Crystallization of Nucleic Acids and Proteins

The Practical Approach Series

SERIES EDITOR

B. D. HAMES
Department of Biochemistry and Molecular Biology
University of Leeds, Leeds LS2 9JT, UK

See also the Practical Approach web site at **http://www.oup.co.uk/PAS**

★ indicates new and forthcoming titles

Affinity Chromatography
Affinity Separations
Anaerobic Microbiology
Animal Cell Culture (2nd edition)
Animal Virus Pathogenesis
Antibodies I and II
Antibody Engineering
Antisense Technology
Applied Microbial Physiology
Basic Cell Culture
Behavioural Neuroscience
Bioenergetics
Biological Data Analysis
Biomechanics—Materials
Biomechanics—Structures and Systems
Biosensors
★ Caenorhabditis Elegans
Carbohydrate Analysis (2nd edition)
Cell-Cell Interactions
The Cell Cycle
Cell Growth and Apoptosis
★ Cell Growth, Differentiation and Senescence
★ Cell Separation
Cellular Calcium
Cellular Interactions in Development
Cellular Neurobiology
Chromatin
★ Chromosome Structural Analysis
Clinical Immunology
Complement
★ Crystallization of Nucleic Acids and Proteins (2nd edition)
Cytokines (2nd edition)
The Cytoskeleton
Diagnostic Molecular Pathology I and II
DNA and Protein Sequence Analysis
DNA Cloning 1: Core Techniques (2nd edition)
DNA Cloning 2: Expression Systems (2nd edition)

DNA Cloning 3: Complex Genomes (2nd edition)
DNA Cloning 4: Mammalian Systems (2nd edition)
★ DNA Microarrays
★ DNA Viruses
Drosophila (2nd edition)
Electron Microscopy in Biology
Electron Microscopy in Molecular Biology
Electrophysiology
Enzyme Assays
Epithelial Cell Culture
Essential Developmental Biology
Essential Molecular Biology I
Essential Molecular Biology II
★ Eukaryotic DNA Replication
Experimental Neuroanatomy
Extracellular Matrix
Flow Cytometry (2nd edition)
Free Radicals
Gas Chromatography
Gel Electrophoresis of Nucleic Acids (2nd edition)
★ Gel Electrophoresis of Proteins (3rd edition)
Gene Probes 1 and 2
Gene Targeting
★ Gene Targeting (2nd edition)
Gene Transcription
Genome Mapping
Glycobiology
★ Growth Factors and Receptors
Haemopoiesis
★ High Resolution Chromatography
Histocompatibility Testing
HIV 1 and 2
★ HPLC of Macromolecules (2nd edition)
Human Cytogenetics I and II (2nd edition)
Human Genetic Disease Analysis
★ Immobilized Biomolecules in Analysis
Immunochemistry 1 and 2
Immunocytochemistry
★ Immundiagnostics
★ *In Situ* Hybridization (2nd edition)
Iodinated Density Gradient Media
Ion Channels
★ Light Microscopy (2nd edition)
Lipid Modification of Proteins
Lipoprotein Analysis
Liposomes
Mammalian Cell Biotechnology
Medical Parasitology
Medical Virology
MHC 1 and 2
★ Molecular Genetic Analysis of Populations (2nd edition)
Molecular Genetics of Yeast
Molecular Imaging in Neuroscience
Molecular Neurobiology
Molecular Plant Pathology I and II

Molecular Virology
Monitoring Neuronal Activity
★ Mouse Genetics and Transfenics
Mutagenicity Testing
Mutation Detection
Neural Cell Culture
Neural Transplantation
Neurochemistry (2nd edition)
Neuronal Cell Lines
NMR of Biological Macromolecules
Non-isotopic Methods in Molecular Biology
Nucleic Acid Hybridization
★ Nuclear Receptors
Oligonucleotides and Analogues
Oligonucleotide Synthesis
PCR 1 and 2
★ PCR 3: PCR In Situ Hybridization
Peptide Antigens
Photosynthesis: Energy Transduction
Plant Cell Biology
Plant Cell Culture (2nd edition)
Plant Molecular Biology
Plasmids (2nd edition)
Platelets
Postimplantation Mammalian Embryos
★ Post-translational Processing
Preparative Centrifugation
Protein Blotting
★ Protein Expression
Protein Engineering
Protein Function (2nd edition)
★ Protein Phosphorylation (2nd edition)
Protein Purification Applications
Protein Purification Methods
Protein Sequencing
Protein Structure (2nd edition)
Protein Structure Prediction
Protein Targeting
Proteolytic Enzymes
Pulsed Field Gel Electrophoresis
RNA Processing I and II
RNA-Protein Interactions
Signalling by Inositides
★ Signal Transduction (2nd edition)
Subcellular Fractionation
Signal Transduction
★ Transcription Factors (2nd edition)
Tumour Immunobiology
★ Virus Culture

Crystallization of Nucleic Acids and Proteins

A Practical Approach

Second Edition

Edited by

ARNAUD DUCRUIX
Laboratoire de Cristallographie et RMN Biologiques, Faculté de Pharmacie Université de Paris V, Paris

and

RICHARD GIEGÉ
Institut de Biologie Moléculaire et Cellulaire du CNRS, Strasbourg

UNIVERSITY PRESS

OXFORD
UNIVERSITY PRESS

Great Clarendon Street, Oxford OX2 6DP

Oxford University Press is a department of the University of Oxford and furthers the University's aim of excellence in research, scholarship, and education by publishing worldwide in

Oxford New York

Athens Auckland Bangkok Bogotá Buenos Aires Calcutta Cape Town Chennai Dar es Salaam Delhi Florence Hong Kong Istanbul Karachi Kuala Lumpur Madrid Melbourne Mexico City Mumbai Nairobi Paris São Paulo Singapore Taipei Tokyo Toronto Warsaw

and associated companies in Berlin Ibadan

Oxford is a registered trade mark of Oxford University Press

Published in the United States
by Oxford University Press Inc., New York

A catalogue record for this book is available from the British Library

Library of Congress Cataloging in Publication Data
Crystallization of nucleic acids and proteins : a practical approach / edited by Arnaud Ducruix, Richard Giegé — [2nd ed.]
(The practical approach series : 210)
Includes bibliographical references and index.
1. Proteins—Analysis. 2. Nucleic Acids—analysis. I. Ducruix. A. (Arnaud) II. Giegé, R. (Richard) III. Series.
QD431.25.A53C79 1999 547.7′5046—dc21 99–15222

ISBN 0–19–963679–6 (Hbk)
0–19–963678–8 (Pbk)

Typeset by Footnote Graphics,
Warminster, Wilts
Printed in Great Britain by Information Press, Ltd,
Eynsham, Oxon.

Preface

With the development of genomics and proteomics, and their applications in basic biological research and the biotechnologies, there is an increasing need of three-dimensional structural knowledge of proteins, nucleic acids, and multi-macromolecular assemblies by X-ray methods. To achieve this aim, crystals diffracting at high resolution are needed. The major aim of the second edition of this book in the *Practical Approach* series is to present an update of the methods employed to produce crystals of biological macromolecules and to outline the newest trends that have entered the field. Since the first edition, which appeared in 1992, the science of crystallogenesis, which was then in its infancy, has grown rapidly balancing between the physics of crystal growth and blind-screen crystallizations. The advances can be appreciated from the Proceedings of the different International Conferences on the Crystallization of Biological Macromolecules (ICCBM 1 to 7) which appear every two years, the latest covering ICCBM-7, held in June 1998 in Granada, and published in *J. Crystal Growth*, Vol. 196, January 1999.

As usual in the series, the emphasis of the present book is to give detailed laboratory protocols throughout the chapters. However, we have not given protocols just as 'recipes', but instead we have intended to always present the methods with reference to the theoretical concepts and principles underlying them. In fact one of the aims of this book was to fight against the fallacious idea according to which crystal growth of biological macromolecules is more an 'art' than a science. Although this is probably sometimes true from a practical point of view, it is certainly incorrect in its principle. Therefore emphasis has been given to the physical parameters involved in crystallization and on the large knowledge on the crystal growth of small molecules, as well as to the particular biochemical and physico-chemical properties of biological macromolecules.

This book is intended to be read by a wide range of scientists. First, by the crystallographers who have to solve three-dimensional structures of macromolecules. Secondly, by all molecular biologists who have access to macromolecules but often do not know how to handle them for crystallization, and who may consider crystallization as an esoteric undertaking, because of lack of basic knowledge about the crystallization process. Thirdly, by the physicochemists and physicists who for other reasons consider biology as an esoteric science. It is our wish that this book will contribute to a better understanding of crystallogenesis by these scientists and to the improved perception of the biological requirements that have to be taken into account for physical studies. Finally, the book should be a laboratory guide for all students and beginners, helping them to avoid making mistakes when entering the field of crystal preparation.

Chapter 1 is an introduction to crystallogenesis of biological macromolecules. It includes a brief historical survey of the subject and introduces the general principles and major achievements of this new discipline. The preparation of biological macromolecules and the concept of 'crystallography-grade purity' are developed in Chapter 2. Chapter 3 is new and introduces the use of molecular biology methods to 'customize' domains for structural biology. It also includes the preparation of protein crystals made of protein molecules containing selenomethionine residues and outlines why they can be used for the multiple anomalous dispersion (MAD) method. Screen-like methods are now widespread but do not provide suggestions when they fail. The answer may then come from statistical methods presented in Chapter 4 which explains their theory and gives practical advice (and a computer program) for protocol design. One of the goals of this book is to give to crystal growers of biomacromolecules the conceptual and methodological tools needed to control crystallization. This is examined in Chapter 5 which includes a description of the classical crystallization methods together with workshop examples. Crystallization in gels is described with theoretical and practical considerations in Chapter 6. This chapter includes a novel section describing the gel acupuncture method; it also contains information on crystal growth under microgravity and hypergravity conditions. Because it is sometimes difficult to reproduce appropriate nucleation conditions, Chapter 7 is devoted to seeding procedures with preformed crystalline material, including micro-, macro-, and cross-seeding with numerous examples.

The special cases of nucleic acids (and their complexes with proteins and nucleoprotein assemblies) and membrane proteins are covered in two individual chapters (8 and 9). Their crystallization is still challenging, but the novel developments in the field have led to a number of recent breakthroughs, that are encouraging for experimenters entering the field. For nucleic acids, Chapter 8 gives emphasis to the strategies for the design and preparation of appropriate DNA or RNA fragments and to the specific features characterizing their crystallization, either free or in complexes with proteins. For membrane proteins, the already published genomes show that a good third of the expressed proteins belong to this category and it is expected that the interest for membrane proteins will expand quickly, especially among structural biologists. It is our wish that the methods described in Chapter 9 will help them to reach this goal.

The link between protein solubility and the physico-chemical parameters governing crystal growth is presented in Chapter 10 with a strong emphasis on the practical issues. Chapter 11 deals with physical methods and gives an introduction to the physics of crystal growth. In particular, the use of light scattering methods to monitor early nucleation events is advocated with examples and a description of the material used.

Chapter 12 is new and covers the expanding field of the two-dimensional crystallization of soluble proteins on planar lipid films. It presents many protocols which may be readily used. Soaking of crystals of biological macromolecules is of great interest for crystallographers, either for resolving a structure (heavy-atom derivatives), or for diffusing inhibitors, activators, or cofactors (eventually photoactivable). As in all previous chapters, practical aspects were the driving force and Chapter 13 is illustrated by a variety of protocols. The editors thought that an introduction to X-ray crystallography should be included in this book. This is done in Chapter 14, that is geared toward biochemists wanting to characterize crystals by themselves rather than explaining how to solve a structure.

It is a great pleasure to acknowledge our gratitude to a number of friends and colleagues. First, our colleagues from Gif-sur-Yvette/Paris and Strasbourg deserve particular thanks for having participated over the years in the development of our studies on crystallogenesis; their enthusiasm was essential and gave us the impetus for the preparation of a book covering this field and for updating it in its second edition. However, without the invaluable help of many friends from both sides of the Atlantic who agreed to cover specialized topics, this venture would not have been possible. We would like to warmly thank all of them. The French Centre National de la Recherche Scientifique (CNRS) and Centre National d'Etudes Spatiales (CNES) are acknowledged for their permanent support in developing biological crystallogenesis and their interest for the physico-chemical aspects of the field.

While preparing this second edition, our dear colleague Roland Boistelle closed his eyes. He was a source of inspiration for all of us and one of the first who geared the biology oriented scientists to the physics of crystal growth. His contribution to the field of macromolecules crystallogenesis was essential and we would like to dedicate this book to his memory.

Paris and Strasbourg — A. D.
June 1999 — R. G.

Contents

Contributors

W. BERGSMA-SCHUTTER
Department of Chemistry / Biophysical Chemistry, University of Groningen, Nijenborgh 4, NL-9747 AG Groningen, The Netherlands.

P. F. BERNE
Rhône Poulenc Rorer, 13, quai Jules Guesde, 94403 Vitry sur Seine Cedex, France.

R. BOISTELLE († DECEASED 1998)
Centre de Recherche sur les Mécanismes de la Croissance Cristalline du CNRS, Campus de Luminy, Case 913, 13288 Marseille Cedex, France.

A. BRISSON
Department of Chemistry / Biophysical Chemistry, University of Groningen, Nijenborgh 4, NL-9747 AG Groningen, The Netherlands.

C. W. CARTER JR
Department of Biochemistry, CB 7260, University of North Carolina at Chapel Hill, Chapel Hill, NC 27599-7260, USA.

A.-C. DOCK-BREGEON
Institut de Génétique et de Biologie Moléculaire et Cellulaire, 1, rue Léon Fries, Parc d'Innovation, BP 163, F-67404 Illkirch Cedex, France.

S. DOUBLIÉ
Department of Microbiology and Molecular Genetics, The Markey Center for Molecular Genetics, University of Vermont, Burlington, VT 05045, USA.

A. DUCRUIX
Laboratoire de Cristallographie et RMN Biologiques, Faculté de Pharmacie, Université de Paris V, 4, Avenue de l'Observatoire, 75270 Paris Cedex 06, France.

J. M. GARCIA-RUIZ
Instituto Andaluz de Ciencias de la Tierra, CSIC-Universidad de Granada, Facultad de Ciencias, 18002-Granada, Spain.

R. GIEGÉ
Institut de Biologie Moléculaire et Cellulaire du CNRS, 15 rue René Descartes, F-67084 Strasbourg Cedex, France.

T. GLEICHMANN
Anorganisch-Chemisches Institut, Westfälische Wilhelms-Universität, Wilhelm-Klemm Str. 8, D-48149 Münster, Germany.

O. LAMBERT
Department of Chemistry/Biophysical Chemistry, University of Groningen, Nijenborgh 4, NL-9747 AG Groningen, The Netherlands.

B. LORBER
Institut de Biologie Moléculaire et Cellulaire du CNRS, 15 rue René Descartes, F-67084 Strasbourg Cedex, France.

D. MORAS
Institut de Génétique et de Biologie Moléculaire et Cellulaire, 1 rue Léon Fries, Parc d'Innovation, BP 163. F-67404 Illkirch Cedex, France.

F. OTALORA
Instituto Andaluz de Ciencias de la Tierra, CSIC-Universidad de Granada, Facultad de Ciencias, 18002-Granada, Spain.

D. PICOT
Institut de Biologie Physico-Chimique, 13 rue P. et M. Curie, F-75005 Paris.

F. REISS-HUSSON
Centre de Génétique Moléculaire du CNRS, Avenue de la Terrasse, F-91190 Gif-sur-Yvette Cedex, France.

M. RIÈS-KAUTT
Laboratoire de Cristallographie et RMN Biologiques, Faculté de Pharmacie, Université de Paris V, 4, Avenue de l'Observatoire, 75270 Paris Cedex 06, France.

M.-C. ROBERT
Laboratoire de Minéralogie-Cristallographie, Universités Pierre et Marie Curie, 4 place Jussieu, F-75252 Paris Cedex 05, France.

L. SAWYER
Structural Biochemistry Group, The University of Edinburgh, Swann Building, King's Buildings, Mayfield Road, Edinburgh EH9 3JR, UK.

E. A. STURA
Dept. d'Ingénierie et d'Etudes des Protéines, Bât. 152, CEA/Saclay, 91191 Gif-sur-Yvette Cedex, France.

M. A. TURNER
X-ray Structure Laboratory, Department of Biochemistry, Hospital for Sick Children, 555 University Avenue, Toronto, Ontario, Canada.

S. VEESLER
Centre de Recherche sur les Mécanismes de la Croissance Cristalline du CNRS, Campus de Luminy, Case 913, F-13288 Marseille Cedex, France.

O. VIDAL
Laboratoire de Minéralogie-Cristallographie, Universités Pierre et Marie Curie, 4 place Jussieu, F-75252 Paris Cedex 05, France.

Abbreviations

BPTI	bovine pancreatic trypsin inhibitor
BTP	bis Tris propane
CMC	critical micellar concentration
DEPC	diethylpyrocarbonate
DiFP	diisopropylfluorophosphate
DLS	dynamic light scattering
DMSO	dimethyl sulfoxide
DOPC	dioleoylphosphatidylcholine
DTT	dithiothreitol
EDTA	ethylenediaminetetraacetic acid
ESI	electronspray ionization mass spectra
GST	glutathione-*S*-transferase
HEW	hen egg white
HEWL	hen egg white lysozyme
HIC	hydrophobic interaction chromatography
IEF	isoelectric focusing
LS	light scattering
MAD	multiple wavelength anomalous dispersion
MALDI	matrix-assisted desorption/ionization
MeTEOS	methyltriethoxysilane
MIR	multiple isomorphous replacement
MPD	2-methyl-2,4-pentane diol
MR	molecular replacement
OD	optical density
OP	osmotic pressure
PBC	periodic bond chain
PC	phosphatidylcholine
PCR	polymerase chain reaction
PEG	polyethylene glycol
pTS	*para*-toluenesulfonate
RPC	reverse-phase chromatography
SANS	small angle neutron scattering
SAXS	small angle X-ray scattering
SIR	single isomorphous replacement
SLS	static light scattering
TEOS	tetraethoxysilane
TLC	thin-layer chromatography
TMOS	tetramethoxysilane

1

An introduction to the crystallogenesis of biological macromolecules

R. GIEGÉ and A. DUCRUIX

'Il y a là des mystères, qui préparent à l'avenir d'immenses travaux et appellent dès aujourd'hui les plus sérieuses méditations de la science'

Pasteur, 1860, in *Leçons de Chimie*.

1. Introduction

The word 'crystal' is derived from the Greek root 'krustallos' meaning 'clear ice'. Like ice, crystals are chemically well defined, and many among of them are of transparent and glittering appearance, like quartz, which was for a long time the archetype. Often they are beautiful geometrical solids with regular faces and sharp edges, which probably explains why crystallinity, even in the figurative meaning, is taken as a symbol of perfection and purity. From the physical point of view, crystals are regular three-dimensional arrays of atoms, ions, molecules, or molecular assemblies. Ideal crystals can be imagined as infinite and perfect arrays in which the building blocks (the asymmetric units) are arranged according to well-defined symmetries (forming the 230 space groups) into unit cells that are repeated in the three-dimensions by translations. Experimental crystals, however, have finite dimensions. An implicit consequence is that a macroscopic fragment from a crystal is still a crystal, because the orderly arrangement of molecules within such a fragment still extends at long distances. The practical consequence is that crystal fragments can be used as seeds (Chapter 7). In laboratory-grown crystals the periodicity is never perfect, due to different kinds of local disorders or long-range imperfections like dislocations. Also, these crystals are often of polycrystalline nature. The external forms of crystals are always manifestations of their internal structures and symmetries, even if in some cases these symmetries may be hidden at the macroscopic level, due to differential growth kinetics of the crystal faces. Periodicity in crystal architecture is also reflected in their macroscopic physical properties. The most straightforward example is given by the ability of crystals to diffract X-rays, neutrons, or electrons, the phenomenon underlying structural chemistry and biology (for introductory texts see refs 1 and 2), and the major aim of this book is to present the methods employed to produce three-dimensional crystals of biological macromolecules, but also two-dimensional crystals (Chapter 12), needed for diffraction studies. Other properties of invaluable practical applications should not be overlooked either, as is the case of optical and electronic properties which are at the basis of non-linear optics and modern electronics (for an introduction to physical properties of molecular crystals see ref. 3). Crystals furnish one of the most beautiful examples of order and symmetry in nature and it is not surprising that their study fascinates scientists (4).

What characterizes biological macromolecular crystals from small molecule crystals? In terms of morphology, one finds with macromolecular crystals the same diversity as for small molecule crystals (*Figure 1*). In terms of crystal size, however, macromolecular crystals are rather small, with volumes rarely exceeding 10 mm^3, and thus they have to be examined under a binocular microscope. Except for special usages, such as neutron diffraction, this is not too severe a limitation. Among the most striking differences between the two families of crystals are the poor mechanical properties and the high content of solvent of macromolecular crystals. These crystals are always extremely fragile and are sensitive to external conditions. This property can be used as a preliminary identification test: protein crystals are brittle or will crush when touched with the tip of a needle, while salt crystals that can sometimes develop in macromolecule crystallization experiments will resist this treatment. This fragility is a consequence both of the weak interactions between macromolecules within crystal lattices and of the high solvent content (from 20% to more than 80%) in these crystals (Chapter 14). For that reason, macromolecular crystals have to be kept in a solvent-saturated environment, otherwise dehydration will lead to crystal cracking and destruction. The high solvent content, however, has useful consequences because solvent channels permit diffusion of small molecules, a property used for the preparation of isomorphous heavy-atom derivatives needed to solve the structures (Chapters 13 and 14). Further, crystal structures can be considered as native structures, as is indeed directly verified in some cases by the occurrence of enzymatic reactions within crystal lattices upon diffusion of the appropriate ligands (5, 6). Other characteristic properties of macromolecular crystals are their rather weak optical birefringence under polarized light: colours may be intense for large crystals but less bright than for salt crystals (isotropic cubic crystals or amorphous material will not be birefringent). Also, because the building blocks composing macromolecules are enantiomers (L-amino acids in proteins—except in the case of some natural peptides—and D-sugars in nucleic acids) macromolecules will not crystallize in space groups with inversion symmetries. Accordingly, out of the 230 possible space groups, macromolecules do only crystallize in the 65 space groups without such inversions (7). While

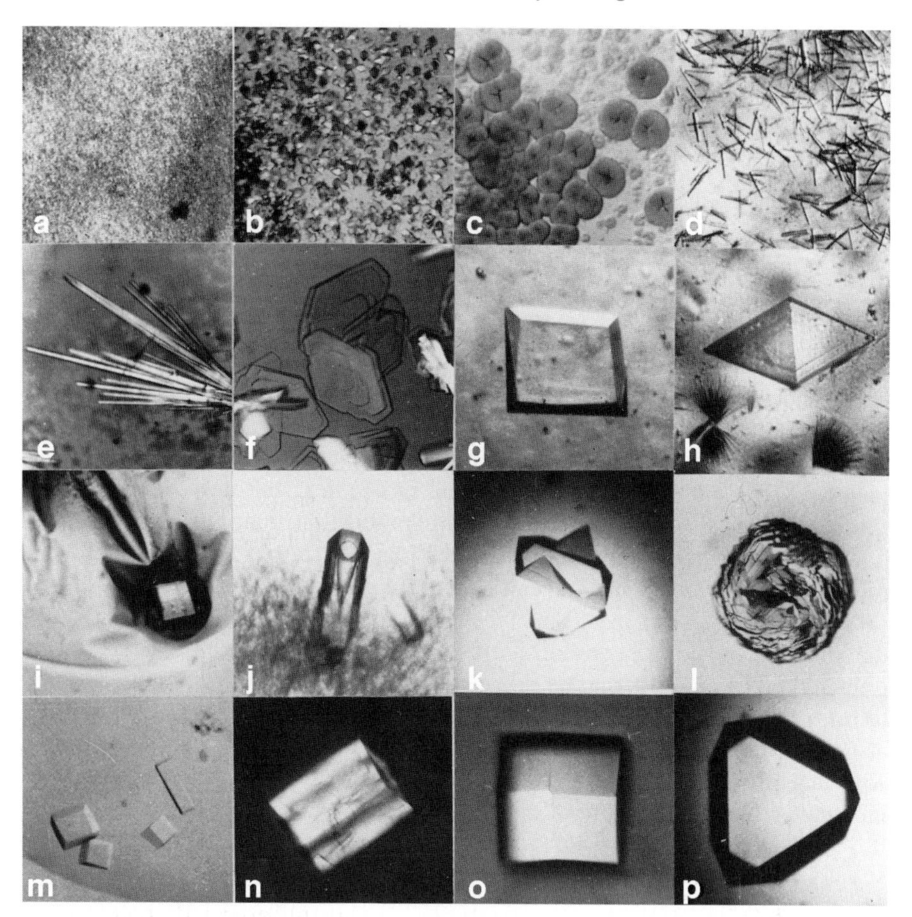

Figure 1. From precipitates to perfect crystals of biological macromolecules. (a) Precipitate of HEW lysozyme; (b) yeast aspartyl-tRNA synthetase microcrystals; (3) spherulites of the complex between yeast aspartyl-tRNA synthetase and tRNA^{Asp}; (d) short needles of tissue inhibitor protein of metalloprotease; (e) aspartyl-tRNA synthetase long needle-like crystals; (f) yeast initiator tRNA^{Met} thin plates with growth defects; (g) plate-shaped crystals of *Hypoderma lineatum* collagenase; (h) tetragonal crystals of aspartyl-tRNA synthetase showing growth defects together with brush-like needle bunches; (i) example of 'skin' of denatured protein around a HEW lysozyme crystal; (j) crystal of mellitin with a hollow extremity; (k, l) twinned and twinned-embedded crystals of collagenase and HEW lysozyme; (m, n, o, p) perfect three-dimensional crystals; (m) polymorphism in a same crystallization drop showing cubic and orthorhombic crystal-habits of aspartyl-tRNA synthetase/tRNA^{Asp} complex; (n) crystals of tRNA^{Asp} with cracks due to ageing; (o) crystals of HEW lysozyme with fourfold symmetry; (p) crystals of *H. lineatum* collagenase.

small organic molecules prefer to crystallize in space groups in which it is easiest to fill space, proteins crystallize primarily in space groups in which it is easiest to achieve connectivity (8). Macromolecular crystals are also characterized by large unit cells with dimensions that can reach up to 1000 Å for virus crystals (9). From a practical point of view, it is important to remember that crystal morphology is not synonymous with crystal quality. Therefore, the final diagnostic of the suitability of a crystal for structural studies will always be the quality of the diffraction pattern which reveals its internal order, as is reflected at first glance by the so-called 'resolution' parameter (Chapter 14).

Crystal growth, which is a very old activity that has always intrigued mankind, and many philosophers and scientists have compared it with the biological process of reproduction, and it has even been speculated that the duplication of genetic material would occur through crystallization-like mechanisms (10). Nowadays, the theoretical and practical frames of crystallogenesis are well established for small molecules, but less advanced for macromolecules, although it can be anticipated that many principles underlying the growth of small molecule crystals will apply for that of macromolecules (11, 12). Until recently, crystallization of macromolecules was rather empirical, and because of its unpredictability and frequent irreproducibility, it has long been considered as an 'art' rather than a science. It is only in the last 15 years that a real need has emerged to better understand and to rationalize the crystallization of biological macromolecules. It can be stated at present that the small molecule and macromolecular fields are converging, with an increasing number of behaviours or features known for small molecules that are now found for macromolecules (13).

2. Crystallization and biology: a historical background

2.1 Before the X-rays

It is often forgotten that some advances in biochemistry and molecular biology have their origin in crystallization data and that empirical crystal growth of biological materials is as old as biochemistry. The first reports on protein crystals were published more than a century ago when Funke, Hünefeld, Lehman, Teichman, and others crystallized haemoglobin from the blood of various invertebrates and vertebrates, and it was prophesied that the study of crystals would shed light on the exact nature of proteinic substances (14–17). This was followed by the crystallization of hen egg white albumin and a series of plant proteins (reviewed in refs 16 and 17). The beauty of crystals certainly fascinated the physiological chemists in these early days, since an atlas with extensive descriptions of the morphologies of haemoglobin crystals was published in 1909 (15). In 1926 Sumner reported the crystallization of urease from jack beans (18), soon followed by Northrop who crystallized pepsin and a series of other proteolytic enzymes (19). It is interesting to note that for technical difficulties, the X-ray structure of urease has only be determined very recently (20). Besides being a method of purification, crystallization experiments established the view that pure enzymes are proteins, a fact not obvious to all at that time (21). Another scientific achievement arose in 1935 when Stanley crystallized tobacco mosaic virus. Influenced by Northrop's conception of enzymes, and using methods developed for proteins, he prepared the virus in pure crystalline state (he believed the virus was an autocatalytic protein) and showed that it retains its infectivity after several recrystallizations (reviewed in ref. 10). The importance and the implications for biology of these discoveries was recognized rapidly and in 1946 the Nobel Prize for Chemistry was awarded to Sumner, Northrop, and Stanley.

2.2 Crystallogenesis and structural biology

The use of crystals in structural biology goes back to 1934, when Bernal and Crowfoot (D. Hodgkin) produced the first X-ray diffraction pattern of a protein, that of crystalline pepsin (22). Since then representatives of most families of macromolecules have been crystallized, but using mainly empirical methods and without rational control of the growth mechanisms. This can be well understood because in the early days of structural biology the interest of scientists was mainly directed at the development of the X-ray methods needed to resolve the structures rather than in that of a rationalization of the crystallization procedures. Therefore, only limited efforts were expended in understanding or improving macromolecular crystallization procedures. At present X-ray methods are well established (2, 7, 23). The overall scheme of the various steps of the resolution of a three-dimensional structure is summarized in *Figure 2*. But production of suitable crystals diffracting at high resolution often remains the bottleneck in structure determination projects. With the rapid development of biotechnologies and the unlimited potential of macromolecular engineering (which requires structural knowledge for site-directed mutagenesis experiments or for design of factitious macromolecules) there is now an increasing need for macromolecular crystals diffracting at high resolution, not only proteins but also nucleic acids and multi-macromolecular assemblies. Thanks to the biotechnological tools it is now possible to obtain rather easily the amounts of pure macromolecules (in most cases several milligrams) needed to start a crystallization project (Chapter 3). In this introductory chapter we present the general trends of the science of crystal growth in biology. More extensive discussions and the practical details will be presented in following chapters.

The first major breakthrough towards better and easier crystallizations was the development, in the 1960s, of crystallization micromethods (e.g. dialysis and vapour phase diffusion) (Chapter 5). It was promoted by structural projects on macromolecules reluctant to crystallize easily and available in limited amounts (24). Further significant improvements came from the discovery of

1- PURIFICATION OF MACROMOLECULES
from wild-type, engineered or overproducing organisms
(possibility of *in vitro* synthesis for nucleic acids and small peptides)

2- CRYSTALLIZATION
by *de novo* crystallization or seeding techniques

3- DATA MEASUREMENTS
characterization of space group and diffraction resolution;
measurements of diffraction intensities on an electronic area detector
(possible use of neutron and frequently of tunable X-ray synchrotron radiation);
frequent data acquisition by cryo-crystallographic methods.

4- PHASE DETERMINATION
using methods based on isomorphous replacement (preparation of heavy atom derivatives),
anomalous scattering, molecular replacement and non-crystallographic symmetry,
or direct calculations (e.g. from maximum entropy)

5- ELECTRON DENSITY MAP COMPUTATION AND INTERPRETATION
interpretation of mini-maps; model building on computer graphic displays

6- MODEL REFINEMENT
least-square refinements; restrained refinements; ...

Figure 2. Steps involved in the resolution of the 3D structure of a biological macromolecule (for more details see Chapter 14).

specific properties of additives to be included in crystallization solvents, such as the polyamines (25, 26) and the non-ionic detergents (27–30) which gave the clue for crystallizing nucleic acids (Chapter 8) or membrane proteins (Chapter 9). Also, a more systematic use of organic cosmotropes (compounds promoting 'order'), like polyhydric alcohols that are stabilizers of protein structure when at high concentration, may facilitate crystallization of flexible proteins (31).

The perception of the importance of purity for growing better crystals (32) was an important achievement in the field (Chapter 2), and the adequate choice of the biological material was an important determinant for the success of many crystallizations. With the ribosome for instance, the preparation of homogeneous particles from halophilic or thermophilic bacteria instead from mesophilic bacteria, considerably improved crystal quality (33, 34). Also, the ease of synthesizing oligonucleotides with automated methods (Chapter 8), or the development of genetic engineering technologies for overexpression of proteins (Chapter 3), explains the increasing number of crystallized nucleic acids or rare proteins.

Today, crystal growth research is stimulated by macromolecular engineering requirements, but also to some extent by space-science projects (crystallization under microgravity conditions, Chapter 6) and its descriptive stage is moving towards a new, more quantitative discipline, biocrystallogenesis, which includes biology, biochemistry, physics, and engineering related aspects. For specialized literature see refs 35–42, and for general reviews 43–46.

3. General principles

3.1 A multiparametric process

Biocrystallization, like any crystallization, is a multiparametric process involving the three classical steps of nucleation, growth, and cessation of growth. What makes crystal growth of biological macromolecules different is, first, the much larger number of parameters than those involved in small molecule crystal growth (*Table 1*) and, secondly, the peculiar physico-chemical properties of the compounds. For instance, their optimal stability in aqueous media is restricted to a rather narrow temperature and pH range. But the main difference from small molecule crystal growth is the conformational flexibility and chemical versatility of macromolecules, and their consequent greater sensitivity to external conditions. This complexity is the main reason why systematic investigations were not undertaken earlier. Furthermore, the importance of some parameters, such as the geometry of crystallization vessels or the biological origin of macromolecules, had not been recognized. It is only recently that the hierarchy of parameters has been perceived. A practical consequence of this new perception was the development of statistical methods to screen crystallization conditions (Chapter 3). For a rational design of growth conditions, however, physical and biological parameters have to be controlled. One of the aims of this book is to give to crystal growers of biological macromolecules the conceptual and methodological tools needed to achieve such control.

3.2 Purity

Because macromolecules are extracted from complex biological mixtures, purification plays an extremely important role in crystallogenesis (Chapter 2). Purity, however, is not an absolute requirement since crystals of macromolecules can sometimes be obtained from mixtures. But such crystals are mostly small or grow as polycrystalline masses, are not well shaped, and are of bad diffraction quality, and thus cannot be used for diffraction studies. However, crystallization of macromolecules from mixtures may be used as a tool for purification (47), especially in industry (48). For the purpose of X-ray

Table 1. Parameters affecting the crystallization (and/or the solubility) of macromolecules[a]

Intrinsic physico-chemical parameters

- Supersaturation (concentration of macromolecules and precipitants)
- Temperature, pH (fluctuations of these parameters)
- Time (rates of equilibration and of growth)
- Ionic strength and purity of chemicals (nature of precipitant, buffer, additives)
- Diffusion and convection (gels, microgravity)
- Volume and geometry of samples and set-ups (surface of crystallization chambers)
- Solid particles, wall and interface effects (e.g. homogeneous versus heterogeneous nucleation, epitaxy)
- Density and viscosity effects (differences between crystal and mother liquor)
- Pressure, electric and magnetic fields
- Vibrations and sound (acoustic waves)
- Sequence of events (experimentalist versus robot)

Biochemical and biophysical parameters

- Sensitivity of conformations to physical parameters (e.g. temperature, pH, ionic strength, solvents)
- Binding of ligands (e.g. substrates, cofactors, metal ions, other ions)
- Specific additives (e.g. reducing agents, non-ionic detergents, polyamines) related with properties of macromolecules (e.g. oxidation, hydrophilicity versus hydrophobicity, polyelectrolyte nature of nucleic acids)
- Ageing of samples (redox effects, denaturation, or degradation)

Biological parameters

- Rarity of most biological macromolecules
- Biological sources and physiological state of organisms or cells (e.g. thermophiles versus halophiles or mesophiles, growing versus stationary phase)
- Bacterial contaminants

Purity of macromolecules

- Macromolecular contaminants (odd macromolecules or small molecules)
- Sequence (micro) heterogeneities (e.g. fragmentation by proteases or nucleases—fragmented macromolecules may better crystallize —, partial or heterogeneous post-translational modifications)
- Conformational (micro) heterogeneities (e.g. flexible domains, oligomer and conformer equilibria, aggregation, denaturation)
- Batch effects (two batches are not identical)

[a] Although all these parameters have not been screened systematically, especially for the crystallization of a given macromolecule, all of them have been evaluated individually in isolated cases.

crystallography, high-quality monocrystals of appreciable size (0.1 mm at least for the dimension of a face) are needed. It is our belief that poor purity is the most common cause of unsuccessful crystallization, and for crystallogenesis the purity requirements of macromolecules have to be higher than in other fields of molecular biology. Purity has to be of 'crystallography grade': the macromolecules not only have to be pure in terms of lack of contaminants, they have also to be conformationally 'pure' (32). Denatured macromolecules, or macromolecules with structural microheterogeneities, adversely affect crystal growth more than do unrelated molecules, especially when structural heterogeneities concern domains involved in crystal packing. On the other hand, the presence of microquantities of proteases (or nucleases) can alter the structure of the macromolecules during storage or the rather long time needed for crystallization. As a consequence, when starting a crystallization project one has to be primarily concerned with purification methodologies and to take all precautions against protease and nuclease action. To have reproducible results, the physiological state of cells should be controlled, because protease (or nuclease) levels may vary as well as the balance between cellular components. As a general rule, batches of macromolecules should not be mixed and crystallization experiments should be conducted on fresh material so that ageing phenomena are limited. For more details see Chapter 2.

In summary, we emphasize the importance of macromolecular purity in biological crystallogenesis, and in cases of unsuccessful experiments we recommend first improvement or reconsideration of the purification procedure of the molecules of interest.

3.3 Solubility and supersaturation

To grow crystals of any compound, molecules have to be brought in a supersaturated, thermodynamically unstable state, which may develop in a crystalline or amorphous phase when it returns to equilibrium. Supersaturation can be achieved by slow evaporation of the solvent or by varying parameters (listed in *Table 1*). The recent use of pressure as a parameter is of note (49). From this it follows that knowledge of macromolecular solubility is a prerequisite for controlling crystallization conditions. However, the theoretical background underlying solubility is still controversial, especially regarding salt effects (50), so that solubility data almost always originate from experimental determinations. Specific quantitative methods permitting such determinations on small protein samples are available (51–53). The main output was the experimental demonstration of the complexity of solubility behaviours, emphasizing the importance of phase diagram determinations for a rational design of crystal growth (Chapter 10).

As to the nature of the salt used to reach supersaturation, one can wonder why ammonium sulfate is so frequently chosen by crystal growers (54). This usage is in fact incidental and results from the practices of biochemists for salting-out proteins. Indeed many other salts can be employed, but their effectiveness for inducing crystallization is variable (52). The practical consequence is that protein supersaturation can be reached (or changed) in a large concentration range of protein and salt, provided that adequate salts are used.

3.4 Nucleation, growth, and cessation of growth

Because proteins and nucleic acids require defined pH and ionic strength for stability and function, biomacromolecule crystals have to be grown from chemically rather complex aqueous solutions. Crystallization starts by a nucleation phase (i.e. the formation of the first ordered aggregates) which is followed by a growth phase. Nucleation conditions are sometimes difficult to reproduce, and thus seeding procedures with preformed crystalline material should not be overlooked as in many cases they represent the only method to obtain reproducible results (55) (Chapter 7). It should be noticed that nucleation requires a greater supersaturation than growth, and that crystallization rates increase when supersaturation increases. Thus nucleation and growth should be uncoupled, which is almost never done consciously but occurs sometimes under uncontrolled laboratory conditions. From a practical point of view, interface or wall effects as well as shape and volume of drops can affect nucleation or growth, and consequently the geometry of crystallization chambers or drops has to be defined. For additional information see Chapter 11.

Cessation of growth can have several causes. Apart from trivial ones, like depletion of the macromolecules from the crystallizing media, it can result from growth defects, poisoning of the faces, or ageing of the molecules. Better control of growth conditions, in particular of the flow of molecules around the crystals, may in some cases overcome the drawbacks as was shown in microgravity experiments (refs 56, 57, and Chapter 6).

3.5 Packing

With biological macromolecules, crystal quality may be correlated with the packing of the molecules within the crystalline lattices, and external crystal morphology with internal structure. As shown by the periodic bond chain (PBC) method, direct protein–protein contacts are essential in determining packing and morphology (Chapter 11). Forces involved in packing of macromolecules may be considered as weak as compared to those maintaining the cohesion of small molecule crystals. They involve salt bridges, hydrogen bonds, Van der Waals, dipole–dipole, and stacking interactions (58–60). It must also be borne in mind that the weak cohesion of macromolecular crystals results from the fact that only a small part of macromolecular surfaces participate in intermolecular contacts (61), the remaining being in contact with the solvent (exceptions may be found for small proteins). This explains the commonly observed polymorphism of biological macromolecular crystals.

4. From empiricism to rationality

4.1 Towards a better understanding of parameters

To date, the major parameters underlying crystallization of macromolecules have been recognized (*Table 1*), and if their respective contributions in the crystallization process are not known with certainty, the theoretical frame needed for explaining their role is well established (11–13, 46, 56). Also, the experimental tools exist that are needed for measuring the contributions of these parameters. This is the case for solubility and aggregation state measurements as well as for monitoring pH and temperature (Chapters 10 and 11), or even the effect of microgravity (Chapter 6). Correlations between the variation of a parameter and the ability of a given macromolecule to crystallize are expected to be found. Finally, and perhaps the most important, was the recognition of the importance of purity. Thus again we emphasize that experimenters should primarily devote their efforts to starting crystallization attempts with molecules of the highest biochemical quality.

4.2 Towards an active control of crystal growth

Only few attempts have been published describing the active control of crystallization experiments. In general the history of experiments is not well known, because crystal growers do not monitor parameters. This is especially the case for temperature, which is almost never known with accuracy, even if experiments are conducted in thermostated cabinets (this may be advantageous because many microconditions may be screened, although at the cost of reproducibility). Also the kinetics of events are practically never monitored. In the following chapters methods to control macromolecular crystal growth will be described (e.g. for the kinetics of evaporation in vapour diffusion crystallization, see Chapter 5; for active video and temperature control, see Chapter 11). Although these different aspects are all in their infancy, and required instrumentation often does not exist, or exists only as prototypes, we believe that user-friendly methodologies will be developed soon, and that laboratories may be equipped with the adequate instrumentation.

4.3 The future of biological crystallogenesis

As mentioned before, modern genetic methods give access to molecules present at very low amounts in cells and help to solve problems linked to structural or conformational heterogeneities of proteins reluctant to crystallize. In particular, engineering of active variants containing compact cores will permit easier crystallizations of proteins with flexible domains. Macromolecular engineering will certainly find many applications in the RNA world, where many ribozymes, pseudo-knots, and other RNA constructs deserve structural investigations (Chapter 8). On the other hand, search of crystal perfection, preparation of large size crystals, together with improvement of data collection methods at synchrotron or neutron sources, are other challenges, notably for the resolution of structures at highest resolution and time-resolved crystallography (62).

For the biologist, studying crystal growth should be correlated with biological problems, and crystallization projects on macromolecular complexes, on membrane proteins (Chapter 9), and especially on engineered proteins, are being developed. For the physicist, growing large monocrystals can be a goal in itself, and one might speculate that exploration of optical, electrical, mechanical, and other physical properties of crystalline arrays made from biological macromolecules or assemblies can lead to novel frontiers in the material science of tomorrow. Finally, for the chemist, usage of chemical and molecular biology tools could lead in future to the design of molecular devices and other nanostructures mimicking macromolecular crystals, as was discussed for artificial self-assembling nucleic acids (63).

In conclusion, the rational approach for prompt crystallizations will demand a synergy between biochemically and physically directed research and usage of automated methods for the control of nucleation and growth, as well as the rapid preparation of high-quality monocrystals.

5. Search for crystallization strategies

No universal answer(s) can be given to the obvious questions about how to start a crystallization project and what kind of strategy would be the most appropriate to be successful. Because of the multiparametric nature of the crystallization process and the diversity of the individual properties of proteins, the advice would be to collect all possible information on the protein one intends to crystallize, so that to hierarchize the variables of the process and if possible to restrict their number. To this aim the questionnaire in *Protocol 1* addresses a number of basic questions.

Protocol 1. Before starting crystallizations, what are the biological and biochemical characteristics of your protein?

If your project concerns nucleic acids or their complexes with proteins see Chapter 8; for membrane proteins see Chapter 9.

A. *Biology and production*

1. What is the biological origin of your protein? From a micro-organism (mesophile or thermophile), a plant, an animal (which tissue)? etc.
2. Is the gene of the protein sequenced?
3. Is the protein cloned? In what expression vector and in what cells was it overexpressed (*E. coli*, yeast, baculovirus, others)?
4. Is it a his-tagged protein? If yes do you plan to crystallize it with or without the tag?

5. Is it a fusion protein? If yes, which protease is used to cleave the protein?

6. How many mg/litre of culture can you produce?

7. How many mg of protein can you obtain per standard purification?

B. *Biochemistry*

1. How long does it take to purify one batch of protein?

2. How do you assess the purity of the protein?

(a) Electrophoresis (native/denaturating conditions).

(b) HPLC (which phase).

(c) Ion spray (mass spectrometry).

(d) Checking of N-terminus.

(e) Activity assay.

2. What are the principal characteristics of the protein?

(a) Molecular weight.

(b) Isoelectric point (calculated or measured).

(c) Glycosylation (yes/no).

(d) Number of free cysteine(s).

(e) Number of disulfide bridges.

(f) Hydrophobicity or hydrophilicity.

(g) What are the ligands?

3. What are the friendly (or unfriendly) solvents?

4. Is the protein monomeric of oligomeric? How did you check it (chromatography, light scattering, others)? Does your protein has a tendency to aggregate?

5. What is the stability of the protein versus time, temperature, or pH?

The better these questions can be answered, the easier will be the design of a crystallization strategy. Good knowledge of the characteristics of the protein, of its availability, will guide the experimenter. Quite often, it helps people to be aware that the premise of crystallization may be as important as the crystallization itself.

References

1. Pickworth Glusker, J. and Trueblood, K. N. (1985). *Crystal structure analysis, a primer.* Oxford University Press, New York.

2. Drenth, J. (1995). *Principles of protein X-ray crystallography.* Springer–Verlag, Berlin.
3. Wright, J. D. (1987). *Molecular crystals.* Cambridge University Press, Cambridge.
4. Lima-de-Faria, J. (ed.) (1990). *Historical atlas of crystallography.* Kluwer, Dordrecht.
5. Hajdu, J., Acharaya, K. R., Stuart, D. I., Barford, D., and Johnson, L. N. (1988). *Trends Biochem. Sci.*, **13**, 104.
6. Mozzarelli, A. and Rossi, G. L. (1996). *Annu. Rev. Biophys. Biomol. Struct.*, **25**, 3430.
7. Blundell, T. L. and Johnson, L. M. (1976). *Protein crystallography.* Academic Press, New York.
8. Wukovitz, W. and Yeates, T. O. (1995). *Nature Struct. Biol.*, **2**, 1060.
9. Usha, R., Johnson, J. E., Moras, D., Thierry, J.-C., Fourme, R., and Kahn, R. (1984). *J. Appl. Crystallogr.*, **17**, 147.
10. Kay, L. L. (1986). *ISIS/J. Hist. Sci. Soc.*, **77**, 450.
11. Feigelson, R. S. (1988). *J. Cryst. Growth*, **90**, 1.
12. Boistelle, R. and Astier, J.-P. (1988). *J. Cryst. Growth*, **90**, 14.
13. Rosenberger, F., Vekilov, P. G., Muschol, M., and Thomas, B. R. (1996). *J. Cryst. Growth*, **168**, 1.
14. Lehman, C. G. (1853). *Lehrbuch der physiologische Chemie.* Leipzig.
15. Reichert, E. T. and Brown, A. P. (1909). *The differentiation and specificity of corresponding proteins and other vital substances in relation to biological classification and evolution: the crystallography of hemoglobins.* Carnegie Institution, Washington DC.
16. Debru, C. (1983). *L'esprit des protéines: histoire et philosophie biochimiques.* Hermann, Paris.
17. McPherson, A. (1990). *J. Cryst. Growth*, **110**, 1.
18. Sumner, J. B. (1926). *J. Biol. Chem.*, **69**, 435.
19. Northrop, J. H., Kunitz, M., and Herriot, R. M. (1948). *Crystalline enzymes.* Columbia University Press, New York.
20. Jabri, E., Carr, M. B., Hausinger, R. P., and Karplus, P. A. (1995). *Science*, **268**, 998.
21. Dounce, A. L. and Allen, P. Z. (1988). *Trends Biochem. Sci.*, **13**, 317.
22. Bernal, J. D. and Crowfoot, D. (1934). *Nature*, **133**, 794.
23. Jones, C., Mulloy, B., and Sanderson, M. R. (ed.) (1996). In *Methods in molecular biology. Crystallographic methods and protocols*, Vol. 114, pp. 1–394. Humana Press, Totowa, NJ, USA.
24. McPherson, A. (1982). *Preparation and analysis of protein crystals.* Wiley, New York.
25. Kim, S. H. and Rich, A. (1968). *Science*, **162**, 1381.
26. Dock, A.-C., Lorber, B., Moras, D., Pixa, G., Thierry, J.-C., and Giegé, R. (1984). *Biochimie*, **66**, 179.
27. Michel, H. (1982). *J. Mol. Biol.*, **158**, 567.
28. Kühlbrandt, W. (1988). *Q. Rev. Biophys.*, **21**, 429.
29. Arnoux, B., Ducruix, A., Reiss-Husson, F., Lutz, M., Norris, J., Schiffer, M., *et al.* (1989). *FEBS Lett.*, **258**, 47.
30. Michel, H. (ed.) (1991). *Crystallization of membrane proteins.* CRC Press, Boca Raton, FL, USA.

31. Jeruzalmi, D. and Steitz, T. A. (1997). *J. Mol. Biol.*, **274**, 748.
32. Giegé, R., Dock, A.-C., Kern, D., Lorber, B., Thierry, J.-C., and Moras, D. (1986). *J. Cryst. Growth*, **76**, 554.
33. Yonath, A., Frolow, F., Shoham, M., Müssig, J., Makowski, I., Glotz, C., *et al.* (1988). *J. Cryst. Growth*, **90**, 231.
34. Trakhanov, S., Yusupov, M., Shirikov, V., Garber, M., Mitschler, A., Ruff, M., *et al.* (1989). *J. Mol. Biol.*, **209**, 327.
35. Feigelson, R. S. (ed.) (1986). Proc. 1st Int. Conf. Protein Crystal Growth, Stanford, CA, USA, 1985. *J. Cryst. Growth*, **76**, 529.
36. Giegé, R., Ducruix, A., Fontecilla-Camps, J., Feigelson, R. S., Kern, R., and McPherson, A. (ed.) (1988). Proc. 2nd Int. Conf. Crystal Growth of Biological Macromolecules, Bischenberg, France, 1987. *J. Cryst. Growth*, **90**, 1.
37. Carter, C. W., Jr. (ed.) (1990). *Methods: a companion to methods in enzymology*, Vol. 1, pp. 1–127.
38. Ward, K. and Gilliland, G. (ed.) (1990). Proc. 3rd Int. Conf. Crystal Growth of Biological Macromolecules, Washington, DC, USA, 1989. *J. Cryst. Growth*, **110**, 1.
39. Stezowsky, J. J. and Littke, W. (ed.) (1992). Proc. 4th Int. Conf. Crystal Growth of Biological Macromolecules, Freiburg, Germany, 1991. *J. Cryst. Growth*, **122**, 1.
40. Glusker, J. P. (ed.) (1994). Proc. 5th Int. Conf. Crystal Growth of Biological Macromolecules, San Diego, CA, USA, 1993. *Acta Cryst.*, **D50**, 337.
41. Miki, K., Ataka, M., Fukuyama, K., Higuchi, Y., and Miyashita, T. (ed.) (1996). Proc. 6th Int. Conf. Crystal Growth of Biological Macromolecules, Hiroshima, Japan, 1995. *J. Cryst. Growth*, **168**, 1.
42. Drenth, J., and Garcia-Ruiz, J. M. (ed.) (1999). Proc. 7th Int. Conf. Crystal Growth of Biological Macromolecules, Granada, Spain, 1998. *J. Cryst. Growth*, **196**, pp. 185–720.
43. Wood, S. P. (1990). In *Protein purification applications: a practical approach* (ed. E. L. V. Harris and S. Angal), pp. 45–58. IRL Press, Oxford.
44. Weber, P. C. (1991). *Adv. Protein Chem.*, **41**, 1.
45. McPherson, A., Malkin, A. J., and Kuznetsov, Y. G. (1995). *Structure*, **3**, 759.
46. Durbin, S. D. and Feher, G. (1996). *Annu. Rev. Phys. Chem.*, **47**, 171.
47. Jakoby, W. B. (1971). In *Methods in enzymology* (ed. W. B. Jakoby), Vol. 22, pp. 248–52. Academic Press, London.
48. Judge, R. A., Johns, M. R., and White, E. T. (1995). *Biotechnol. Bioeng.*, **48**, 316.
49. Lorber, B., Jenner, G., and Giegé, R. (1996). *J. Cryst. Growth*, **103**, 117.
50. Von Hippel, P. H. and Schleich, T. (1969). In *Structure and stability of biological macromolecules* (ed. S. N. Timasheff and G. D. Fasman), Vol. 2, pp. 417–574. Dekker.
51. Mikol, V. and Giegé, R. (1989). *J. Cryst. Growth*, **97**, 324.
52. Ries-Kautt, M. and Ducruix, A. (1989). *J. Biol. Chem.*, **264**, 745.
53. Cacioppo, E., Munson, S., and Pusey, M. L. (1991). *J. Cryst. Growth*, **110**, 66.
54. Gilliland, G. L., Tung, M., Blakeslee, D. M., and Ladner, J. E. (1994). *Acta Cryst.*, **D50**, 408.
55. Thaller, C., Weaver, L. H., Eichele, G., Wilson, E., Karlson, R., and Jansonius, J. N. (1981). *J. Mol. Biol.*, **147**, 465.
56. Giegé, R., Drenth, J., Ducruix, A., McPherson, A., and Saenger, W. (1995). *Prog. Cryst. Growth Charact.*, **30**, 237.
57. McPherson, A. (1997). *Trends Biotechnol.* **15**, 197.

58. Bergdoll, M. and Moras, D. (1988). *J. Cryst. Growth*, **90**, 283.
59. Salemme, F. R., Genieser, L., Finzel, B. C., Hilmer, R. M., and Wendolosky, J. J. (1988). *J. Cryst. Growth*, **90**, 273.
60. Wang, A. H. J. and Teng, M. K. (1988). *J. Cryst. Growth*, **90**, 295.
61. Carugo, O. and Argos, P. (1997). *Protein Sci.*, **6**, 2261.
62. Chayen, N. E., Boggon, T. J., Cassetta, A., Deacon, A., Gleichmann, T., Habash, J., *et al.* (1996). *Q. Rev. Biophys.*, **29**, 227.
63. Seeman, N. C. (1991). *Curr. Opin. Struct. Biol.*, **1**, 653.

2

Biochemical aspects and handling of macromolecular solutions and crystals

B. LORBER and R. GIEGÉ

1. Introduction

The quality and quantity of the macromolecular samples are important prerequisites for successful crystallizations. Proteins and nucleic acids extracted from living cells or synthesized *in vitro* differ from small molecules by additional properties intrinsic to their chemical nature and their larger size. They are frequently difficult to prepare at a high degree of purity and homogeneity. Besides traces of impurities, harsh treatments may decrease their stability and activity through different kinds of alterations. Consequently, the quality of biomacromolecules depends on the way they are prepared and handled. As a general rule purity and homogeneity are regarded as conditions *sine qua non*. Accordingly, purification, stabilization, storage, and handling of macromolecules are essential steps prior to crystallization attempts. Other difficulties in crystal growth may come from the source of the biological material. It is advisable to have at disposal a few milligrams of material when starting first crystallization trials although structures were solved with submilligram quantities of protein (1). Once crystals suitable for X-ray analysis can be produced, additional material is often needed to improve their quality and size and to prepare heavy-atom derivatives. It is thus essential that isolation procedures are able to supply enough fresh material of reproducible quality. Similar situations are encountered with multi-macromolecular assemblies (e.g. viruses, nucleosomes, ribosomal particles, or their subunits).

This chapter discusses biochemical methods used to prepare and characterize macromolecules intended for crystallization assays. Practical aspects concerning manipulation and qualitative analyses of soluble proteins will be emphasized. The cases of nucleic acids and membrane proteins are described in more detail in Chapters 8 and 9. Peculiar aspects of molecular biology that are important for crystallogenesis are presented in Chapter 3. They include the design of engineered macromolecules with new physical properties or modified to simplify purification or crystallographic analysis. Finally, methods for identification of macromolecular content of crystals and measurements of their density are presented as well.

2. The biological material

2.1 Sources of biological macromolecules

Many biological functions are sustained by classes of proteins and nucleic acids universally present in living organisms so that the source of macromolecules may seem unimportant. In fact, better crystallization conditions or better diffracting crystals are frequently found by switching from one organism to another. Variability in sequences between heterologous macromolecular species may lead to different conformations and consequently to different crystallization behaviours. Also, differences in crystal quality may result from addition or suppression of intermolecular contacts because of the high solvent content (50–80%) (2) and the existence of relatively few contacts in macromolecular crystal lattices. In practice, proteins isolated from eukaryotes are frequently more difficult to crystallize than their prokaryotic counterparts due to the presence of additional flexible domains. In eukaryotes, post-translational modifications are often responsible for structural and conformational microheterogeneity. Proteins isolated from thermophilic micro-organisms are more stable at higher temperatures than those from other organisms and may be more amenable to crystallization. Components of the protein biosynthesis machinery are examples (3, 4). Proteins from halophilic micro-organisms are alternative candidates whose stability is optimal in the presence of high salt concentrations close to those needed to reach supersaturation but their purification presents methodological difficulties (5). Finally, the physiological state of the cells and the 'freshness' of the starting material may be important. Certain proteins from unicellular organisms are isolated in their native state only when cells are in exponential or pre-stationary growth phase (6–8). Since catabolic processes are predominant in tissues of dead organisms, they should not be stored before use unless they are frozen immediately *post mortem*. Plant material should be processed immediately after harvest or quick-frozen for storage.

2.2 Macromolecules produced in host cells or *in vitro*

In the past, macromolecules that were most abundant, easy to isolate, and most stable were the first crystallized. Today, many researchers deal with biological molecules that are only present in trace amounts and the preparation of the quantities needed for crystallization assays can become a limiting step. In a number of cases this problem can be circumvented owing to the advancement of genetic engineering methods which make it possible to clone and overexpress genes in bacterial or eukaryotic cells (9, 10). In recombinant bacteria with multicopy plasmids non-toxic proteins can accumulate and in exceptional cases reach a quarter of total proteins (more typically a 100 mg of protein are isolated with a good yield from less than a 100 g of bacterial host cells). High intracellular concentration of certain proteins may lead to growth inhibition (11) or formation of inclusion bodies of denatured, aggregated, or even pseudo-crystallized material that requires adapted isolation procedures (12). Therefore, proper vectors and host cells must be chosen to optimize over-production levels and to retain native conformation and functional activity of the proteins (13). Rare codons may reduce translation rate and efficiency due to limiting concentrations of minor tRNA isoacceptors or amino acid misincorporation (14, 15). Separation of foreign macromolecules from endogenous ones can be difficult in the absence of specific biological assays (16). Finally, the presence of ligands or chaperones may be indispensable to maintain the native conformation of certain proteins (17). All these factors have to be kept in mind when planning a purification strategy.

Overexpressed proteins can be hypomodified because maturation enzymes responsible for the co- or post-translational modifications may not work with the same efficiency on recombinant macromolecules. Protein glycosylation, unknown in prokaryotes, occurs to various extents in eukaryotes. Mammalian proteins are glycosylated in yeast and in insect cells infected by baculovirus (18) but different modification patterns lead to structural variants (19) especially when modification enzymes are limiting. Continuous cell-free translation systems are potential alternatives to produce natural or factitious proteins for crystallization (20). Similarly, modification of nucleotides may be partial when RNA genes are overexpressed in host cells (21) and are absent when transcribed *in vitro*. RNAs produced *in vitro* by polymerases (from bacteriophages SP6 or T7) are frequently heterogeneous in length because transcription does not always terminate at a unique position (Chapter 8). Changes introduced in genes by site-directed mutagenesis (22) may result in proteins with altered conformations, decreased stability and activity, perturbed folding, changed modified post-synthetic modifications and solubility, or unforeseen degradations. Amino acid substitutions to increase stability or solubility, deletion of flexible domains, fusion with sequences binding to immobilized ligands in affinity chromatography or with polypeptides may facilitate crystallization. Mutants that are more suitable for the preparation of heavy-atom derivatives and selenomethionine containing proteins for anomalous scattering studies are now of general use (Chapter 3). Crystallization of protein–RNA complexes also profits from advantages of protein engineering methods (23).

3. Isolation and storage of pure macromolecules

Methods for the purification of proteins (24) and nucleic acids (Chapter 8) have been reviewed extensively. A few practical points are discussed here.

Readers are encouraged to consult manufacturer's or supplier's catalogues and application notes for updated technical information.

3.1 Preparative isolation methods

3.1.1 Generals

The manifold properties of macromolecules makes it difficult to give a general scheme to facilitate purification of an unknown biomolecule. Optimized protocols are usually achieved by trial-and-error approaches but more systematic procedures have been described (25). In such protocols, the sequence of events is important because macromolecules in crude extracts may be protected by interaction with ligands or other macromolecules which will probably be eliminated during isolation. Harmful compounds, like hydrolases, must be separated as early as possible. In no case they should be enriched or co-fractionated. *Table 1* lists major purification methods together with the appropriate equipment.

Table 1. Methods and equipment for the purification of biological macromolecules

- **Cell culture**
  Fermentors, culture flasks and plates, thermostated cabinets
  High capacity centrifuges or filtration devices for cell recovery
- **Cell disruption**
  Mechanical disruption devices (grinders, glass bead mills, French press)
  Chemical treatments (e.g. phenolic extraction of small RNAs)
  Biochemical treatments (e.g. cell lysis by enzymes)
  Others (e.g. sonication, freezing/thawing)
- **Centrifugation**
  Low speed centrifuge (to remove cell debris or recover precipitates)
  High speed centrifuge (to fractionate subcellular components)
- **Dialysis and ultrafiltration**
  Dialysis tubing (hollow fibres or membranes of various porosities and sizes)
  Concentrators (from 0.5 ml to a few litres with high flow rate low macromolecule-binding membranes of various cut-offs)
- **Chromatography** (prefer metal-free systems)
  Low pressure equipment for fast separation (FPLC, hyperdiffusion, perfusion)
  High pressure equipment (HPLC)
  Columns of various capacities filled with various matrices (particle size 10–30 mm)
  Pumps, programmer, on-line absorbance detector, fraction collector, recorder
- **Preparative electrophoresis and isoelectric focusing**
  Electrophoresis apparatus for large rod or slab gels
  Preparative liquid IEF apparatus (column or horizontal cells)
  Power supplies
- **Detection, characterization, and quantitation**
  Spectrophotometer, fluorimeter
  pH meter, conductimeter, refractometer (to monitor chromatographic elution)
  Liquid scintillation counter (for radioactivity detection)
  Analytical gel electrophoresis, capillary electrophoresis, and IEF equipment

The preparation of a cellular extract and fractionation of its components are the two stages common to most purification protocols (except for macromolecules secreted in culture media). Intracellular macromolecules are released using physical, chemical, or biological disruption methods and extracts are clarified by centrifugation or ultrafiltration. Membrane proteins and proteins with hydrophobic surfaces are solubilized with detergents or sulfobetaines (26) (Chapter 9). Extracellular compounds and macromolecules synthesized *in vitro* may be recovered either by ultrafiltration, centrifugation, flocculation, or liquid–liquid partitioning.

3.1.2 Proteins

Gross fractionation includes one or several precipitations induced either by addition of salts (e.g. ammonium sulfate), organic solvents (e.g. acetone), or organic polymers (e.g. PEG). Temperature or pH variation are applied to decrease solubility or stability of unwanted macromolecules. Fractionation between two liquid phases and selective precipitation (e.g. of nucleic acids by protamine) are additional methods. The next steps involve more resolutive methods, generally a combination of column chromatographies. These are based on separation by charge (adsorption, anion or cation exchange, chromatofocusing), hydrophobicity (hydrophobic interaction (HIC) or reverse-phase (RPC) chromatographies), size (exclusion chromatography), peculiar structural features (e.g. affinity for heparin, antibodies, metal ions, or thiol groups), or activity (affinity for catalytic sites, receptors, or biomimetic compounds). HPLC yields higher resolution than standard techniques because of the monodispersity and small size of the spherical matrix particles (27) and new matrices take advantage of hyperdiffusion (Beckman) or perfusion (PerSeptive Biosystems) to accelerate elution. Preparative IEF is carried on in gels with free or immobilized ampholytes (Immobiline®, Pharmacia) (28) in rotating cells divided in compartments by permeable nylon grids (Rotofor®, Bio-Rad) or in multichamber units holding fixed pH membranes (Isoprime™, Pharmacia). Differential centrifugation and free flow electrophoresis (29) are other methods. Monitoring of specific activities during the purification procedure helps to identify unsatisfactory steps in which macromolecules are lost or inactivated. Guidelines for effective protein purification may be summarized as follows:

- work in the cold room (i.e. at 4 °C) with chilled equipment and solutions if the protein is unstable at higher temperature
- use precipitation steps to speed up fractionation
- limit the number of chromatographies (to three or four)
- prefer quick assays to characterize macromolecules
- use short and efficient non-denaturing intermediary treatments (e.g. repeated dialysis)
- add stabilizing agents and protease inhibitors (see Section 5.5).

In summary, success in crystallization is often dependent on rapid purification. To reach this aim advanced equipment and chromatographic systems enabling high flow rate (e.g. advanced HPLC and perfusion chromatography) are recommended.

3.1.3 Nucleic acids

Purification of nucleic acids requires specific methods. For tRNAs, phenol extraction precedes counter-current distribution or chromatography on benzoylated DEAE–cellulose (also on Sepharose® or other matrices). Further purification is based on anion exchange, adsorption, reverse-phase, mixed-mode, hydrophobic interaction, perfusion, or affinity chromatographies. Intermediary treatments include precipitation by ethanol, dialysis, and concentration by evaporation under vacuum. HPLC on coated silica substituted by short aliphatic chains (C_4) gives separations with good resolution. Oligo-DNA or RNA are synthesized chemically on solid phase supports or enzymatically *in vitro*. Abortive sequences are eliminated by denaturing gel electrophoresis (Prepcell™, Bio-Rad) or HPLC (30–32). For further details see Chapter 8.

3.2 Specific preparation methods

Re-chromatography of samples prior to crystallization removes minor contaminants (e.g. degradation or ageing products), aggregates appearing during storage or small molecules (e.g. additives like glycerol) (33), and favours crystallization. HIC (at a 1–100 mg scale) can bring proteins and nucleic acids directly into solutions containing crystallizing agents (e.g. salt, MPD, or PEG) (34, 35). Purification on hydroxyapatite was cardinal for crystallizing some proteins. Microscale concentration devices are useful to remove small size contaminants and exchange buffers prior to crystallization. Additional advice is given in Section 5.5.

3.3 Stabilization and storage

Macromolecules extracted from cells must be kept in solutions having properties close to those of the cellular medium in order to maintain native conformations. Storage under improper conditions spoils the precious material obtained after long and hard work. Buffers whose pK is only weakly affected by temperature (36) are recommended to avoid pH variations in samples stored frozen and assayed at room temperature. Macromolecules prone to aggregation require a minimal ionic strength to stay soluble but all ions may not be compatible with their native structure or activity. Denaturation is minimized by avoiding pH or temperature extremes as well as contact with organic solvents, chaotropic agents, or oxidants (37, 38). Thiol groups in proteins require a reducing agent (e.g. DTE, DTT, 2-mercaptoethanol, or glutathione). Finally, diluted proteins may adsorb onto the walls of glass or plastic containers (39).

Structure and stability of globular proteins depend upon intramolecular hydrogen bonds. Interaction with ions of the Hofmeister series affects stability (40) (Chapter 10). Ice formation during freezing or freeze-drying induce partial unfolding (41). Therefore, do not freeze or lyophilize unprotected macromolecules (the latter process removes bound water belonging to the macromolecular solvation shell). Glycerol at high concentration (e.g. 50–60%, v/v) stabilizes proteins (42) and stays liquid at –20 °C (its high viscosity reduces diffusion by about two orders of magnitude). Trehalose, sucrose (43), and other cosmotropic agents (44) preserve biomaterials by favouring more compact structures and by reducing backbone thermal fluctuation (45). Storage as a suspension in an ammonium sulfate solution is efficient but sometimes less convenient since after some time it generates heterogeneities due to amidation of glutamate and aspartate side chains. Ligands enhance protein stability. Bactericidal or fungicidal agents (like *highly toxic* sodium azide, ethylmercurithiosalicylate, or volatile thymol) should be added for storage or crystallization. Nucleic acids stored dry or as alcoholic precipitates should be free of phenol which leads to alkaline-type hydrolysis. RNA molecules are chemically and structurally stable at slightly acidic pH (4.5–6.0) and in the presence of Mg^{2+} (Chapter 8).

3.4 Ageing

Properties of macromolecular samples change with time as explicitly illustrated for lysozyme crystallization (46). Ageing results from the action of contaminants present or introduced in samples or of modifications generated by oxidants. In the example of lysozyme, changes in crystallizability are due to the presence of fungi that multiply in the stored protein solution (46). Water molecules or metal ions induce slow hydrolysis in RNAs (Chapter 8).

Self-cleaving macromolecules, like certain proteases and ribozymes, pose specific problems. For protease crystallization the problem can be solve by storing and co-crystallizing the protease with an active site inhibitor. For instance, the three-dimensional structure of a human protease of the ICE type participating in apoptosis could be solved because it was crystallized as a covalent complex with a tetrapeptide inhibitor (47). For hammerhead ribozyme, self-cleavage was prevented by introducing modified bases into the molecule (48).

4. Characterization and handling of macromolecules

Numerous analytical tools are available to detect, characterize, and quantitate macromolecules. This section deals with general methods of particular interest for crystal growers and gives practical advice for handling pure macromolecules.

4.1 Analytical biochemical methods

Gel electrophoresis quickly visualizes the macromolecular content of a sample. Procedures for proteins are well known (49) and *Protocol 1* is adapted for small nucleic acids. Electrophoresis gives an estimate of the apparent size of proteins (49) and nucleic acids (50). IEF in polyacrylamide gels containing free or immobilized pH gradients separates proteins with differences in isoelectric points smaller than 0.01 pH unit (51). Capillary electrophoresis and mass spectrometry give rapid purity and homogeneity diagnostics (52, 53). Electronspray ionization mass spectra (ESI) inform about the counterion distribution around a protein (54). ESI and MALDI (matrix-assisted desorption/ionization) are the most sensitive and rapid methods to detect microheterogeneities in macromolecular samples. HPLC is useful to separate small amounts of macromolecules for further analysis (28). Sequence analysis is complementary to amino acid composition determination (55).

Protocol 1. Gel electrophoresis of nucleic acids (up to 150-mers)

Caution! Acrylamide, *N,N'*methylene bisacrylamide, 'Stains all', and ethidium bromide are highly toxic. Wear gloves when working with their solutions. Also protect your eyes by wearing glasses or a facial screen when working with UV light.

Equipment and reagents

- PAGE equipment (e.g. Mini-Protean, Bio-Rad) sold for protein electrophoresis
- Acrylamide-bisacrylamide solution: 38% (w/v) acrylamide, 2% (w/v) *N,N'*methylene bisacrylamide
- 20 × Tris-borate buffer: 243 g Tris base, 110 g boric acid, 18.6 g EDTA for 1 litre, pH 8.3
- TEMED
- Fresh 5% (w/v) solution of ammonium peroxodisulfate
- Urea (ultrapure or electrophoresis grade)
- Saccharose or glycerol
- Bromophenol blue
- Ultrafiltration membrane (0.45 μm pore size) (Millipore)
- 'Stains all' dye powder (Eastman Kodak, Aldrich)
- Ethidium bromide

Method

1. Prepare a stock solution to make gels having a total acrylamide concentration T = 8% (w/v), a cross-linker concentration C = 5% (w/w), and containing 8 M urea by mixing:
   - 20 ml acrylamide-bisacrylamide solution
   - 5 ml of 20 × Tris-borate buffer
   - 50 g urea
   - distilled water up to 100 ml
2. Filter the above stock solution on a 0.45 μm pore size membrane.

3. To 10 ml solution add 10 μl TEMED and 100 μl fresh 5% (w/v) ammonium peroxodisulfate solution. Mix and pour in the mould. Polymerization occurs in about 30 min.
4. Denature nucleic acids in a solution containing 8 M urea, 20% (w/v) saccharose (or 20% (v/v) glycerol), and 0.025% (w/v) bromophenol blue. Load samples onto the gel and run electrophoresis under appropriate voltage.
5. Stain nucleic acids by soaking gels in the dark in a solution containing 30 mg 'Stains all', 100 ml dimethylformamide, and distilled water up to 1 litre. Destain in the light. Another technique employs electrophoresis buffer containing 0.5 μg/ml ethidium bromide. Wait 10 min and view the gel in UV light (254 nm). Silver stain techniques for proteins are also suitable to visualize nucleic acids (49).

Methods to quantitate proteins are dictated by sample size (volume and concentration) and required degree of specificity (56). Spectrophotometry is non-invasive and accurate when extinction coefficients are known. Theoretical molar absorption coefficients ε of polypeptides can be calculated from tryptophan and tyrosine content using:

$$\varepsilon_{280\ \mathrm{nm}} = 5690\ n_x + 1280\ n_y,$$

where 5690 and 1280 are the molar absorption coefficients at 280 nm of tryptophan and tyrosine, and n_x and n_y the numbers of tryptophan and tyrosine residues, respectively (57). Hence, protein concentrations are obtained from:

$$\mathrm{c\ (mg/ml)} = \mathrm{A}_{280\ \mathrm{nm}} \times \varepsilon_{280\ \mathrm{nm}}/\mathrm{M}_r.$$

If the amino acid composition of a protein is unknown, the ε coefficient can be determined from ponderal, spectrophotometric, or refractive index measurements, as well as from colorimetric dye binding assays, but results are skewed when its composition deviates from that of the reference protein or when contaminants interfere (57). In mixtures, protein concentrations are estimated by empirical formulas eliminating the contribution of nucleic acids. For quartz cuvettes of l cm optical pathway (58, 59):

$$\mathrm{c\ (mg/ml)} = (1.55\ \mathrm{A}_{280\ \mathrm{nm}} - 0.76\ \mathrm{A}_{260\ \mathrm{nm}})/l,$$

or for a better estimate

$$\mathrm{c\ (mg/ml)} = 0.3175\ (\mathrm{A}_{228.5\ \mathrm{nm}} - \mathrm{A}_{234.5\ \mathrm{nm}})/l.$$

Active site titration monitors the functionality of individual enzyme molecules (60) but it cannot detect altered molecules whose activity is unaffected. Immunological properties may be used to assess the ability of antibodies to recognize conformational states or parts of molecules.

Approximate concentrations of RNA are obtained assuming that 1 $A_{260\ nm}$ unit (for 1 cm path length) corresponds to 0.040 mg/ml. Values are more accurate when extinction coefficients are known (61) from absorbance and phosphorus content measurements (62).

4.2 Prevention of macromolecular damage

Pure macromolecules require special care to prevent damage or loss. Following practical advice can contribute to the success of a crystallization project:

(a) Concentrate samples by ultrafiltration in devices using pressure or centrifugal force (cylindrical cells or parallel flow plates) or by dialysis against hygroscopic compounds (e.g. dry, high M_r PEG or gel filtration matrices). Choose membranes with low binding capacity for stirred pressure cells. Optimize stir-rate to impede denaturation through shearing or adsorption onto membranes. Do not create oxidizing environments (foam or air bubbles) to avoid formation of disulfide bonds in proteins.

(b) Centrifuge aggregates forming as a consequence of pH decrease, oxidation, or increase of salt or protein concentration.

(c) Concentrate biomolecules by precipitation (e.g. add ammonium sulfate) and dissolve the precipitate in a small volume, or adsorb them on a chromatographic matrix and elute them at higher ionic strength.

(d) Beware of techniques concentrating contaminants (e.g. proteases and nucleases).

(e) Exchange buffers and concentrate macromolecules over membranes with appropriate cut-offs to eliminate small M_r compounds.

(f) Solubilize proteins with mild non-ionic detergents (e.g. octylglucoside) or non-detergent sulfobetaines (26) (Chapter 9).

(g) Avoid high concentrations of denaturing agents (e.g. guanidinium chloride, urea, or chaotropic detergents) which inactivate or unfold macromolecules.

(h) Prepare buffers freshly with ultra-pure water and high-grade chemicals and adjust their pH after mixing all ingredients (pH may change after dilution or in the presence of other compounds). Purify suspicious chemicals by crystallization, distillation, or chromatography.

(i) Add bactericidal or fungicidal agents (e.g. sodium azide, sodium ethylmercurithiosalicylate, or thymol at 0.02%, w/v) in solutions (some buffers, like *toxic* cacodylate ions, have bactericidal properties).

(j) Experiment on aliquots to limit handling of stock solutions and avoid repeated freezing/thawing of macromolecules.

(k) Remove undesired molecules by dialysis, ultrafiltration, or size exclusion chromatography.

(l) Prepare macromolecules with or without their ligands (e.g. coenzyme, metal ions) or try additives (e.g. ions, reducing agents, chelators) to search for conformers crystallizing more readily.

(m) Sterilize glass- or plasticware in contact with nucleic acids. Wear gloves during manipulations; fingers are always contaminated by nucleases (63), proteases, and bacteria.

(n) Store solutions in air-tight bottles to prevent contamination by airborne micro-organisms.

5. The problem of purity and homogeneity

5.1 The concept of 'crystallography-grade' quality

The concept of purity takes a peculiar meaning in biological crystallogenesis (36). Not only must molecules be pure, i.e. deprived of unrelated macromolecules or undesired small molecules, but they must be 'pure' in terms of structure and conformation. In other words crystallization trials should be done with homogeneous populations of conformers. This concept is based on the fact that the best crystals can only be grown from solutions containing well-defined entities with identical conformations and physico-chemical properties. Over the years it has been refined with the improvements of analytical biochemical and biophysical micromethods.

The importance of purity may appear exaggerated and contradictory with earlier views because crystallization can be used as a purification method in chemistry and biochemistry (64). For structural studies, however, the aim is to prepare monocrystals diffracting at high resolution with a good mosaicity and a prolonged stability in the X-ray beam. It is thus understandable that contaminants may compete for sites on the growing crystals and generate lattice errors leading to internal disorder, dislocations, irregular faces and secondary nucleation, twinning, poor diffraction, or early cessation of growth (65). Dynamic light scattering of contaminated protein solutions and *in situ* atomic force microscopy (AFM) on growing crystals have provided convincing evidence that these phenomena occur for biomolecules (66, 67). Because of the high number of molecules in a single crystal ($\sim 10^{20}$ per mm^3), p.p.m. amounts of contaminant may induce formation of non-specific aggregates, alter macromolecular solubility, or interfere with nucleation and crystal growth (66, 67). The effects of impurities are reduced in gelified media (68, 69, and Chapter 6). Successful crystallization of rare proteins and nucleic acids support the importance of purity and homogeneity (1, 4, 33, 70).

5.2 Purity of samples

The level of confidence for the purity of a macromolecule depends upon the resolution, specificity, and sensitivity of the methods used to identify contaminants. Protein or nucleic acid samples are seldom analysed for contamination by other classes of molecules. Although most of the contaminants are eliminated along the purification steps through which the proteins or nucleic acids have gone, traces of polysaccharides, lipids, proteases, or nucleases may be sufficient to hinder crystallization. Small molecules, like peptides, oligonucleotides, amino acids, carbohydrates, or nucleotides as well as uncontrolled ions should also be considered as contaminants. Buffer molecules remaining from a former purification step can be responsible for irreproducible crystallization (e.g. phosphate ions are relatively difficult to remove and may crystallize in the presence of other salts). Counterions play a critical role in the packing of biomolecules. Often macromolecules do not crystallize or yield different habits in the presence of various buffers adjusted at the same pH. Consequently, 'purity' means also that reagents used with pure macromolecules (e.g. precipitants, buffers, detergents, or additives) should be of the highest grade. This is especially true for precipitants present at molar concentrations. Contamination of a 2 M ammonium sulfate solution by only 0.001% (w/w) (1 mM) Pb^{2+} equals a stoichiometry of one molecule of impurity per molecule of a protein of M_r 50000 present at 5 mg/ml (note that such contamination level may promote crystallization; for instance Cd^{2+} is needed for ferritin crystallization because it is involved in packing contacts) (71). Purification techniques for common precipitants are listed in *Table 2*. Commercial detergents should also be repurified (75, 76). Chemicals in which contaminants do not exceed a few p.p.m. are commercially available but the label 'ultra pure' is sometimes exaggerated. Molecules released from non-inert chromatography matrices (e.g. Sephadex, celluloses) by enzymatic digestion or by desorption of organic compounds (e.g. organic phases bound to silica matrices) fall in the category of impurities.

5.3 Microheterogeneity of samples

Microheterogeneity in pure macromolecules is only revealed by very resolutive methods (Section 5.4). Although its causes are multiple, the most

Table 2. Techniques for the purification of major crystallization agents

Chemicals	Major contaminants in commercial batches	Purification techniques and references
Ammonium sulfate	Ca^{2+}, Fe ions, Mg^{2+}, $PbSO_4$,[a] CN^-, NO^{3-}	Recrystallization
PEG[b]	Cl^-, F^-, NO^{3-}, PO_4^{2-}, SO_4^{2-}, peroxides, aldehydes	Column chromatography[b] (72, 73), recrystallization[b]
MPD	Cl^-, K^-, Na^+, SO_4^{2-}	Distillation under vacuum, column chromatography (74)

[a] Non-soluble species.
[b] See Chapter 5, *Protocol 1*.

Table 3. Frequent sources of microheterogeneity in pure proteins and ribonucleic acids

Variation in primary structure (genetic variations, synthesis errors, hydrolysis)
Variation in secondary structure (misfolding or partial unfolding)
Variation in tertiary structure (conformers)
Variation in quaternary structure (oligomerization)
Molecular dynamics (flexible domains)
Incomplete post-transcriptional or post-translational modifications
Partial binding of ligands or foreign molecules
Aggregation (specific or non-specific)
Fragmentation (i.e. chemical or enzymatic hydrolysis)
Chemical alterations (e.g. partial oxidation of sulfhydryl groups, deamidation)
Others

common ones are uncontrolled fragmentation and post-synthetic modifications (*Table 3*). It must be emphasized that the strict identity in sequence of *in vivo* or *in vitro* produced polypeptides or nucleic acids with that of natural molecules should be confirmed.

5.3.1 Structural microheterogeneity

Proteolysis normally takes part in many physiological processes (e.g. maturation, regulation of enzymatic activity, and catabolism) (77–82), and represents a major difficulty to overcome during protein isolation because proteases (M_r 20 000–800 000) are localized in various cellular compartments or secreted in the extracellular medium. Proteases are distinguished by the structure of their catalytic site containing either a serine, aspartic acid, or cysteine residue, or a metal ion. Not all can be inhibited by commercial compounds (*Table 4*). Upon cell disruption, cellular compartments are mixed with extracellular proteases and control over proteolysis is lost. Decrease of protein size or stability, modification of their charge or hydrophobicity, partial or total loss of activity or of immunological properties are signs of proteolysis. Hydrolysis of nucleic acids by nucleases is frequently detected on sequencing gels (83). Traces of proteases or nucleases may not be detectable even when overloading electrophoresis gels but they can cause damage during concentration or storage of samples. Fragmentation of RNA by chemical hydrolysis is catalysed by metal ions and enhanced at alkaline pH (84) (Chapter 8).

Co- or post-translational enzymatic modifications generate microheterogeneities in proteins when different groups (e.g. oligosaccharide chains) occupy all sites or when correct modifications are unevenly distributed over the polypeptide chains (e.g. when all sites are not substituted). Over a hundred modifications are known of which some are listed in *Table 5*. Most of them require special methods for their analysis (85, 86). Only some modifications are reversible (e.g. phosphorylation) but not glycosylation or methylation. Heterogeneity in carbohydrate chains, either *N*-linked at asparagine or *O*-linked at serine or threonine residues, is frequent in eukaryotic proteins

Table 4. Some commercially available protease or nuclease inhibitors

Proteases or nucleases	Inhibitors[a]
All protease classes	Possibly α_2-macroglobulin or DEPC
Serine proteases	DIFP, PMSF, Pefabloc® SC,[b] aminobenzamidine, 3,4-dichloro isocoumarin, antipain, chymostatin, elastinal, leupeptin, boronic acids, cyclic peptides, trypsin inhibitors (e.g. aprotinin, peptidyl chloromethyl ketone)
Aspartic acid proteases	Pepstatins and statin-derived inhibitors
Cysteine proteases	All thiol binding reagents, peptidyldiazomethanes, epoxysuccinyl peptides (e.g. E-64), cystatins, peptidyl chloromethanes
Metalloproteases	Chelators (e.g. EDTA, EGTA), phosphoramidon and phosphorus containing inhibitors, bestatin, amastatin and structurally related inhibitors, thiol derivatives, hydroxamic acid
Ribonucleases	RNasin® (Promega), ribonucleoside–vanadyl complexes, DEPC
Deoxyribonucleases	DEPC, chelators (e.g. EDTA, EGTA)

[a] These compounds are toxic and must be manipulated with caution. For more inhibitors, see ref. 96.
[b] According to the manufacturer (Pentapharm Ltd.), Pefabloc® SC or AEBSF (4-(2-aminoethyl)-benzensulfonyl fluoride) is a non-toxic alternative to PMSF and DIFP.

Table 5. Some co- or post-translational modifications of proteins

Amino acid residues or chemical groups	Chemical group added or modification
Amino terminal -NH^{3+}	Formyl-, acetyl-, glycosyl-, aminoacyl-, cyclization of Gln
Carboxy terminal -COOH	Amide, amino acyl-
Arg	ADP ribosyl-, methyl-, ornithine
Asp, Glu	Carboxyl-, methyl-
Asn, Gln	Glycosyl-, deamination
Cys	Seleno-, heme, flavin-
His	Flavin-, phospho-, methyl-
Lys	Glycosyl-, pyridoxyl-, biotinyl-, phospho-, lipoyl-, acetyl-, methyl-
Met	Seleno-
Phe	Hydroxyl-
Pro	Hydroxy-
Ser	Phospho-, glycosyl-, ADP ribosyl-
Thr	Phospho-, glycosyl-, methyl-
Tyr	Iodo-, hydroxy-, bromo-, chloro-

(87). Microheterogeneities may appear during storage, e.g. by deamidation of asparagine or glutamine residues.

Similarly, nucleotides in RNA and especially in tRNAs are modified or hypermodified co- or post-transcriptionally (88). Purification from bulk and subsequent crystallization of individual tRNA species is affected by the physiological state of cells (21) as well as by the amplitude of the modifications which modulates their overall charge and hydrophobicity.

5.3.2 Functional versus conformational heterogeneity

Pure macromolecules can be fully functional in a biochemical activity assay even though they are microheterogeneous. Conformational heterogeneity may have several origins: binding of ligands, intrinsic flexibility of molecular backbones, oxidation of cysteine residues, or partial denaturation. In the first case, macromolecules should be prepared in both forms, the one deprived of and the other saturated with ligands (89). In the second case, controlled fragmentation may be helpful. In the last one, oxidation of a single cysteine residue leads to a complex mixture of molecular species for which the chances of growing good crystals are low (90); reducing agents reverse such redox effects (Section 5.5).

5.4 Probing purity and homogeneity

Although macromolecules may crystallize readily in an impure state (63), it is always preferable to achieve a high level of purity before starting crystallization trials. Biochemical quality controls can be performed at relatively little expense in comparison to time-consuming crystallizations. It is recommended to combine several independent analytical methods (e.g. those described in Section 3.1) to assert the absence of contaminants or of microheterogeneities in samples (91).

Spectrophotometry and fluorimetry give information about the quality of samples if macromolecules or their contaminants have special absorbance or emission properties. As illustrated in *Figure 1*, a sharp peak in HPLC is insufficient evidence for the quality of a product. SDS–PAGE indicates the size of protein contaminants (visualized by staining, autoradiography, or immunodetection) but not that of non-protein contaminants. Gel IEF gives an estimate of the pI of protein components in a mixture and electrophoretic

Table 6. Selected techniques to detect structural and conformational heterogeneity

Techniques	Amount required (μg)	Information on
Activity assay	<1	Biological activity
Active site titration	<1	Ligand binding, affinity
Gel electrophoresis	<1–10	Mobility, size, charge
Gel filtration	<10–100	Size, shape
IEF/titration curve	<1–10	Charge, pI, mobility
Capillary electrophoresis/IEF	<1	Size, charge
Immunological titration	<1–10	Antigenic determinants
Scattering methods (light, X-rays, neutrons)	100–1000	Size, shape
Spectrophotometry, fluorimetry	<10–100	Absorption, emission
Ultracentrifugation	<100–1000	Size, shape
Mass spectrometry	<1	Mass

titration shows the mobility of individual proteins as a function of pH (92). The latter method can also suggest the type of chromatography (i.e. anion or cation exchange, chromatofocusing) suitable for further purification or guide toward other chromatographies (adsorption, size exclusion, hydrophobic interaction, or affinity). Capillary electrophoresis is well adapted for purity analysis (52). Amino acid composition and sequencing of N- and C-termini verify in part the integrity of primary structure (54). ESI and MALDI mass spectrometries are powerful tools in recombinant protein chemistry (93). Homogeneity of nucleic acids is probed by electrophoresis in gels containing urea (50, 89). Radioactive end-labelling enables detection of low levels of cleavage in ribose-phosphate chains (84). NMR detects small size contaminants and gives structural information on biomolecules (94). Useful methods for detecting conformational heterogeneity are given in *Table 6*.

5.5 Improving purity and homogeneity in practice

Some difficulties in crystallization (e.g. absence of crystals or poor diffraction) may be overcome by reconsidering the purification protocol. Start with fresh

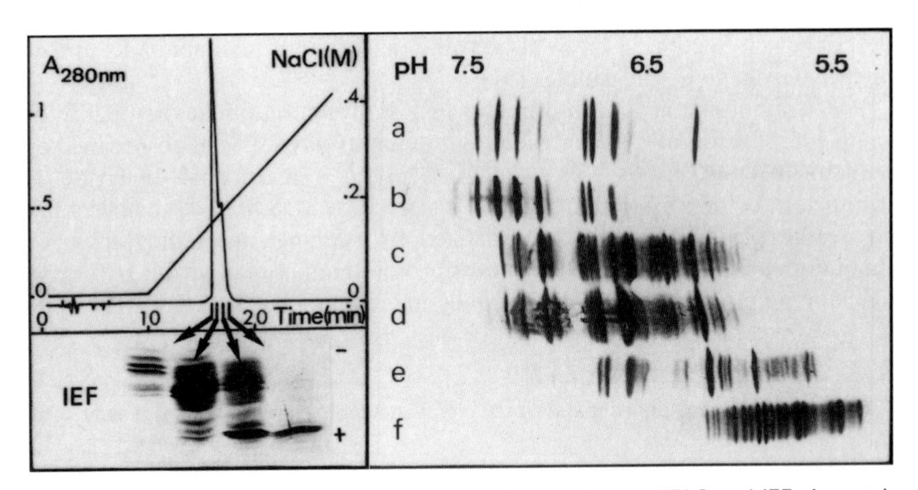

Figure 1. (*Left*) Comparison of the resolution of ion exchange HPLC and IEF. Aspartyl-tRNA synthetase from yeast (250 μg, pure according to standard ion exchange chromatography) was fractionated by anion exchange HPLC on a Mono Q column (i.d. 5 mm × length 5 cm, v = 1 ml, Pharmacia) in 50 mM Tris–HCl buffer pH 7.5, and was eluted at 0.5 ml/min with increasing NaCl concentration. IEF was performed on aliquots (3 μg protein) of the fractions. The polyacrylamide gel was 10 × 10 cm^2 (thickness 0.5 mm) and contained 2% (w/v) ampholytes (pH range 4–7). Staining with Coomassie Blue R-250 reveals several protein populations differing by charge. (*Right*) Batch-dependent variation in the microheterogeneity of pure aspartyl-tRNA synthetase. Six batches of protein purified according to a standard procedure and having the same specific activity, were compared by IEF under native conditions (samples of 5 μg protein, a dimer with a subunit M_r of 60 000, were analysed. Differences in charge result from uncontrolled proteolysis between positions 14 and 33 in the polypeptide chain.

material, change the sequence of events (by inverting chromatographic steps) or the steps themselves (by using other chromatographic matrices). To avoid cross-contamination never mix batches of pure macromolecules even when they look apparently identical. A small shift in the elution from a chromatography column or a preparation done on the same columns but at another scale or temperature can introduce other contaminants in active fractions. Such variability can sometimes be detected by IEF (*Figure 1*). Clean and sterilize by filtration (e.g. over 0.22 μm porosity membranes) all solutions in contact with pure macromolecules. Use chemically inert and autoclavable chromatography matrices which do not release molecules (e.g. Trisacryl, IBF Biotechnics, or TSK gels, Merck).

Macromolecules can be rendered more homogeneous in various ways. Addition of protease inhibitors is generally effective (*Table 4*) (95, 96). Assays to detect proteases by solubilization of clotted protein or by degradation of labelled peptides are commercially available (e.g. Peptag™, Boehringer). A cocktail of inhibitors should contain at least one specific for each protease class; an example is given in *Protocol 2*. On a small scale, chromatography over a column of immobilized inhibitors (e.g. α_2-macroglobulin) or substrate analogues (like arginine or benzamidine) may trap proteases. The major drawback of inhibitors lies in their possible binding to or inactivating of the proteins they should protect. Over-production in strains deprived of harmful proteases is a common solution to proteolysis (97).

Protocol 2. Preparation of buffered solutions of protease inhibitors and stabilizing agents

Caution! DIFP is a powerful inhibitor of human acetylcholine esterase. It must be handled with extreme caution. See manufacturer's safety data sheet. Other protease inhibitors are also very toxic. Wear gloves when handling solutions.

Reagents

- Buffer solutions containing 10% (v/v) glycerol and 10^{-3} M EDTA
- 0.1 M stock solution of DIPF (Sigma) prepared by diluting a 1 g commercial sample (about 1 ml) in 50 ml cold anhydrous isopropanol; always keep this solution at –20 °C
- 10^{-3} M stock solutions of peptidic inhibitors[a] (pepstatin, bestatin, and E-64 from Sigma) in ethanol:water (50:50)
- Reducing agents (2-mercaptoethanol and DTE or DTT) as stock solutions at 10^{-1} M

Method

1. Add DIPF, peptidic inhibitors, and 2-mercaptoethanol (DTE or DTT)[b] in buffer solutions just prior to use (final concentrations 5×10^{-4} M, 5×10^{-6} M, and 5×10^{-3} M, respectively).

Protocol 2. *Continued*

2. Add inhibitors afresh before cell disruption and at each step of the isolation procedure.

[a] Experimenters should be aware of low solubility, limited stability, affinity, and reversibility of inhibitors.
[b] Final concentration 10^{-3} to 10^{-4} M.

The action of nucleases is minimized by the addition of non-specific inhibitors, e.g. ribonucleoside–vanadyl complexes (98) (*Table 4*). Diethylpyrocarbonate (DEPC) reacts with histidyl residues at the surface of proteins and blocks the catalytic site of nucleases. RNasin® (Promega), a protein of M_r 51 000 isolated from human placenta, inactivates RNases by stoichiometric non-covalent and non-competitive binding (99). When added in transcription media to protect *in vitro* synthesized RNAs, it must be removed prior to crystallization. Nucleases show affinity for blue dextran (100) and 5′-(4-aminophenyl)-uridine-(2′,3′) phosphate (101) bound to agarose and chromatography on these media can remove them. Metal ions responsible for chemical hydrolysis of RNAs are eliminated by ion exchange or chelation (e.g. with EDTA, EGTA, or chelators immobilized on agarose beads). Magnesium is an exception and stabilizes RNAs at physiological pH (Chapter 8).

To enhance compactness and homogeneity undesirable parts of multidomain macromolecules may be removed by controlled fragmentation. Indeed, some proteins yield crystals of better quality after limited proteolysis (102). Unsuspected contamination by proteases may lead to a similar result. Mammalian glycoproteins must be deprived of their heterogeneous flexible carbohydrate moieties unless they are over-produced in prokaryotes. Treatment with specific endoglycosidases leaves only a single or a few sugar groups at each site (103–105). Crystals of deglycosylated proteins can diffract X-rays to high resolution (106). After controlled cleavage, the macromolecular core must be purified in order to remove proteases, glycosidases, peptides, or oligosaccharides. To be reproducible, enzymatic tools must be free of contaminant proteases that could introduce undesirable cleavages. The isolation of a single molecular species is essential to grow large monocrystals. In many instances additional purification by HPLC or IEF yielded better crystals (107–111).

6. Characterization and handling of crystals

6.1 Analysis of crystal content

Once crystals have been obtained, their macromolecular content must be verified. An X-ray diffraction pattern is convincing proof (Chapter 14) but simple analytical biochemical and biophysical methods give complementary information. *Table 7* lists two classes of biochemical and biophysical methods

Table 7. Selected methods for biochemical and biophysical characterization of crystals

Analytical methods[a]	Expected information and references
Crystalline material	
Mechanical stability test (with glass needle)	Softness of macromolecular crystals
Soaking with selective dyes[b]	Chemical nature of components
Enzymic digestion	Biochemical nature of components
X-ray diffraction[c]	Space group, resolution, unit cell parameters
Density measurements[d]	Solvent content, molecular mass of protomer
Microscopy (EM, AFM)	Crystallinity, surface topology (65)
Microspectrophotometry	Chemical composition (112)
Laser Raman spectroscopy	Conformational characteristics (113)
Soaking with ligands	Binding affinity (crystals may crack!)
In situ catalysis (for enzymes)	Catalytic activity (114)
Dissolved crystals	
Spectrophotometry/fluorimetry	Characterization and quantitation of molecules
Gel or capillary electrophoresis	Characterization of macromolecules, size
Gel or capillary IEF (for proteins only)	Characterization and charge
Column chromatography (microscale)	Characterization and quantitation (115)
Activity assays	Biological activity
Mass spectrometry	Molecular mass of components, detection of microheterogeneities, of counterions, sequencing (52–54, 93)
Sequencing (protein, nucleic acid)	Integrity of primary structure

[a] Other methods may be employed in particular cases, e.g. dichroism (116), analytical ultracentrifugation (117).
[b] For proteins, most stains used in light microscopy. For nucleic acids, see *Protocol 1*.
[c] See Chapter 14.
[d] See Section 6.

for the analysis of crystals. Some are applicable to the crystalline material itself whereas others require solubilized molecules. Since crystals contain only micrograms of macromolecules most methods must be scaled down. In all cases, the aims are:

(a) To verify that crystals contain the desired macromolecules and in the right stoichiometry in the case of co-crystals.

(b) To ensure that macromolecules are in an active conformation within crystals (for enzymes, this can be asserted by *in situ* catalytic assays provided active sites are accessible and ligands can diffuse harmlessly within the crystalline lattice).

(c) To compare the macromolecules in the crystalline state with those in solution (e.g. by laser Raman spectroscopy).

Prior to any analysis, uncrystallized macromolecules or amorphous material (present within the mother liquor or deposited onto the crystal faces) must be washed away. This is done by transferring crystals several times in large volumes of mother liquor.

6.2 Crystal density

Unlike crystals of small molecules made of densely packed matter, macromolecular crystals contain non-negligible amounts of solvent (2). The density of a crystal not only reflects the way molecules and solvent occupy the three-dimensional lattice but also provides structural information about the molecules contained in the unit cells. In the past, crystal density was used to estimate molecular weights (118) but sometimes wrong estimations have led to erroneous structure models.

In macromolecular crystals, the solvent represents on average 30–80% of the unit cell volume (2). It is made of free solvent (mother liquor) and bound solvent (water, ions, but also precipitant or buffer molecules, additives) forming the solvation (hydration) shell of the crystallized macromolecule (119). The knowledge of crystal density may be of importance for crystals of complexes (either of protein–ligand, protein–protein, or protein–nucleic acid nature) or co-crystals. The number n of macromolecules (or protomers) contained in one unit cell of volume V (cm^3) can be derived from crystal density r_c (g/cm^3) using the relation:

$$n = NV\,(r_c - r_s)/[M\,(1 - v_p\,r_s)]$$

where N is Avogadro's number (6.02×10^{23}), r_s the density of free and bound solvent (g/cm^3), M the protomer molar weight (g/mol), and v_p the partial specific volume of the dry protein (cm^3/g). Hence, the solvent fraction:

$$F_s = [(1/v_p) - r_c]/[(1/v_p - r_s)] = 1 - (nv_pM/NV).$$

In the case of proteins, the mass per asymmetric unit M_p is obtained with good approximation with the simplified formula (120):

$$M_p\,[\mathrm{Da}] = 2.324\,(V/n)\,[\text{Å}^3]\,(r_c\,[g/cm^3] - 1)$$

assuming the density of the solvent equals that of water ($r_w = 1.0$) and $v_p =$ 0.74 cm^3/g. In practice, the volume of the unit cell is calculated from cell parameters measured on X-ray diffraction patterns (Chapter 14). The molar weight of the macromolecule is determined by biochemical or biophysical methods (*Table 6*). Partial specific volumes of proteins (v_p) are either assumed to be equal to the inverse of their density, computed from partial specific volumes of individual amino acids (121) and the amino acid composition, or approximated as above. Partial specific volumes of nucleic acids can be approximated to 0.54 cm^3/g (122). The densities of solvent and crystals are determined experimentally. The resulting number of protomers is rounded to the nearest integer.

Two common methods for the measurement of crystal densities use either organic solvents or solutions of large polymers in which the sedimentation of the crystals is compared with internal references. Both are applicable to proteins, nucleic acids, and viral or ribosomal particles. They have their advantages and limitations. Other methods are described separately below. The densities of most protein crystals range from 1.10–1.60 g/cm^3 (123).

6.2.1 Organic solvent and Ficoll™ methods

The first method, adapted from the small molecule field, uses mixtures of water saturated carbon tetrachloride or chloroform and xylene. Experimental details are given in *Protocol 3*, part A. Measurements are sensitive to the presence of mother liquor around the crystals and difficulties may arise when trying to remove the excess solution. Readers are referred to ref. 123 for further details.

Protocol 3. Experimental determination of crystal density (adapted from ref. 126)

Caution! Organic solvents are toxic, especially carbon tetrachloride. Manipulate them under a fume-hood. Wear gloves and glasses.

Equipment and reagents

- Transparent glass tubes of 0.7–1 mm inner diameter
- Pycnometer or electronic densimeter equipped with a vibrating tube (DMA, Anton Paar)
- Glass micropipettes
- Centrifuge (reaching 3000 *g*)
- X-ray diffraction equipment
- Water saturated xylene, carbon tetrachloride, chloroform, toluene
- Salt (e.g. sodium phosphate) solutions
- Ficoll™ (Pharmacia)

A. *Organic solvent method*

1. Prepare gradients, in transparent glass tubes of 0.7–1 mm inner diameter, by combining various proportions of water saturated xylene and carbon tetrachloride (or chloroform).
2. Calibrate the gradients with salt (e.g. sodium phosphate) solutions whose densities are measured with a pycnometer or an electronic densimeter equipped with a vibrating tube.
3. Deposit one or two crystals with a minimal volume of mother liquor onto the gradient with a glass micropipette.
4. Estimate the final position of the crystals and compare with references.

B. *Ficoll™ method*

1. Prepare a series of solutions (from 30–60%, w/w) by mixing appropriate

Protocol 3. *Continued*

amounts of Ficoll™ powder and water. Heat at 55°C to dissolve. Cooled solutions are yellowish and viscous.

2. Work at constant temperature. Prepare gradients in transparent glass tubes of 0.7–1 mm inner diameter. Use a gradient maker or proceed by layers of decreasing density. Centrifuge after each layer (5 min, 1000 *g*) and end with a longer centrifugation (e.g. 1 h at 3000 *g*) to smooth the gradient.
3. Calibrate gradients with drops of carbon tetrachloride (or chloroform) and toluene mixtures (all saturated with water). The densities of these organic solutions are measured as in part A.
4. Deposit one or two crystals and a droplet of mother liquor onto the gradient with a glass micropipette. Centrifuge at 1000 *g* for 1–60 min and compare the displacement of the crystals with that of calibrated drops.
5. Take X-ray diffraction pictures to verify that unit cell parameters are unchanged.

Designed for macromolecular crystals, the second method uses a large sucrose polymer (Ficoll™, M_r 400 000, Pharmacia) as the only solute (124). The density of Ficoll solutions varies linearly with concentration. Density gradients are prepared by centrifugation and calibrated with mixtures of organic solvents obtained as in the previous method. Technical details are summarized in *Protocol 3*, part B. The method may lead to overestimation of crystal densities presumably due to the slow diffusion of polymer in the crystal lattice (120). Extrapolation of successive measurements to time zero should give a good estimate of density.

6.2.2 Other methods

Other methods can be used to estimate the solvent content of a crystal. Cross-linking may be helpful when crystals are too fragile. It is carried out by soaking them for a few hours in mother liquor containing either 1% or 2% (v/v) glutaraldehyde or formaldehyde to bind proteins or nucleic acids and proteins, respectively (125). For proteins or nucleic acids whose M_r and sequence are known, the molar absorption coefficient ε can be calculated and used to measure the amount contained in single crystals. To be applicable, the volume of the crystal must be known precisely (the task is easier when crystals have a regular geometrical shape). Then the crystal is dissolved in a small volume of buffer solution (e.g. 50 ml) and its absorption spectrum recorded. (A rough estimate of protein or RNA amounts in a crystal can also be obtained by comparison with known quantities analysed in parallel by gel electrophoresis.) From the weight amount of macromolecules in the crystal and the unit cell parameters, the fraction volume occupied by the solvent can be estimated and an approximate density can be extrapolated. More accurate results are expected when molar absorption coefficients are high. For crystals containing more than one type of macromolecule, protein and nucleic acid or a macromolecule and a smaller molecule (e.g. a protein and a ligand), their stoichiometry can be calculated when molecules differ by their spectroscopic properties.

References

1. Wierenga, R. K., Lalk, K. H., and Hol, W. G. J. (1987). *J. Mol. Biol.*, **198**, 109.
2. Matthews, B. V. (1968). *J. Mol. Biol.*, **33**, 491.
3. Garber, M., Davydova, N., Eliseikina, I., Fomenkova, N., Grysaznova, O., Gryshkovskaya, I., *et al.* (1996). *J. Cryst. Growth*, **168**, 301.
4. Thygesen, J., Krumholz, S., Levin, I., Zaytzev-Bashan, A., Harms, J., Bartels, H., *et al.* (1996). *J. Cryst. Growth*, **168**, 308.
5. Zaccaï, G. and Eisenberg, H. (1990). *Trends Biochem. Sci.*, **15**, 333.
6. Kern, D., Giegé, R., Robbe-Saul, S., Boulanger, Y., and Ebel, J.-P. (1975). *Biochimie*, **57**, 1167.
7. Adelman, R. C. and Dekker, E. E. (ed.) (1985). *Modification of proteins during aging*. Alan R. Liss, New York.
8. Lorber, B. and DeLucas L, J. (1990). *FEBS Lett.*, **241**, 14.
9. Sambrook, J., Fritsch, E. F., and Maniatis, T. (ed.) (1989). *Molecular cloning: a laboratory manual*, 3 Vols. Cold Spring Harbor Laboratory Press.
10. Goeddel, D. V. (ed.) (1991). *Methods in enzymology*, Vol. 185, pp. 1–599. Academic Press, London.
11. Kurland, C. G. and Dong, H. (1996). *Mol. Microbiol.*, **21**, 1.
12. Schein, C. H. (1990). *Biotechnology*, **8**, 308.
13. Miroux, B. and Walker, J. E. (1996). *J. Mol. Biol.*, **260**, 289.
14. Zahn, K. and Landy, A. (1996). *Mol. Microbiol.*, **21**, 69.
15. Calderone, T. L., Stevens, R. D., and Oas, T. G. (1996). *J. Mol. Biol.*, **262**, 407.
16. Carter, C. W. (1988). *J. Cryst. Growth*, **90**, 168.
17. Horwich, A. L., Neupert, W., and Hartl, F. U. (1990). *Trends Biotech.*, **8**, 126.
18. Richardson, C. D. (ed.) (1995). *Methods in molecular biology*, Vol. 39, pp. 1–420. Humana Press, Totowa, NJ.
19. Marston, F. A. O. (1986). *Biochem. J.*, **240**, 1.
20. Spirin, A. S., Baranov, V. I., Ryabova, L. A., Ovodov, S. Y., and Alakhov, Y. B. (1988). *Science*, **242**, 1162.
21. Perona, J. J., Swanson, R., Steitz, T. A., and Söll, D. (1988). *J. Mol. Biol.*, **202**, 121.
22. McPherson, M. J. (ed.) (1991). *Directed mutagenesis: a practical approach*, pp. 1–288. IRL Press, Oxford.
23. Oubridge, C., Io, N., Teo, C. H., Fearnley, I., and Nagai, K. (1995). *J. Mol. Biol.*, **249**, 409.
24. Harris, E. L. V. and Angal, S. (ed.) (1990). *Protein purification applications: a practical approach*, pp. 1–192. IRL Press, Oxford.
25. Yin, Y. and Carter, C. W., Jr. (1996). *Nucleic Acids Res.*, **24**, 1279.

26. Vuillard, L., Baalbaki, B., Lehmann, M., Norager, S., and Legrand, P. (1996). *J. Cryst. Growth*, **168**, 150.
27. Oliver, R. W. A. (ed.) (1989). *HPLC of macromolecules: a practical approach*, pp. 1–252. IRL Press, Oxford.
28. Righetti, P. G. (1983). *Isoelectric focusing: theory, methodology and application*. Elsevier, Amsterdam.
29. Wagner, H. (1989). *Nature*, **341**, 669.
30. Ott, G., Dorfler, S., Sprinzl, M., Muller, U., and Heinemann, U. (1996). *Acta Cryst.*, **D52**, 871.
31. Wahl, M. C., Ramakrishnan, B., Ban, C., Chen, X., and Sundaralingam, M. (1996). *Acta Cryst.*, **D52**, 668.
32. Cate, J. H., Gooding, A. R., Podell, E., Zhou, K., Golden, B. L., Kundrot, C. E., *et al.* (1996). *Science*, **273**, 1678.
33. Aoyama, K. (1996). *J. Cryst. Growth*, **168**, 198.
34. Gulewicz, K., Adamiak, D., and Sprinzl, M. (1985). *FEBS Lett.*, **189**, 179.
35. Giegé, R., Dock, A.-C., Kern, D., Lorber, B., Thierry, J.-C., and Moras, D. (1986). *J. Cryst. Growth*, **76**, 554.
36. Stoll,V. S. and Blanchard, J. S. (1990). In *Methods in enzymology* (ed. M. P. Deutscher), Vol. 182, pp. 24–38. Academic Press, London.
37. Yang, A. S. and Honig, B. (1993). *J. Mol. Biol.*, **231**, 459.
38. Darby, N. J. and Creighton, T. E. (ed.) (1989). *Protein structure: a practical approach*, pp. 1–120. IRL Press, Oxford.
39. Norde, W. (1995). *Cells Mater.*, **5**, 97.
40. Baldwin, R. L. (1996). *Biophys. J.*, **71**, 2056.
41. Strambini, G. B. and Gabellieri, E. (1996). *Biophys. J.*, **70**, 971.
42. Sousa, R. (1995). *Acta Cryst.*, **D51**, 271.
43. Crowe, L. M., Reid, D. S., and Crowe, J. H. (1996). *Biophys. J.*, **71**, 2087.
44. Jeruzalmi, D. and Steitz, T. A. (1997). *J. Mol. Biol.*, **274**, 748.
45. Buttler, S. L. and Falke, J. J. (1996). *Biochemsitry*, **35**, 10595.
46. Chayen, N. E., Radcliffe, J. W., and Blow, D. M. (1993). *Protein Sci.*, **2**, 113.
47. Mittl, P. R. E., Di Marco, S., Krebs, J. F., Bai, X., Karanewsky, D. S., Priestle, J. P., *et al.* (1997). *J. Biol. Chem.*, **272**, 6539.
48. Scott, W. G., Finch, J. T., Grenfell, R., Fogg, J., Smith, T., Gait, M. J., *et al.* (1995). *J. Mol. Biol.*, **250**, 327.
49. Hames, B. D. and Rickwood, D. (ed.) (1990). *Gel electrophoresis of proteins: a practical approach*, pp. 1–250. IRL Press, Oxford.
50. Rickwood, D. and Hames, B. D. (ed.) (1990). *Gel electrophoresis of nucleic acids: a practical approach*, pp. 1–336. IRL Press, Oxford.
51. Righetti, P. G., Gianazza, E., and Gelfi, C. (1988). *Trends Biochem. Sci.*, **13**, 333.
52. Karger, B. L. and Hancock, W. S. (ed.) (1996). *Methods in enzymology*, Vol. 270, pp. 1–611 and Vol. 271, pp. 1–543. Academic Press, London.
53. Muddiman, D. C., Bakhtiar, R., Hofstaler, S. A., and Smith, R. C. (1997). *J. Chem. Edu.*, **74**, 1288.
54. Riès-Kautt, M., Ducruix, A., and Van Dorsselaer, A. (1994). *Acta Cryst.*, **D50**, 366.
55. Matsudaira, P. (ed.) (1993). *A practical guide to protein and peptide purification and microsequencing*. Academic Press, Inc., San Francisco.
56. Stoscheck, C. M. (1990). In *Methods in enzymology* (ed. M. P. Deutscher), Vol. 182, pp. 50–68. Academic Press, London.

57. Pace, C. N., Vajdos, F., Fee, L., Grimsley, G., and Gray, T. (1995). *Protein Sci.*, **4**, 2411.
58. Warburg, O. and Christian, W. (1941). *Biochem. Z.*, **310**, 384.
59. Ehresmann, B., Imbault, P., and Weil, J.-H. (1973). *Anal. Biochem.*, **54**, 454.
60. Fersht, A. (1977). *Enzyme structure and mechanism*. Freeman & Company, Reading.
61. Guéron, M. and Leroy, J.-L. (1978). *Anal. Biochem.*, **91**, 691.
62. Ames, B. N. and Dubin, D. T. (1960). *J. Biol. Chem.*, **235**, 769.
63. Holley, R. W., Apgar, J., and Merrill, S. H. (1961). *J. Biol. Chem.*, **236**, PC42.
64. Judge, R. A., Johns, M. R., and White, E. T. (1995). *Biotech. Bioeng.*, **48**, 316.
65. Vekilov, P. G. and Rosenberger, F. (1996). *J. Cryst. Growth*, **158**, 540.
66. Skouri, M., Lorber, B., Giegé, R., Munch, J.-P., and Candau, S. J. (1996). *J. Cryst. Growth*, **152**, 209.
67. McPherson, A., Malkin, A. J., Kuznetsov, Y. G., and Koszelak, S. (1996). *J. Cryst. Growth*, **168**, 74.
68. Provost, K. and Robert, M.-C. (1995). *J. Cryst. Growth*, **156**, 112.
69. Hirschler, J., Charon, M.-H., and Fontecilla-Camps, J. C. (1995). *Protein Sci.*, **4**, 2573.
70. Doudna, J. A., Grosshans, C., Gooding, A., and Kundrot, C. E. (1993). *Proc. Natl. Acad. Sci. USA*, **90**, 7829.
71. Lawson, D. M., Artymiuk, P. J., Yewdall, S. J., Smith, J. M. A., Livingstone, J. C., Treffry, A., *et al.* (1991). *Nature*, **349**, 541.
72. Ray, W. J. and Puvathingal, J. (1985). *Anal. Biochem.*, **146**, 307.
73. Jurnak, F. (1986). *J. Cryst. Growth*, **76**, 577.
74. Bello, J. and Nowoswiat, E. F. (1965). *Biochim. Biophys. Acta*, **105**, 325.
75. Kuhlbrandt, W. (1988). *Q. Rev. Biophys.*, **21**, 429.
76. Lorber, B., Bishop, J. B., and DeLucas, L. J. (1990). *Biochim. Biophys. Acta*, **1023**, 254.
77. Achstetter, T. and Wolf, D. H. (1985). *Yeast*, **1**, 139.
78. Barrett, A. J. and Salvesen, G. (ed.) (1986). *Research monographs in cell and tissue physiology*, Vol. 12, pp. 1–661. Elsevier, Amsterdam.
79. Dalling, M. J. (ed.) (1986). *Plant proteolytic enzymes*, 2 Vol. CRC Press, Boca Raton, FL.
80. Bond, J. S. and Butler, P. E. (1987). *Annu. Rev. Biochem.*, **56**, 333.
81. Arfin, S. M. and Bradshaw, R. A. (1988). *Biochemistry*, **27**, 7979.
82. Wandersman, C. (1989). *Mol. Microbiol.*, **3**, 1825.
83. Howe, C. J. and Ward, E. S. (ed.) (1989). *Nucleic acids sequencing: a practical approach*, pp. 1–254. IRL Press, Oxford.
84. Moras, D., Dock, A.-C., Dumas, P., Westhof, E., Romby, P., Ebel, J.-P., *et al.* (1985). *J. Biomol. Struct. Dyn.*, **3**, 479.
85. Wold, F. and Moldave, K. (ed.) (1984). *Methods in enzymology*, Vol. 106, pp. 1–574 and Vol. 107, pp. 1–688. Academic Press, London.
86. Kendall, R. L., Yamada, R., and Bradshaw, R. A. (1990). In *Methods in enzymology* (ed. D. V. Goeddel), Vol. 185, pp. 398–408. Academic Press, London.
87. Rudd, P. M. and Dwek, R. A. (1997). *Crit. Rev. Biochem. Mol. Biol.*, **32**, 1.
88. Grosjean, H. and Benne, R. (ed.) (1998). *Modification and editing of RNA*. ASM Press, Washington DC.
89. Lorber, B., Giegé, R., Ebel, J.-P., Berthet, C., Thierry, J.-C., and Moras, D. (1983). *J. Biol. Chem.*, **258**, 8429.

90. Van der Laan, J. M., Swarte, M. B. A., Groendijk, H., Hol, W. G. J., and Drenth, J. (1989). *Eur. J. Biochem.*, **179**, 715.
91. Rhodes, D. G. and Laue, T. M. (1990). In *Methods in enzymology* (ed. M. P. Deutscher), Vol. 182, pp. 555–66. Academic Press, London.
92. Lorber, B., Kern, D., Mejdoub, H., Boulanger, Y., Reinbolt, J., and Giegé, R. (1987). *Eur. J. Biochem.*, **165**, 409.
93. Andersen, J. S., Svensson, B., and Roepstorff, P. (1996). *Nature Biotech.*, **14**, 449.
94. Wüthrich, K. (1995). *Acta Cryst.*, **D51**, 249.
95. Beynon, R. J. and Bond, J. S. (ed.) (1989). *Proteolytic enzymes: a practical approach*, pp. 1–278. IRL Press, Oxford.
96. Beynon, R. J. (1989). In *Protein purification methods: a practical approach* (ed. E. L. V. Harris and S. Angal), pp. 40–51. IRL Press, Oxford.
97. Gottesman, S. (1990). In *Methods in enzymology* (ed. D. V. Goeddel), Vol. 185, pp. 119–29. Academic Press, London.
98. Berger, S. L. and Birkenmeier, C. S. (1979). *Biochemistry*, **18**, 5143.
99. Scheele, G. and Blackburn, P. (1979). *Proc. Natl. Acad. Sci. USA*, **76**, 4898.
100. Thompson, S. T., Cass, K. H., and Stellwagen, E. (1975). *Proc. Natl. Acad. Sci. USA*, **72**, 669.
101. Wierenga, K., Huizinga, J. D., Gaastra, W., Welling, G. W., and Beintema, J. J. (1973). *FEBS Lett.*, **31**, 181.
102. Bergfors, T., Rouvinen, J., Lehtovaara, P., Caldentey, X., Tomme, P., Claeyssens, M., *et al.* (1989). *J. Mol. Biol.*, **209**, 167.
103. Kalisz, H. M., Hecht, H. J., Schomburg, D., and Schmid, R. D. (1990). *J. Mol. Biol.*, **213**, 207.
104. Stura, E. A., Chen, P., Wilmot, C. M., Arevalo, J. H., and Wilson, I. A. (1992). *Proteins*, **12**, 24.
105. Baker, H. M., Day, C. L., Norris, G. E., and Baker, E. N. (1996). *Acta Cryst.*, **D50**, 380.
106. Nitanai, Y., Satow, Y., Adachi, H., and Tsujimoto, M. (1996). *J. Cryst. Growth*, **168**, 280.
107. Anderson, W. F., Boodhoo, A., and Mol, C. D. (1988). *J. Cryst. Growth*, **90**, 153.
108. Bentley, G. A., Boulot, G., Riottot, M. M., and Poljak, A. (1988). *J. Cryst. Growth*, **90**, 213.
109. Spangler, B. D. and Westbrook, E. M. (1989). *Biochemistry*, **28**, 1333.
110. Moreau, H., Abergel, C., Carrière, F., Ferrato, F., Fontecilla-Camps, J. C., Cambillau, C., *et al.* (1992). *J. Mol. Biol.*, **225**, 147.
111. Weber, W., Wenisch, E., Gunther, N., Marnitz, U., Betzel, C., and Righetti, P. G. (1994). *J. Chromatogr.*, **A679**, 181.
112. Kirsten, H. and Christen, P. (1983). *Biochem. J.*, **211**, 427.
113. Chen, M., Lord, R., Giegé, R., and Rich, A. (1975). *Biochemistry*, **14**, 4385.
114. Mozzarelli, A. and Rossi, G. L. (1996). *Annu. Rev. Biophys. Biomol. Struct.*, **25**, 3430.
115. Lorber, B. and Giegé, R. (1985). *Anal. Biochem.*, **146**, 402.
116. Ford, R. C., Picot, D., and Garavito, R. M. (1987). *EMBO J.*, **6**, 1581.
117. Schuster, T. M. and Toedt, J. M. (1996). *Curr. Opin. Struct. Biol.*, **6**, 650.
118. Palmer, K. J., Ballantyne, M., and Galvin, J. A. (1948). *J. Am. Chem. Soc.*, **70**, 906.
119. Salemme, F. R., Genieser, L., Finzel, B. C., Hilmer, R. M., and Wendoloski, J. J. (1988). *J. Cryst. Growth*, **90**, 273.

120. Bode, W. and Schirmer, T. (1985). *Biol. Chem. Hoppe-Seyler*, **366**, 287.
121. Cohn, E. J. and Edsall, J. T. (ed.) (1943). In *Proteins, amino acids and peptides*, p. 370. Van Nostrand Reinhold, Princeton, NJ.
122. Cohen, G. and Eisenberg, H. (1968). *Biopolymers*, **6**, 1077.
123. Matthews, B. W. (1985). In *Methods in enzymology* (ed. H. W. Wyckoff, C. H. W. Hirs, and S. N. Timasheff), Vol. 114, pp. 176–87. Academic Press, London.
124. Westbrook, E. M. (1985). In *Methods in enzymology* (ed. H. W. Wyckoff, C. H. W. Hirs, and S. N. Timasheff), Vol. 114, pp. 187–96. Academic Press, London.
125. Sewell, B. T., Bouloukos, C., and Von Holt, C. (1984). *J. Microsc.*, **136**, 103.
126. Mikol, V. and Giegé, R. (1992). In *Crystallization of nucleic acids and proteins: a practical approach* (ed. A. Ducruix and R. Giegé), 1st edn, pp. 219–40. IRL Press, Oxford.

3

Molecular biology for structural biology

P. F. BERNE, S. DOUBLIÉ, and C. W. CARTER, JR

1. Introduction

The number of published 3D structures has increased exponentially in the last decade and the resulting mass of structural data has contributed significantly to the understanding of mechanisms underlying the biology of living cells. However, these mechanisms are so complex that structural biologists face still greater challenges, such as the study of higher-order functional complexes. As an example, we can mention the protein complexes that assemble around activated growth factor receptors to allow the transduction of extracellular signals through the membrane and inside the cell (1).

Because of their diverse intrinsic properties, proteins exhibit variable difficulty for structural biology studies. Before the rise of recombinant expression methods, only a minority of protein structures were determined, representing mainly favourable cases: proteins of high abundance in their natural source which could be purified and crystallized, in contrast to rare proteins that were often refractory to crystallization. The advent of methods for recombinant protein overexpression was a breakthrough in this area. It was followed by an increasing number of publications describing the crystallization of proteins, not under their native form, but in modified versions after sequence engineering.

First we will consider the classical use of molecular biology applied to optimize the expression system for a recombinant protein for structural biology, without modification of its sequence. In the second part, we will deal with molecular biology procedures aimed at engineering the properties of a protein through sequence modifications in order to make its crystallization possible. In the last part we will give an example where molecular biology can help solve a crystallographic problem, namely that of phase determination by introducing anomalous scatterers (e.g. selenium atoms) into the protein of interest.

2. Molecular biology to optimize protein expression systems

Whenever extraction of a protein from its natural source appears unsuitable for structural studies, molecular biology resources can be brought in, initially aiming at choosing and setting up an appropriate expression system. This initial approach could involve comparing various expression hosts and vectors and deciding if the protein is to be produced as a fusion to facilitate its purification.

2.1 Expression systems

Structural biology, whether NMR or crystallography, consumes large amounts of protein (up to 100 mg or more for a complete study). When a protein of major interest is discovered, it is often the case that the natural source does not express it at a sufficient level to allow structural studies. If this protein represents a small fraction of the total cellular proteins, important quantities of cell extracts have to be treated, and multiple purification steps are generally needed, therefore decreasing the overall yield of recovery.

Provided a gene encoding the protein to crystallize is available, expression systems can be used for large scale production of recombinant proteins. By 'expression system' one should understand the host, vector, and all procedures necessary to allow expression of a particular protein in a non-natural host. The most frequently used expression host is the bacterium *Escherichia coli*, followed by insect (*Spodoptera frugiperda*) cells infected by a baculoviral vector. Mammalian cells (for example the CHO cell line) and yeast strains are used less frequently and will be less detailed. The choice of an expression system (see *Table 1* for commercially available vectors) should take into account the nature of the protein and the experimental constraints specific to each system.

2.1.1 Bacteria

The bacterial host is by far the easiest to manipulate, particularly when dealing with large culture volumes, and it often yields a high level of overexpression (up to 300 mg/litre of low density culture and 50% of the total cellular proteins). The ease of manipulating bacterial expression systems facilitates approaches involving several mutants, for example to engineer variants in order to investigate the crystallization properties of a protein or to perform structure/function studies. However, a major drawback of the *E. coli* system lies in the fact that overexpressed proteins often misfold in the bacterium and aggregate into inclusion bodies (see Section 2.3). Additionally, the lack of post-translational modifications may be detrimental to protein activity.

2.1.2 Insect cells

The baculoviral system is used increasingly in many laboratories, facilitated by the availability of commercial tools that make this eukaryotic expression system more user-friendly (2). In this system, a viral vector carrying the gene of interest is used to infect insect cells. These host cells provide a more favourable environment for the folding of most eukaryotic proteins than does the cytoplasm of bacteria. In addition, cytoplasmic, membrane-localized, transmembrane, or secreted proteins will each be directed to their respective compartment, provided that the signals for localization are included in the vector. Insect cells have the machinery to perform most of the usual post-translational modifications, including the glycosylation of extracellular domains. Expression levels can vary from one particular construction to another, and overexpression yields of up to 50% of total proteins have been reported. Tropomyosin is a typical example of high expression (30 mg purified protein per litre of culture) in Sf9 insect cells that allowed crystallization (3). However, compared to the bacterial system, baculovirus systems require additional steps, such as the purification of a viral clone and the amplification of the virus by several infection cycles.

2.1.3 Mammalian cells

Expression of a protein in mammalian cells for structural studies is not frequent (4), because the set-up of a cell line overexpressing a given protein is time-consuming and the level of expression is never very high. Nevertheless, one should keep in mind that it could be the only system allowing expression of a mammalian protein in a context where it will be correctly processed and fully functional.

2.1.4 Yeast

Yeast can also be a potential host and has proved to be extremely efficient for expressing certain secreted proteins (up to 4 g per litre of culture). However, it should be limited to naturally secreted proteins because the vast majority of other proteins would undergo abnormal glycosylation if secreted by yeast cells. The *Pichia pastoris* strain has been advantageous for producing several complex eukaryotic proteins (5) and a *Pichia* kit is commercially available from Invitrogen. Yeast cells are less efficient tools for intracellular expression and require special equipment to be disrupted (6). However, there are examples of successful intracellular expression in *P. pastoris*, *Saccharomyces cerevisiae*, or *Schizosaccharomyces pombe* (7) for structural studies.

2.2 Factors influencing the level of expression

Expression vectors dedicated to the transfer and expression of a particular gene of interest into an expression host have been created, in order to optimize the level of expression, by including a strong promoter and the required elements for transcription and translation by the host machinery. However, for a single expression vector, it is well known that expression levels can vary from one protein to another, and there are no rules to predict these variations.

For expression in *E. coli*, the signals required for the expression vector (promoter, Shine–Dalgarno sequence for ribosome binding, transcription terminator) are well characterized (8), but the influence of gene-dependent factors is not fully understood (9). Codon usage is assumed not to play a significant role (10), except when particularly rare codons for arginine and/or isoleucine are repeated (11). The secondary structure of the messenger RNA can be critical (9) for its stability and for initiation efficiency (accessibility of the Shine–Dalgarno sequence). In addition, DNA hairpin structures can result in premature transcription termination. It is not clear why large proteins (i.e. > 50 kDa) are expressed on average at lower levels.

In *E. coli* expression vectors, the short number of nucleotides between the Shine–Dalgarno sequence and the initiation codon has to be strictly respected. This requires that the gene of interest be inserted using a restriction site overlapping with the initiation codon (except when making N-terminal fusion proteins). In contrast, there is no equivalent sequence in eukaryotic vectors. The first ATG codon found downstream from the promoter is used as the initiation codon. The sequence found around this codon seems to influence to some extent the efficiency of translation (12).

2.3 Inclusion bodies in *E. coli*

Of the many specific problems encountered when recombinant proteins are expressed in bacteria, the most frequent is the formation of inclusion bodies (8, 13). Inclusions can be seen as aggregates of misfolded protein. Their formation could result, at least in some cases, from the intrinsic insolubility of the protein of interest in the cytoplasm of *E. coli*. This can be the case for aberrant proteins, such as truncated variants or proteins mutated in structurally important residues. In other cases, however, formation of inclusion bodies may arise because the high rate of synthesis results in high concentrations of transient states in the folding pathway. Because such folding intermediates can expose surfaces that ultimately come together to stabilize the native conformation, they can lead to intermolecular interactions and hence to aggregation and lower stability. A fraction of the synthesized protein may escape this aggregation, fold properly, and be perfectly soluble in *E. coli*. This explains why some of the protein can be found soluble and active in the cytoplasm of *E. coli* together with inclusion bodies.

Some 'recipes' used to reduce the fraction of protein trapped into inclusion bodies in *E. coli* involve manipulating various factors that may contribute to this phenomenon. Lowering the culture temperature (30 °C, 25 °C, or even 20 °C) is the most widely used procedure for this purpose (13). Lower temperature reduces the rate of synthesis and may destabilize hydrophobic interactions in the aggregated protein, increasing its stability and permitting a higher fraction to fold properly. Improvements have also been obtained occasionally by limiting the level of overexpression or by the use of additives in the culture medium (sorbitol or betain) (14). Lack of the appropriate chaperones, post-translational modifications, or appropriate partner interacting with the protein in the natural host suggest other possible remedies including co-expression of the protein with chaperones (15, 16) or co-expression of both subunits of a heterodimeric protein in the same host (17).

There are several examples for which none of the above procedures prevented the formation of inclusion bodies. It is then possible to extract the protein from the inclusion bodies using a denaturing agent, such as urea or guanidinium chloride. This can be used as a selective purification step, because inclusion bodies are often highly enriched in the desired protein, and can be separated from the membranous fraction (18). Subsequent renaturation of the protein works best if refolding is allowed to take place on a solid support provided, for example, by a hydrophobic interaction resin (19). The procedure in *Protocol 1* was adapted by Sharon Campbell (personal communication) and has been shown to be effective in a number of cases. The specific hydrophobic column support is not critical; it apparently serves a function analogous to that provided by chaperones in the natural host cell, allowing an orderly progression of intermediate structures to form with minimal intermolecular interactions.

Protocol 1. Renaturation of proteins isolated from inclusion bodies

Equipment and reagents

- Hipropyl hydrophobic interaction column

Method

1. Resolubilize the insoluble protein in 6–8 M guanidinium hydrochloride.
2. Dilute it to ~ 100 μg/ml in a buffer with ~ 1.5 M NaCl, 4.5 M guanidinium hydrochloride, 10% glycerol, 10 mM DTT, 5% ethylene glycol, and other components necessary for the activity of the protein. Use a weakly buffered solution compatible with activity (i.e. ~ 40 mM acetate).
3. Apply the mixture to a hipropyl hydrophobic interaction column pre-equilibrated with the same buffer used in step 2, plus 2 M NaCl, 10–20 mM DTT, additional components as before, 10% glycerol, and 5% ethylene glycol. For a $7.75 \times 100\ mm^2$ semi-preparative column, load ~ 1.0 mg of diluted protein. For a preparative $2.2 \times 10\ cm^2$ column, load ~ 5.0 mg.
4. Elute with a gradient of the starting buffer mixed progressively with the same buffer without NaCl.

Alternatives to renaturation trials, when a protein is produced in the bacteria as a totally insoluble form, are to turn to other expression hosts, to try to express the protein as a fusion protein, or to have it secreted by *E. coli*, which will provide a different environment for the folding of the protein. This is done by fusing the protein to an N-terminal signal peptide. One commercially available vector is suitable for this purpose (*Table 1*).

The response surface methodology described in Chapter 4 offers a rational, coherent way to evaluate and optimize the variety of factors influencing the proportion of soluble protein.

2.4 Tagged proteins to facilitate purification

A protein can be modified in order to facilitate its purification for structural studies. This is done by fusing its N- or C-terminus with another protein or with a peptidic tail allowing for a purification by affinity chromatography (20). *Table 2* summarizes a selection of fusion tags commercially available. This method can be applied, in principle, to proteins expressed in any host and is especially useful in two cases: when the level of expression is low and when there is a need to produce in parallel several variants of the protein for crystallization. Although each of them could be preferred for specific reasons, it is obvious from a literature survey of proteins successfully expressed and crystallized in the last years that the polyhistidine tag has encountered a much broader success than any other. This is probably due to the availability of nickel chelate resins (Ni-NTA, Qiagen) which have higher capacities than most other affinity resins. This allows for the relatively cheap purification of 10–100 mg of protein.

Besides facilitating its purification, the fusion of the target protein with a partner protein, such as thioredoxin, may be done to allow the soluble expression of a protein otherwise insoluble (21). However this approach remains largely empirical: in contrast to a generally accepted idea, fusion of a protein with a very soluble bacterial protein, such as glutathione-*S*-transferase (GST), does not always result in an improved solubility of the fusion product, compared to its non-fusion counterpart (P. F. B., unpublished data).

A major drawback of working with fusion proteins is that recovery of the native protein requires a cleavage step. When dealing with a short peptidic tail, it can be envisaged to leave it for crystallization (22), but the tag could affect the proper folding of the protein or prevent crystallization itself (23). Fusions with proteins like GST (25 kDa) usually have to be dissociated by proteolytic cleavage before crystallization, although there are exceptions. Usually, the expression vectors for fusion proteins encode a cleavage site for a specific protease between the two fusion partners. However, the cleavage step happens to be costly for large quantities and can generate heterogeneity. Indeed, it is difficult to control the proteolytic reaction to get a full cleavage at the desired site without secondary cleavages. It should also be kept in mind that even after removal of the fusion partner, there remains generally a small part of the linker at the extremity of the cleaved protein. This can be minimized by an appropriate cloning strategy.

2.5 Use of molecular biology to design an appropriate expression vector

In common language, molecular biology technology includes all the methods for gene manipulation. They are currently used by an increasing number of laboratories and this availability was greatly facilitated by the development of standard techniques and the commercialization of the required enzymes and reagent kits. In contrast to proteins, which exhibit a great diversity of physical and chemical behaviour, DNA fragments are easily modified under standard conditions, independently of base composition. Most basic techniques of DNA manipulation, as well as the basic techniques for the expression of proteins in *E. coli* and mammalian cells, can be found in the classical protocol manual by Maniatis *et al.* (18).

Our purpose is to give strategies to construct new DNA vectors appropriate to the expression of a target protein rather than detailed protocols. Using an expression system requires, as a preliminary condition, the gene encoding the

Table 1. Selection of commercial expression vectors

Name (family)	System	Promoter	Supplier
pET series	*E. coli*	T7	Novagen
pTrc99	*E. coli*	trc	Pharmacia
pLEX	*E. coli*	pL	Invitrogen
pET12, pET20	*E. coli* secretion	T7	Novagen
pVL1392	Baculovirus	Polyhedrin	Pharmingen
pAcSG2	Baculovirus	Polyhedrin	Pharmingen
pAcUW1	Baculovirus	p10	Pharmingen
pFastBac[a]	Baculovirus	Polyhedrin	Gibco BRL
pBluebac	Baculovirus	Polyhedrin	Invitrogen
pBac	Baculovirus	Polyhedrin or gp64	Novagen
pBacPAK	Baculovirus	Polyhedrin	Clontech
pAC5	*Drosophila* cells	Actin 5c (constitutive)	Invitrogen
pMT	*Drosophila* cells	Metallothionein (inducible)	Invitrogen
pPIC	Yeast *P. pastoris*	Alcohol oxidase	Invitrogen
pYEUra3	Yeast	GAL1	Clontech
pcDNA3	Mammalian	CMV	Invitrogen
pZeoSV2	Mammalian	SV40	Invitrogen
pMSG	Mammalian	MMTV-LTR	Pharmacia

[a] This baculoviral system is the only one where the recombination with the viral DNA takes place in *E. coli*, instead of inside insect cells. Therefore, the purification and amplification of a viral clone is not necessary, which shortens the delays, especially for large scale production.

Table 2. Fusion tags for purification

(a) Bacterial vectors

Vector name	Fusion partner	Affinity	Supplier
pET-15, pET-28	Poly-His	Metal chelate	Novagen
pQE	Poly-His	Metal chelate	Qiagen
pTrcHis	Poly-His	Metal chelate	Invitrogen
pGEX	Glutathione-*S*-transferase	Glutathione	Pharmacia
pCAL	Calmodulin binding domain	Calmodulin	Stratagene
pMAL	MalE	Maltose	New England Biolabs
pTrxFus	Thioredoxin	Phenylarsine oxide	Invitrogen
pET-32	Poly-His + thioredoxin	Metal chelate	Novagen
pET-29	S-tag	S-protein	Novagen
pEZZ	Protein A domain	IgG	Pharmacia
pET34–38	Cellulose binding domain	Cellulose	Novagen
Impact system	Chitin binding domain	Chitin	New England Biolabs
FLAG system	FLAG peptide	Anti-FLAG monoclonal antibody	Sigma

(b) Baculovirus transfer vectors

Vector name	Fusion partner	Affinity	Supplier
pBacPAKHis	Poly-His	Metal chelate	Clontech
pAcHLT	Poly-His	Metal chelate	Pharmingen
pFastBacHT	Poly-His	Metal chelate	Gibco BRL
pBlueBacHis	Poly-His	Metal chelate	Invitrogen
pAcGHLT	Glutathione-*S*-transferase	Glutathione	Pharmingen
pBac7-11	Cellulose binding domain	Cellulose	Novagen

protein of interest. It is beyond the scope of this chapter to describe the techniques available for cloning new genes (see, for example, ref. 18) and it is assumed that the gene encoding the protein of interest is available to the reader.

Given the gene of interest, the initial task may be to switch from one expression vector and/or host system to another. *Protocol 2* gives a general 'strategy for subcloning' a gene of interest and *Table 1* contains a selection of common, commercially available expression vectors. The procedure can be applied to transfer a gene (or part of a gene if one wishes to reduce the size of a protein) from one expression vector to another, including vectors to generate tagged proteins. The use of PCR allows this work to be done independently of the presence of compatible restriction sites on the vectors because any chosen site may be introduced using PCR.

Protocol 2. General subcloning strategy using PCR

Equipment and reagents

- Oligonucleotides (see Chapter 8)
- Appropriate restriction enzymes, plasmid, expression vector
- Kits for DNA extraction, purification (e.g. Qiagen or Promega)
- Bacterial strains
- PCR thermocycler and current molecular biology materials
- DNA sequencing facility

Method

1. Perform a standard PCR reaction (25 cycles: 30 sec at 94 °C, 1 min at 60 °C, 1 min per kb amplified at 72 °C) with the selected primers using 1 ng of a plasmid carrying your gene as template. Include control reactions with individual primers.
2. Isolate the PCR product using a phenol:chloroform extraction, followed by ethanol precipitation or using one of the commercially available kits.
3. Digest the acceptor vector (1 μg) and your PCR product with the selected restriction enzymes. Isolate the fragments of interest using agarose gel electrophoresis. Under long UV illumination (365 nm), cut out gel pieces containing the digested vector and PCR insert. Use one of the commercial kits for the extraction of DNA out of the gel.
4. Perform a standard ligation reaction in the presence of about 1 nM of the linearized vector and 2 nM of the PCR insert (concentration is not critical). Include a ligation of the vector alone as control.
5. Transform competent bacteria (e.g. TG1) strain with 2 μl of ligation.
6. Prepare a small amount of DNA (commercial miniprep kit) of a few or several clones and check for the presence of the insert using restriction enzymes.
7. Sequence a few positive clones to make sure that no error has been introduced by the PCR reaction. The complete region amplified by PCR has to be checked.

2.5.1 Choice of restriction sites on the vector

Expression vectors contain host-specific elements required for efficient expression, that have to be preserved, and a location delimited by restriction sites (sometimes multiple ones) dedicated to the proper insertion of the target gene. On a map of the vector, identify this location, and the promoter upstream from it. The coding sequence has to be inserted in the proper orientation between an upstream site, closer to the promoter, and a downstream site. The cloning will be much facilitated if two different, protruding, and non-compatible restriction sites are used.

Note that the upstream site of bacterial expression vector is normally an imposed one, *Nde*I or *Nco*I, because the initiation codon has to be precisely six or seven bases downstream from the Shine–Dalgarno sequence (to locate on the map downstream from the promoter). In contrast, in eukaryotic systems, there is no such sequence and protein synthesis starts at the first ATG following the transcription start. Therefore, any of the multiple sites could be chosen.

A particular case is the insertion of the target gene in a vector designed for fusion proteins. These vectors often contain several cloning sites but it is only by choosing the most upstream one (N-terminal tag) or the most downstream one (C-terminal tag) that one minimizes the number of unrelated amino acids at the junction.

In all cases, check that the intended sites are not present in the target gene. If this is not the case, the vector has to be cut in two pieces for a ligation with three partners (as described in Section 2.5.3) or another expression vector has to be chosen.

2.5.2 Design of oligonucleotide primers

The sense (upstream) primer should contain, in the following order:

- a 4 base extension, for example TAGC, to make an overhang allowing cleavage by the restriction enzyme
- the selected upstream restriction site
- the initiation codon ATG (overlapping with the restriction site in *E. coli* vectors), except in case of fusion with an N-terminal tag
- 0 to 2 added bases to preserve the reading frame
- a 23 base sequence identical to the beginning of the sequence or, if one wishes to reduce the size of the protein, of the chosen coding region.

The antisense (downstream) primer should contain, in the following order:

- a 4 base extension, for example TAGC, to make an overhang allowing cleavage by the restriction enzyme
- the selected downstream restriction site
- a TTA, CTA, or TCA sequence, complementary to a stop codon, except in case of fusion with a C-terminal tag
- 0 to 2 added bases to preserve the reading frame (in case of C-terminal tag)
- a 23 base sequence complementary to the end of the sequence, or, if one wishes to reduce the size of the protein, to the end of the chosen coding region.

2.5.3 Case of large inserts (> 1500 bp)

In principle *Protocol 2* could function with any size of coding sequence. However, the yield of the PCR amplification will decrease with the size of the fragment, and the risk of introducing mutations will increase. Also, the whole PCR-amplified fragment has to be checked by sequencing. For these reasons, there might be a better strategy in the case of a large insert (> 1500 bp) which consists in amplifying only one part of the sequence, up to an internal restriction site, and then constructing the vector by the ligation of three pieces or, alternatively, in two subcloning steps.

2.5.4 From expression vector to expression strain

Once a new expression vector is constructed, it has to be introduced into the host and tested for protein expression. The various stages of this process (cell transformation, transfection or infection, clone selection, cell cultivation, and induction of protein expression) are host- and vector-dependent and it is not our purpose to detail them. We can recommend handbooks treating *E. coli* (18), yeast (24), baculovirus (2), and mammalian (18) cell expression and to refer to the technical tips accompanying commercial expression vectors.

3. Engineering physical properties of macromolecules

Direct modification of the properties of a protein by modifying its sequence is another strategy for preparing suitable samples for 3D structure determination. Such an approach is generally not considered first, but rather when difficulties arise. If this strategy is chosen, make sure that the biological activity of the wild-type protein is conserved in the variants generated for crystallogenesis.

3.1 Problems encountered and possible solutions

Various problems can be encountered at different stages of a structure determination. They are generally detected by one of the following signs:

(a) During purification, the protein tends to aggregate or exhibits solubility problems: these will likely become more pronounced when the protein is concentrated for crystallization.

(b) The purified protein is heterogeneous, contaminated by slightly differing variants, which are often proteolysis products (see Chapter 2).

(c) Changes in a protein's characteristics (e.g. solubility in the presence of precipitating agents), during crystallization trials indicate protein instability, typically the most critical case being the spontaneous formation of irreversible precipitates in the concentrated protein solution.

(d) Diffraction patterns exhibit poor resolution, although crystals are treated under optimal conditions.

(e) No useful heavy-atom derivative can be obtained by a traditional screening approach.

Molecular biology may offer solutions in each of these cases. In most cases, understanding the phenomenon may help in choosing the appropriate changes in the protein sequence. First analyse the nature of the problem by all available experimental tools: electrophoresis, isoelectric focusing, gel filtration, native electrophoresis, mass spectrometry, activity and stability assays, dynamic light scattering (see Chapter 2).

Proteins can suffer various types of chemical degradation, and appropriate alterations could reduce these instabilities. One frequently recognized mechanism is oxidation of sulfhydryl groups, which can lead to formation of oligomers, and/or to aggregation of the protein. In all cases, oxidation produces some degree of heterogeneity in the protein solution, detrimental to the crystallization efforts. Adding reducing agents is recommended to slow down this phenomenon but, in some cases, it could be beneficial to mutate a particularly sensitive cysteine residue to a serine in order to produce or improve crystals.

When a protein exhibits a tendency towards aggregation, however, it is generally not so simple to identify the residues responsible for this behaviour. When the formation of disulfide bridges can be ruled out, it is often thought that protein molecules associate through hydrophobic residues that are located near the surface, or which become exposed under partially denaturing conditions (freezing and thawing, warming, addition of chaotropes). Denaturation being avoided, one can stabilize a protein by changing hydrophobic amino acids into more hydrophilic ones using mutagenesis. This type of strategy was successfully used to allow the crystallization of the HIV integrase (25). The authors faced a high tendency of the integrase to aggregate. They systematically replaced every hydrophobic residue, except those strictly conserved among the family of retroviral integrases, which where supposed to play a critical structural role. As a general rule, the less conserved hydrophobic amino acids are within a family of proteins, the better their chance to be localized at the surface and be responsible for aggregation. An alignment of hydrophobic cluster analysis (HCA) plots (26) can be used to identify such residues, which are prime candidates for mutagenesis.

Current characterization methods, as mentioned above, allow detection of the main problems of stability and heterogeneity. Mass spectrometry reveals other potential sources of microheterogeneity, such as the presence of post-translational modifications. An example of post-translational modification that hampers crystallization is autophosphorylation of the protein, as found with a tyrosine kinase (27). The authors decided to mutate the relevant tyrosine residues into phenylalanine and succeeded in crystallizing the protein.

There are, however, cases where lack of crystallization is not linked to a problem of protein heterogeneity or stability, but to the impossibility of the protein to establish the adequate intermolecular interactions required for nucleation and crystal growth. Mutations can be created that favour these processes, as was done for fibronectin crystallization (28). Here, engineering the protein to have an isoleucine at the C-terminus allowed its easy crystallization and it was later observed that this isoleucine was involved in crystal packing. However, our current understanding of the mechanisms of protein crystallization does not allow predictions on which mutations will favourably affect the interactions leading to crystallization.

Finally, when difficulties arise only in the phasing stage of structure determination, specific approaches can be employed, like the production of selenomethionyl proteins, which is the subject of Section 4, or cysteine mutagenesis, recently discussed by Martinez-Hackert *et al.* (29).

3.2 Defining the optimal size of the molecule to be crystallized

Although one should always first attempt to crystallize full-length proteins, useful structural data have been obtained from systems that are hard to crystallize by cutting out regions that interfere with crystal growth. This approach may consist of limiting the study to a single functional domain (typically a binding domain in the case of the crystallization of a complex) or shortening the size by eliminating a dispensable extremity (e.g. dimerization, nuclear localization, or membrane anchorage domains).

There are several methodological reasons for reducing the size of the studied protein. Over-production of high molecular weight proteins is generally not very efficient. Truncating a protein often decreases its flexibility, a property which is in general detrimental to crystallization. This is true for proteins consisting of two or more functional domains joined by a flexible link as well as for proteins carrying flexible segments at one extremity. An additional reason to eliminate flexible parts in proteins is that they are generally good substrates for proteases. Therefore, a residual proteolytic activity present in the sample could easily generate heterogeneity, as illustrated by the crystallization of the interferon γ receptor (30), which required the deletion of eight amino-terminal residues. A flexible extremity might even be responsible for the poor diffraction quality of crystals (31).

Even though proteolysis under specific conditions has been used in some cases for the preparation of a shorter variant for crystallization (32), genetic engineering of a protein is a more secure way to obtain the same result. The major difficulty of this approach consists in choosing the borders of the domain to be expressed. Ideally, this domain should correspond to a structural and functional domain and be devoid of terminal parts dispensable for activity. Identifying a functional domain in a protein can result from sequence alignments although the corresponding structural domain is likely to overhang on both sides the region of homology. Secondary structure and/or other predictive methods (e.g. HCA diagrams) can be used to define the position of the junction between two domains, which should correspond to a

Table 3. Commonly used proteases

Protease	**Supplier**	**Storage condition (2–10 mg/ml)**	**Incubation buffer**
Trypsin	Sigma Ref. T8642	1 mM HCl, 20% glycerol, –80 °C	20 mM Tris, 10 mM $CaCl_2$ pH 8
Chymotrypsin	Merck Ref. 2307	1 mM HCl, 20% glycerol –80 °C	20 mM Tris, 10 mM $CaCl_2$ pH 8
Subtilisin	Boehringer Mannheim Ref. 165 905	20 mM Tris pH 8, 10 mM $CaCl_2$, 20% glycerol, –80 °C	20 mM Tris, 10 mM $CaCl_2$ pH 8
Thermolysin	Boehringer Mannheim Ref. 161 586	20 mM Tris pH 8, 10 mM $CaCl_2$, 20% glycerol, –80 °C	20 mM Tris, 10 mM $CaCl_2$ pH 8
Endoproteinase Glu-C	Boehringer Mannheim Ref. 791 156	20 mM sodium phosphate pH 7.8, 20% glycerol, –80 °C	25 mM sodium phosphate pH 7.8
Papain	Boehringer Mannheim Ref. 108 014	Suspension, 4 °C	20 mM sodium phosphate pH 8

gap in the secondary structure. More reliable information should be obtained from experimental data such as limited proteolysis (see *Protocol 3*) or expression and characterization of truncated species (33, 34). Such a strategy led to the successful crystallization of the kinase domain of the insulin receptor (27) and of the catalytic domain of type II adenylyl cyclase (31).

From a practical point of view, it is quite easy for a molecular biologist to modify an existing expression vector and to express a shorter form of a protein. For this purpose, we suggest following the same general strategy as described in *Protocol 2* (general subcloning protocol).

Protocol 3. Limited proteolysis assays

Equipment and reagents

- Proteases and protease inhibitors (see *Table 3*)
- Water-bath
- Equipment for SDS gel electrophoresis
- Polypeptide sequencing facility

Method

1. Prepare, as substrate for the assays, a stock solution of your protein (2–10 mg/ml) in a buffer in which it is stable.
2. Select proteases exhibiting broad specificity (with several potential sites inside proteins, e.g. trypsin, chymotrypsin, subtilisin, thermolysin, papain, endoproteinase Glu-C). Prepare stock solutions (2–3 mg/ml) and store them as frozen aliquots in appropriate buffers (see *Table 3*) containing 20% glycerol.[a]

3. Prepare on ice serial dilutions (0.1 mg/ml down to 10 ng/ml) of the proteases in their specific incubation buffers. In a first series of experiments, serial tenfold dilutions might be done and the conditions may then be refined in subsequent experiments.
4. Mix a small volume of the concentrated protein solution (2–5 μl) with the diluted protease solution (40 μl for example) in order to reach a final concentration of about 0.2 mg/ml.
5. Incubate for 1 h at room temperature or 37 °C. Stop the reaction by addition of a specific inhibitor. We recommend PMSF (1 mM final concentration) for all the mentioned serine proteases and E-64 for papain (cysteine protease) (see also Chapter 2).
6. Analyse the reaction products by migration on SDS–PAGE. Try to identify the products of proteolysis at preferred sites. These products correspond to a unique cleavage site at low protease concentration, whereas a higher protease concentration leads to non-specific cleavage at multiple sites.
7. For each protease, optimize the protease concentration that leads to the limited cleavage at the preferred site. Once these conditions have been established, perform a larger scale reaction in order to isolate a larger amount of the proteolysis products.
8. Isolate the gel band of interest using PAGE followed by transfer on a polyvinylidene difluoride (PVDF) membrane (Millipore). Give the samples to a specialized service for N-terminal sequence analysis.[b]
9. The sequence information originating from various proteases may pinpoint areas in the protein that are especially sensitive to proteolysis. These positions probably correspond to connections between various domains in the protein.

[a] Appropriate storage conditions have to be investigated for other proteases. Most importantly, the activity has to be strictly reproducible from one experiment to another, and this is the reason for using frozen aliquots (see also Chapter 2).
[b] Alternatively, the protein species might be analysed by mass spectrometry, as in ref. 27.

3.3 Site-directed mutagenesis

Whatever the envisaged change in the amino acid sequence (amino acid replacement, deletion, or insertion) it is possible to use a general and simple method based on PCR amplification, named the overlap extension method (35). The main requirement of this technique is the synthesis of four oligonucleotides, used as primers in the PCR amplifications, two primers carrying the appropriate DNA modification, and two 'external' primers. This is generally not a limitation, since synthesis of oligonucleotides has become cheap and accessible to all laboratories.

Protocol 4. Site-directed mutagenesis using PCR

Equipment and reagents

- As for *Protocol 2*

Method

1. Design two mutagenic primers, a sense primer (primer a), encoding the target sequence, and an antisense primer (primer b), strictly complementary to primer a. The sense primer could include one or a few modified bases, an insertion, or a deletion, provided that these modifications are surrounded by two 13 base segments strictly complementary to the starting sequence. These segments are necessary to ensure that the primers will recognize the template in spite of the sequence changes.
2. In order to design two external primers (20- to 22-mers with 50–60% GC content), identify two unique restriction sites located each side of the mutation, separated by 200–1500 base pairs. Choose a sense primer (primer c) overlapping with or immediately upstream from the upstream site and an antisense primer (primer d) overlapping with or immediately downstream from the downstream site.
3. Perform two independent PCR reactions using the starting vector[a] as template (10 ng) and standard PCR conditions. The first reaction uses primers a and d, the second reaction uses primers b and c. In this way, the PCR products will be two fragments overlapping over the segment encoded by the mutagenic primers.
4. Isolate the two PCR products using agarose gel electrophoresis.
5. Perform a third PCR reaction using the two external primers (c and d) and, as a template, a mixture of the two previous PCR products ($\sim$ 10 ng each). Using gel electrophoresis, check an aliquot for the presence of the expected final PCR product.[b]
6. Follow *Protocol 2*, steps 3–7.

[a] If available, use as template a vector containing your gene but different from the final vector. That will facilitate the identification of the mutated versus wild-type clones. The template vector should, of course, contain all the region between the two external primers and be recognized by these primers.
[b] During the hybridization step, some hybrids will appear carrying one strand of each fragment annealed through their overlapping segment. These hybrids will be complemented to double-strand DNA by the polymerase and will constitute an efficient template for further amplification by the external primers.

The overlap extension method is described in *Protocol 4*, adapted from ref. 35. Note that there are simpler strategies for mutagenesis in some particular cases (e.g. mutations close to an existing restriction cleavage site). The method proposed here has the advantage of being general. It is cheap and relatively easy to perform, and one can envisage constructing several directed mutations in parallel, if required. In this case, the external primers are common to all mutations and only two oligonucleotides are required for each specific mutant. Additionally, it is often possible to combine directed mutagenesis with subcloning in another expression vector, as described in *Protocol 2*, by designing appropriately the external primers.

3.4 Random mutagenesis

3.4.1 Background

As mentioned above, a great number of mutations at hydrophobic residues were generated to obtain a more soluble variant of HIV integrase (25). It is possible through random mutagenesis to generate an even broader range of variants. This strategy can be used to improve a particular property of a protein (level of expression, solubility, tendency towards aggregation). It is especially suited when the mutations that have a chance to succeed cannot be predicted. The less precisely the encountered problem can be localized in terms of protein sequence, the larger will be the number of mutations that have to be generated and tested.

3.4.2 Strategy

In comparison to classical chemical methods for random mutagenesis, a more recent method (36) produces a broader spectrum of mutations. This method relies on amplification of the gene — or piece of gene — of interest by *Taq* DNA polymerase. This enzyme is usually employed in PCR reactions because of its high thermostability. Although this enzyme is relatively accurate under normal PCR conditions, it tends to introduce random mutations under special conditions, i.e. in the presence of Mn^{2+}. *Protocol 5* (according to ref. 36) gives a guideline for this procedure.

Protocol 5. Random mutagenesis

Equipment and reagents

- As for *Protocol 2*
- *Taq* DNA polymerase

Method

1. Choose the part of the protein DNA sequence that you want to mutagenize randomly. It should be included between two unique restriction sites. It may be necessary to first introduce these restriction sites by directed mutagenesis.

Protocol 5. *Continued*

2. Design oligonucleotides as primers for a PCR amplification. Choose a sense primer immediately upstream from the first restriction site and an antisense primer immediately downstream from the second site.
3. Perform the PCR reaction under specific conditions in order to favour the misincorporation of deoxyribonucleotides by *Taq* DNA polymerase. The PCR reaction should include, in addition to the usual buffer, 0.5 mM $MnCl_2$, 7 mM $MgCl_2$, and modified concentrations of dNTPs (0.2 mM of dATP and dGTP, 1 mM of dTTP and dCTP). Use a very low initial concentration of the plasmid template (e.g. 1 pg in a reaction volume of 50 μl), as the error rate of the PCR will increase if the amplification factor is increased.[a]
4. Follow *Protocol 2*, steps 4–7.
5. Sequence several clones to determine the average number of mutations per clone. An average rate of one per clone seems reasonable for most applications. It might be necessary to sequence a series of clones coming from PCR reactions starting with various amounts of DNA template in order to identify the conditions that generate the appropriate rate of mutation.[b]
6. Using the ligation mixture selected according to step 5, transform an appropriate expression strain, e.g. BL21.
7. Grow small volume cultures of several clones to characterize potential protein variants.

[a] It might be necessary to adjust the amount of polymerase, the annealing temperature, the length of elongation, the number of cycles, in order to get a good yield of amplification, depending on the length of the amplified fragment and on the choice of the primers.

[b] The rate of various mutation types might be adjusted by varying the nucleotide concentration, as described by Fromant *et al.* (63).

3.5 Selection of variants following mutagenesis

When generating multiple protein variants in order to improve crystallization, an obvious question is which criteria to choose to discriminate among different constructs. The ultimate goal is to obtain diffraction-quality crystals. However it is time-consuming to purify large amounts of several variants and to carry on crystallization trials with each of them. There is therefore a need for preliminary selection criteria that can be applied to many variants at an early stage of the process.

Various criteria for the selection of variants can be retained, depending on the nature of the problem encountered. Expression level or solubility in *E. coli* can be tested easily in parallel on multiple clones using electrophoresis or, even simpler if available, an immunodetection method. To assess the chances of succeeding in crystallizing a protein, one could utilize methods that analyse its aggregation status. This can be done using native electrophoresis, analytical gel filtration chromatography, or dynamic light scattering. These various techniques require 10–100 μg of purified protein. In practice, the comparative test of several variants using these techniques requires that the protein be well overexpressed and purified selectively in one step. *E. coli* as expression host and a fusion with a purification tag are, therefore, almost prerequisites.

Protocol 6 provides a guideline for screening variants generated by random mutagenesis. This strategy has been used in our laboratory (P. F. B., personal communication), but has to be adapted to other cases. The studied protein exhibited multiple oligomerization states and we looked for a more homogeneous variant. The phenomenon was clearly detectable using native electrophoresis. We took advantage of the high level of expression in bacteria and of an easy and selective purification step. Growth of bacteria, lysis, and a small scale purification could be performed in 96-wells plates, so that 2000 clones could be analysed in a few months. A few of them were stabilized as dimers and looked homogeneous in native electrophoretic analysis.

Although there are still few reported cases of random mutagenesis to engineer a protein for crystallization, it should be considered as a possible approach for proteins particularly reluctant to crystallize. The development of tools for the parallel micropurification of several proteins, such as Ni-NTA magnetic agarose beads (Qiagen) for polyhistidine tagged proteins, will facilitate such approaches.

Protocol 6. Screening for mutants derived from random mutagenesis

Equipment and reagents

- As for *Protocol 2*
- Lysozyme, DNase, ammonium sulfate solution
- Equipment for gel electrophoresis

Method

1. Add 100 μl of a bacterial growth medium, supplemented with the appropriate antibiotics, to each well of a microtitre plate.
2. With a tip, pick up colonies out of a Petri dish originating from a transformation made the day before. Inoculate each well with a unique colony and strike the tip on a new dish at an identified position to keep a replicate of each clone.
3. Let the cultures grow by incubating the plate for 6 h at 25 °C. In our case, the gene was cloned into vector pET3©, which exhibited a sufficient level of expression without IPTG under these conditions. When working with a vector with a more tightly regulated promoter, it could be necessary to induce the expression in a more specific way.

Protocol 6. *Continued*

4. Collect the cells by centrifuging the plate for 10 min at 2000 *g* (Sigma centrifuges, for instance, have rotors that accept microtitre plates). Suck in carefully the medium. The cells can be frozen at this stage for future use.
5. This step and the following are done easily using a multichannel pipette. Lyse the cells by resuspending the pellets in 100 μl of a buffer containing 50 mM Tris, 20 mM NaCl, 0.1% Triton X-100, 0.5 mM EDTA, 10 μg/ml lysozyme pH 8.4. Incubate for 20 min at room temperature.
6. Add 10 μl of a DNase solution (10 μg/ml in the same buffer as above plus 20 mM $MgCl_2$) and homogenize each well. Incubate for 20 min at room temperature.
7. Centrifuge the plate for 20 min at 2000 *g*. Carefully transfer the supernatant to another plate.
8. Perform a purification step, which could be an incubation with an affinity resin. In our case, the protein could be precipitated selectively by the addition of 0.6 M ammonium sulfate and recovered after centrifugation and elimination of the supernatant.
9. Analyse the protein by native PAGE electrophoresis. This gives a good idea of the state of the protein under native conditions. Ideally, the protein is expected to migrate as a unique and sharp band.

4. Preparation of selenomethionyl protein crystals

4.1 Background

Since the first edition of this book in 1992, the use of selenomethionyl proteins for phase determination has increased dramatically. Selenomethionine substitution now accounts for about two-thirds of all structures solved by multiple wavelength anomalous dispersion (MAD) (37). Use of MAD has been introduced to circumvent non-isomorphism problems that can occur when using heavy metal derivatization as a phasing method (38, 39). The MAD method exploits the presence of anomalous scatterers, such as copper or iron in metalloproteins, heavy metal derivatizing agents, or selenium in selenomethionyl proteins. All measurements relevant to determining a single phase can be made on the same crystal so isomorphism is exact, and electron density maps are very often of high quality (40, 41). With the increased number of synchrotron beam lines dedicated to MAD experiments, the MAD method has rapidly become a general method for phase determination.

The pioneering work of Hendrickson and co-workers showed that selenium is a useful anomalous scatterer (42). Selenomethionine can totally replace methionine in *E. coli* (43). Substitution of methionine by selenomethionine thus offers a general method for introducing anomalous scatterers into cloned proteins. Moreover, most selenomethionyl protein crystals are isomorphous to their native counterpart. As a result, the difference of 18 e^- between selenium and sulfur can be used in a conventional isomorphous replacement structure determination. In addition to its use as an anomalous scatterer and isomorphous derivative, selenomethionine presents the additional advantage of pinpointing methionines, therefore aiding model building.

Preparation and crystallization of selenomethionine-substituted proteins are straightforward procedures (42, 44). In brief, one needs to:

- express the cloned protein in a strain auxotrophic for methionine or more simply, block methionine biosynthesis
- ferment this strain in a medium in which methionine is replaced by selenomethionine
- avoid oxidation during purification of the substituted protein
- crystallize the substituted protein under conditions similar to those used for the native protein.

4.2 Expression and cell growth in a prokaryotic system

4.2.1 Expression in a methionine auxotroph strain

i. Transformation

Historically, the procedure used to engineer selenomethionyl proteins has consisted of transforming an existing methionine auxotroph strain (met^-) with a plasmid containing the cloned gene of the protein of interest and growing the resulting strain in a medium devoid of methionine and containing selenomethionine. *E. coli* met^- strains differ in their tolerance to selenomethionine. LeMaster studied the selenomethionine tolerance of several met^- strains and constructed a strain, DL41, which grows well on selenomethionine-containing media and therefore can be used as a general host for plasmid transformation (42). Note: the strain DL41 can be obtained from the *E. coli* Genetic Stock Center, Yale University School of Medicine, New Haven, CT 06510 USA (`http://cgsc.biology.yale.edu`). Plasmid transformation is carried out by standard procedures (18).

ii. Cell growth

Although selenomethionine is recognized almost equally well as methionine by methionyl-tRNA synthetase (Doublié and Carter, unpublished observations), cells grow more slowly in selenomethionine and generally reach stationary phase at a lower final cell density. Furthermore, cells grown in selenomethionine tend to stay in stationary phase. A low percentage of LB in the starter culture of a met^- strain will provide sufficient methionine to revive the cells from stationary phase, as well as thiamine and biotin. Hydrolysed LB contains 5 mg/litre of methionine, which will be incorporated preferentially to selenomethionine during fermentation (45). To minimize the amount of

Table 4. Starter medium for met⁻ strains

Ingredients	Concentration
1. Minimal medium[a]	Minimal medium A[b] or M9 (18), with carbon source 5 g/litre
2. All amino acids except methionine[a]	40 mg/litre
3. Selenomethionine[c]	20–60 mg/litre
4. LB (12)	x% (v/v) to be determined[d]

[a] One can also use LeMaster's medium (60) instead of 1 and 2.
[b] Ausubel, F. M., Brent, R., Kingston, R. E., Moore, D. D., Seidman, J. G., Smith, J. A., and Struhl, K. (ed.) (1987). *Current protocols in molecular biology.* Greene Publishing Associates and Wiley Interscience, New York, NY.
[c] L-selenomethionine can be purchased from Fisher/Acros or Sigma.
[d] As an example, Yang *et al.* used a 100 ml starter medium containing 5% (v/v) LB for a 20 litre fermenter (43).

Table 5. Fermentation medium for met⁻ strains

Ingredients	Concentration
1. Minimal medium[a]	Minimal medium A[b] or M9 (12), with carbon source 5 g/litre
2. All amino acids except methionine[a]	40 mg/litre
3. Selenomethionine	20–60 mg/litre
4. Thiamine	2 mg/litre
5. Biotin (if needed)	2 mg/litre

[a] One can also use LeMaster's medium (60) instead of 1 and 2.
[b] Ausubel, F. M., Brent, R., Kingston, R. E., Moore, D. D., Seidman, J. G., Smith, J. A., and Struhl, K. (ed.) (1987). *Current protocols in molecular biology.* Greene Publishing Associates and Wiley Interscience, New York, NY.

residual methionine in the purified protein, the dilution factor for the final inoculation must be adjusted according to the amount of LB used in the starter inoculum. This amount will be a compromise between a better growth rate and complete selenomethionine substitution. For each particular strain, one will have to determine the amount of rich medium in the starter culture, as well as the optimal concentration of selenomethionine throughout the fermentation (see *Protocol 7*).

Protocol 7. Cell growth of met⁻ strains

Equipment and reagents

- 100 ml of starter medium (*Table 4*)
- Fermentation medium (*Table 5*)

Method

1. Isolate single colonies by streaking an LB plate (18) supplemented with antibiotics with strain of interest. Incubate at 37 °C.

2. Ferment the cells in medium containing all appropriate antibiotics.
   (a) Inoculate 100 ml of starter medium with a single colony. Shake at 37°C.
   (b) Inoculate a 10–20 litre fermenter[a] containing pre-warmed fermentation medium with the 100 ml starter inoculum in mid-log phase.
   (c) Monitor cell growth in order to identify times for induction and harvest.
3. Induce, if necessary.
4. Harvest the cells in mid- to late log phase.[b] Resuspend the harvested cells in the appropriate lysis buffer and quick-freeze in dry ice or liquid nitrogen. Store at –80°C.

[a] Cell growth should be done in a fermenter because regulated temperature and pH improve the yield.
[b] We have noticed cell lysis shortly after cells reached late log phase. Care must be taken to harvest cells as quickly as possible.

4.2.2 Methionine pathway inhibition

In prokaryotes, an alternative to using a met$^-$ strain is simply to block methionine biosynthesis (46). Aspartate is a precursor in the biosynthesis pathway of lysine, threonine, and homoserine (an intermediate in the formation of methionine). Lysine and threonine block the methionine biosynthesis pathway in *E. coli* by inhibiting the enzymes that phosphorylate aspartate (aspartokinases). Moreover, phenylalanine and leucine are known to act in synergy with lysine. One can therefore produce selenomethionyl protein by growing a non-auxotroph *E. coli* strain in the absence of methionine but with ample amounts of selenomethionine and of the amino acids known to block methionine biosynthesis. This procedure has been successfully applied to a number of proteins, among them FKBP12 (46), UDP-*N*-acetylenopyruvylglucosamine (47), T7 DNA polymerase (48), and human 3-methyladenine DNA glycosylase (49). This procedure is more straightforward than the one

Table 6. Fermentation medium for methionine pathway inhibition

Ingredients	Concentration
1. Minimal medium	Minimal medium A[a] or M9 (18), with carbon source 5 g/litre
2. Thiamine	2 mg/litre
3. Biotin (optional)	2 mg/litre
4. Lysine, phenylalanine, and threonine	100 mg/litre
Isoleucine, leucine, and valine	50 mg/litre
5. Selenomethionine[b]	60 mg/litre

[a] Ausubel, F. M., Brent, R., Kingston, R. E., Moore, D. D., Seidman, J. G., Smith, J. A., and Struhl, K. (ed.) (1987). *Current protocols in molecular biology.* Greene Publishing Associates and Wiley Interscience, New York, NY.
[b] L-selenomethionine can be purchased from Fisher/Acros or Sigma.

involving a met$^-$ strain and should be applicable to any prokaryotic strain (see *Protocol 8*).

Protocol 8. Methionine pathway inhibition

Equipment and reagents

- Current molecular biology reagents
- LB medium (18), minimum medium (*Table 6*)

Method

1. Isolate single colonies by streaking an LB plate (18) supplemented with antibiotics with strain of interest. Incubate at 37 °C.
2. Inoculate 1 ml of LB medium with a single colony. Grow overnight.
3. Spin down cells (2 min at 1300 *g* in a microcentrifuge) and resuspend in 1 ml of supplemented minimum medium (items 1, 2, and 3 in *Table 6*). Inoculate 1 litre of the same, pre-warmed medium.
4. Add all seven amino acids at mid-log phase.
5. Induction is done 15 min after addition of the amino acids, if necessary.
6. Harvest the cells in mid- to late log phase. Resuspend the harvested cells in the appropriate lysis buffer and quick-freeze in dry ice or liquid nitrogen. Store at –80 °C.

4.3 Eukaryotes

Animal organisms are naturally auxotrophic for methionine and cells can grow in a selenomethionine containing medium with good incorporation of the modified amino acid. There are to date only a few examples of selenomethionyl protein production in eukaryotic cells. Selenomethionyl human chorionic gonadotropin (hCG) was produced in insect cells (Sf9) (50, 51) as well as in Chinese hamster ovary (CHO) cells (52, 53). The selenomethionyl variant of a functional fragment of sialoadhesin was also produced in CHO cells (54). The substitution rates reported for these three proteins range from 84–92%.

4.4 Purification

Introduction of selenomethionine into proteins has two consequences that impact on purification. The altered chemistry of selenium makes substituted proteins more sensitive to oxidation than natural proteins. Moreover, if selenium atoms are solvent exposed, they can alter protein solubility and behaviour on chromatography resins. These properties require the following modifications to the normal purification as shown in *Protocol 9*.

Protocol 9. Purification of selenomethionyl proteins

Equipment and reagents

- Chromatographic equipment
- Dithiothreitol (DTT)
- Ethylenediaminetetraacetic acid (EDTA)

Method

1. Purify as quickly as possible, with modifications to avoid oxidation.
   (a) Degas all buffers by boiling or evacuation.
   (b) Include a reducing agent such as DTT and a chelator such as EDTA to remove traces of metals that could catalyse oxidation (55) (see also Chapter 2). Use 0.2–1 mM EDTA and 5–20 mM DTT.
2. Expect selenomethionyl proteins to be slightly less soluble than their natural counterparts.
   (a) Anticipate lower optimal ammonium sulfate concentrations in trituration protocols.
   (b) Anticipate increased retention in some chromatography procedures.
3. Store purified protein in an oxygen-free environment, at –80 °C in the presence of glycerol, or if possible, as frozen droplets at –180 °C.
4. Mass spectroscopy is the most accurate method to quantitate selenomethionine incorporation. If this technique is not available, one can undertake an amino acid analysis to check the percentage of substitution. Selenomethionine is destroyed under the acid hydrolysis conditions used in amino acid analysis, so that it is the disappearance of methionine that is monitored.

4.5 Crystallization

Experience to date suggests that selenomethionyl proteins crystallize in conditions that are very similar to those used with native proteins (42, 45, 48, 49). As a consequence of the lowered solubility of selenomethionyl proteins, either the protein or the precipitant concentration should be slightly reduced to achieve comparable degrees of supersaturation. It is often the case that growth of selenomethionyl protein crystals require microseeding with a crushed wild-type protein crystal. Selenomethionine oxidation can lead to aberrant X-ray fluorescence spectra in which the position and shape of the K-edge are altered and the white line intensity decreased (37). This can be avoided by maintaining the crystals in a solution containing DTT and EDTA. Crystals should be stored in an oxygen-free environment such as an anaerobic chamber if possible (see Chapter 5 for a crystallization method suitable for oxygen-sensitive proteins). They should be irradiated as soon as possible or flash-frozen and stored in liquid nitrogen while they await data collection. Selenomethionine incorporation does not appear to alter diffraction limits and selenomethionyl protein crystals are generally isomorphous with native crystals. However, they can be more sensitive to radiation damage.

There are now several large protein structures (> 90 kDa) that have been solved by the MAD method using solely the anomalous signal of selenium (48, 56; J. L. Smith, personal communication). It is also clear that large numbers of selenium sites (15 or more) can be readily located with direct methods programs. This realization should increase even further the widespread use of selenomethionyl proteins for phase determination. There are also encouraging results regarding the incorporation of telluromethionine into proteins (57, 58). Even though tellurium cannot be used as an anomalous scatterer (its K-edge (0.389 Å) and L-edges (> 2.5 Å) correspond to wavelengths not usually reachable at synchrotron facilities), its 36 e^- difference with sulfur has been successfully used in conventional isomorphous replacement methods (58). This procedure will reach its true potential when telluromethionine becomes commercially available.

4.6 Warning

Selenium is an essential element for most animal and bacterial life (59), but it is also a very *toxic* compound because of its ability to replace sulfur. In mammals (e.g. protein crystallographers), ingested methionine is a source of sulfur. As a result, selenomethionine can be *harmful* or even *fatal* if inhaled, swallowed, or absorbed through the skin. Experiments should always be done in a hood and the experimenter should be sure to wear gloves. Experimenters should contact the Health and Safety office at their institution and inquire about proper disposal of selenomethionine containing media.

5. Conclusion

Structural studies require homogeneous concentrated solutions of the protein of interest. Some proteins cannot be obtained in this form and optimizing the expression system can sometimes solve the problem. However, even an apparently perfect protein solution might be reluctant to crystallize, perhaps because of surface residues that could form destabilizing intermolecular contacts. In some cases, such a protein can be stabilized by the addition of specific agents like glycerol, detergents, zwitterions, or amino acids (61, 62) (see also Chapter 2). An alternative approach consists in identifying residues responsible for the instability of the protein and mutating them. Chances are that these residues, being located at the surface, will not be crucial for keeping the overall 3D folding of the protein. This chapter gives an overview of the genetic engineering tools that one can exploit to optimize expression systems and to design variants by systematic or random mutagenesis. These modifications can lead to improved crystallization or to novel physical properties useful for structure determinations by X-ray diffraction studies. Such macromolecular engineering strategies may find routine use in modern structural biology.

References

1. Pawson, T. (1995). *Nature*, **373**, 573.
2. Gruenwald, M. D. and Heitz, M. S. (1993). In *Baculovirus expression vector system: procedures and methods.* A manual published by Pharmingen.
3. Miegel, A., Sano, K. I., Yamamoto, K., Maeda, K., Maeda, Y., Taniguchi, H., *et al.* (1996). *FEBS Lett.*, **394**, 201.
4. Marchot, P., Ravelli, R. B. G., Raves, M. L., Bourne, Y., Vellom, D. C., Kanter, J., *et al.* (1996). *Protein Sci.*, **5**, 672.
5. Cregg, J. M., Vedvick, T. S., and Raschke, W. C. (1993). *Biotechnology*, **11**, 905.
6. Kern, D., Dietrich, A., Fasiolo, F., Renaud, M., Giegé, R., and Ebel, J.-P. (1977). *Biochimie*, **59**, 453.
7. Weijland, A., Williams, J. C., Neubauer, G., Courtneidge, S. A., Wierenga, R. K., and Superti-Furga, G. (1997). *Proc. Natl. Acad. Sci. USA*, **94**, 3590.
8. Friehs, K. and Reardon, K. F. (1993). *Adv. Biochem. Eng. Biotech.*, **48**, 53.
9. Olins, P. O. and Lee S. C. (1993). *Curr. Opin. Biotech.*, **4**, 520.
10. Dale, G. E., Broger, C., Langen, H., D'Arcy, A., and Stüber, D. (1994). *Protein Eng.*, **7**, 933.
11. Kim, K. K., Yokota, H., Santoso, S., Lerner, D., Kim, R., and Kim, S. H. (1998). *J. Struct. Biol.*, **121**, 76.
12. Kozak, M. (1997). *EMBO J.*, **16**, 2482.
13. Schein, C. H. (1989). *Biotechnology*, **7**, 1141.
14. Blackwell, J. R. and Horgan, R. (1991). *FEBS Lett.*, **295**, 10.
15. Caspers, P., Stieger, M., and Burn, P. (1994). *Cell. Mol. Biol.*, **40**, 635.
16. Yasukawa, T., Kanei-Ishii, C., Maekawa, T., Fujimoto, J., Yamamoto, T., and Ishii, S. (1995). *J. Biol. Chem.*, **270**, 25328.
17. Mayer, M. P., Prestwich, G. D., Dolence, J. M., Bond, P. D., Wu, H., and Poulter, C. D. (1993). *Gene*, **132**, 41.
18. Maniatis, T., Fritsch, E. F., and Sambrook, J. (ed.) (1989). *Molecular cloning: a laboratory manual*, 2nd edn. Cold Spring Harbor Press, Cold Spring Harbor, NY.
19. Geng, X. and Chang, X. (1992). *J. Chromatogr.*, **599**, 185.
20. Ford, C. F., Suominen, I., and Glatz, C. E. (1991). *Protein Expression Purification*, **2**, 95.
21. Stoll, V. S., Manohar, A. V., Gillon, W., Macfarlane, E. L. A., Hynes, R. C., and Pai, E. F. (1998). *Protein Sci.*, **7**, 1147.
22. Zhang, F., Robbins, D. J., Cobb, M. H., and Goldsmith, E. J. (1993). *J. Mol. Biol.*, **233**, 550.
23. Fields, B. A., Ysern, X., Poljak, R., Shao, X., Ward, E. S., and Mariuzza, R. A. (1994). *J. Mol. Biol.*, **239**, 339.
24. Campbell, I. and Duffus, J. H. (ed.) (1989). *Yeast: a practical approach.* IRL Press, Oxford.
25. Jenkins, T. M., Hickman, A. B., Dyda, F., Ghirlando, R., Davies, D. R., and Craigie, R. (1995). *Proc. Natl. Acad. Sci. USA*, **92**, 6057.

26. Woodcock, S., Mornon, J.-P., and Henrissat, B. (1992). *Protein Eng.*, **5**, 629.
27. Wei, L., Hubbard, S. R., Hendrickson, W. A., and Ellis, L. (1995). *J. Biol. Chem.*, **270**, 8122.
28. Dickinson, C. D., Gay, D. A., Parello, J., Ruoslahti, E., and Ely, K. R. (1994). *J. Mol. Biol.*, **238**, 123.
29. Martinez-Hackert, E., Harlocker, S., Inouye, M., Berman, H. M., and Stock, A. M. (1996). *Protein Sci.*, **5**, 1429.
30. Windsor, W. T., Walter, L. J., Syto, R., Fossetta, J., Cook, W. J., Nagabhushan, T. L., *et al.* (1996). *Proteins: Structure, Function, Genetics*, **26**, 108.
31. Zhang, G., Liu, Y., Qin, J., Vo, B., Tang, W. J., Ruoho, A. E., *et al.* (1997). *Protein Sci.*, **6**, 903.
32. Bergfors, T., Rouvinen, J., Lehtovaara, P., Caldentey, X., Tomme, P., Claeyssens, M., *et al.* (1989). *J. Mol. Biol.*, **209**, 167.
33. Cohen, S. L., Ferré-D'Amare, A. R., Burley, S. K., and Chait, B. (1995). *Protein Sci.*, **4**, 1088.
34. Pfuetzner, R. A., Bochkarev, A., Frappier, L., and Edwards, A. M. (1997). *J. Biol. Chem.*, **272**, 430.
35. Ho, S. N., Hunt, H. D., Horton, R. M., Pullen, J. K., and Pease, L. R. (1989). *Gene*, **77**, 51.
36. Cadwell, C. and Joyce, G. F. (1992). *PCR Methods Applications*, **2**, 28.
37. Smith, J. L. and Thompson, A. (1998). *Structure*, **6**, 815.
38. Hendrickson, W. A. (1988). In *Crystallographic computing*, Vol. 4 (ed. N. W. Isaacs and M. R. Taylor), pp. 97–108. Oxford University Press, Oxford, UK.
39. Hendrickson, W. A. (1985). *Trans. Am. Crystallogr. Assoc.*, **21**, 11.
40. Hendrickson, W. A., Pähler, A., Smith, J. L., Satow, Y., Merrit, E. A., and Phizackerley, R. P. (1989). *Proc. Natl. Acad. Sci. USA*, **86**, 2190.
41. Yang, W., Hendrickson, W. A., Crouch, R. J., and Satow, Y. (1990). *Science*, **249**, 1398.
42. Hendrickson, W. A., Horton, J. R., and LeMaster, D. M. (1990). *EMBO J.*, **9**, 1665.
43. Cowie, D. B. and Cohen G. N. (1957). *Biochim. Biophys. Acta*, **26**, 252.
44. Doublié, S. (1997). In *Methods in enzymology* (eds C. W. Carter and R. Sweet), Academic Press, London. Vol. 276, p. 523.
45. Yang, W., Hendrickson, W. A., Kalman, E. T., and Crouch, R. J. (1990). *J. Biol. Chem.*, **265**, 13553.
46. Van Duyne, G. D., Standaert, R. F., Karplus, P. A., Schreiber, S. L., and Clardy, J. (1993). *J. Mol. Biol.*, **229**, 105.
47. Benson, T. E., Filman, D. J., Walsh, C. T., and Hogle, J. M. (1995). *Nature Struct. Biol.*, **2**, 644.
48. Doublié, S., Tabor, S., Long, A. M., Richardson, C. C., and Ellenberger, T. (1998). *Nature*, **391**, 251.
49. Lau, A. Y., Schärer, O. D., Samson, L., Verdine, G. L., and Ellenberger, T. (1998). *Cell*, **95**, 249.
50. Chen, W. Y. and Bahl, O. P. (1991). *J. Biol. Chem.*, **266**, 8192.
51. Chen, W. Y. and Bahl, O. P. (1991). *J. Biol. Chem.*, **266**, 9355.
52. Wu, H., Lustbader, J. W., Liu, R. E., Canfield, W., and Hendrickson, W. A. (1994). *Structure*, **2**, 545.
53. Lustbader, J. W., Wu, H., Birken, S., Pollak, S., Gawinowicz Kolks, M. A., Pound, A. M., *et al.* (1995). *Endocrinology*, **136**, 640.

54. May, A. P., Robinson, R. C., Aplin, R. T., Bradfield, P., Crocker, P. R., and Jones, E. Y. (1997). *Protein Sci.*, **6**, 717.
55. Scopes, R. K. (1982). *Protein purification*. Springer–Verlag, New York, NY.
56. Turner, M. A., Yuan, C. S., Borchardt, R. T., Hersfield, M. S., Smith, G. D., and Howell, P. L. (1998). *Nature Struct. Biol.*, **5**, 369.
57. Boles, J. O., Lewinski, K., Kunke, M., Odom, J. D., Dunlap, B. R., Lebioda, L., *et al.* (1994). *Nature Struct. Biol.*, **1**, 283.
58. Budisa, N., Karnbrock, W., Steinbacher, S., Humm, A., Prade, L., Neuefeind, T., *et al.* (1997). *J. Mol. Biol.*, **270**, 616.
59. Stadman, T. C. (1980). *Annu. Rev. Biochem.*, **49**, 93.
60. LeMaster, D. M. and Richards, F. M. (1985). *Biochemistry*, **24**, 7263.
61. Sousa, R., Lafer, E. M., and Wang, B. C. (1991). *J. Cryst. Growth*, **110**, 237.
62. Jeruzalmi, D. and Steitz, T. A. (1997). *J. Mol. Biol.*, **274**, 748.
63. Fromant, M., Blanquet, S., and Plateau, P. (1997). In *Genetic engineering with PCR* (ed. R. M. Horton and R. C. Tait), p. 39. Horizon Scientific Press, Wymondham, UK.

4

Experimental design, quantitative analysis, and the cartography of crystal growth

C. W. CARTER, JR

1. Introduction

This chapter is about practical uses of mathematical models to simplify the task of finding the best conditions under which to crystallize a macromolecule. The models describe a system's *response* to changes in the independent variables under experimental control. Such a mathematical description is a *surface*, whose two-dimensional projections can be plotted, so it is usually called a 'response surface'.

Various methods have been described for navigating an unknown surface. They share important characteristics: experiments performed at different levels of the independent variables are scored quantitatively, and fitted implicitly or explicitly, to some model for system behaviour. Initially, one examines behaviour on a coarse grid, seeking approximate indications for multiple crystal forms and identifying important experimental variables. Later, individual locations on the surface are mapped in greater detail to optimize conditions. Finding 'winning combinations' for crystal growth can be approached successively with increasingly well-defined protocols and with greater confidence. Whether it is used explicitly or more intuitively, the idea of a response surface underlies the experimental investigation of all multivariate processes, like crystal growth, where one hopes to find a 'best' set of conditions. The optimization process is illustrated schematically in *Figure 1*.

In general, there are three stages to this quantitative approach:

(a) *Design*. One must first induce variation in some desired experimental result by changing the experimental conditions. Experiments are performed according to a plan or *design*. Decisions must be made concerning the experimental variables and how to sample them.

(b) *Experiments and scores*. Each experiment provides an estimate for how the system behaves at the corresponding point in the experimental space.

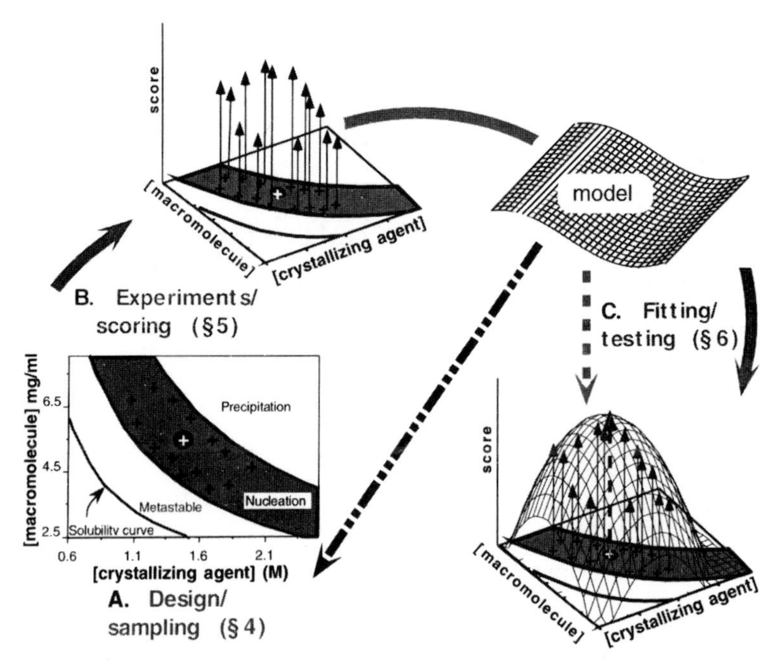

Figure 1. Three stages of a response-surface experiment aimed at locating an optimal point in the solubility diagram for macromolecular crystal growth. (A) Design of the experiment involves decisions regarding which variables to test, the resources (number of individual tests, amount of protein, etc.) to be devoted, and explicit descriptions of the experiments in the experimental matrix. (B) Performing the experiments involves quantitative measurement of one or more 'responses'. (C) Fitting and testing the model involves taking a generic model (the gauze) and fitting it to the observed data by adjusting model parameters. Sections of the chapter devoted to each stage are indicated in the grey labels.

When these estimates are examined together as a group, patterns often appear. For example, a crystal polymorphism may occur only in restricted regions of the variable space explored by the experiment.

(c) *Fitting and testing models.* Imposing a mathematical model onto such patterns provides a way to predict how the system will behave at points where there were no experiments. The better the predictions, the better the model. Adequate models provide accurate *interpolation* within the range of experimental variables originally sampled; occasionally a very good model will correctly predict behaviour *outside* it (1). Quadratic polynomial models are particularly useful for optimization, because they can possess 'stationary points', where their gradient vanishes, and which may represent optima (2).

It is increasingly important for structural molecular biology to establish the repertoire of different molecular conformations accessible to a given macromolecule, in order to understand their functional roles (3–6). Perhaps the most valuable aspects of the quantitative procedures described in this chapter have to do with building evidence about relationships between physico-chemical parameters and crystal growth. Crystal growth depends on many physico-chemical factors that also influence protein conformation (7–13). Understanding relationships between conformational equilibria and crystal growth conditions may benefit from establishing the reliability of anecdotal evidence about crystal growth (14). Statistical analysis of models entails an inherent management of 'signal' and 'noise' from the experiments, providing an appropriate framework of confidence limits within which inferences about such relationships can be drawn (15–18).

2. Response surfaces and factorial design

2.1 Mathematical models and inference

By *quantitative analysis*, is meant estimating and interpreting parameters for an appropriate mathematical model for crystal growth that minimizes the sum of squared differences between observed results, Q_{obs}, and predictions of the model, Q_{calc}. A model provides a way to coordinate the sometimes complex interplay between the experimental variables, making sense of how they affect crystal growth. Sometimes, a good model can suggest novel ways to interpret experimental behaviour. Models can be useful in three distinct phases of a crystal growth project: *screening*, *characterization*, and *optimization*.

2.2 What is a response surface?

Response-surface *models* use a small number of parameters to describe system behaviour; therein lies their economy. Good model parameters predict approximate values for the system response at any set of values for the independent, experimental variables. Linear one-dimensional response-surface models, $y = \beta_1 x + \beta_0$, are familiar in many contexts. Quadratic polynomial models describe more complex behaviour, and hence can indicate the location of optimal conditions.

A response-surface *experiment* works in the opposite direction, using the system behaviour at a defined sample of points representing the experimental space to estimate the parameters, $\{\beta_i\}$, which are the coefficients of the model. This reciprocal nature of a response-surface model, endows it with dual abilities to estimate parameters from experimental observations and to predict experimental results based on the parameters (*Figure 2*).

2.3 Factorial experimental design

Because they coordinate the influence of all important experimental variables, response-surface models are closely linked to *factorial experimental*

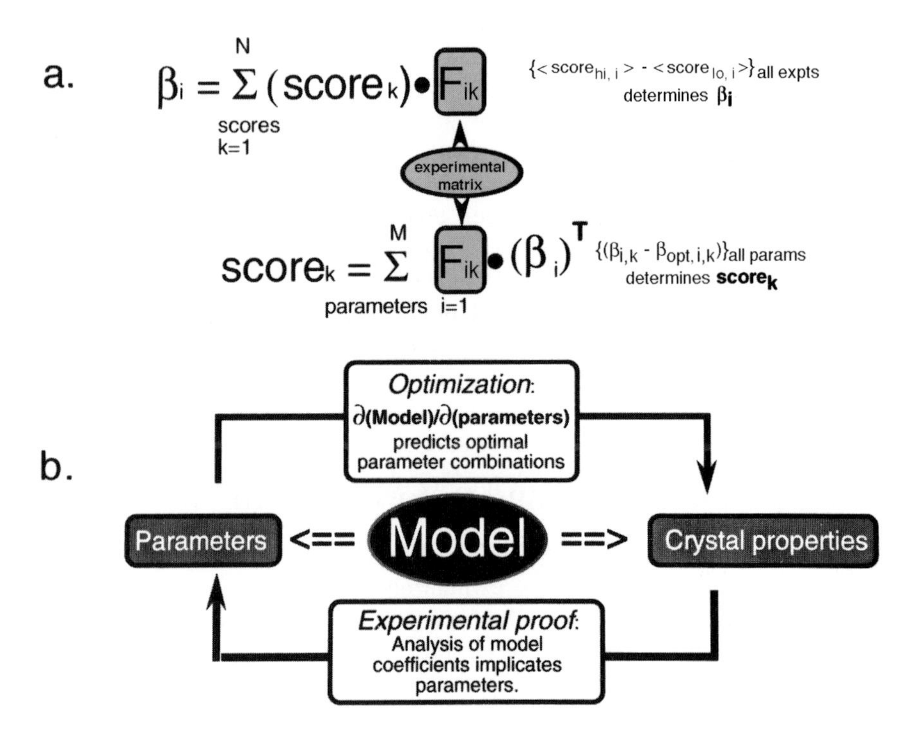

Figure 2. Mathematical models provide a two-way link between crystal properties and experimental effects. (a) The mathematical relationships linking response-surface parameters and experimental scores via the design matrix, $\mathbf{F}_{ik}$. ($score_k$) is a row vector of the N experimental scores, $(\beta_i)^T$ is a column vector of the model coefficients. (b) Schematic presentation of the reciprocity between parameters and scores described in the text.

design. Factorial design makes the experimental dependencies explicit in the form of an *experimental matrix*, denoted here by $\mathbf{F}_{ik}$, whose elements, $\{F_{ik}\}$, indicate the level of experimental factor i used in the kth experiment.

The 'causes' of particular processes, like crystallization of a macromolecule, frequently comprise only a small number of particular factors, or combinations of factors. These specific factors, the 'explanation' as well as the 'recipe' for the effect, usually act somewhat independently of one another. Rarely do all contributing factors interact so intimately as to require simultaneous fine-tuning of each one. In the jargon of experimental design, the contributing factors themselves are called *main effects*, while combinations of main effects are called *n-factor interactions*. Thus a factorial design is a set of N experiments intended to identify and map important main effects and interactions simultaneously for M experimental variables or *factors*. The N $\times$ M matrix, $\mathbf{F}_{ik}$, assigns different levels of each experimental variable to each experimental test, *simultaneously changing values systematically for each factor*. When all possible combinations are tested, the design is called a *full factorial* and contains

$$N = \prod_{i_{\text{var}\,s}}^{M} (j_{levels})_{i\text{factors}}\ \textit{experiments.} \qquad [1]$$

2.3.1 Simultaneous variation of several variables

Factorial designs change several variables simultaneously. An example is given in *Figure 3*, where a 6 × 3 experimental matrix and observed scores are shown in reverse contrast. The levels are conveniently coded as numbers, $-1 < [F_{ik}] < 1$, which are referred to as the assigned 'treatments'. $[F_{ik}] = 0$ represents the mean value of the range studied, and the point [0,0,0....0] represents the 'centre' of the experiment. Other choices ($1 < [F_{ik}] < 2$) can be used. Unlike traditional 'one-at-a-time' designs, simultaneous variation of several variables allows each experiment to contribute consistently to the estimation of all effects and interactions, doing several jobs at the same time. *Figure 3* shows how averaging three results at each pH can be achieved by using two experiments at one temperature and one at the other temperature.

2.3.2 Contrasts

This example shows the rudimentary relationships between experimental matrix, observed and calculated scores, and model parameters. This particular

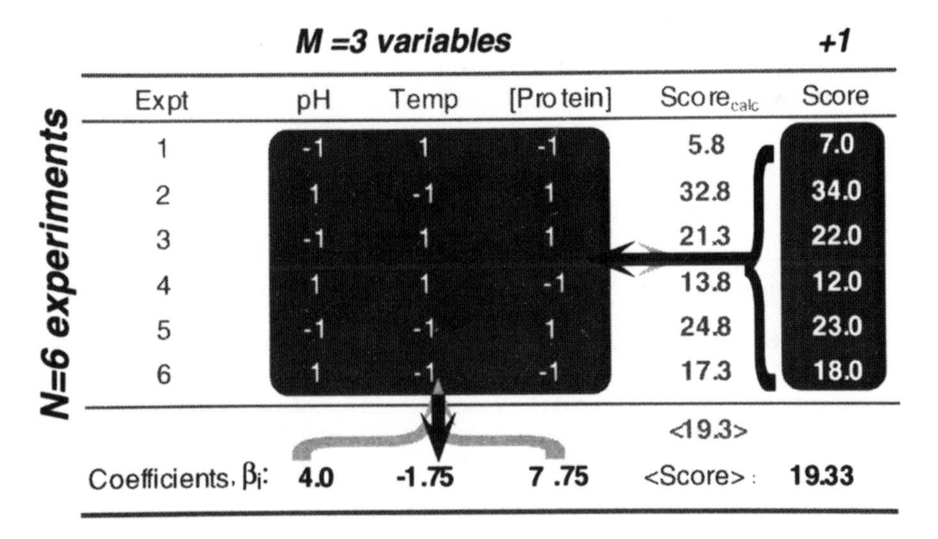

Figure 3. Example of an incomplete factorial design matrix, showing how scores and contrasts (model parameters) are connected to each other via the experimental matrix (shaded). Here, six experiments are used to define a model with four parameters, which include the constant term and the three main effect contrasts.

model states that the expected score for the kth experiment, $Score_{calc,k}$, is given by the average observed score, $\beta_0 = 19.33$, plus or minus a constant value, β_i, from each of the three columns:

$$Score_{calc,k} = \beta_0 + \sum_{i=1}^{M_{params}} F_{ik} \times \beta_i,$$

where β_0 is the average score. Thus, for the first experiment, $Score_{calc,1} = 19.33 + [(-1 \times 4.0) + (1 \times -1.75) + (-1 \times 7.75)] = 5.8$. Similarly, the three coefficients of the model, $\{\beta_i\}$, are obtained from the average sum of the products of each score and the matrix element from the appropriate column:

$$\beta_i = \frac{\sum_{k=1}^{N_{expts+}} Score_{kcalc} \times F_{ik}}{N(+)} + \frac{\sum_{k=1}^{N_{expts-}} Score_{kcalc} \times F_{ik}}{N(-)}, \quad [2]$$

where the sums over experiments at the two levels are kept separately, in case there are not the same numbers of experiments at each level. *Equation 2* is also called the *contrast* between the high and low levels for the ith factor.

Other contrasts can be calculated, and may be important. Most important are those for the two-way interactions between pairs of factors. Consider the data for the full-factorial design in *Table 1*, and in which the incomplete factorial sample from *Figure 3* is indicated by bold face. The effects of three variables are each tested at two levels. Contrasts for all possible n-factor interactions are shown, in addition to those for the main effects. There are three two-factor interactions and one three-factor interaction. Entries for each interaction are the products of the entries for the respective main effects. Contrasts for higher-order interactions are generally smaller than those for main effects.

In summary, the following aspects of the entries in *Table 1* should be noted.

(a) All possible combinations for the three variables are tested at two levels by the 8 (~ 2^3) experiments in the full design. This means that the

Table 1. Extended matrix and contrasts for a three-factor, two-level full-factorial design

Exp't	pH	Temp	[Prt]	pH × Temp	pH × [Prt]	Temp × [Prt]	pH × Temp × [Prt]	Score
1	–1.000	–1.000	–1.000	1.000	1.000	1.000	–1.000	8.7
2	**–1.000**	**–1.000**	**1.000**	**1.000**	**–1.000**	**–1.000**	**1.000**	**23.0**
3	**–1.000**	**1.000**	**–1.000**	**–1.000**	**1.000**	**–1.000**	**1.000**	**7.0**
4	**–1.000**	**1.000**	**1.000**	**–1.000**	**–1.000**	**1.000**	**–1.000**	**22.0**
5	**1.000**	**–1.000**	**–1.000**	**–1.000**	**–1.000**	**1.000**	**1.000**	**18.0**
6	**1.000**	**–1.000**	**1.000**	**–1.000**	**1.000**	**–1.000**	**–1.000**	**34.0**
7	**1.000**	**1.000**	**–1.000**	**1.000**	**–1.000**	**–1.000**	**–1.000**	**12.0**
8	1.000	1.000	1.000	1.000	1.000	1.000	1.000	30.0
Contrasts	4.16	–1.59	7.91	–0.91	0.59	0.34	0.16	19.338

experiments are necessarily *uniformly distributed* among the different possible combinations.

(b) Each column is a *different linear combination* of the eight scores. This is true both for the full-factorial design and the 6-experiment sample. Designs for which any two F_{ij} columns are the same are said to involve *confounding* or *aliasing* of the effects denoted by identical columns. It is impossible to distinguish which of the confounded columns is responsible for the contrast in the experimental scores without additional experiments specifically designed to distinguish between the multiple possibilities (19).

(c) The seven columns all have equal numbers of 1s and –1s. This is true for the complete factorial and for all but the temp x [Prt] column of the 6-experiment sample. Both the full design and the sampled design are therefore balanced with respect to the main effects.

2.3.3 Balance

If each level is tested by the same number of experiments, the design is said to be *balanced*. Balance is important. The standard deviation of an average value, $<x>$, is given by

$$\sqrt{\frac{\sum_{i=1}^{n}(x_i - <x>)^2}{n-1}},$$

where x_i refers to an observation. If the design in *Figure 3* had only two experiments at the low pH and four at the high pH, then the standard deviation of the average score for the latter four experiments would tend to smaller by a factor of $\sqrt{(2-1)/(4-1)} = \sqrt{1/3} = 0.57$. The estimate would tend to be more precise, and probably more accurate than that for the other two experiments. For this reason, balanced designs are preferred for quantitative multivariate problems, because testing each level the same number of times distributes both signal and noise as evenly as possible among the different experiments.

2.3.4 Resolution

The dot products between all of the columns in the full (but not the in-complete) factorial design all equal zero. Thus, they form a set of orthogonal basis vectors for the experimental space. This property means that in a full-factorial design, the experimental treatments for all effects and interactions are completely uncorrelated, which enhances the ability to separate their impact on the scores. This 'separability' property is related to what is called the *resolution* of a sampling design.

The optimal balance, resolution, and freedom of confounding of full-factorial experiments make them especially useful whenever a phenomenon is already suspected to depend on the effects and interactions of a small number of factors. An example was the demonstration of how pH, temperature, and two low molecular substrates influenced the crystal growth of tryptophanyl-tRNA synthetase (TrpRS) (2). TrpRS shows both pH- and ligand-dependent conformational changes, so the main effects (pH) and two-factor interactions (pH $\times$ substrate and substrate $\times$ substrate) identified from a quantitative analysis of the crystal growth behaviour in a 2^4 factorial experiment were biochemically significant. Strategies for sampling full-factorial designs discussed in Section 3 differ from one another in the degree to which each compromises one or more of these properties.

3. Sampling appropriate subsets from a full-factorial design

Sampling is always an issue. No experiment provides more than a sample of how a system responds to changes in the conditions that vary from experiment to experiment. A full-factorial design in M factors yields just enough observations to uniquely define all of the main effects and all n-factor interactions. However, for M > 3–5, it becomes increasingly difficult to carry out all such experiments even at only two levels per factor, and additional sampling is required. Sampling sacrifices some of the advantages of the full-factorial design in return for the economy of doing fewer experiments. The particular selection of experiments actually performed can determine whether or not a design achieves a desired goal, and the appropriate sampling strategy depends on the experimental context.

3.1 Screening versus optimization

As the broad distinction between 'screening' and 'optimization' suggests, one often proceeds by first identifying 'nodes' of the response surface where crystallization occurs, and then mapping finer details of individual nodes. Neither process is typically done with full-factorial designs. Different sampling strategies can be used. The appropriate quantitative analytical methods are generally similar, but use different types of models which, in turn, dictate how the experimental sampling points should be selected. The sampling strategies are illustrated schematically in *Figure 4*. Each sacrifices some of the strengths of the full-factorial design, making different compromises with respect to balance, confounding, and resolution.

3.2 Subsets for screening

3.2.1 Incomplete factorial sampling

Efficient covering of the entire experimental space is a key requirement for screening. Incomplete factorial designs were introduced to detect the most important factors and their interactions from screening experiments (20).

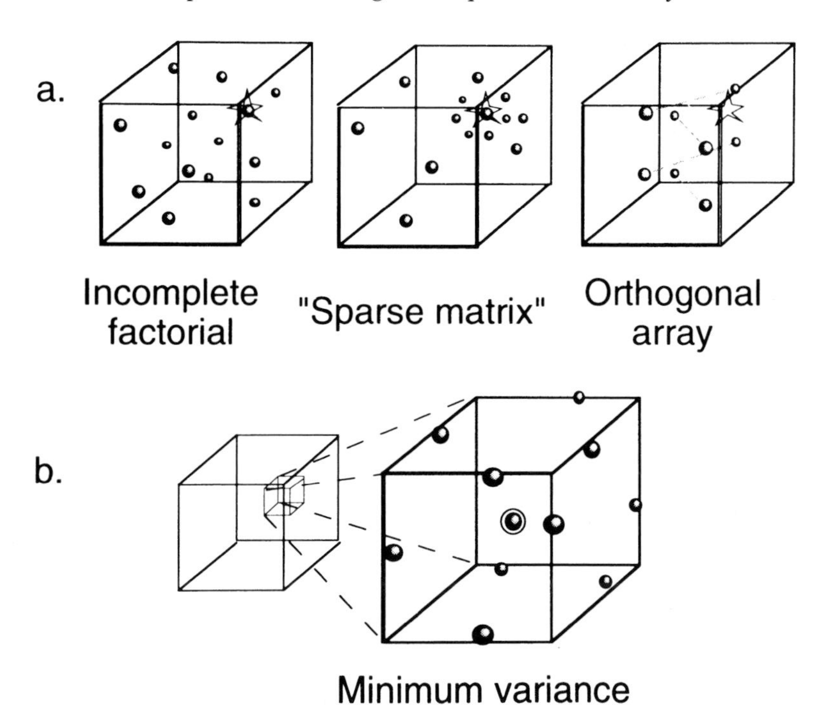

Figure 4. Sampling strategies. (a) Screening. Incomplete factorial designs sample parameter space randomly and uniformity is enforced on the sampling of main effects and two-dimensional interactions. 'Sparse matrix' sampling involves an intentional bias toward combinations that have worked previously. Orthogonal arrays impose strict orthogonality on the columns of the experimental matrix, and are thus less flexible. (b) Optimization. Hardin-Sloane design for minimum variance parameter estimation distribute experiments non-uniformly, with one or two at the centre and the others at the boundary of the search region.

Their use has been described previously (15–18). Factor levels are chosen randomly and then balanced to achieve nearly uniform sampling (*Figure 4*). Levels for each main effect are sampled the same number of times, and all two-factor interactions are sampled as uniformly as possible. This strategy preserves the ability to detect large main effects and two-factor interactions with minimal confounding (18). This process leads to a very flexible sampled factorial design which has given superior performance in a wide variety of contexts (15, 21–24).

The number of experiments in a design should be chosen relative to the size, N, of the full-factorial design, which, in turn, depends on the number of variables, and their levels according to *Equation 1*. Incomplete factorial designs with a sampling density as coarse as roughly $\sqrt{N/2}$ can be found that are quite evenly balanced with respect to main effects and two-factor interactions, and entirely free of explicit confounding. This rule of thumb is more useful for larger factorial designs; designs with smaller numbers of factors are less efficient, as they require more experiments for adequate signal-to-noise.

3.2.2 'Sparse matrix' recipes

Factorial design strongly influenced the development of 'sparse matrix kits' now used for screening in an overwhelming majority of crystal growth efforts. These unbalanced, pre-packaged samples from a broad, factorial space, emulate factorial design without actually capturing their essential features (*Figure 4*) (25–29). The conceptual basis for such designs is to bias the choices of combinations in favour of conditions that have previously produced crystals for other proteins, RNA, or membrane proteins, as found, for example, in the Biological Crystallization Database (30) to select conditions that have previously produced crystals. A recent extension of this approach uses the biases in that database to weight sampling strategies for particular classes of proteins (31).

The success of kits, arising from their ease of use and commercial accessibility, speaks for itself, but comes at significant cost. Biasing is necessarily done at the expense of balance. Thus, scoring is nearly useless for quantitative model building and analysis. Biased searches may also miss useful crystal polymorphs. Moreover, reinforcing existing biases during crystallization also biases the database of solved macromolecular crystal structures in quite subtle ways. For example, the high salt structure of haemoglobin A, known as the R structure (32), differs from those obtained under near physiological conditions, and is almost certainly an artefact of crystallization at high salt (33–36).

The attractions of sparse matrix kits are virtues only because the physical chemistry of pure, homogeneous protein solutions strongly favours crystallization over a considerable range of solubility. When kits work, one still must optimize conditions suggested by preliminary screening. When they fail one must re-screen, from new combinations of conditions. Since little can be learned from quantitative analysis of results obtained using kits, the focus here is on experimental designs that can be analysed.

3.2.3 Orthogonal arrays and fractional factorial designs

Other sampling strategies place a higher priority on separating effects of a particular subset of factors as completely as possible; what was termed the 'resolution' in Section 2.3.4. Resolution is achieved by accepting a higher degree of confounding between main effects and higher-order interactions, which are usually less significant. For full-factorial designs, higher-order interactions can provide estimates for the experimental variance. 'Fractional factorial' designs (37) intentionally alias higher-order interactions with treatments of additional main effects, thereby increasing the number of factors tested. 'Orthogonal arrays' preserve much of the geometrical symmetry of full-factorial designs and have been developed and exploited for screening (38). Here, the treatment effects are completely uncorrelated and can hence be more completely resolved. Orthogonal subsets of the two level, three-factor design in *Table 1* each have four experiments. As noted in Section 6, they provide poor quantitation that is improved substantially by increasing the number of experiments from four to six by the use of an incomplete factorial design.

3.3 Minimum variance sampling: Hardin–Sloane designs

Response-surface experiments are sampled by different criteria from those used in screening (*Figure 4b*). Optimization requires a non-uniform sampling strategy because the objective is to construct an accurate analytical approximation to how the system actually responds to the input variables. If one already knows something about where the best result might be obtained, it is no longer sensible to scatter the experimental test points uniformly throughout the space. Rather, experimental points are selected for maximal impact on the accuracy of the parameters of the response-surface function, and hence on the coordinates of its stationary points. Designs that minimize errors in parameter estimation are called 'minimum-prediction variance' designs (39).

The strategy of Hardin–Sloane designs can be visualized in a one-dimensional example where the goal is to distinguish as accurately as possible between a linear relation, $y = ax + b$, and a parabola, $y = ax^2 + bx + c$. Three groups of experimental points have maximal impact on this distinction: those near the suspected maximum value of the parabola and those at upper and lower limits of x for which the parabolic approximation may be appropriate. Distributing several experiments near the suspected maximum value and the remaining ones at low and high values of x gives the most accurate (averaged over multiple experimental measurements) values for these crucial points of the response surface. For multivariate response-surface models the same strategy applies; a small number of experiments should be done near the centre of the design and the remainder evenly distributed around the *perimeter* of the experimental space (*Figure 4b*). Hardin–Sloane design matrices are given in *Tables 2* and *3*.

3.4 Computer programs to generate designs for special purposes

Because they require efficient sampling, screening and response-surface designs are often selected from a large number of potential designs generated by a computer program according to specific criteria. Such programs include *INFAC* (`http://russell.med.unc.edu/~carter/designs`), a program to generate incomplete factorial designs, and *GOSSET* (`http://www.research.att.com/~njas/gosset`), a more general and much more powerful program to generate minimum-variance designs for response-surface determination. Other programs for experimental design also have

Table 2. Hardin–Sloane minimum integrated variance design matrix for four factors, 20 experiments[a]

Exp't	Variable 1	Variable 2	Variable 3	Variable 4
1	0.000	−0.056	0.000	−0.250
2	0.000	−0.056	0.000	−0.250
3	0.000	1.000	0.000	−0.250
4	0.000	−1.000	0.000	−1.000
5	1.000	−0.007	0.116	1.000
6	−1.000	−0.007	−0.116	1.000
7	0.210	0.108	−1.000	−1.000
8	−0.210	0.108	1.000	−1.000
9	−1.000	−1.000	1.000	−0.250
10	−1.000	1.000	−1.000	−0.250
11	1.000	−1.000	−1.000	−0.250
12	1.000	1.000	1.000	−0.250
13	0.492	−1.000	1.000	1.000
14	−0.492	−1.000	−1.000	1.000
15	−1.000	1.000	0.577	−1.000
16	1.000	1.000	−0.577	−1.000
17	0.669	1.000	−1.000	1.000
18	−0.669	1.000	1.000	1.000
19	−1.000	−1.000	−1.000	−1.000
20	1.000	−1.000	1.000	−1.000

[a] This design was prepared specifically for use in the experiments reported here by N. J. A. Sloane, using *GOSSET* (39). Matrix entries should be interpreted as: 0 = the centre, −1 = the low end, and 1 = the high end of the variable range. The same design has been used repeatedly in different contexts, by assigning the matrix entries to different parameters and/or ranges.

been described (40). *INFAC* designs have optimal coverage and minimal aliasing of main effects with each other and with two-way interactions. The program is entirely interactive, prompting for all necessary information. The *GOSSET* interface is less intuitive and requires explicit description of variable ranges and the type of response-surface function to be fitted. Much can be obtained from the examples in the users manual, but the overview in *Protocol 1* of how to generate a design like that in *Table 2* should be helpful.

Protocol 1. Generating a Hardin–Sloane design using *GOSSET*

Method

1. Once launched, *GOSSET* requests the user to select a working directory. Description of the design is done without prompts, and must be preceded by line numbers, e.g. 10 range × y z −1 1.
2. Enter the range to be considered for each of the variables, including

those that will take on discrete values, e.g. 20 discrete T 4 14 21 can be used to specify three discrete values for temperature.

3. Enter the model for which the design will be used, e.g. 30 model (1+x+y+z+T)^2 will specify a quadratic polynomial model in four variables, x, y, z, and T. Such a model has (4+1)*(4+2)/2 = 15 parameters, so the design must have at least that number of experiments.
4. Compile the above 'program' using: compile (from this point, no line numbers are entered!).
5. Compute the matrix of 'experimental moments' using: moments n=1000000. These moments are used internally to represent the impacts of experimental points on the predicted variance (39).
6. Ask for the design. The command: 'design' will generate a design with the minimum required number of experiments for the specified model. Various modifications include:
   (a) design runs=24 n=20. This forces the program to generate 24 experiments and to find the best design from 20 different starting points.
   (b) design type=I extra=5. This forces the program to generate an I-optimal design (the default choice, which optimizes the **I**ntegrated prediction variance) with 5 more than the minimum number of experiments. The resulting design will have 20 experiments in this case.
7. Generate a formatted file with the design: interp >20expt_xyzT.design. This converts the output file into a formatted table in the file 20expt_xyzT.design.

4. Screening with factorial designs

Surprisingly little thought has been devoted to characterizing the dimensions of 'crystallization space' for any protein. Nevertheless, recent studies of how model proteins crystallize (41–50) have led to new insights (see Chapter 10) that change how one should think about screening and optimization of crystal growth. These insights provide a rudimentary, but essential guide to experimental design.

4.1 Selecting experimental variables

Crystallization requires an appropriate balance of variables that influence:

- *materials*: composition and homogeneity of the unit cell

Table 3. Hardin–Sloane minimum integrated variance design matrix for five factors, 30 experiments[a]

Exp't	[Macromol]	Supersaturation	[Additive]	pH	Ligand
1	−1.00	−1.00	0.03	−1.00	−1.00
2	0.04	0.04	0.10	0.07	0.03
3	−0.82	−1.00	1.00	−1.00	0.86
4	−0.04	1.00	−1.00	−0.11	−1.00
5	1.00	−1.00	−1.00	1.00	1.00
6	−1.00	−0.45	−1.00	−0.18	−0.20
7	1.00	1.00	−0.12	−1.00	1.00
8	0.18	−0.17	0.03	−0.12	1.00
9	−1.00	1.00	−0.08	1.00	−1.00
10	1.00	0.00	−0.63	1.00	−1.00
11	−0.04	0.01	−0.07	1.00	0.05
12	−0.68	0.45	−1.00	−1.00	1.00
13	−1.00	1.00	1.00	−0.23	1.00
14	0.20	1.00	0.73	1.00	1.00
15	−0.61	−1.00	−1.00	1.00	−1.00
16	1.00	1.00	1.00	0.91	−0.94
17	1.00	−0.12	−1.00	−1.00	−1.00
18	1.00	−1.00	−0.02	−0.30	−0.55
19	1.00	1.00	−1.00	0.37	0.31
20	0.36	1.00	1.00	−1.00	−1.00
21	0.21	−1.00	1.00	0.29	−1.00
22	−1.00	−0.87	1.00	1.00	0.19
23	−1.00	0.07	1.00	−0.11	−1.00
24	0.04	0.04	0.10	0.07	0.03
25	−1.00	−1.00	−0.22	0.44	1.00
26	1.00	−0.20	1.00	−1.00	0.42
27	0.37	−1.00	−1.00	−1.00	0.39
28	1.00	−1.00	1.00	1.00	1.00
29	−1.00	1.00	−1.00	1.00	1.00
30	−1.00	1.00	0.04	−1.00	−0.16

[a] This five-variable design was prepared using the *GOSSET* program (39). It was designed to compensate for the failure of one of the experiments to produce a score, and is called J-optimal. It contains 30 experiments, which are nine more than the 21 experiments required to estimate parameters for a five-variable quadratic model. Although variable names have been suggested, this matrix can be used in any desired context.

- *relative interaction potentials and free energies*: interparticle potentials, solubility, and supersaturation
- *relative rates of various processes*: nucleation, diffusive transport, and interfacial deposition.

Screening therefore must sample four different types of factors that intervene between an experimental design and crystal growth (*Figure 5*), including

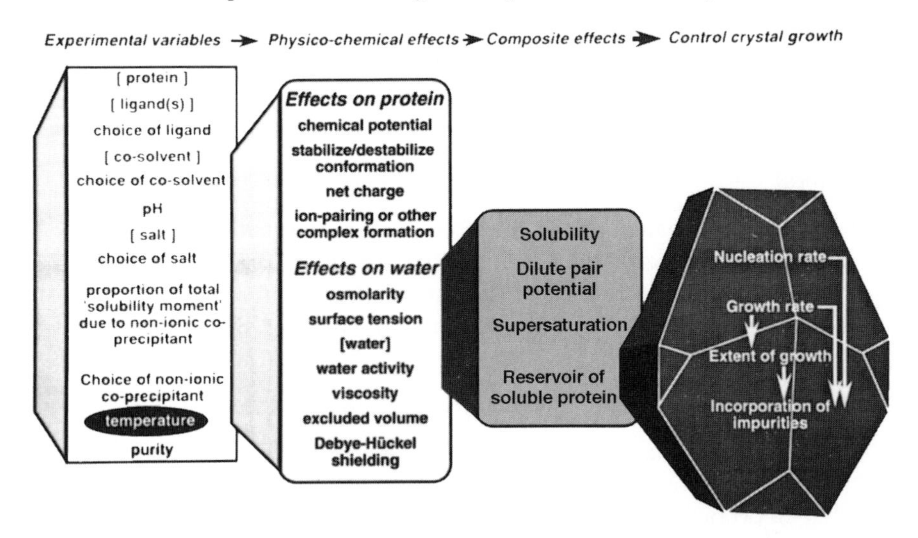

Figure 5. Crystal growth variable space. Variables are organized hierarchically, from those directly under explicit experimental control on the left, to those that dictate crystal growth behaviour on the right. To challenge one experimental design in screening and/or optimization of crystal growth is to find an efficient parameterization of variables in this space that will permit *post-hoc* quantitative analysis.

those under direct experimental control (column I) and composite properties that ultimately determine the physical system evolution (column IV). These variables can influence both macromolecule and solvent (column II), changing the conformation, net charge, and quaternary structure of what is to be crystallized while reducing solubility and promoting crystal growth. Proteins have significant net charge, except at their isoelectric points (45), and are accompanied by stoichiometric amounts of counterions, whose concentration increases with the protein concentration (51,52) and these ions may incorporate in, and influence crystal properties. Solvent properties, notably the activity of water and the dielectric constant, are also sensitive to variables in column I. The balance between thermodynamic and kinetic effects, including rates of equilibration (53,54), will determine the extent of growth and hence the crystal size. They also influence the rate at which impurities are incorporated (50, 55–57). The challenges in designing experiments to study crystallogenesis arise from the need to sample the effects suggested in *Figure 5* as effectively as possible.

4.2 Preparing the experimental matrix

The experimental matrix specifies how each individual experiment or 'test' is to be carried out. *Protocol 2* summarizes a procedure for constructing an incomplete factorial experimental matrix.

Protocol 2. Preparation of the experimental matrix for an incomplete factorial design

Method

1. Gather information about the protein to be crystallized from all involved with expression and purification; from previous screening experiments; and from databases (Biological Macromolecule Crystallization Database): (`http://ibm4.carb.nist.gov:4400/bmcd/bmcd.html`) (30, 58).
2. List factors that might influence crystal growth.
   (a) Estimate the variation of net charge with pH from the amino acid composition (Chapter 10). Verify the pI by isoelectric focusing. Use a pH range with values above and below the pI.
   (b) Factors required for stability and monodispersity (59, 60).
   (c) Ligands and other factors likely to influence the conformation of the macromolecule.
4. Choose from this list the factors to be screened in the current experiment and the levels to test.
5. Choose the number of tests. In general, this number should be somewhat more than the number of factors to be screened. There is no hard and fast rule; experience from diverse sources (20, 61) suggests using $\sim \sqrt{N/2}$ tests if the full-factorial design requires N tests.
6. Compile the experimental matrix itself (the computer program *INFAC* `http://russell.med.unc.edu/~carter/designs` will do steps a–e interactively).
   (a) Choose factor levels at random, working down each column, and from column to column.
   (b) Balance each column, readjusting levels to equilibrate the numbers of tests at each level.
   (c) Balance each two-factor interaction by compensating readjustments to two columns. In this case, it is adequate to ensure that each combination is represented by at least one test.
   (d) Verify the balance of all columns and two-factor interactions.
   (e) Examine the experimental treatments for possible confounding. Confounding is indicated whenever two different effects have

identical patterns of level assignments. If a confounded effect turns out to be large, this knowledge is useful in further experimentation to distinguish between the two possibilities. (Designs generated by *INFAC* are selected to minimize confounding.)

4.3 'Floating' variables, initial values, and sampling intervals

Crystal nucleation can be achieved by changing either the solubility or the macromolecular concentration, or both, via changes in [crystallizing agent], temperature (62, 63), or pH. One of these must be chosen as a 'floating variable'; the appropriate choice is system-dependent (45). Temperature and pH (net protein charge) have reduced influence on solubility at high ionic strength, and ionic strength has reduced influence on solubility close to the isoelectic point. Thus, for example, proteins that crystallize near their isoelectric points will almost certainly be insensitive to ionic strength, leaving pH or temperature as possible floating variables. Alternatives include finding an ion that interacts with the protein, changing its pI.

4.3.1 Scanning for supersaturation

Since solubility behaviour is difficult to determine a priori, bootstrap methods must be used in order to exploit prior knowledge of solubility behaviour in choosing concentrations of protein and crystallizing agent. An upper limit to the solubility in the absence of crystallizing agents can be estimated by concentrating the protein in its usual buffer to the greatest extent possible for storage. Screening should sample fractions of this value, say 1.0, 0.5, and 0.3, and crystallizing agent concentrations should reduce solubility to ~ 10–30% of its initial value. A range of crystallizing agent concentrations can be scanned using the same sample by dialysis. Vapour diffusion uses less protein, permitting tests at different [crystallizing agent]. This is the basis of the 'footprint' method (64, 65), in which four concentrations are tested for six reagents in a Linbro plate. Footprinting can be effectively combined with streak seeding (66) (Chapter 5) because it provides protein samples equilibrated with supersaturated, but non-nucleated solutions, which are ideal for seeding. Moreover, coverslips can be transferred to reservoirs of lower vapour pressure, permitting simultaneous increases of [macromolecule] and [crystallizing agent]. The simplex procedure described in Section 7.1 offers another way to find the approximate macromolecule and crystallizing agent concentrations for the nucleation zone.

Supersaturation is so critical to controlling crystal nucleation that an approach called 'reverse screening' (65), presumes that crystals should grow with almost any agent if the appropriate supersaturation can be found, and examines solubility behaviour for two or three crystallizing agents. The importance of solubility data, and the *ad hoc* nature of screening reinforce the importance of utilizing early screening experiments to construct experimental solubility databases when working with a new problem. Mother liquors should be separated routinely from all solid phases, and assayed for the concentration of soluble macromolecule. Parameters of the Cohn–Green equation, $\ln C_{sol} = \beta - K_s$ [crystallizing agent] can be estimated from two or three different determinations of the protein concentration in the liquid phases of initial experiments. K_s changes significantly with the net charge of the protein, the position of the anion in the Hofmeister series, as indicated in Section 4.3.

4.3.2 Sampling intervals

How finely sampled a design should be depends on its purpose. For screening, the high level should 'titrate' out any specific binding interactions that might influence crystal growth. For optimization experiments, which require scores for nearly all experiments, the following guidelines may be useful:

(a) Variables whose effects are related to chemical potentials (e.g. pH) are logarithmic; changes of $\sim \pm 0.5$ will effect a tenfold variation in equilibrium concentrations.

(b) Protein concentration and supersaturation exert higher-order effects on nucleation rates, proportional to n ln[protein], where n is the (generally unknown) order of the nucleation step (67, 68). Finer sampling of ln[protein], say by $\pm 0.5/n$, should change nucleation rates by an order of magnitude and may be more appropriate.

4.4 Design for initial screening of variables for crystallizing a new protein

The experimental variables in *Figure 5* suggest that screening designs can be made more effective in producing crystals and more useful sources of evidence about the factors governing crystal growth. The design in *Table 4* is

Table 4. Incomplete factorial screening design in ten variables

Exp't	A	B	C	D	E	F	G	H	I	J
1	2	2	4	1	1	3	2	3	1	1
2	4	5	5	1	2	3	1	1	1	2
3	5	1	4	2	1	1	3	1	2	2
4	1	2	3	1	2	2	3	3	2	1
5	3	5	2	3	2	1	1	2	1	2
6	4	3	1	1	1	3	1	3	2	1
7	5	4	1	3	3	1	1	2	2	2
8	1	4	3	3	1	2	2	2	2	2
9	2	1	2	1	3	2	2	3	1	1

Table 4. *continued*

Exp't	A	B	C	D	E	F	G	H	I	J
10	3	3	5	3	3	3	3	3	1	2
11	1	1	5	2	3	2	1	1	2	1
12	2	1	1	2	1	1	3	2	1	1
13	3	1	2	2	2	3	2	2	2	1
14	5	1	4	3	2	2	3	1	1	1
15	4	5	3	2	3	1	2	1	1	2
16	5	2	2	3	1	1	3	1	2	2
17	4	4	2	2	2	2	3	2	1	2
18	2	5	1	3	3	1	2	3	2	1
19	3	5	4	2	1	2	2	2	1	1
20	1	4	3	2	3	2	1	2	2	2
21	5	5	5	1	1	3	1	3	2	2
22	2	4	5	3	2	1	1	1	1	2
23	1	3	4	3	1	3	2	1	2	1
24	3	4	1	2	2	3	3	3	1	1
25	4	2	1	1	3	3	3	3	1	2
26	5	3	3	1	2	3	2	3	2	2
27	4	1	3	3	3	3	1	2	2	1
28	3	2	5	2	1	1	3	2	1	1
29	2	3	5	2	3	1	1	1	1	1
30	1	2	3	1	1	2	2	1	2	2
31	5	4	4	1	3	2	1	3	2	1
32	3	3	1	3	2	2	3	1	2	2
33	2	2	4	3	1	3	1	1	1	2
34	1	5	2	1	2	1	2	2	1	2
35	4	3	2	1	1	2	3	1	1	1
36	5	2	5	1	2	1	1	2	2	1
37	3	4	2	2	1	1	3	3	1	1
38	4	5	2	3	3	1	2	2	2	1
39	2	1	5	3	2	3	2	3	2	2
40	1	3	3	3	3	3	3	2	1	2
41	4	2	2	1	3	2	2	1	2	2
42	2	5	5	3	2	2	1	3	1	1
43	5	1	1	2	2	3	3	1	2	1
44	3	2	1	1	2	2	2	3	1	2
45	1	4	5	1	3	3	2	1	1	2
46	1	2	4	2	3	1	3	2	1	1
47	5	4	4	1	1	3	1	1	2	1
48	4	3	2	2	2	1	2	2	2	2
49	3	3	4	3	1	1	1	2	1	2
50	2	4	3	2	3	2	3	3	2	2
51	3	1	2	3	2	1	3	2	2	1
52	5	5	4	2	1	2	1	3	1	1
53	2	3	1	3	3	1	1	1	1	2
54	1	5	1	2	3	3	2	3	2	1
55	4	1	4	1	3	2	1	2	2	2
56	3	5	3	3	1	2	3	3	2	1
57	4	4	1	1	2	2	2	2	1	1
58	2	2	3	1	1	1	2	1	1	1
59	5	3	5	2	1	3	3	1	2	2
60	1	1	3	2	2	3	1	3	1	2

motivated by that goal. It provides a balanced incomplete factorial screen covering a similar sample of the conditions from *Figure 5* to that represented in the Hampton kit(s) (25), and is informed by the underlying physical chemistry. It can be used either with a standard set of pHs or centred on a known or estimated pI value.

4.4.1 Physico-chemical parameterization

As far as possible, this design associates each variable with a single physico-chemical property from column II in *Figure 5*. The Hofmeister series (42, 44, 45) likely reflects the kosmotropy or lyotropy of the salt. Its effect on solubility and hence on crystal growth is likely to be related, via the surface tension (69) to solubility and to the relative interactions of the various ion pairs in the protein–solvent interface (70). The series of polyols with increasing molecular weight samples a range of excluded volumes. Different glycerol concentrations sample osmotic pressure exerted on water molecules in intramolecular crevices and hence possible conformational changes (71) are screened explicitly. The design was implemented in *Protocol 3* according to these and similar considerations.

4.4.2 The 'organic moment'

A distinctive feature of the Jancarik and Kim screen (25) is that organic crystallizing agents like PEG are combined with significant amounts of salts. One rationale for doing this is that different proteins may have different proportions of polar and non-polar surface textures, whose solubilities depend in different ways on volume exclusion mediated by PEG and electrostatic shielding by ionic crystallizing agents. An extreme example, for which there is considerable literature (27, 72), are the integral membrane proteins, which explicitly demand simultaneous manipulation of ionic and polymer excluding crystallizing agents. The design in *Table 4* allows for the interaction of salts with organic and/or polymeric crystallizing agents at three levels, e.g. 0%, 35%, 70%.

If solubility constants are available, the interaction can be quantified explicitly by a 'moment',

$$M = \frac{(Ks_O[\text{Organic co-precipitant}])}{(Ks_A[\text{Ionic precipitant}] = Ks_O[\text{Organic co-precipitant}])}.$$

M approximates the relative impact on solubility for each crystallizing agent, via the molar concentration of each together with its solubility coefficient, K_s, which usually must be determined (Section 4.3.1). If these are unavailable, M can be approximated for screening purposes using molar concentrations alone for all agents except the high molecular weight PEGs, whose K_s values are generally an order of magnitude greater than those for salts and low molecular weight organic reagents.

Protocol 3. Example: incomplete factorial screening design (*Table 4*)

Method

Identify columns in Table 4 with factors such as the following:

A. pH: five levels, 4.5, 5.5, 6.5, 7.5, 8.5, cover the range normally observed for crystallization of proteins. Alternately, if the pI is known five levels can be centred on the pI.

B. Use five ionic crystallizing agents, with the following anions: sulfate, phosphate, acetate, chloride, nitrate from the Hofmeister series.

C. Organic crystallizing agent: five levels, isopropanol, methylpentane-diol, PEG 4000, PEG 8000, PEG 20 000.

D. Protein concentration: three levels, Max, Max/2, Max/3. Three levels are chosen because this variable may not be linear, and may have a maximum. Some effort is placed on estimating the curvature of its behaviour.

E. Temperature: three levels, for instance 4°C, 14°C, 22°C.

F. 'Moment' of non-polar reagent, M: 0%, 35%, 70%, as defined in Section 4.4.2.

G. Divalent cation: three levels, none, Mg^{2+}, and either Ca^{2+}, Cd^{2+}, or Mn^{2+}. Salts with acetate, nitrate, sulfate, and chloride are sufficiently soluble. All are problematical with phosphate. This is a constant weakness of all such plans. Phosphate experiment(s) calling for level three of this variable can use Mn^{2+}.

H. Additive: three levels, none, arginine, βOG. These suggested choices are based on the selection of additives that either stabilize proteins (arginine ~ 50–100 mM) (59, 60) or destabilize weak, non-specific contacts (βOG 0.1–0.3%, w/v).

I. Glycerol: two levels, 0%, 10% (v/v). This variable samples additional osmotic pressure exerted by the solution over and above that produced by the crystallizing agent(s) themselves.

J. Ligand: two levels, presence or absence of a substrate or inhibitor.

4.5 Experimental set-ups

Experiments in the design should be carried out by systematically approaching supersaturation, manipulating the floating variable to ensure that each experiment is taken to completion. This should be done repeatedly for conditions producing precipitates, to verify that precipitation was not due simply to excessive precipitant concentrations. Microdialysis buttons have obvious advantages from this point of view, but vapour diffusion can be used if desired. Vapour diffusion is essential for some experimental conditions, such as organic crystallizing agents and polymers. Factorial experiments preclude using the same buffer for any two experiments, so some thought should be given to rational preparation of stock solutions. We have found no good alternative to simply making up each buffer separately, in order to keep the ionic compositions consistent with design requirements.

5. Quantitative scoring

Previous quantitation and statistical analysis of crystal growth experiments (1, 2, 17, 73) are sufficiently compelling to justify a more intensive investigation of practical requirements and procedures for using quantitative analysis of crystal growth more routinely. We presume throughout this chapter that experiments can be quantitatively scored in a variety of different ways. This section addresses problems associated with scoring crystallization experiments, and suggests protocols for solving those problems.

5.1 Hierarchical evaluation: interrogating nature by experimental design

Quantitative scoring can be integrated into the study of crystal growth most effectively in the context of a hierarchical framework of specific questions (*Protocol 4*). *Figure 6* shows how we 'interrogate' crystal growth experiments progressively, first establishing which kind of quantitation may be appropriate. Several qualitatively different 'scores' give purpose to the interrogation, increasing the likelihood of success by directing attention to the appropriate context and question, before attempting to assign scores. Further statistical analysis is appropriate for many of the scores.

Scores should reflect the consequences of changing the experimental variables. Several pre-conditions help assure that they do, and should be satisfied first; the questions in the scheme provide a guide to assuring that these are met before trying to build models for any particular scores. Above all, it is essential to have a sufficient number of scores for analysis. Hence, the initial question concerns how many tests actually produced crystals. The answer to this question determines whether to pursue the left-hand or right-hand branches of *Figure 6*.

5.1.1 Scoring crystal properties

If most tests produced crystals, then two kinds of scores representing crystal properties can be developed, depending on whether or not the crystals correspond to the same or to different polymorphs. Screening experiments provide qualitatively more diverse information about the location of station-

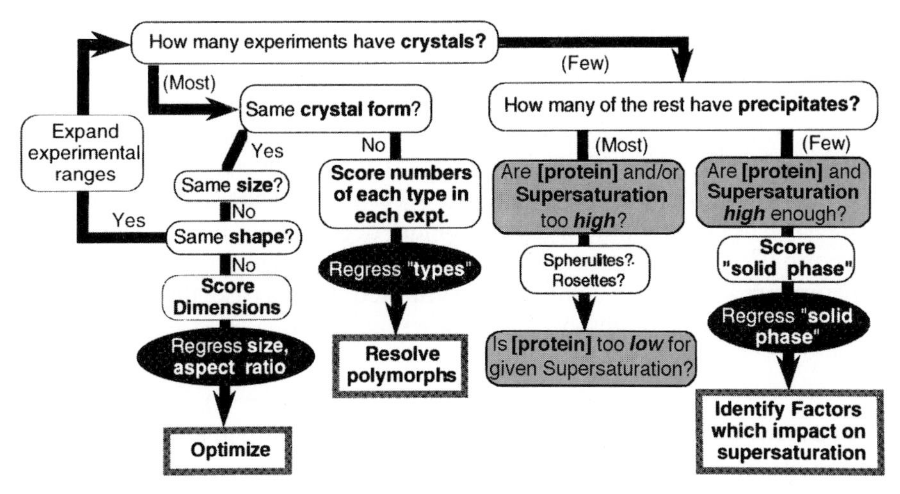

Figure 6. A hierarchical scheme for evaluating designed crystallization experiments. A nested set of questions is posed of the experiments in the design. Answers to these questions determine the appropriate action to take. Questions in lightly shaded boxes can be answered categorically, either with 'yes' or 'no' or comparative answers. Questions in mid-grey boxes can be addressed by obvious subsequent experiments. Directives to 'score' a particular parameter imply that a quantitative score can be used, leading to multivariate regression of those scores against experimental factors for purposes suggested in rectangles with large grey borders.

ary points than do optimization experiments. In general, the solid phases of screening experiments may be microcrystalline or amorphous precipitates, and there may be different crystalline polymorphs in the ensemble. The scoring system should therefore make use of this broad range of information.

Polymorphism frequently occurs in crystallization, and probably reflects, in part, underlying structural polymorphisms that may be associated with function. One should distinguish cases where the different polymorphs appear in different tests from a screen from those where the polymorphism is evident within the same sample. In the former case, subsequent experiments should be directed toward optimizing the different nodes. The latter situation often arises in the early stages of optimization from a screen that produced conditions where response surfaces for more than one polymorph overlap. Scoring the proportions of each polymorph in mixtures can help resolve conditions under which the different polymorphs can be grown uniquely (73) (Section 8). Finally, if all crystals belong to the same polymorph, it makes sense to determine if their sizes and habits vary enough to justify optimizing those properties by response-surface methods (1).

5.1.2 Troubleshooting supersaturation

When few of the experiments in a design produced crystals, as depicted on the right-hand side of *Figure 6*, other types of questions can lead to alternative scoring aimed at finding a 'sense of direction' from the current position. This is the case when seeking initial conditions with a screening experiment. One searches in this case for an *unknown* point in a multidimensional solubility diagram where discrete crystal nucleation occurs, but similar considerations underlie the interrogation. Unless one has the happy result of generating many crystals from an initial screen, it is necessary to make sense of variations in the type of phase separation. Prior binary decisions are nevertheless useful here, too. If few experiments have a solid phase, the culprit is likely that the experiments are not sufficiently supersaturated. In this case, the score can be simply the presence or absence of a phase separation.

Protocol 4. Scoring crystallization experiments (see *Figure 6*)

Method

1. Determine and score how many experiments in the design have some kind of phase separation.
   (a) If a majority are still clear, adjust the experiments to increase supersaturation. Regression of this first score against the experimental variables can reveal which changes to make in experimental conditions to increase the proportion of experiments producing a solid phase.
   (b) If a majority have a solid phase, proceed to step 2.
2. Determine and score, according to Table 5, how many experiments in the design have crystals.
   (a) If most experiments have crystals, how many different morphologies are represented? If there is only a single crystal habit, proceed to step 4. Otherwise, proceed to step 5.
   (b) If most experiments have precipitates, oils, and spherullites, streak seeding (Chapter 7) can be helpful.
3. Examine a 'scatterplot matrix' showing how each score depends on each independent variable for obvious trends and/or non-linearities.
4. Measure crystal sizes and shapes.
   (a) Two dimensions, and occasionally a third can usually be measured using a microscopic ruler. Enter dimensions for representative samples from each experiment.
   (b) Calculate sizes (surface area or volume) and shapes (aspect ratio, or width/length).

(c) Determine the maximum, minimum, mean value, for each score. Note significant variations.

5. Count the numbers of crystals with the same habit in each experiment. Enter separate scores for each distinct morphology. These scores can help resolve polymorphs (Section 8).

Table 5. Scale of crystal quality

Result	Score, Q
Cloudy/amorphous precipitates	1.0
Gelatinous/particulate precipitates	2.0
Oils	3.0
Spherulites	4.0
Needles	5.0
Plates	6.0
Prisms	7.0

5.2 Rating different solid phases

When no proper crystals are obtained, sense often can be made from variations in the quality of the solid or denser phase. Precipitates differ somewhat from one another. Fluffy, cloudy, or filamentous precipitates have little likelihood of being crystalline, but uniform, granular, and/or particulate precipitates often are microcrystalline. Similarly, among different crystalline samples, one can readily distinguish spherulites (radially symmetric aggregates of microscopic needles) from needles, plates, and prisms, with obvious preference for the latter over the former. An intuitive sense of this progression (15) provides a basis for the quantitative scoring protocol in *Table 5*. Exploration of the lysozyme phase diagram into regions where a liquid–liquid phase separation occurs (49) provide a solid rationale for scoring of pathological behaviour in this way. Under appropriate conditions even lysozyme can be induced to produce all of the pathological solid phases observed with many proteins which seem refractory to crystal growth. The authors proposed a 'generic phase diagram' applicable to all or many proteins, in which the types of solid phases listed in *Table 5* (spherullites, oils, and various kinds of precipitate) assume a coherent progression as one approaches some optimal value of supersaturation. Relationships between these and more optimal locations are probably somewhat reproducible from protein to protein.

Further support for quantitative scoring came when detailed examination of the behaviour of monoclinic tryptophanyl-tRNA synthetase (1) provided evidence for a 'sweet-spot' in the solubility diagram where crystal volume and shape were optimized simultaneously, suggesting that growth in all three dimensions progressed unimpeded and at similar rates. Interestingly, secondary nucleation, which had been a difficult problem, was also minimal for the same, optimal conditions, whereas various pathologies, including elongation (needles) and multiple nucleation appeared away from the optimal supersaturation, which was higher than that suggested by intuition.

A problem with the scale in *Table 5* is the tendency to confuse microcrystalline and amorphous precipitates (16, 17). The following methods can help confirm the rank-ordering of scores:

(a) Streak seed from precipitates to assay for microcrystallinity (Chapter 7).

(b) Examine precipitates for birefringence in a glass depression slide to avoid plastic/air interfaces, with reservoir solutions in the depressions surrounding the sample and a large glass slide as a coverslip over the entire dish to avoid crystallization of the crystallizing agent.

5.3 Size and shape

Optimization experiments show how the properties of a particular crystal form depend on smaller variations of the important parameters. Thus, the scoring must reflect this variation. Crystal dimensions can be scored directly by microscopic inspection.

5.4 Scoring the best result or the 'average' from a given test?

Intuition and experience suggest different answers to this question. Intuitively, averaging seems safer, because it helps to insulate the scores from fluctuations, giving a truer account of given experimental conditions. On the other hand, an exceptional crystal among many smaller or misshapen crystals is often an 'existence proof' that one is close to an optimum. Averaging properties of the good crystal with those of the many poorer samples will tend to dilute the influence of that one experimental combination, submerging the signal in the analysis. Our experience is that using the properties of the best crystals in a drop has been considerably more successful than focusing on average properties. Replication (see Section 7.2.3) helps to separate spurious from meaningful results.

6. Regression, the analysis of variance, and analysis of models

Key to all factorial response-surface methods is the quantitative analysis of the different behaviours observed as experimental conditions are varied.

Quantitative analysis begins with some quantitative '*observed* score', given here by Q_{obs}. Two related kinds of analysis can be done (74):

(a) Estimation of main effect and interaction contrasts.

(b) Construction and testing of linear models for calculating the scores, Q_{calc}.

Before using these procedures, carry out a preliminary analysis as given in *Protocol 5* to assess which scores and/or variables deserve more careful analysis.

Protocol 5. Preliminary analysis of scores

Method

1. Prepare a file with the experimental matrix and all scores. Most statistics programs provide separate modules for data entry, statistical model building and analysis, and presentation graphics. For factors whose levels are attributes (e.g. different ions), create a separate column for each attribute, giving values of 0 or 1 for its presence or absence. It may also be useful to approximate monotonic behaviour like that of the ions of the Hofmeister series in a separate column.
2. Determine the maximum, minimum, mean value, for each score (this may be programmed in *EXCEL;* in *SYSTAT,* use the Stats/Stats command). Identify scores with significant variation.
3. Examine the data graphically. A matrix of two-way scatterplots for each score versus each experimental variable is a useful, visual presentation of the data, showing obvious correlations.
4. Try to express such trends as linear models explaining the variation in scores (*Protocol 6*).

6.1 Analysis of contrasts

Contrasts (Section 2) compare averages of specific, balanced sets of experiments treated at different levels of a particular factor for information about whether or not that factor is a *significant source of the variation* observed in the results. The null hypothesis is that these averages, also called the treatment means, are equal within the error of the measurements. Unless all variables are converted to 'standard' variables with all means = 0 and standard deviations = 1, the contrasts themselves can be misleading because the units of different variables may be on quite different scales. Thus, as the italicized phrase suggests, the simple contrast sum calculations presented in Section 2.3.2, *Equation 2*, can be misleading, and should be verified by a full analysis of variance whenever possible.

6.1.1 Averaging results from replicated experiments

Averaging is a well-understood source of confidence in making inferences. The signal-to-noise of a repeated experimental result increases as the square root of the number of times it is repeated. For example, a 24-experiment design provides 12 experiments at each level of a two-level variable, increasing the signal-to-noise by ~ 2.5 times over that from two replicates.

6.1.2 Averaging experiments that are not exact replicates

The efficiency of factorial experiments comes from contrasts between ensembles of experiments that are not exact replicates of one another. Interestingly, the standard deviations of averages from different experiments in a design depend strongly only on the precision of individual measurements and only weakly on the number of different factors being varied. Balanced designs average-out the effects of other variables, highlighting the bona fide sources of variation in scores from all experiments. In these cases, the average of all experiments treated at one level of a particular factor, relative to the average of experiments treated at another level remains sensitive to the influence of that factor.

6.2 Analysis of models by multiple regression and the analysis of variance

Multiple regression provides β_i values that minimize the sum of squared differences $\sum_{i=1}^{n}(Q_{obs,i} - Q_{pred,i})^2$. This predictive model can provide an estimate, Q_{calc}, for the experimental result, based on contributions from the different factors. If there are K adjustable parameters in the model, they can be estimated by minimizing the sum of the squares of differences between Q_{obs} and Q_{calc} over all the experiments in a design of $N > K$ experiments.

Analysis of designed experiments is a broad and well-established field, and no attempt will be made here to summarize what can be found in excellent monographs on the subject (19, 75–79). Procedures necessary for rudimentary statistical and graphic analyses are available in standard data analysis programs (*EXCEL*, *KALEIDAGRAPH*) for personal computers. Regression analysis of linear models is available only in the more comprehensive statistical packages. Particularly useful are the two programs *SYSTAT* (80) (now *SPSS SYSTAT*) and *JMP* (80). *MATHEMATICA* is a useful adjunct for examining linear models, locating stationary points, and plotting two-dimensional level surfaces (81). Templates are available from the author at `http://russell.med.unc.edu/~carter/designs`.

The type of design matrix and mathematical model used for fitting the observed data are associated with specific objectives typically encountered in a crystal growth or other experimental project. Three types of models are summarized in *Table 6*, together with definitions for some terms used in the text, and how they are most often used to identify suitable conditions, characterize effects, and optimize. Models (II) and (III) are derived from the first by adding specific terms, which are indicated in bold face. The linear model used in our original study (15), included only main effects (*Table 6*-I). For the full-factorial design in *Table 2* we added all the multi-factor interaction terms (*Table 6*-II). Quadratic models (*Table 6*-III) supplement the general linear model with all possible squared terms and two-factor interactions. By substituting new variable names for the squared and interaction terms, these models become special cases of the general linear model (I).

For all models, ϵ is the residual error to be minimized, and the constant value, β_0, is the mean score for all experiments in the design. Each β_i coefficient in the linear model (I) is the average amount by which the presence of factor, F_i, raises or lowers the score from the overall average. Higher-order β_{ij} and β_{ii} coefficients have similar meaning for models II and III. Since many important variables have non-linear effects, it is not surprising that many processes, including crystal growth, can be modelled more effectively by multivariate

Table 6. Designed experiments for different contexts

Factorial design	**N experiments with simultaneous variation of M < N factors**
Quantitative analysis.	Minimizing $\left\{ \sum_{j=1}^{M(\text{expts})} \lvert Q^j_{obs} - Q^j_{calc} \rvert \right\}$
Objective	**Design/model**
Detection of important main effects and interactions.	Incomplete factorial $Q_{calc} = \beta_0 + \sum_{j=1}^{N(\text{factors})} \beta_i F_i = \varepsilon$ (I)
Verification of these inferences.	Replicated, full factorial $Q_{calc} = \beta_0 + \sum_{i}^{N(\text{factors})} \beta_i F_i + \sum_{j=1}^{N-1} \sum_{j>1}^{N} \beta_{ij} F_i F_j \ldots\text{higher terms} + \varepsilon$ (II)
Optimization of crystal growth conditions.	Response surface $Q_{calc} = \beta_0 + \sum_{i}^{N(\text{factors})} \beta_i F_i + \sum_{i=1}^{N-1} \sum_{j>1}^{N} \beta_{ij} F_i F_j + \sum_{i=1}^{N} \beta_{ii} F_i^2 + \varepsilon$ (III)

quadratic functions. These are the simplest functions that assume maxima, minima, and saddle points, within the range of independent variables. Once a model has been fitted, its stationary points can be determined analytically by partial differentiation with respect to all the independent variables and equating the gradient to zero. Stationary point coordinates provide estimates for the factor levels giving the best result. This is the basis of the response-surface method (19).

6.2.1 Selecting, fitting, and evaluating models

Protocol 6 outlines the search for a good regression model. Statistical analysis is used to gauge how much of the scatter in scores can be attributed to experimental changes, relative to what must be attributed to noise. It is worth repeating that if all experiments have nearly the same scores, then one has a smaller chance of identifying significant gradients. A full quadratic response-surface model with N variables takes the general form of a polynomial with $(N + 1)(N + 2)/2$ coefficients (*Table 6*, model III).

Protocol 6. Data analysis: finding a good regression model

Method

1. Examine the mean value, range, and standard deviation statistics for each score. Identify scores with significant variations.
2. Examine scatterplot matrices of each score for each independent variable.
   (a) Look for and note any obvious trends.
   (b) Identify any categorical variables that might usefully be re-ordered. These include cases where scores are clustered within a category, but where their mean values show no pattern. Re-order numerical assignments for categories, if doing so would create a physically sensible monotonic or parabolic series.
   (c) Rationalize patterns created in step 2(b) in terms of physically reasonable effects.
3. Decide whether or not the variation in mean scores (step 1) and suggested patterns (step 2) are sufficient to support meaningful regression against the experimental variables.
4. Build and test trial models as described in *Protocol 7*, testing all main effects first, then including two-way interactions and quadratic terms.

The analysis involves two interdependent tasks: model selection and parameter estimation. The terms of a model represent the calculable effects of the experimental factors, $\beta_i F_i$, their interactions, $\beta_{ij} F_i F_j$, and their squares, $\beta_{ii} F_i^2$, plus the intrinsic variation or noise, ε, associated with the experimental set-up and scoring. It is essential to discard terms that do not contribute significant information about the response, and use the extra degrees of freedom to improve the estimate for the residual error, thereby reducing the variances of parameter estimates. Incorrect fitting of questionable parameters can lead to model bias. Similarly, finding the best 'interpretation' for a response-surface experiment can be haphazard because coefficients and their statistical significance change when the model itself changes. Choosing which coefficients should be retained in the response-surface model is therefore a challenging task (19, 82). Once the best set of predictors has been identified, their coefficients are estimated by multiple regression least squares.

These tasks are the job of a full statistics program. We have used two such programs. *SYSTAT* (80) (now *SPSS/SYSTAT*) and *JMP* (80). Both provide a powerful multiple regression module, with the appropriate statistical calculations, and graphing tools. The following illustrations present output from the *SYSTAT MGLH* (**M**ultiple regression, **G**eneral **L**inear **H**ypothesis) module. This module cannot evaluate partial derivatives or solve for stationary points, but it is easy to use, intuitive, and fast, and the former tools are available in *Mathematica* (81). The *JMP* user interface is well-developed, and *JMP* may be easier to use.

Protocol 7. Identifying and fitting model parameters

Method

1. Define (on a command line or in a dialog box) a linear model for a single dependent variable (the score) as a function of a set of independent variables.
2. Identify the best subset of terms by using 'stepwise multiple regression'. As the name suggests, this algorithm is an automated procedure for choosing the best subset of coefficients. Generally, start with a complete model and gradually eliminate insignificant contributors. It is sometimes useful to work forwards, finding the most significant terms first. Stepping can be carried out with a variety of different tolerance and threshold criteria for including or eliminating terms, as appropriate, until the model appears stable and sensible. Depending on the algorithm, terms may be recycled if they appear to regain or lose significance as the stepping proceeds.

Protocol 7. *Continued*

3. Think about what the model is saying; add/delete coefficients that might/might not make sense.
4. Compare different models according to three types of criteria:
   (a) The squared multiple correlation coefficient, R^2. This indicates the percentage of the variation that can be 'explained' by the model. It should be as high as possible.
   (b) The probability of the F-ratio test under the null hypothesis that the model has no predictive value. This value should be very small.
   (c) Individual Student t-test probabilities for each coefficient. These should be as small as possible.
5. Verify coefficients indicating two-way interactions; calculate and examine the average scores for all four of the combinations ($--$, $+-$, $-+$, $++$) in the two-way matrix.
6. Expect that useful models of optimum behaviour will have positive linear and negative quadratic coefficients. Look for models with these characteristics.
7. Plot two-dimensional projections of the model surface. Superimposing these plots onto the observed scores is a good way to get a feel for what the model has to tell about the system.
8. Verify that plots actually reflect the data. Discrepancies usually mean errors in entering data, scores, and/or model coefficients, but can point to unexpected effects.
9. Generally, the residuals, ($[Q_{obs,i} - Q_{calc,i}]$), can be saved and analysed in the same ways used for the scores (*Protocols 5–8*). Examine them for clues about where the model may be deficient.

Statistics programs output a summary table with the relevant information. Models first of all should make physical sense. Different models then are compared using two statistical properties indicated by bold face in *Table 7*, which summarizes a model based on the data in *Table 1* and *Figure 3*. First is how well they account for the deviation of scores from their mean value (the 'variation'). The squared multiple correlation coefficient, R^2, represents the percentage of this variation predicted by the model, and should be as close as possible to 1.0. By itself, however, it is insufficient to discriminate between good and bad models. An incomplete, but valid model may have a low value for R^2, (0.3–0.7). High values also can result from models with too many parameters.

Table 7. Multiple regression and analysis of variance for the illustration in *Figure 3*

A. Data from the full-factorial experiment from *Table 1*

Squared multiple R: 0.994		F-ratio = 128.7		P = 0.001
Variable	**Coefficient**	**Std error**	**T**	**P(2 tail)**
Constant	19.4	0.402	48.1	0.00002
pH	4.2	0.402	10.4	0.002
Temp	−1.6	0.402	−4.0	0.029
Protein	7.9	0.402	19.7	0.00029
pH*Temp	−0.9	0.402	−2.3	0.108

B. Data from incomplete factorial subset in *Figure 3*

Squared multiple R: 0.999		F-ratio = 443.1		P = 0.036
Variable	**Coefficient**	**Std error**	**T**	**P(2 tail)**
Constant	18.9	0.22	87.2	0.007
pH	4.0	0.25	16.0	0.040
Temp	−1.8	0.25	−7.0	0.090
Protein	7.8	0.25	31.0	0.021
pH*Temp	−1.4	0.22	−6.4	0.099

C. Data from orthogonal array A

Squared multiple R: 0.951		F-ratio = 9.78		P = 0.221
Variable	**Coefficient**	**Std error**	**T**	**P(2 tail)**
Constant	19.175	2.175	8.816	0.072
pH	3.825	2.175	1.759	0.329
Protein	8.825	2.175	4.057	0.154

D. Data from orthogonal array B

Squared multiple R: 0.986		F-ratio = 34.6		P = 0.119
Variable	**Coefficient**	**Std error**	**T**	**P(2 tail)**
Constant	19.500	1.000	19.500	0.033
pH	4.500	1.000	4.500	0.139
Protein	7.000	1.000	7.000	0.090

A second kind of statistic, the 'P values', give the probability of obtaining equally good models by random processes, i.e. under the 'null hypothesis' that the variation is uncorrelated with changes in the experimental variables. This information is available for the Student t-tests of individual coefficients and the overall F-ratio. The Student t value is the ratio of a coefficient to its standard error, so it is a statistic about the signal-to-noise of an *estimated coefficient*. The overall F-ratio is the squared distance between calculated and *average* scores for all experiments divided by the squared distance between calculated and *observed* scores. It can be considered an estimate of the signal-to-noise of the *model*. Useful models can have P-values as high as 10^{-3}, our best models are better than 10^{-11}. *Protocol 8* and the associated statistical criteria apply equally to the investigation of any model.

Try models using main effects one or two at a time, to see how much of the variation is explained (multiple R^2), and at what cost in terms of the F-ratio probability. How much better does the model get by adding another factor? Since this procedure is the most difficult, it requires some intuitive feel, which comes from practice. This approach is illustrated for data from *Table 1* in *Table 7*, which illustrates the trade-off that necessarily accompanies sampling. All models have quite high R^2 values, and satisfy the criterion of predictive power. In other respects, however, the quality of the models depends on the amount of available data. The complete ensemble of eight experiments affords the best model by all criteria: it supports a model with more significant parameters, while at the same time giving the best F-ratio and t-test probabilities. The incomplete factorial subset performs nearly as well, giving nearly the same parameter estimates. Their statistical significance, however, is degraded by about two orders of magnitude. Deleting additional data, as with the two different orthogonal arrays, degrades the models well beyond the point where they are useful. None of the parameters is statistically significant.

Usually, the default stepwise regression will produce a reasonable representation of what is in the data. That model can occasionally be improved using different tricks. Check individual t-tests and try deleting the worst one (with the highest probability). Omit the constant term only when its t-test is poor (has a high probability). It is recommended to retain the main effect in any model that involves a higher-order interaction, even if this makes the model worse. There are exceptions to all of these guidelines. A trade-off must always be made between the decreased variance of the model parameters, achieved by reducing their number, and the potential loss of real information about the response surface that occurs when a 'true' parameter with a large variance is deleted from the model. A more detailed description, with examples, is provided in ref 83). The ultimate test is a model's usefulness.

6.2.2 Stationary points

Quadratic and higher-order polynomial models are selected, fitted, and evaluated in the same way. However, these models can be sufficiently curved to possess maxima and minima, and require another level of analysis concerning their stationary points. Stationary points of a function occur where its gradient goes to zero with respect to all the independent parameters, as illustrated for a one-dimensional case in *Figure 7* and in *Figure 8b*. They are solutions to the simultaneous equations obtained by equating the partial derivatives to zero, and are determined as in *Protocol 8*. Partial derivatives, by definition, estimate changes in a dependent variable (the response) induced by small fluctuations in the independent variables (the experimental con-

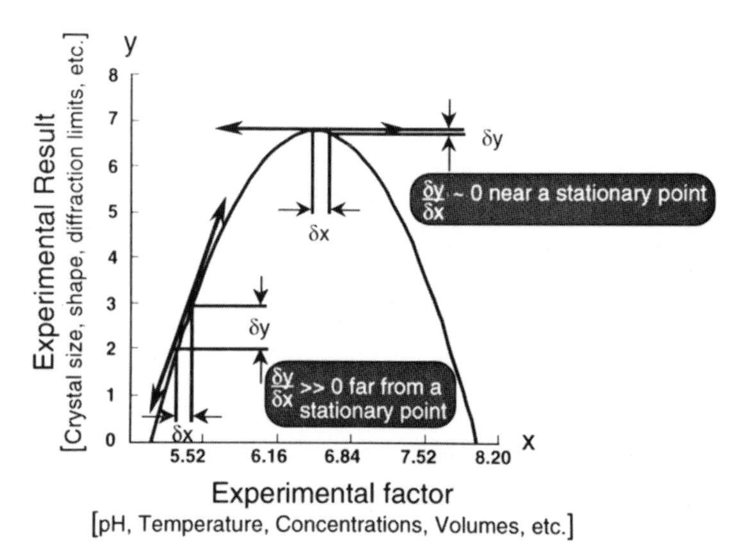

Figure 7. Stationary points of polynomial functions and experimental reproducibility. Stationary points occur where the partial derivatives of a function all vanish. A one-dimensional parabola is illustrated to show the relationship between its derivative and its maximum value, and to emphasize that far from the maximum value, the function has increasingly steep slopes. These steep slopes translate into experimental fluctuations, which have greater impact far from the stationary point.

ditions). If the function describes the system behaviour adequately, the stationary point coordinates will specify an optimal set of experimental conditions.

Protocol 8. Identifying stationary points

A. *Determination and characterization of stationary point coordinates*

These steps can be programmed in *MATHEMATICA*. Templates are available at `http://russell.med.unc.edu/~carter/designs`).

1. Calculate the partial derivatives of the model function with respect to each of the independent variables. The set of partial derivatives constitutes the gradient of the function.
2. Solve the simultaneous equations for the stationary point by equating the partial derivatives to zero.
3. Check that the stationary point coordinates correspond to experimentally sensible values.

Protocol 8. *Continued*

4. Determine the nature of the stationary point by examining the signs of the second derivatives, and/or by plotting two-dimensional level surfaces with constant values for all but two variables (1).

B. *Verification of the behaviour at and near a stationary point*

1. Set up duplicate experiments under conditions given by the stationary point.
2. Compare the experimental results with the predictions that they should be 'optimal'.

6.2.3 Analysis and verification

Identification of a stationary point does not guarantee optimality. One must first examine the behaviour nearby to determine whether it corresponds to a maximum, a minimum, or to a saddle point. This can be done by evaluating the second partial derivatives: negative curvature in all variables implies a local maximum, whereas positive curvature implies a minimum, and mixed second partial derivatives imply a saddle point. An accessory strategy in such cases is to examine two-dimensional level surface plots in all subspaces. Examples of these level surfaces are shown in *Figure 8*.

Frequently, the dominant feature of a response surface is not a stationary point, but a 'ridge', along which the value of the function increases, but normal to which it decreases (*Figure 8a*). It is hard to overemphasize the importance of using this kind of feedback to iterate the search for improvements; an illustration is provided in Section 7.

A model is only as good as its valid predictions. Consequently, any predictions regarding optimum behaviour of the system must be verified with replicated experiments at the stationary point.

6.2.4 Advantages of working with stationary points

Stationary points may be determined for any desirable crystalline property which can be 'scored' precisely enough for optimization, including volume, shape, diffraction limits, stability, and relative freedom from secondary nucleation (1). Finding and using conditions close to stationary points of analytical response surfaces has several important advantages.

(a) *Optimization*. The first real benefit is the obvious one: conditions at a convex stationary point produce crystals that are in some sense optimized (84, 85).

(b) *Reproducibility*. Stationary points are not only optima; they also represent points where the response is most reproducible. Even the most carefully

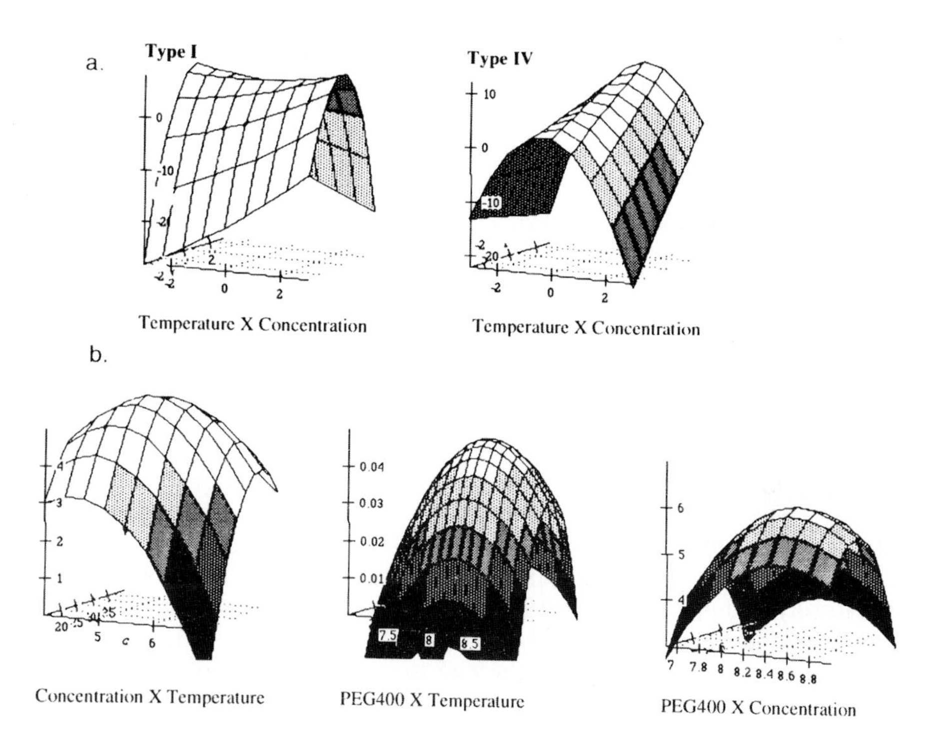

Figure 8. (a) Ridges observed in the temperature × concentration level surface for response surfaces determined for two different TrpRS crystal polymorphs (1). The ridgeline represents combinations of temperature and [protein] which produce the same supersaturation level, suggesting that supersaturation is a natural search variable. (b) Level surfaces for a third TrpRS polymorph determined using as one of the search variables an approximation to supersaturation (87). Here, the three level surfaces involving [protein] all show optima.

performed experiments can suffer from a frustrating level of irreproducibility. One source of variability is intimately connected to the partial derivatives of the underlying, multidimensional response surface (*Figure 7*). Working at stationary points helps insulate crystal growth from experimental errors in pH determination, pipetting errors affecting concentrations of various components, temperature fluctuations, and so on.

(c) *Insight*. Examining plots of the response surface is often a powerful aid to the interpretation of crystal growth experiments. We discovered a useful new search direction for response-surface experiments when graphs of two-dimensional level surfaces (*Figure 8a*) revealed *ridges*, where the same result was obtained for many combinations of factors. Ridges were conspicuous in the temperature × concentration level planes. Temperature and concentration affected the protein's solubility coordinately, the ridge corresponding to a constant value of supersaturation. An important inference was that it would be better to sample simultaneously for the effects of protein concentration and supersaturation. Using [protein] and [protein] × [crystallizing agent] eliminated the ridges in the resulting level surfaces (*Figure 8b*). This observation led to the approximations to supersaturation described in Section 7.2.1.

Empirical response surfaces provide scientific documentation about crystallogenesis that is otherwise difficult to achieve, including conclusions with important and interesting biochemical relevance (2). Particularly interesting as evidence is the statistical support for the stationary point coordinates.

7. Optimization

The term 'optimization' is used frequently in discussions of crystal growth. Usually, it refers to variation of some experimental variables with the aim of empirically finding 'better' crystals from among the conditions tested. Given sufficient time and materials, and a fortunate choice of experimental variables to explore, any search strategy can lead to better crystals. Often, however, both time and materials are limited. In such cases, there are two ways in which the search for optimal conditions can be made more systematic and efficient. One uses either a line search or a more elaborate variant called 'simplex' optimization (86), the other uses response surfaces. The two approaches are complementary; the former may actually be more appropriate if existing conditions are far from an optimum.

7.1 Steepest ascent and simplex optimization

Conceptually, these two closely related methods provide the most intuitive algorithms for optimization. Sensing improvement along a direction, one marches toward that direction which gives the greatest improvement. This single search direction is specified by the gradient of the function to be optimized, which may depend on several variables. Generally, this approach is appropriate far from a stationary point, where experiments are available to fit a plane or ridge function. An example is given in Section 8. A simplex is a polytope having one more point than the dimension of the space in which it is conceived; the two-dimensional simplex is a triangle. Simplices provide a model-independent basis for an optimization search. One deletes the worst point from the current simplex, and reflects that point across the face to which it forms a perpendicular, forming a new simplex. The gradient of the underlying surface is detected indirectly, by comparing scores at each point of the simplex (86).

7.2 Optimization using quadratic polynomial models

A more powerful optimization procedure is to fit quadratic polynomial models to data from a Hardin–Sloane response-surface design. In our experience, quadratic models with statistical significance orders of magnitude better than the familiar 95% confidence limit can be obtained for a variety of different types of scores (1, 24, 83). *Protocol 9* outlines the set-up for a response-surface experiment.

Protocol 9. Setting up a response-surface crystal growth optimization

Method

1. A least four variables should be sampled simultaneously to use quadratic polynomial models.
2. Two variables should always represent the solubility diagram, irrespective of the other variables.
   (a) Protein concentration and supersaturation are more nearly orthogonal than protein concentration and precipitant concentration, facilitating sampling of the nucleation zone (*Figure 9*).
   (b) A product, either [protein] × [crystallizing agent] (87) or ln [protein] + [crystallizing agent] are useful approximations to supersaturation or ln[supersaturation] when the latter are unknown.
3. Choose additional variables based on any available prior information. Regression analyses of an incomplete factorial screening experiment often provide indications of the most significant main effects (24).
4. Centre the experiment close to the best known set of conditions.
   (a) Exploitation of response-surface experiments is most successful when quantitative and reliable scores have been obtained for all or nearly all experiments.
   (b) Designs not centred on conditions known to produce crystals are effectively screening experiments, and should be carried out using qualitatively different experimental matrices.
5. Choose sufficiently large ranges for each variable to induce significant variation in the score without losing the response itself. Follow the guidelines in Section 4.3.

7.2.1 Identifying critical variables

It is essential before optimizing a system that the critical variables be identified. This is one of the most important reasons to analyse screening experiments quantitatively. Two of the most important search variables can usually be chosen at the outset: the protein concentration and some variable representing supersaturation. These two variables 'orthogonalize' the irregularly shaped nucleation zone of the solubility diagram (*Figure 9*). We have used the product {[protein] × [crystallizing agent]} to approximate this variable (87). Logarithmic sampling based on Green's approximation to solubility: **S** = [protein]·exp(k[crystallizing]), leads to ln**S** ~ {ln[protein] + k[precipitating agent]}, which gives a more manageable range of [crystallizing agent], and should be preferred. Other important variables include the pH (implicitly the net charge), the position of the anions in the Hofmeister series (43, 45), and the presence of any important ligands. The range and mean values for these values are often critical to success: too narrow a range leads to too small a variation in the results for the regression to attribute effects to anything but fluctuations. By the same token, the ranges must be small enough that all or most experiments actually produce crystals. Some guidelines are given in Section 4.3.2.

7.2.2 Replication

The variances of replicate experiments done at a random combination of parameters near a stationary point should vary inversely with the distance of that point from the stationary point (*Figure 7*). Our experience suggests that this is indeed the case. An important consequence is that the estimates of the variance obtained from an ensemble of unreplicated experiments may fail to capture the information about stationary points that can be obtained from replicated sets of experiments, which would provide an approximate map of the variances over the experimental space. For this reason, it is useful to carry out each experiment in a design twice.

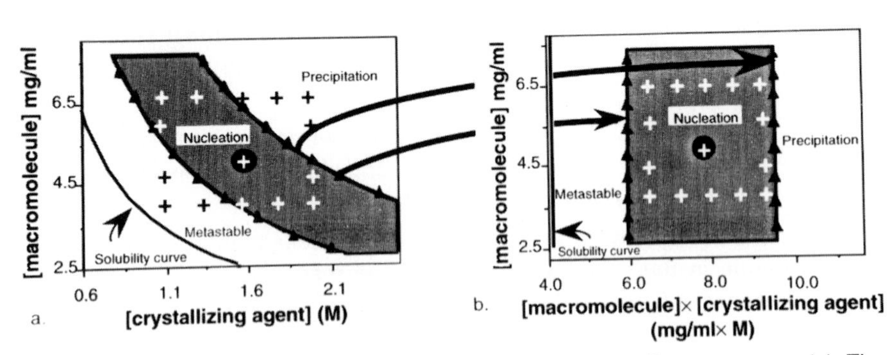

Figure 9. Orthogonalization of the nucleation zone for sampling purposes. (a) The nucleation zone is shown as the curved figure bounded by the metastable and precipitation zones. (b) The product [macromolecule] × [crystallizing agent] is constant along the rectangular hyperbolae in (a). Thus, the nucleation zone becomes a rectangle which can be sampled evenly on a regular grid (87).

7.2.3 An example

A response-surface experiment for monoclinic tryptophanyl-tRNA synthetase (TrpRS) crystals is presented step-by-step to illustrate the process of building and analysing the model. This crystal form had always previously given long and thin crystals that had to be grown bigger by repeated reseeding (84, 85). Intense efforts to improve the size of the initial crystals by systematic variation of protein concentration had led only to modest improvements. After some initial successes with other forms of TrpRS, we proceeded as follows with the monoclinic form (1):

(a) We selected [protein], {[prot] × [ppnt]}, temperature, and the concentration of the additive, PEG 400, as the independent variables. A range of values (+\- 9–23%) was chosen to surround the best conditions we had previously achieved. Values of these four variables were assigned to 20 experiments of the Hardin–Sloane matrix in *Table 2*. These experiments were each done twice and scored using three different criteria: volume, the ratio of the smallest dimension to the largest, and a subjective assessment of their uniformity and freedom from satellite crystals. These scores are more objective and quantitative than the subjective scale we used previously to score screening experiments and should be easier to use. They were input to a *SYSTAT* data file together with the H–S matrix.

(b) Using the *MGLH* (**M**ultiple Regression, **G**eneral **L**inear **H**ypothesis) module in the statistics program *SYSTAT* for the Macintosh (82), tests were carried out first using a model containing all 15 terms of equation (III) (a constant, four each of the main effects and the quadratic terms, and the six two-factor interactions). The initial and subsequent models were evaluated according to criteria described in Section 6.2.1. This complete model had an F-ratio probability of 10^{-4}. R^2 was 0.87, indicating that all but about 13% of the variation in observed scores could be attributed to the model. Nevertheless, in several respects the model needed adjustment. In particular, three coefficients had t-test probabilities > 0.05, indicating that they were without significance and should be removed.

(c) The full model was pruned by backward stepwise regression, eliminating three of the 15 parameters. The final model was obviously very significant; its F-ratio probability, P, was 10^{-11}, R^2 was 0.95; and t-test probabilities of the 11 coefficients of the model were nearly all below the 5% confidence limit (1). Two were around 0.1 and of questionable significance. However, removing them resulted in a serious deterioration, causing P to decrease by an order of magnitude, and four additional factors with significant t-tests in the best model became completely insignificant.

Prediction and verification. Partial derivatives of the calculated score with respect to all variables were evaluated from the model expression. Equating the four derivatives to zero and solving for the coordinates ([protein]$_{opt}$, [pro_ppnt]$_{opt}$, [PEG]$_{opt}$, and Temp$_{opt}$) of the optimum predicted that crystals grown at these values would be better any of those observed in the H–S experiment. Crystals grown at the optimum point were two orders of magnitude larger and of sufficient volume for diffraction experiments. Similar analysis for two other scores (volume and uniformity) showed that the three optima were essentially in the same place, simultaneously optimizing all three scores (1).

8. Resolution of polymorphs

A recurring problem with screening experiments is that they sample conditions that may be far from optimal for a particular crystal form. Attempts to optimize such cases using Hardin–Sloane designs can give rise to surfaces like that in *Figure 10*, where instead of an optimum, the surface represents a saddle. Sometimes, this problem is confounded by the appearance of multiple crystal forms in the same experiments. Indeed, the appearance of saddle points is often diagnostic of the superposition of overlapping response surfaces for different poymorphs. Since different crystal polymorphs rarely, if ever, have the same dependence on all experimental variables, finding optimal stationary points for different polymorphs can be a useful way to 'purify' them away from one another.

Figure 11 shows how a line search helped relocate the centre for a Hardin–

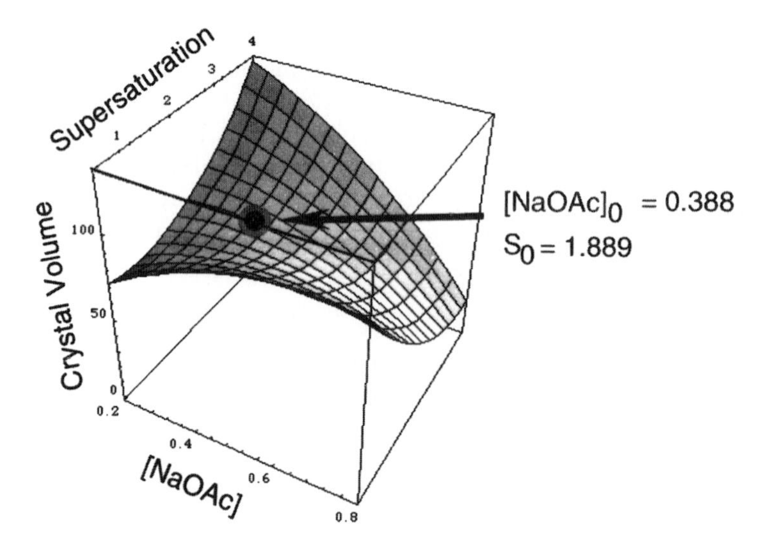

Figure 10. Response surface determined for *E. coli* cytidine deaminase at the neighbourhood of a 'hit' from a kit (73) showing that the conditions are nearly at a saddle point and hence far from an optimum. The dot indicates experiment number 28 of Jancarik and Kim (25).

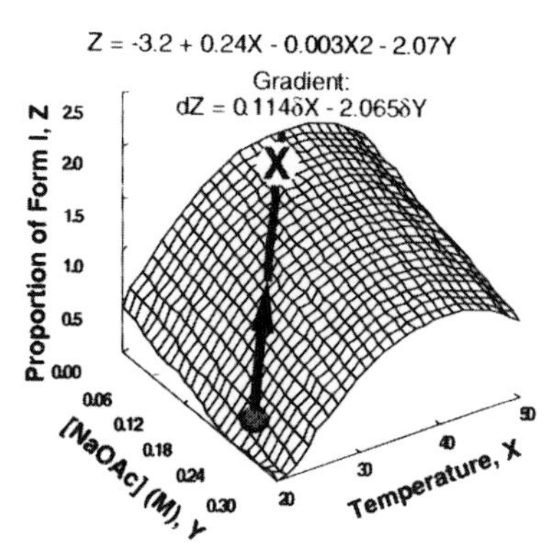

Hardin-Sloane

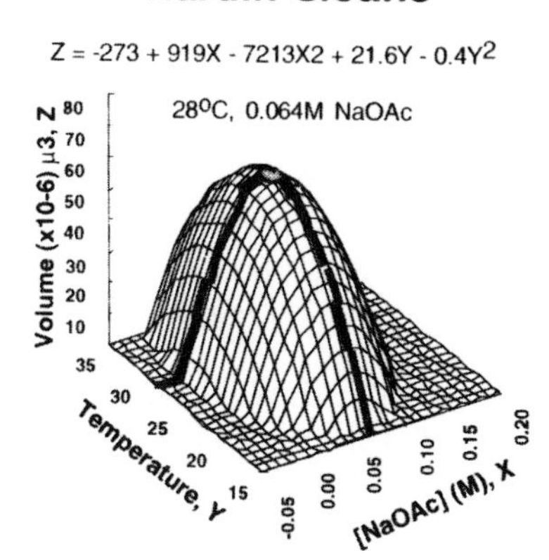

Figure 11. Use of steepest ascent (line search) to resolve polymorphs of *E. coli* cytidine deaminase crystals. (a) The proportion of the desired form (Form I) fitted to the ridge function shown above the figure. Tests selected to fall on the line as described in the text showed that this proportion increased along that search, until at the X, only Form I was observed. (b) A Hardin–Sloane design was then centred on the X, giving the optimum shown (73).

Sloane design away from the saddle point in *Figure 10*, ultimately locating optimal conditions for a new polymorph of *E. coli* cytidine deaminase (73). Experiments at higher temperature in the original Hardin–Sloane design had variable amounts of a second polymorph. The proportion of this second polymorph was used as a score and fitted to the ridge surface in *Figure 11a*. The initial conditions from the Hampton screen lay near the bottom of a steeply sloped ridge in two variables, temperature and [Na Acetate]. The gradient at that point indicated that the proportion of this form increased by 0.114 for every degree the temperature was raised, and by 2.07 as the concentration of acetate was reduced by 1 mole. Combining the two indications gives that for each degree of temperature increase, the acetate concentration should be reduced by 0.114/2.07 = 0.055 M. Experiments stepped out along this gradient showed the expected increase in the fraction of the desired crystal form. When the proportion was equal to 100%, a new Hardin–Sloane design was performed, giving the surface in *Figure 11b*.

The first step in separating out the polymorphs was to identify the variables critical to the different polymorphs. This was inferred from how the polymorphs behaved with respect to changes in all variables in the first Hardin–Sloane design. A subset was identified by model fitting and analysis, and the surface used to identify the gradient for a steepest ascent search to resolve the different polymorphs from one another. Subsequent, re-centred response-surface experiments can locate the stationary points associated with each polymorph. The potential advantages of using response surfaces in this way also include the evidence provided about the influence of each of the factors on the polymorphism.

References

1. Carter, C. W., Jr and Yin, Y. (1994). *Acta Cryst.*, **D50**, 572.
2. Carter, C. W., Jr., Doublié, S., and Coleman, D. E. (1994). *J. Mol. Biol.*, **238**, 346.
3. Schutt, C. E., Lindberg, U., Myslik, J., and Strauss, N. (1989). *J. Mol. Biol.*, **209**, 735.
4. Anderson, C. M., Zucker, F. H., and Steitz, T. A. (1979). *Science*, **204**, 375.
5. Sack, J. S., Saper, M. A., and Quiocho, F. (1989). *J. Mol. Biol.*, **206**, 171.
6. Bullough, P. A., Hughson, F. M., Skehel, J. J., and Wiley, D. C. (1994). *Nature*, **371**, 37.
7. Colombo, M. F., Rau, D. C., and Parsegian, A. V. (1992). *Science*, **256**, 665.
8. Rand, R. P., Fuller, N. L., Butko, P., Francis, G., and Nicholls, P. (1993). *Biochemistry*, **32**, 5925.
9. Rand, R. P. (1992). *Science*, **256**, 618.
10. Reid, C. and Rand, R. P. (1997). *Biophys. J.*, **72**, 1022.
11. Parsegian, V. A., Rand, R. P., and Rau, D. C. (1995). In *Methods in enzymology* (eds M. L. Johnson and G. Ackers), Academic Press, Inc., London. Vol. 259, p. 43.
12. Leikin, S. and Parsegian, V. A. (1994). *Proteins: Structure, Function, Genetics*, **19**, 73.
13. Royer, W. E., Pardnamani, A., Gibson, Q. H., Peterson, E. S., and Friedman, J. M. (1996). *Proc. Natl. Acad. Sci. USA*, **93**, 14526.
14. Ladner, J. E., Finch, J. T., Klug, A., and Clark, B. F.C. (1972). *J. Mol. Biol.*, **72**, 99.
15. Carter, C. W., Jr. and Carter, C. W. (1979). *J. Biol. Chem.*, **254**, 12219.
16. Carter, C. W., Jr., Baldwin, E. T., and Frick, L. (1988). *J. Cryst. Growth*, **90**, 60.
17. Carter, C. W., Jr. (1990). *Methods: a companion to methods in enzymology*, **1**, 12.
18. Carter, C. W., Jr. (1992). In *Crystallization of proteins and nucleic acids: a practical approach* (ed. R. Giegé and A. Ducruix), p. 47. IRL Press, Oxford University Press, Oxford, UK.
19. Box, G. E. P., Hunter, W. G., and Hunter, J. S. (1978). *Statistics for experimenters.* Wiley Interscience, New York.
20. Carter, C. W. (1995). *Qual. Eng.*, **8**, 181.
21. Bell, J. B., Jones, M. E., and Carter, C. W., Jr. (1991). *Proteins: Structure, Function, Genetics*, **9**, 143.
22. Betts, L., Frick, L., Wolfenden, R., and Carter, C. W., Jr. (1989). *J. Biol. Chem.*, **264**, 6737.
23. Doublié, S., Xiang, S., Bricogne, G., Gilmore, C. J., and Carter, C. W. J. (1994). *Acta Cryst* **A50**, 164.
24. Yin, Y. and Carter, C. W., Jr. (1996). *Nucleic Acids Res.*, **24**, 1279.
25. Jancarik, J. and Kim, S.-H. (1991). *J. Appl. Cryst.*, **24**, 409.
26. Doudna, J. A., Grosshans, C., Gooding, A., and Kundrot, C. (1993). *Proc. Natl. Acad. Sci. USA*, **90**, 7829.
27. Song, L. and Gouaux, E. (1997). In *Methods in enzymology*, ibid ref. 54. Vol. 276, p. 60.

28. Kundrot, C. E. (1997). In *Methods in enzymology*, Vol. 276, p. 143.
29. Scott, W. G., *et al.* (1995). *J. Mol. Biol.*, **250**, 327.
30. Gilliland, G. L., Tung, M., Blakeslee, D. M., and Ladner, J. E. (1994). *Acta Cryst.*, **D50**, 408.
31. Rosenberg, J. M., *et al.* (1997). In *Annual Meeting* 56 (American Crystallographic Association, St. Louis).
32. Shaanan, B. (1983). *J. Mol. Biol.*, **175**, 159.
33. Smith, F. R., Lattman, E. E., and Carter, C. W., Jr. (1991). *Proteins: Structure, Function, Genetics*, **10**, 81.
34. Janin, J. and Wodak, S. J. (1993). *Proteins*, **15**, 1.
35. Srinivasan, R. and Rose, G. D. (1994). *Proc. Natl. Acad. Sci. USA*, **91**, 11113.
36. Smith, F. R. and Simmons, K. C. (1994). *Proteins: Structure, Function, Genetics*, **18**, 295.
37. Plackett, R. L. and Burman, J. P. (1946). *Biometrika*, **33**, 305.
38. Kingston, R. L., Baker, H. M., and Baker, E. N. (1994). *Acta Cryst.*, **D50**, 429.
39. Hardin, R. H. and Sloane, N. J. A. (1993). *J. Stat. Plan. Inference*, **37**, 339.
40. Audic, S., Lopez, F., Claverie, J.-M., Poirot, O., and Abergel, C. (1997). *Proteins: Structure, Function, Genetics*, **29**, 252.
41. Ferrone, F., Hofrichter, J., and Eaton, W. A. (1985). *J. Mol. Biol.*, **183**, 611.
42. Ducruix, A. F. and Riès-Kautt, M. (1990). *Methods: a companion to methods in enzymology*, **1**, 25.
43. Riès-Kautt, M. and Ducruix, A. (1992). In *Crystallization of nucleic acids and proteins: a practical approach* (ed. A. Ducruix and R. Giegé), p. 195. IRL Press, Oxford, UK.
44. Carbonnaux, C., Riès-Kautt, M., and Ducruix, A. (1995). *Protein Sci.*, **4**, 2123.
45. Riès-Kautt, M. and Ducruix, A. (1997). In *Methods in enzymology* ibid ref. 54. Vol. 276, p. 23.
46. Ataka, M. and Tanaka, S. (1986). *Biopolymers*, **25**, 337.
47. Ataka, M. and Michihiko, A. (1988). *J. Cryst. Growth*, **90**, 86.
48. Rosenberger, F. (1996). *J. Cryst. Growth*, **166**, 40.
49. Muschol, M. and Rosenberger, F. (1997). *J. Chem. Phys.*, **107**, 1953.
50. Rosenberger, F., Muschol, M., Thomas, B. R., and Vekilov, P. G. (1996). *J. Cryst. Growth*, **168**, 1.
51. Retailleau, P. (1997). Thesis Ph. D. Université Paris XI Orsay.
52. Retailleau, P., Riès-Kautt, M., and Ducruix, A. (1997). *Biophys. J.*, **73**, 2156.
53. Luft, J. R., *et al.* (1994). *J. Appl. Cryst.*, **27**, 443.
54. Luft, J. R. and DeTitta, G. T. (1997). In *Methods in enzymology* (eds C. W. Carter and R. Sweet), Academic Press, Inc., London. Vol. 276, p. 110.
55. Rosenberger, F., Vekilov, P. G., Lin, H., and Alexander, J. I.D. (1997). *Microgravity science and technology* **10**, 29.
56. Vekilov, P. G., Thomas, B. R., and Rosenberger, F. (1998). *J. Phys. Chem.*, **2**, 5208.
57. Vekilov, P. G. and Rosenberger, F. (1998). *J. Cryst. Growth*, **186**, 251.
58. Gilliland, G. and Bickham, D. M. (1990). *Methods: a companion to methods in enzymology*, **1**, 6.
59. Guilloteau, J. P., *et al.* (1996). *Proteins: Structure, Function, Genetics*, **25**, 112.
60. Timasheff, S. N. (1992). In *Pharmaceutical biotechnology* (ed. T. J. Ahera and M. C. Manning), p. 265. Plenum, New York.
61. Bricogne, G. (1993). *Acta Cryst.*, **D49**, 37.

62. Rosenberger, F. (1986). *J. Cryst. Growth*, **76**, 618.
63. Rosenberger, F. and Meehan, E. J. (1988). *J. Cryst. Growth*, **90**, 74.
64. Stura, E. and Wilson, I. (1990). *Methods: a companion to methods in enzymology*, **1**, 38.
65. Stura, E. A., Satterthwait, A. C., Calvo, J. C., Kaslow, D. C., and Wilson, I. A. (1994). *Acta Cryst.*, **D50**, 448.
66. Stura, E. (1998). In *Crystallization of nucleic acids and proteins: a practical approach* (ed. A. Ducruix and R. Giegé). IRL Press, Oxford, UK.
67. Hofrichter, J., Ross, P. D., and Eaton, W. A. (1974). *Proc. Natl. Acad. Sci. USA*, **71**, 4864.
68. Hofrichter, J., Ross, P. D., and Eaton, W. A. (1976). *Proc. Natl. Acad. Sci. USA*, **73**, 3035.
69. Collins, K. D. and Washabaugh, M. W. (1985). *Q. Rev. Biophys.*, **18**, 323.
70. Collins, K. D. (1997). *Biophys. J.*, **72**, 65.
71. Sousa, R. (1996). In *Methods in enzymology* Vol. 276, p. 131.
72. Scarborough, G. (1994). *Acta Cryst.*, **D50**, 643.
73. Kuyper, L. and Carter, C. W., Jr. (1996). *J. Cryst. Growth*, **168**, 155.
74. Edwards, A. L. (1979). *Multiple regression and the analysis of variance and covariance*. W. H. Freeman and Company, San Francisco.
75. Milliken, G. A. and Johnson, D. E. (1989). *Analysis of messy data: Volume 2, nonreplicated experiments*, pp. 1–199. Chapman and Hall, New York.
76. Milliken, G. A. and Johnson, D. E. (1992). *Analysis of messy data: Volume 1, designed experiments*, pp. 1–473. Chapman and Hall, New York.
77. Neter, J. and Wasserman, W. (1974). *Applied linear statistical models*, pp. 1–842. Richard D. Irwin, Homewood, IL.
78. Morrison, D. F. (1990). *Multivariate statistical methods*. McGraw Hill, New York.
79. Mardia, K. V., Kent, J. T., and Bibby, J. M. (1979). *Multivariate analysis*. Academic Press, London.
80. SAS Institute, I. (1989). SAS Institute, Inc., Cary, NC.
81. Wolfram, S. (1994). *Mathematica, the student book*. Addision Wesley, Reading, MA.
82. Wilkinson, L. (1987). SYSTAT, Inc., Evanston, IL 60601.
83. Carter, C. W., Jr. (1997). In *Methods in enzymology* Vol. 276, p. 74.
84. Thaller, C., *et al.* (1981). *J. Mol. Biol.*, **147**, 465.
85. Thaller, C., *et al.* (1985). In *Methods in enzymology* Vol. 114, p. 132.
86. Wilson, L. (1998). In *SpaceBound 1997* (ed. J. Sygusch). Canadian Space Agency, Montreal, Quebec, Canada.
87. Carter, C. W., Jr. (1996). *Acta Cryst.*, **53**, 647

5

Methods of crystallization

A. DUCRUIX and R. GIEGÉ

1. Introduction

There are many methods to crystallize biological macromolecules (for reviews see refs 1–3), all of which aim at bringing the solution of macromolecules to a supersaturation state (see Chapters 10 and 11). Although vapour phase equilibrium and dialysis techniques are the two most favoured by crystallographers and biochemists, batch and interface diffusion methods will also be described.

Many chemical and physical parameters influence nucleation and crystal growth of macromolecules (see Chapter 1, *Table 1*). Nucleation and crystal growth will in addition be affected by the method used. Thus it may be wise to try different methods, keeping in mind that protocols should be adapted (see Chapter 4). As solubility is dependent on temperature (it could increase or decrease depending on the protein), it is strongly recommended to work at constant temperature (unless temperature variation is part of the experiment), using commercially thermoregulated incubators. Refrigerators can be used, but if the door is often open, temperature will vary, impeding reproducibility. Also, vibrations due to the refrigerating compressor can interfere with crystal growth. This drawback can be overcome by dissociating the refrigerator from the compressor. In this chapter, crystallization will be described and correlated with solubility diagrams as described in Chapter 10.

Observation is an important step during a crystallization experiment. If you have a large number of samples to examine, then this will be time-consuming, and a zoom lens would be an asset. The use of a binocular generally means the presence of a lamp; use of a cold lamp avoids warming the crystals (which could dissolve them). If crystals are made at 4°C and observation is made at room temperature, observation time should be minimized.

2. Sample preparation

2.1 Solutions of chemicals

2.1.1 Common rules

Preparation of the solutions of all chemicals used for the crystallization of biological macromolecules should follow some common rules:

- when possible, use a hood (such as laminar flux hood) to avoid dust

- all chemicals must be of purest chemical grade (ACS grade)
- stock solutions are prepared as concentrated as possible with double distilled water.

Solubility of most chemicals are given in Merck Index. Filter solutions with 0.22 μm minifilter. If you use a syringe, do not press too hard as it will enlarge the pores of the filter. Filters of 0.4 μm will retain large particles whereas 0.22 μm filters are supposed to sterilize the solution. Label all solutions (concentration, date of preparation, initials) and store at 4°C. Characterize them by refractive index from standard calibrated solutions. Use molar units (mole per litre) in preference to percentage. This avoids confusion between weight to weight (w/w), weight to volume (w/v), and volume to volume (v/v). Quite often crystallization articles refer to percentage without any information, making the results difficult to reproduce. As an example, a 20% (w/v) stock solution twice diluted will give a 10% solution whereas this would not be the case if starting from a 20% (w/w) solution.

2.1.2 Buffer

The chemical nature of the buffer is an important parameter for protein crystal growth. It must be kept in mind that the pH of buffers is often temperature-dependent; this is particularly significant for Tris buffers. Buffers, which must be used within one unit from their p*K* value, are well described in standard text books (4).

2.1.3 Purification of PEG

PEG is available in a variety of polymeric ranges; the most commonly used are compounds with mean molecular weights of 2000, 4000, and 6000. These are polydisperse mixtures and their composition around the mean may vary from one producer to the other; it is better to always use the same brand. Molecular weights higher than 10000 are rarely used because of excessive viscosity of their solutions. Reproducibility and quality of crystals may depend on PEG molecular weight.

The optimal range of PEG concentration for crystallization of a given protein depends on PEG molecular weight and may be very narrow (about 1.5%, w/v). As the viscosity of PEG solutions may lead to pipetting errors, it is better to routinely verify the concentration of PEG in reservoirs for dialysis or vapour diffusion by measuring the refractive index with a Abbe refractometer. A reference curve is established on known amounts of PEG dissolved in the same buffer. From a practical point of view, commercial PEG does contain contaminants, either ionic (5) or derived from peroxidation. Repurification as shown in *Protocol 1* is strongly recommended before use (6).

Protocol 1. Purification of PEG[a]

Equipment and reagents

- Bio-Rad AG501X8
- $Na_2S_2O_4.5H_2O$
- PEG

Method

1. Pour a column (2.5 × 10 cm^2) with a mixed-bed strong ion exchange resin in the H^+-OH^- form (e.g. Bio-Rad AG501X8). Wash with 300 ml methanol:water (3:7, v/v) then with 500 ml water.
2. Dissolve 200 g PEG in water (500 ml final volume). Measure the refractive index of the solution. Degas for 30 min under vacuum (water aspirator) with gentle magnetic stirring. Add 1.24 g $Na_2S_2O_4.5H_2O$ and let stand for 1 h.
3. Pass the solution through the column at a flow rate of 1 ml/min. Discard the first 30 ml and collect the following eluate.
4. Check the concentration of PEG by refractometry and store frozen in small aliquots at –20 °C.
5. Before use, an antioxidant can be added (para-hydroxyanisole, stock solution in isopropanol, 1.3 mg/ml; add 1 μl per ml PEG stock solution).

[a] See Chapter 2, *Table 2* for another method.

2.1.4 Mother liquor

Mother liquor is defined as the solution containing all crystallization chemicals (buffer, salt, crystallizing agent, and so on) except protein or nucleic acid at the final concentration of crystallization.

2.2 Preparing samples of biological macromolecules

2.2.1 Removing salts

Proteins and nucleic acids often contain large amount of salts of unknown composition when first obtained. Thus it is wise to dialyse a new batch of a macromolecule against a large volume of well-characterized buffer of given pH, to remove unwanted salts and to adjust the pH. Starting from known conditions helps to ensure reproducibility.

Commercially available dialysis tubing are generally composed of cellulose or polyacetate. They should be prepared as described in *Protocol 2*. Molecular cut-off (i.e. the pore size limit) is given by manufacturers for spherical particles. As most proteins and nucleic acids are better described as ellipsoids or cylinders, a cut-off far enough from the molecular weight should be chosen.

As an example, a 12000 cut-off is not appropriate for lysozyme (M_r 14305) and if used will allow the protein to leak slowly through the membrane, thus diluting it in the external chamber. You can check the impermeability of the membrane towards the protein by placing a dialysis tube containing the macromolecule at a given concentration in a beaker (outer reservoir) containing a small volume (a few millilitres). After 24 h, check the biological macromolecule concentration in the outer reservoir. In practice, it should not contain more than 1% of the amount of the macromolecule.

Depending on the choice of the commercial membrane, it should be either prepared and demetallized (following *Protocol 2*) or washed with distilled water. The membrane must be kept cold. If kept for a long period (even in the form of dry tubes), contamination problems may arise leading to leaks of protein. Dialysis membranes are fragile and it is quite easy to puncture them with nails; so do wear plastic gloves, remembering to rinse them because they are often treated with talc.

Protocol 2. Preparation of dialysis tubing

Equipment and reagents

- Dialysis tubing
- Bunsen burner
- Solution A: 5% (w/v) $NaHCO_3$ (50 g/litre) and 50 mM EDTA (18.6 g/litre)

Method

1. Boil the tubing for 30 min in solution A. Avoid puncture at this stage when mixing with glass rods or magnetic stirrers.
2. Rinse several times with distilled water.
3. Store in 50% (v/v) ethanol solution.
4. Check each tubing integrity for possible puncture.
5. Prior to crystallization, rinse membranes several times with distilled water then with buffer.

2.2.2 Concentration

Whatever the crystallization method used, it requires high concentrations of biological macromolecules as compared to normal biochemistry conditions. Before starting a crystallization experiment, a concentration step is generally needed. Keep pH and ionic strength at desired values, since pH may vary when the concentration of the macromolecule increases. Also, low ionic strength could lead to early precipitation (see Chapter 2 for further practical advice). It could be very frustrating when the macromolecule precipitates irreversibly or adsorbs on concentration apparatus membrane and/or support. Many commercial devices are available; they are based on different principles and operate:

(a) Under nitrogen pressure.
(b) By centrifugation (e.g. Centricon).
(c) By lyophilization (because it may denature some proteins, test first on a small amount). Non-volatile salts are also lyophilized and will accumulate.

Choice of the method of concentration depends on the quantity of macromolecule available. Dialysis against high molecular PEG proved to be successful in our hands. We use a dialysis chamber (volume 50–500 μl), the top of which is covered by a glass coverslip which is sealed to the plastic chamber with grease. *Figure 1* describes the apparatus. This allows for an easy access from the top of the dialysis chamber.

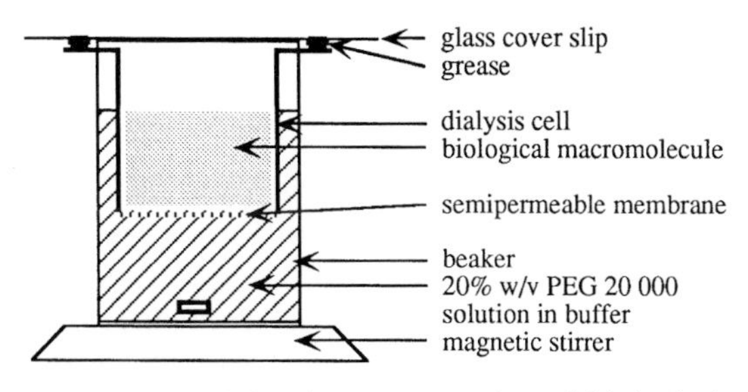

Figure 1. Dialysis apparatus used for the concentration of biological macromolecules. Prepare a solution of 20% (w/v) of PEG 20 000 in an appropriate buffer and dialyse your biological macromolecule against it. Check the macromolecule concentration using optical absorbance, colorimetric or enzymatic assay on a small aliquot.

2.2.3 Removing solid particles

Before a crystallization experiment, solid particles such as dust, denatured proteins, and solids coming from purification columns (beads) or lyophilization should be removed. This could be achieved by centrifugation or filtration, depending on the available quantity.

2.2.4 Measuring concentration of biological macromolecules

The most common method to measure macromolecular concentrations is to sample an aliquot, dilute it with buffer, and measure absorbance at 280 nm or 260 nm (for proteins or nucleic acids, respectively) within the linear range of a spectrophotometer. Proper subtraction with the reference cell should be made especially when working with additives absorbing in the 260–300 nm wavelength range. When working with enzymes, an alternative method to measure the concentration of protein is to perform activity tests, otherwise, colorimetric methods can be used. This can be done either by a modification (7) of the assay described by Winterbourne (8) or by the modification (9) of the reagent assay developed by Bradford (10). See also Chapter 2, Section 4.2 and Chapter 8, Section 2.2.

3. Crystallization by dialysis methods

3.1 Principle

These methods allow for an easy variation of the different parameters which influence the crystallization of biological macromolecules. Different types of dialysis cells are used but all follow the same principle. The macromolecule is separated from a large volume of solvent by a semi-permeable membrane which gives small molecules (ions, additives, buffer, and so on) free passage but prevents macromolecules from circulating. The kinetics of equilibrium will depend on the membrane cut-off, the ratio of the concentration of crystallizing agent inside and outside the protein chamber, the temperature, and the geometry of the cell.

3.2 Examples of dialysis cells

3.2.1 Macrodialysis

The most simple technique is to use a dialysis bag. Large crystals are occasionally obtained. It is very convenient for successive recrystallization. Commercially available dialysis tube such as Spectrapor of inner diameter 2 mm can be used to limit the amount of protein. However, each assay requires about 100 μl at least per sample.

3.2.2 Microdialysis

i. Zeppenzauer cells

Crystallization by dialysis was first adapted to microvolumes by Zeppenzauer (11). The miocrodialysis cells are made from capillary tubes closed either by dialysis membranes or polyacrylamide gel plugs. Those cells require only 10 μl or less of macromolecule solution per assay. A modified version of the Zeppenzauer cell was described by Weber and Goodkin (12).

ii. Dialysis buttons

Commercially available, microdialysis cells are made of transparent Perspex (*Figure 2a*). They can be obtained from your local workshop. The protein chamber should be filled so that it forms a dome. The membrane is then placed over the button and held by an O-ring of appropriate diameter. Installing the membrane has a reputation of difficulty, and beginners often trap air bubbles between the protein solution and the membrane. To avoid the problem, one can use either a piece of plastic (*Figure 2b*) having the same diameter as the button, or when working with flat buttons use a plastic cork. The cell is then immersed in a vial and an inexpensive way is to use transparent scintillation counting vials.

Observation with a binocular or microscope through the membrane is easy. However, if you use cross-polarizer, the membrane will depolarize light. For crystal mounting, O-ring and membrane should be removed gently. Problems

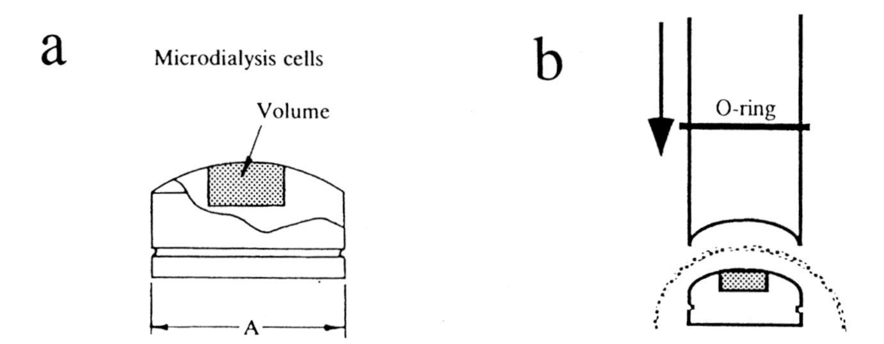

Figure 2. Dialysis button. (a) Diameter of the buttons (A) generally varies between 10–20 mm and volume of the biological macromolecule chamber is 5–350 μl. (b) To install a dialysis membrane, use a pipetter tip of diameter adapted to the concave shape of the dialysis button.

occur when crystals stick to the wall of the chamber. In this case a whisker can be used to gently free the crystal.

iii. Microcap dialysis

The technique, described in *Figure 3* and adapted from ref. 13, was useful for membrane proteins (see Chapter 9). Although it is more difficult to observe crystal growth with this method it is very convenient for storing, and microcaps are disposable. The method is quite easy to use (*Protocol 3*) and you can play with the ratio of the diameter versus height of the microcap to influence the kinetics of crystallization. It should be noted that when the macromolecule does not entirely fill the microcap chamber, the presence of air (which is compressible) allows osmotic pressure to develop, and thus modifies the macromolecule concentration.

Protocol 3. Crystallization by microcap dialysis

Equipment and reagents

- Microcaps
- Hamilton syringe
- Tygon tubing of 1.3 and 3 mm diameter
- 1.5 ml Eppendorf tubes
- Low melting wax
- Dialysis membrane cut in small squares
- Low temperature soldering

Method

1. Commercial microcaps are cut with a glass saw (for instance a 50 μl cap is cut in three parts to fit in an Eppendorf tube of 1.5 ml).
2. Wrap a piece of dialysis membrane around one end (the one which is smooth) and secure with a piece of tubing of diameter 1.3 mm.

Protocol 3. *Continued*

3. Load the biological macromolecule with a Hamilton syringe.
4. Shake the assembly to bring the biological macromolecule solution in contact with the membrane.
5. Seal the free microcap end with wax molten by soldering bit.
6. Split a second ring of tubing of diameter 3 mm and superpose it to the first one. The aim is to prevent the membrane from touching the bottom of the Eppendorf tube which would limit the exchange with the reservoir.
7. Insert in an Eppendorf tube (volume 1.5 ml) containing 1 ml of the crystallizing solution.
8. Close cap and wrap top of Eppendorf tube with Parafilm (American Can Company).

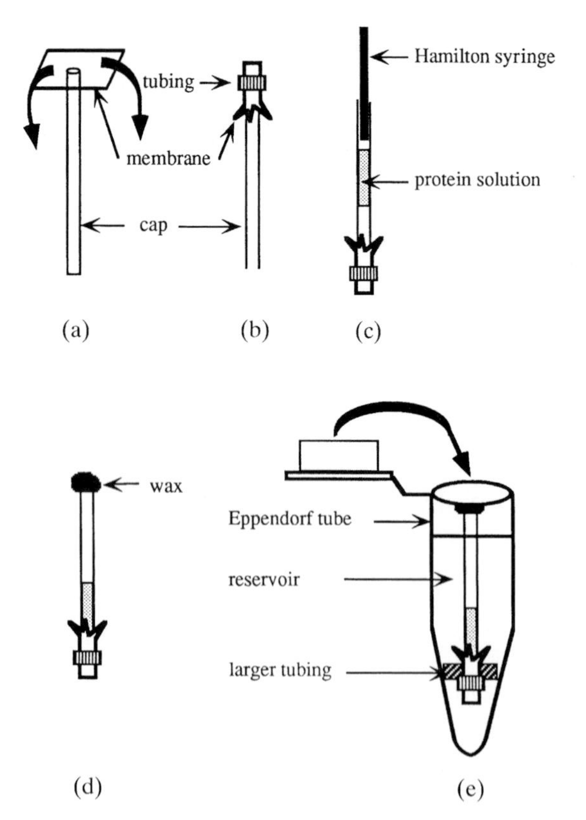

Figure 3. Crystallization by microcap dialysis. (a) Place the dialysis membrane on the microcap; (b) secure with Tygon ring; (c) load the protein; (d) close the extremity with wax; (e) fill up a 1.5 ml Eppendorf tube with crystallizing agent and insert microcap.

3.2.3 Double dialysis

The purpose of double dialysis (14) is to reduce the rate of equilibration and therefore to provide a better control of crystal growth. The apparatus is shown in *Figure 4*. Large crystals of delta toxin of *Staphylococcus aureus* were obtained this way (14). Equilibration time is rather long (could be several weeks) as the gradient concentration of crystallizing agent is low. It is therefore more geared toward production of large crystals than screening. *Protocol 4* describes the methodology. One can manipulate the different parameters (membrane cut-off, distance between dialysis membranes, relative volumes, and so on) to optimize crystallization.

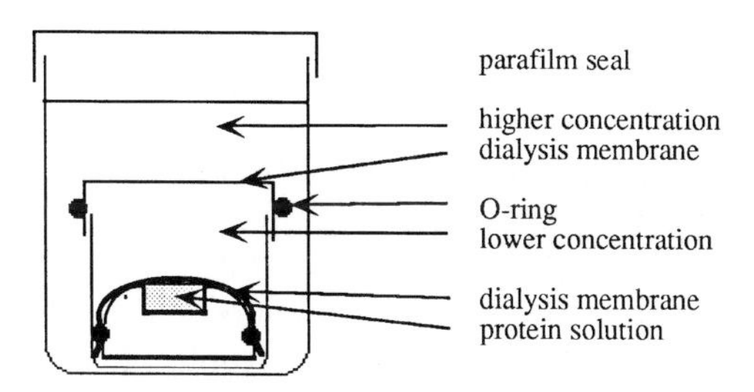

Figure 4. Double dialysis set-up (adapted from ref. 14). Macromolecule is contained in a conventional dialysis button placed in a second dialysis set-up. The equilibration rate depends upon the volumes of buffers in the different compartments.

Protocol 4. Crystallization by double dialysis

Equipment and reagents

- Dialysis button
- O-ring
- Dialysis membrane cut in small squares

Method

1. Prepare the dialysis button as in Section 3.2.2 with a solution of crystallizing agent at a concentration in which the biological macromolecule is undersaturated. This is called the 'inner compartment'.
2. Insert the conventional dialysis button in a vial (about 10 ml) called the 'middle compartment' containing a solution of crystallizing agent at a concentration in which the biological macromolecule is supersaturated.
3. Cover with a dialysis membrane maintained by an O-ring.

Protocol 4. *Continued*

4. Place in a larger vial (for instance a beaker of 50 ml) containing the solution of crystallizing agent at a concentration in which the biological macromolecule will precipitate completely. This is the 'outer compartment'.
5. Cover with Parafilm or a stopper.

4. Crystallization by vapour diffusion methods

Among the crystallization micromethods, vapour diffusion techniques are probably the most widely used throughout the world. They were first used for the crystallization of tRNA (15).

4.1 Principle

The principle of vapour diffusion crystallization is indicated in *Figure 5*. It is very well suited for small volumes (down to 2 μl or less). A droplet containing the macromolecule to crystallize with buffer, crystallizing agent, and additives, is equilibrated against a reservoir containing a solution of crystallizing agent at a higher concentration than the droplet. Equilibration proceeds by diffusion of the volatile species (water or organic solvent) until vapour pressure in the droplet equals the one of the reservoir. If equilibration occurs by water exchange from the drop to the reservoir, it leads to a droplet volume decrease. Consequently, the concentration of all constituents in the drop will increase. For species with a vapour pressure higher than water, the exchange occurs from the reservoir to the drop. In such a 'reverse' system, the drop volume will increase as well as the concentration of the drop constituents. This last solution, less widely used, has led to the crystallization of $tRNA^{Asp}$ (16) and of several proteins (17, 18). The same principle applies for hanging drops, sitting drops, and sandwich drops.

Glass vessels in contact with macromolecular solutions should be treated

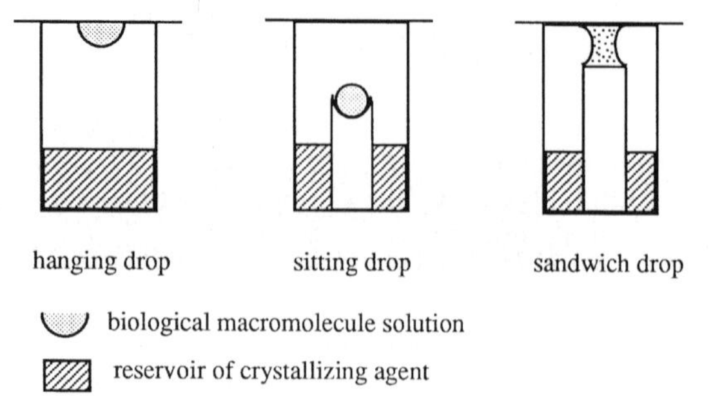

Figure 5. Schematic representation of hanging drop, sitting drop, and sandwich drop.

in a way to obtain an hydrophobic surface. Coated glass coverslips are commercially available but can be prepared following *Protocol 5*.

Protocol 5. Preparation of glass coverslips[a]

Equipment and reagents

- Coverslips
- Temperature controlled water-bath
- Toluene
- Dimethyldichlorosilane
- Soap solution
- Distilled water
- Ethanol

A. *Silanization*

1. Place coverslips in a bath of toluene containing 1% dimethyldichlorosilane at 60°C for 10 min.
2. Coverslips are then washed with soap solution and rinsed with distilled water and ethanol.
3. The same procedure is used for Pyrex plates.
4. Dry overnight at 120°C to sterilize vessels.

B. *Siliconization*

1. This can be achieved with commercially available reagent solutions (e.g. Sigmacoat). Coverslips are washed in the solution, and dried overnight at 120°C.

[a] All operations can be performed under vacuum when dealing with narrow vessels like capillaries.

A device shown in *Figure 6* helps to treat coverslips. If made in Perspex, the device is only used for drying coverslips; if Teflon made, it can be used for all the silanization process.

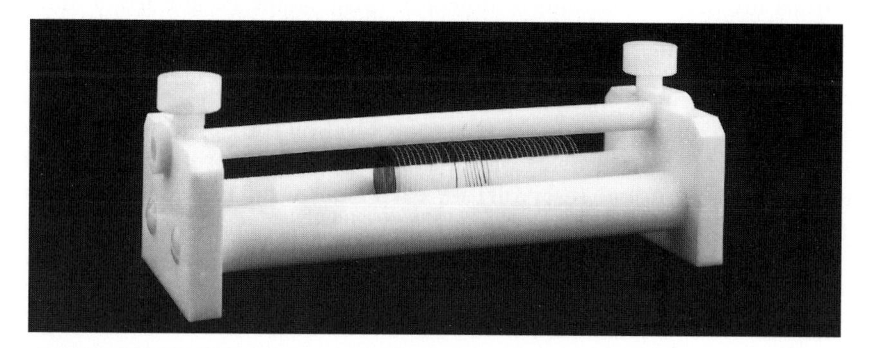

Figure 6. A device for treating coverslips (Perspex or Teflon made). The set-up displayed (in Teflon) can be manufactured in the laboratory workshop. Coverslips are held in the threading of the two bottom axis; a smaller unthreaded axes secures the coverslips. The set-up shown is about 20 cm long and permits handling of about 60 coverslips.

4.2 Experimental set-ups

4.2.1 Hanging drops in Linbro boxes

Commercially available Linbro boxes are plastic boxes (*Figure 7*) normally used for tissue culture (they may be replaced by Costar or VDX plates, supplied by Hampton Research). Plastic boxes will depolarize light; so unless for a particular orientation of a crystal, no birefringence will be observed. Boxes contain 24 wells labelled A, B, C, D vertically, and 1–6 horizontally. It is convenient, to avoid confusion, to always use them in the same orientation. Each box must be carefully labelled (date, experiment number, operator, and so on). Each well has a volume of approximately 2 ml and an inner diameter of 16 mm. There is a small rim which will be used for sealing the system. Each well will be covered by a glass coverslip of 22 mm diameter treated as described in *Protocol 5*.

Figure 7. A Linbro box for vapour diffusion crystallization in hanging drops. The photograph shows the box with its cover which is held by Plasticine in the corners. For a better display, two drops were prepared with dyes (at the left).

Drops are set up following *Protocol 6*. Most of the people use a 'magic' ratio of two between the concentration of the crystallizing agent in the reservoir (well) and in the droplet. This is conveniently achieved by mixing a droplet of protein at twice the desired final concentration with an equal volume of the reservoir at the proper concentration (other ratios can be used as well). Avoiding local over-concentration can be achieved by placing the two drops (protein and reservoir) on each side of an Eppendorf tube and vortexing it quickly.

Protocol 6. Crystallization in Linbro boxes

Equipment and reagents

- Silicone grease
- Linbro boxes
- Coverslips
- Pair of brussel
- Plasticine

Method

1. Grease rims with silicone grease.[a]
2. Fill up reservoir with 1 ml of filtered (0.22 μm) crystallizing agent.
3. Spray glass coverslip with antidust.
4. Mix a 2–10 μl drop of filtered (0.22 μm) biological macromolecule solution with an equivalent volume of reservoir.
5. Layer the drop on the 22 mm diameter coverslip (do not touch the coverslip with the extremity of the tip of the pipettor or it will spread) so that a nearly hemispherical drop is formed.[b]
6. Return the coverslip with a pair of brussel (or fingers). First train yourself with water!
7. Set on the grease rim and gently press to seal the well with the grease. Do not press too firmly, otherwise the coverslip will break.
8. Check the sealing by inspecting the rim in an azimuthal way. If sealing is not properly done, the drop will concentrate as well as the reservoir. Crystallization will occur eventually, but will be very difficult to reproduce.
9. Adjust glass coverslips tangent to each other otherwise they overlap.
10. Put Plasticine in the corners to avoid the contact between the cover and grease, otherwise slips get stuck at the cover.

[a] To dispense the grease, fill up a syringe and replace the needle by an Eppendorf yellow tip.
[b] You may layer several microdrops on a coverslip.

When no crystal or precipitate is observed, either supersaturation is not reached or one has reached the metastable region (see Chapter 10 for definition). In the latter case, changing the temperature by a few degrees is generally sufficient to initiate nucleation. For the former, the concentration of crystallizing agent in the reservoir must be increased. In the former case, gently rotate the coverslip in the plane of the rim to ease the grease (which becomes 'stiff' with time), then lift it, suck the reservoir entirely, and replace it by a more concentrated solution. More grease is added and the coverslip sealed. The volume of the drop will decrease again and all constituent concentrations in the drop increase.

i. Problems

For membrane proteins (see Chapter 9), the presence of detergent tends to spread out the drops and lower the surface tension. In all cases gravity will tend to sink the drops containing the macromolecule in the reservoir for volumes exceeding 25 μl. Shaking Linbro boxes will give the same results. Boxes must be transported horizontally and carefully. It is always painful for beginners to ruin an experiment when transporting a box. If you prepare boxes at room temperature and transfer them to 4^°C, condensation will occur on the surface of the coverslip. Water droplet will surround the macromolecule drop. If it mixes, the protein will dilute and probably stay in an undersaturated state. To avoid this problem, set up boxes at the final experiment temperature and cover boxes with polystyrene sheets.

It is possible to recrystallize macromolecules using hanging drops as described in *Protocol 7*.

Protocol 7. Recrystallization of macromolecules

Equipment and reagents

- Minifilters
- Coverslips
- Crystallizing agent
- Grease

A. *For purification*

1. Remove mother liquor.
2. Wash crystals with fresh buffer.
3. Redissolve in fresh crystallizing agent solution.
4. Recrystallize.

B. *For crystal growth*

1. Redissolve crystals by replacing crystallizing agent in the reservoir by buffer. Depending on the biological macromolecule, it may take a few hours or a few days. The drop volume will increase; so control the process, otherwise the drop will sink.
2. Filter the drop in a minifilter (e.g. Costar, Millipore) by centrifugation. **Warning**: the dead volume is at least 5 μl.
3. Place the drop on a clean glass coverslip.
4. Add some grease to the rim.
5. Fill the reservoir with crystallizing agent and recrystallize.

4.2.2 Crystallization with ACA CrystalPlates®

The American Crystallographic Association (ACA) sponsored a vapour diffusion plate called CrystalPlate® (manufactured by ICN Flow). Although dedicated for use with automated systems (19), it is equally useful for manual crystallization. As shown in *Figure 8*, it may be used to set up crystallization by hanging drops, sitting drops, or sandwich drops, depending on the thickness of the lower glass slip and the volume of the drop (*Protocol 8*). Each box contains 15 wells. The glass coverslips should be prepared as described in *Protocol 5*. If desired, the breakaway plug in the bottom of a reservoir — see

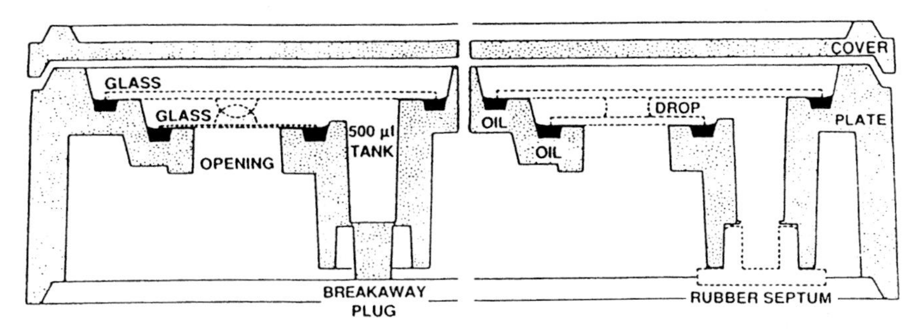

Figure 8. ACA CrystalPlate® (courtesy of ICN Flow). This is a versatile system for vapour diffusion crystallization allowing individual experiments on sitting, hanging, or sandwiched drops under classical or automated conditions.

tank in *Figure 8* — may be removed with pliers and a rubber septum inserted so that the reservoir solution may be changed with an hypodermic syringe. A crystallization set-up based on the same versatile concept as the CrystalPlates® is the so-called Q-Plate™ (supplied by Hampton Research).

i. Advantages

Because of the glass windows, when looking at drops under polarized binocular, birefringence of crystals can be observed. Drops with macromolecules are no longer above the reservoir thus eliminating sinking. Large drops can be prepared with the sandwich method.

Warning: if you use a rubber septum, plates stick when you translate them during observation, eventually provoking disasters.

Protocol 8. Crystallization with ACA CrystalPlates®

Equipment and reagents

- ACA CrystalPlates® and glass slips
- Grease

Method

1. Prepare the plate by filling up the upper and lower troughs of each well with ordinary hydrocarbon vacuum pump oil or grease.
2. To dispense oil or grease, fill up a syringe and replace the needle by an Eppendorf yellow tip.
3. Put 0.5 ml of crystallizing agent into each reservoir.
4. Position one of the 14×14 mm^2 glass coverslips over the hole in each well. The coverslips should seal quickly if there is enough oil or grease.
   (a) For hanging or sitting drops use $14 \times 14 \times 0.2$ mm^3 glass coverslips.

Protocol 8. *Continued*

(b) For sandwich drops (5–25 μl) use 14 × 14 × 1 mm^3 glass coverslips.

(c) For sandwich drops (15–75 μl) use 14 × 14 × 1.5 mm^3 glass coverslips.

5. Put a drop of the biological macromolecule solution in the centre of the lower (for sitting or sandwich drops) or upper (for hanging drops) glass coverslips and then set one of the 24 × 30 × 1 mm^3 glass coverslip in position on the upper trough.

4.2.3 Other systems

Among the first used sitting drop set-ups are systems using Pyrex plates (e.g. Corning Glass 7220) with three or more depressions placed in different type of boxes, with the reservoir solution below the plate (20, 21). Cryschem MVD24 plates are commercially available (manufactured by Cryschem Inc. or C. Supper Company). Each well contains a plastic post in the centre to hold the protein sitting drop. The cup has been designed (*Figure 9*) to provide maximum surface area for free diffusion during equilibration. The reservoir solution is held within the narrow moat surrounding the support port. The plates are sealed with an adhesive tape which is supplied by the manufacturer. An ingenious way to seal the plates with the tape and to set hanging drops is

Figure 9. Crystal growth multi-chamber plate for vapour diffusion crystallization on individual sitting drops (courtesy of C. Supper Company).

to use the HANGMAN framework. Here the protein droplets are first installed on the tape which is than inverted in a second step over the plate (22).

A variety of other set-ups have been designed in many laboratories, allowing for instance Linbro or VDX boxes to be used for sitting drop experiments (with the depression on a small plastic bridge (23) or on a glass rod as on the Oxford or Perpetual Systems Corporation set-ups, respectively); or doing vapour phase equilibration in capillaries (24, 25), or even directly in X-ray capillaries as was described for ribosome crystallizations (26) or in the gel acupuncture method. For extremely fragile crystals, when transfer from crystallization cells to X-ray capillaries (see Chapter 14) can lead to internal damage and mechanical cracks of crystals, this last method may be well adapted.

4.3 Varying parameters

Although unique in this respect, vapour diffusion methods permit easy variation of physical parameters during crystallization, and many successes were obtained by modifying supersaturation by temperature or pH changes (27, 28) (see also Section 5). With ammonium sulfate as the crystallizing agent, it has been shown that the pH in the droplets is imposed by that of the reservoir (29). Consequently, varying the pH of the reservoir permits gentle adjustment of that in the droplets. From another point of view, sitting drops are well suited for attempting epitaxial growth of macromolecule crystals on appropriated mineral matrices (30).

In vapour diffusion crystallizations, the contamination by micro-organisms can be prevented by placing a small grain of thymol, a volatile organic compound, in the reservoir (see also Chapter 2). Thymol, however, can have specific effects on crystallization, as shown with glucose isomerase (31), and may thus represent an useful additive to assay in crystallization screenings.

4.4 Kinetics of evaporation

4.4.1 Final concentrations

Calculating the final concentration of constituents in an equilibrated drop is often a source of misunderstanding in protocols. Sometimes it refers to the final concentration in the drop before the vapour diffusion process is initiated, sometimes it describes the final concentration in the drop at equilibrium at the end of the process. So, except for the species of vapour pressure higher than that of water, at equilibrium, if the ratio of crystallizing agent concentration between reservoir and drop is two, final concentrations and volumes are as follows:

- final drop volume = 1/2 initial volume
- final concentration of all constituents of the drop (protein, additive, and so on) equals twice the initial concentration.

Many other ratios can be used. Varying the volume of the droplet will influence the kinetics of crystallization and so the protein crystal size.

4.4.2 Equilibration kinetics

The kinetics of water evaporation determines the kinetics of supersaturation and accordingly affects nucleation rates. Evaporation rates from hanging drops have been determined experimentally in the presence of ammonium sulfate, PEG, MPD (32), and NaCl (33–35) as crystallizing or dehydrating agents. The main parameters which determine the rate of water equilibration are temperature, initial drop volume (and initial surface to volume ratio of the drop and its dilution with respect to the reservoir), water pressure of the reservoir, and the chemical nature of the crystallizing agent (*Figure 10a*). Theoretical modelling (35) has shown in addition the pivotal role of the drop to reservoir distance (d), but effect of this parameter is negligible in classical set-ups, e.g. in Linbro boxes (32), and becomes only noticeable in special experimental arrangements (34, 35) when $d > 2$ cm (*Figure 10b*). Note: the presence of macromolecule does not seem to affect the water evaporation rate.

From the practical point of view, the time for water equilibration to reach 90% completion can vary from about 25 hours to more than 25 days, the fastest equilibration occurring in the presence of ammonium sulfate, that in the presence of MPD being slower, and that in the presence of PEG by far the slowest (*Figure 10a*). Estimates of the minimal duration of equilibration under several standard experimental conditions can be obtained from an empirical model (36). Equilibration rates are significantly slowed down by increasing appropriately the distance between the drop and the reservoir (*Figure 10b*). This can be done using the Z/3 plate design (34) or just simple test-tubes as reservoirs (35). An alternative solution to decrease equilibration rates is to layer oil over the reservoir (37).

The particularly slow equilibration rates observed with PEG may explain crystallization successes using this precipitating agent (38). Indeed crystal growth may be favoured when supersaturation is attained very slowly. This fact is corroborated by independent experiments in which the terminal crystal size was significantly increased by reducing the vapour pressure of the reservoir, i.e. the evaporation rate, as a function of time (37, 39, 40).

5. Crystallization by batch methods

5.1 Classical methods

The biological macromolecule to be crystallized is mixed with the crystallizing agent at a concentration such that supersaturation is instantaneously reached. This can be achieved with all methods previously described. For hanging drop or sitting drops, the reservoir no longer acts to concentrate the drop but is

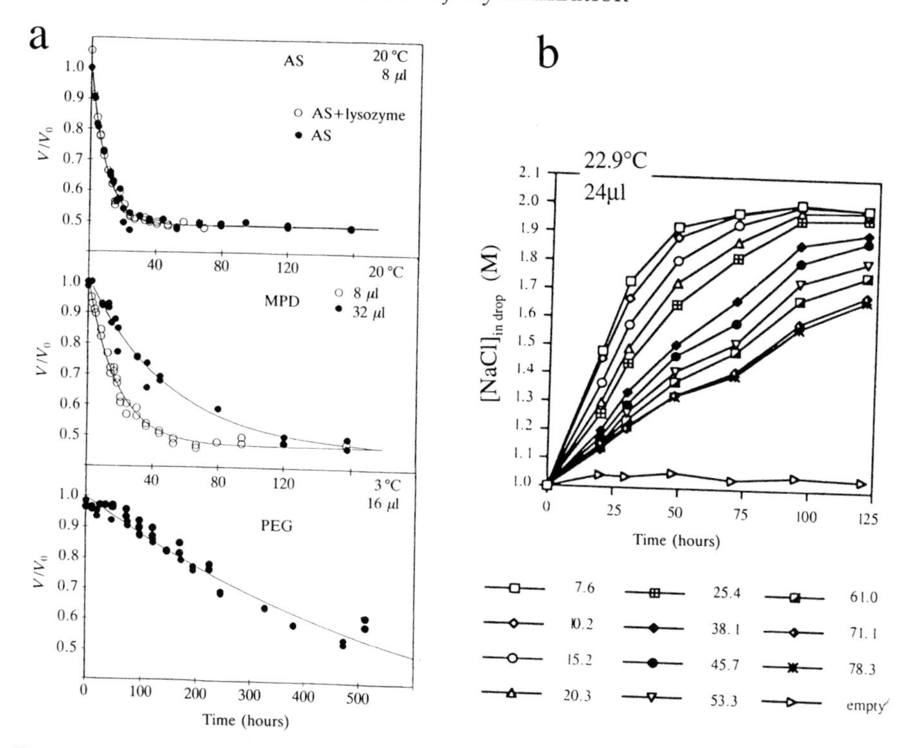

Figure 10. Water evaporation kinetics in the presence of ammonium sulfate (AS), MPD, PEG, and NaCl as dehydrating agents. (a) Measurements done in drops set in Linbro boxes with AS, MPD, and PEG. The data also show the influence of protein (in AS experiment), and of initial drop volume (in MPD experiment) on final drop volume. V_o is the initial volume of the drop; experiments were conducted with a concentration of crystallizing agent in the reservoir twice that in the drop, time at zero. Adapted from ref. 32. (b) Measurements done on hanging drops (24 μl) set over test-tubes with distances between the drop and reservoir varying from 7.6–78.3 mm. Experiments were conducted at 22.9°C with 1.0 M NaCl as initial concentration in the drop and 2.0 M NaCl in the reservoirs. Adapted from ref. 35.

only present to maintain constant vapour pressure. Because one starts from supersaturation, nucleation tends to be too large. However, in some cases fairly large crystals can be obtained when working close to the metastable region. This is illustrated in Chapter 10. If supersaturation is too high, a precipitate may develop in a batch crystallization vessel; do not discard such experiments because crystals can eventually grow from the precipitate by Ostwald ripening (41). Notice, however, that the growth kinetics under such circumstances are decreased.

An automated system for microbatch macromolecule crystallizations and screening has been described (42) allowing the set up of samples of less than 2 μl. Reproducibility of experiments is guaranteed because samples are dispensed and incubated under paraffin oil, thus preventing evaporation and uncontrolled concentration changes of the components in the micro-droplets. Using silicone oils that are slightly soluble in water, or appropriate mixtures of paraffin and silicone oils, results in gradual protein concentration in the droplets like in vapour diffusion experiments (43).

A variation of classical batch crystallization is the sequential extraction procedure of Jakoby (44), based on the property that many proteins (not all) are more soluble in concentrated salt (e.g. ammonium sulfate) when lowering the temperature. The method can be adapted for microassays and was successfully applied for the crystallization of a proteolytic fragment of methionyl-tRNA synthetase from *Escherichia coli* (45).

5.2 Advanced methods

Solubility and supersaturation of proteins is influenced by hydrostatic pressure. Advantage has been taken of this fact to crystallize proteins at high pressure (46, 47). Effects become significant for pressures higher than 50 Mpa (500-fold atmospheric pressure) as shown for lysozyme crystals that can be grown in the range 50–250 Mpa. Such crystals exhibit habits different from controls grown at atmospheric pressure and diffract at high resolution (47). Experimental set-ups that can contain up to 24 samples of 80 μl crystallizing solution each and that can be pressurized up to 400 Mpa have been designed (47). Although the method has only been occasionally used in the macromolecular field, in the future it may represent an interesting alternative to obtain new crystalline forms of proteins.

The floating drop method enables crystallization of biological macromolecules under conditions where the crystallizing solution has no contact with the container walls (48, 49). Drops (5–100 μl) are placed at the interface between two layers of inert and non-miscible silicone fluids contained in square glass or plastic cuvettes. The density of the fluids can be such that drops containing the most common crystallizing agents can be floated (*Figure 11*).

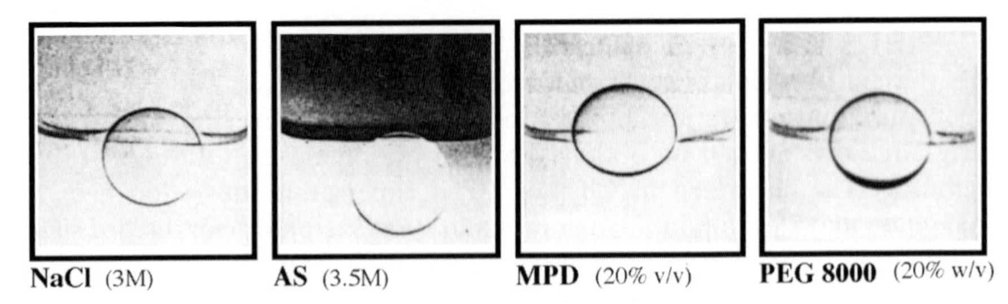

Figure 11. Crystallization in floating drops. Droplets (40 μl) of four different crystallizing agent solutions placed at the interface of two silicone fluids are displayed. The two silicone fluids at 20 °C (800 μl each with low density fluid PS037 layered over high density PS181 from Hüls America, Inc.) (silicone oils may be provided by Hampton Research) are placed in square glass or polystyrene spectrophotometer cuvettes. Adapted from ref. 48.

Several proteins and a spherical plant virus were crystallized in the temperature range 4–20 °C using this method (48). Its main advantage is to reduce the nucleation rate. Thus crystallization in floating drops provides a means to obtain a small number of larger crystals in an homogeneous liquid medium. Because drops are not in contact with air, the method may be convenient to crystallize proteins sensitive to oxidation. Further, when implemented in a thermostated device, the method provides a simple and convenient way for kinetic measurements of macromolecule crystal growth (48).

Other advanced methods useful to prepare crystals for diffraction studies (e.g. in gelified media, under microgravity, and the gel acupuncture method) are described in Chapter 6.

6. Crystallization by interface diffusion

This method was developed by Salemme (50) and used to crystallize several proteins. In the liquid/liquid diffusion method, equilibration occurs by diffusion of the crystallizing agent into the biological macromolecule volume. To avoid rapid mixing, the less dense solution is poured very gently on the most dense (salt in general) solution. Sometimes, the crystallizing agent is frozen and the protein layered above to avoid rapid mixing.

One generally uses tubes of small inner diameter in which convection is reduced. This could be achieved more easily by using gels as described in Chapter 6. It follows the same diffusion method without the inconvenience of the metastability of two liquids sitting on the top of each other. This method gained new attention because of microgravity experiments (see Chapter 6).

7. Correlations with solubility diagrams

Even if it is not possible to determine the solubility (or phase) diagram for each biological macromolecule, it is important to understand the correlation between solubility diagrams and the method used to reach supersaturation and crystallization, using schematic diagrams (for more details see Chapter 10).

7.1 Dialysis

In the case of dialysis buttons, if one considers that stretching of the membrane is negligible, the macromolecule concentration will remain constant. However, if the macromolecule solution does not fill the chamber entirely, leaving room for air, it is no longer exactly true since the macromolecule concentration may vary (increase or decrease depending on the situation). The initial concentration of the crystallizing agent in the reservoir (this could be buffer) leaves the macromolecule in an undersaturated state.

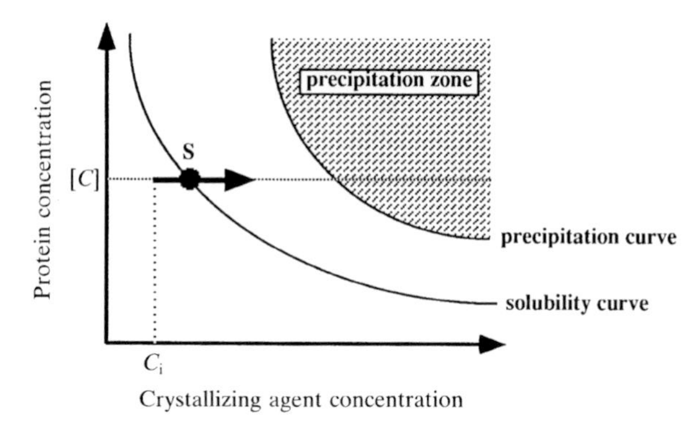

Figure 12. Schematic solubility diagram and correlation between macromolecule and crystallizing agent concentrations in a crystallization experiment using a dialysis set-up. *Ci* is the initial concentration of crystallizing agent and *C* the constant protein concentration. The area between the precipitation and solubility curves is the supersaturated region where crystallization can occur. Precipitation and solubility curves can be determined experimentally, although for the latter one crystals should be obtained first. For more details see Chapter 10.

As shown in *Figure 12*, when increasing the crystallizing agent concentration, the supersaturation state is reached after passing through point S which is the equilibrium point on the solubility curve. Then, depending on the final crystallizing agent concentration, it will crystallize or precipitate.

7.2 Vapour diffusion

In a classical case when the concentration of crystallizing agent in the reservoir is twice the one in the drop, the protein will start to concentrate from an undersaturated state A (at concentration C_i) to reach a supersaturated one B (at concentration C_f) with both protein and crystallizing concentrations increasing by a factor two. Two hypothetical cases are represented in *Figure 13a* and *13b* corresponding to experiments not leading (*Figure 13a*) or leading (*Figure 13b*) to crystals. Since no crystals are obtained in *Figure 13a*, the equilibrium at point B will be located in the metastable region; when the first crystals appear (at the break of the arrow) the trajectory of equilibration is more complex. In that case the remaining concentration of protein in solution will converge towards point C located on the solubility curve.

7.3 Batch crystallization

In batch crystallization using a closed vessel, three cases can be considered as shown in *Figure 14*. If the protein concentration is such that the solution is undersaturated (point A), crystallization will never occur (unless another parameter such as temperature is varied). The protein concentration may belong to the supersaturated region between solubility and precipitation

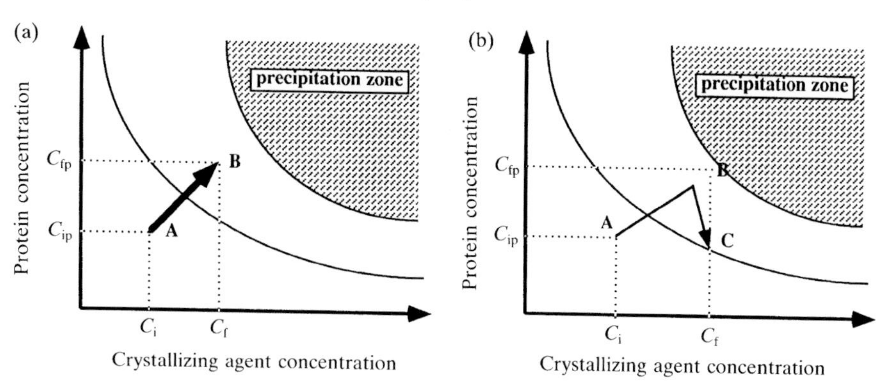

Figure 13. Schematic solubility diagram and correlation between macromolecule and crystallizing agent concentrations in crystallization experiments using vapour diffusion set-ups. Situation without (a) and with (b) crystallization. See legend to *Figure 12* and text for further explanations.

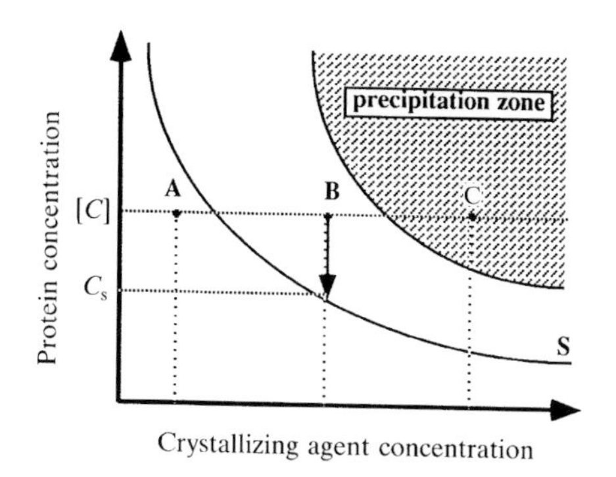

Figure 14. Schematic solubility diagram and correlation between macromolecule and crystallizing agent concentrations in crystallization experiments using a batch method (in closed vessels). See legend to *Figure 12* and text for further explanations.

curves (point B). In that case the arrow describes the variation of the remaining concentration of protein in solution. In the last case (point C) the protein will precipitate immediately because supersaturation is too high. In some cases, however, crystals may grow from the precipitates (41).

8. Practising crystallization

It is a good exercise to train oneself with a cheap easy accessible protein. Lysozyme, thaumatin, thermolysin, and BPTI are good candidates which are commercially available from various manufacturers and which crystallize readily. Examples of crystallization using various methods in hanging drops are given in *Protocols 9* and *10*.

Protocol 9. Testing lysozyme crystallization using hanging drops

Equipment and reagents

- Linbro box
- 50 mM sodium acetate pH 4.5
- 3 M NaCl
- 40 mg/ml lysozyme in 50 mM acetate pH 4.5
- 0.22 μm microfilter
- Coverslips

Method

1. Prepare stock solutions of 3 M NaCl and 40 mg/ml (2.74 mM) lysozyme in 50 mM acetate pH 4.5 and buffer stock solution (50 mM sodium acetate pH 4.5). Filter all solutions with a 0.22 μm microfilter.
2. Prepare a Linbro box as described in Section 4.2.1.
3. Fill up reservoirs of row A with solutions of NaCl ranging from 0.5–1.5 M in steps of 0.2 M.
4. On a coverslip, mix 4 μl of protein stock solution with 4 μl of reservoir. Flip it and set it on the greased rim.
5. Fill up reservoirs of row B with solutions of NaCl ranging from 0.8–1.8 M in steps of 0.2 M. Repeat the experiment on row B after diluting the protein stock solution by a factor two to obtain a new one of 20 mg/ml (1.37 mM).
6. Fill up reservoirs of row C with solutions of NaCl ranging from 1.5–2.5 M in steps of 0.2 M. Repeat the experiment after diluting the protein stock solution by a factor two.
7. Use row D for duplicate or testing particular parameters (e.g. volume of drops to see the influence of kinetic effects on growth).
8. Store the experiments at 18°C.
9. Observe the experiments once a day for a week.
10. Train yourself in mounting crystals (see Chapter 14).

Protocol 10. Testing thaumatin crystallization using a dialysis button

Equipment and reagents

- Dialysis membrane
- Dialysis button
- Linbro box
- Stock solutions (see *Protocol 9*)

Method

1. Prepare stock solutions as in *Protocol 9*.
2. Fill the protein chamber of a dialysis button with protein stock solution

diluted to 10 mg/ml. The solution must form a dome above the entry. Install the dialysis membrane of appropriate cut-off as described in Section 3.2.2.

3. Fill up with 1 ml of reservoir solution the reservoir of a Linbro box and drop the button in it with the aperture of the protein chamber on the top.
4. Store the experiments at 18°C.
5. Increase the concentration in the reservoir by 200 mM of crystallizing agent every day.
6. Observe the next day.
7. Repeat steps 5 and 6 until crystallization occurs.

9. Concluding remarks

In this chapter we have described the most common crystallization methods; all of them have advantages and drawbacks. Most of crystallographers favour vapour phase diffusion which provides an easy way to practise crystallization. It is also the method of choice for robotics (51). Dialysis presents the advantage that macromolecular concentrations remain constant, so that only one parameter varies at a time and nature of buffer or crystallizing agent can be changed easily. It differs with a classical vapour phase equilibrium crystallization experiment where all constituents in the drop are concentrated.

Beside the classical and advanced crystallization methods described in this chapter, less standard methods may be useful in particular cases. These can be methods based on old ideas not yet well explored, as pulse-diffusion of precipitant combining dialysis and free diffusion in capillaries (52), or combinations of dialysis and electrophoresis (53). Also, crystallization in particular environments should be considered, such as under levitation (54), in centrifuges (55, 56), or in magnetic (57, 58) or electric (59) fields. Particular attention should be given to crystallization methods where convection is reduced; e.g. the gel acupuncture method, crystallization in gels which is becoming popular in the macromolecule field, and crystallization in microgravity (Chapter 6). Methods based on temperature diffusion, which are widely used in material sciences (60), may be adapted under certain conditions for macromolecule crystallization (61, 62). Finally, the use of novel types of crystallization cells may represent an interesting alternative for growing better crystals. In particular, cells based on principles developed for microgravity experiments may be appropriate (see refs 63–65). It is our hope that the methods and ideas discussed in this chapter will help readers, not only to solve their crystallization problems, but also to improve existing methods, and even to develop new crystallization methodologies.

References

1. McPherson, A. (1985). In *Methods in enzymology*, (eds Wickoff, H. W., Hirs, C. H., and Timasheff, S. N.), Academic Press, London. Vol. 114, p. 112.
2. Giegé, R. (1987). In *Crystallography in molecular biology* (ed. D. Moras, J. Drenth, B. Strandberg, D. Suck, and K. Wilson), NATO ASI Series, Vol. 126, pp. 15–26. Plenum Press, New York and London.
3. McPherson, A. (1990). *Eur. J. Biochem.*, **189**, 1.
4. Perrin, D. D. and Dempsey, B. (1974). *Buffer for pH and metal ion control.* Chapman and Hall Ltd., London and New York.
5. Jurnak, F. (1986). *J. Cryst. Growth*, **76**, 577.
6. Ray, W. J. and Puvathingal, J. (1985). *Anal. Biochem.*, **146**, 307.
7. Chayen, N., Akins, J., Campbell-Smith, S., and Blow, D. M. (1988). *J. Cryst. Growth*, **90**, 112.
8. Winterbourne, D. J. (1986). *Biochem. Soc. Trans.*, **14**, 1179.
9. Mikol, V. and Giegé, R. (1989). *J. Cryst. Growth*, **97**, 324.
10. Bradford, M. M. (1976). *Anal. Biochem.*, **72**, 248.
11. Zeppenzauer, M. (1971). *Methods in enzymology*, ibid ref. 1. Vol. 22, p. 253.
12. Weber, B. H. and Goodkin, P. E. (1970). *Arch. Biochem. Biophys.*, **141**, 489.
13. Pronk, S. E., Hofstra, H., Groendijk, H., Kingma, J., Swarte, M. B. A., Dorner, F., *et al.* (1985). *J. Biol. Chem.*, **260**, 13580.
14. Thomas, D. H., Rob, A., and Rice, D. W. (1989). *Protein Eng.*, **2**, 489.
15. Hampel, A., Labananskas, M., Conners, P. G., Kirkegard, L., RajBhandary, U. L., Sigler, P. B., *et al.* (1968). *Science*, **162**, 1384.
16. Giegé, R., Moras, D., and Thierry, J.-C. (1977). *J. Mol. Biol.*, **115**, 91.
17. Richard, B., Bonneté, F., Dym. O., and Zaccai, G. (1995). In *Archaea, a laboratory manual*, pp. 149–54. Cold Spring Harbor Laboratory Press.
18. Jeruzalmi, D. and Steitz, T. A. (1997). *J. Mol. Biol.*, **274**, 748.
19. Zuk, W. M. and Ward, K. B. (1991). *J. Cryst. Growth*, **110**, 148.
20. Kim, S.-H. and Quigley, G. (1979). In *Methods in enzymology*, Vol. 59, p. 3.
21. Dock, A.-C., Lorber, B., Moras, D., Pixa, G., Thierry, J.-C., and Giegé, R. (1984). *Biochimie*, **66**, 179.
22. Luft, J. R., Cody, V., and DeTitta, G. T. (1992). *J. Cryst. Growth*, **122**, 181.
23. Harlos, K. (1992). *J. Appl. Crystallogr.*, **25**, 536.
24. Phillips, G. N., Jr. (1985). *Methods in enzymology*, ibid ref. 1. Vol. 104, p. 128.
25. Luft, J. R. and Cody, V. (1989). *J. Appl. Crystallogr.*, **22**, 386.
26. Yonath, A., Müssig, J., and Wittmann, H. G. (1982). *J. Cell. Biochem.*, **19**, 145.
27. McPherson, A. (1985). In *Methods in enzymology*, Vol. 114, p. 125.
28. McPherson, A. (1995). *J. Appl. Crystallogr.*, **28**, 362.
29. Mikol, V., Rodeau, J.-L., and Giegé, R. (1989). *J. Appl. Crystallogr.*, **22**, 155.
30. McPherson, A. and Shlichta, P. (1988). *Science*, **239**, 385.
31. Chayen, N. E., Lloyd, L. F., Collyer, C. A., and Blow, D. M. (1989). *J. Cryst. Growth*, **97**, 367.
32. Mikol, V., Rodeau, J.-L., and Giegé, R. (1990). *Anal. Biochem.*, **186**, 332.
33. Luft, J. R., Arakali, S. V., Kirisits, M. J., Kalenik, J., Wawzzak, I., Cody, V., *et al.* (1994). *J. Appl. Crystallogr.*, **27**, 443.
34. Arakali, S. V., Easley, S., Luft, J. R., and DeTitta, G. T. (1994). *Acta Cryst.*, **D50**, 472.

35. Luft, J. R., Albrigth, D. T., Baird, J. K., and DeTitta, G. T. (1996). *Acta Cryst.*, **D52**, 1098.
36. Fowlis, W. W., DeLucas, L. J., Twigg, P. J., Howard, S. B., Meehan, E. J., and Baird, J. K. (1988). *J. Cryst. Growth*, **90**, 117.
37. Chayen, N. E. (1997). *J. Appl. Crystallogr.*, **30**, 198.
38. McPherson, A. (1976). *J. Biol. Chem.*, **251**, 6300.
39. Gernert, K. M., Smith, R., and Carter, D. C. (1988). *Anal. Biochem.*, **168**, 141.
40. Przybylska, M. (1989). *J. Appl. Crystallogr.*, **22**, 115.
41. Ng, J., Lorber, B., Witz, J., Théobald-Dietrich, A., Kern, D., and Giegé, R. (1996). *J. Cryst. Growth*, **168**, 50.
42. Chayen, N. E., Shaw Stewart, P. D., Maeder, D. L., and Blow, D. M. (1990). *J. Appl. Crystallogr.*, **23**, 297.
43. d'Arcy, A., Elmore, C., Stihle, M., and Johnston, J. E. (1996). *J. Cryst. Growth*, **168**, 175.
44. Jakoby, W. B. (1971). In *Methods in enzymology*, Vol. 22, p. 248.
45. Waller, J. P., Risler, J.-L., Monteilhet, C., and Zelwer, C. (1971). *FEBS Lett.*, **16**, 186.
46. Visuri, K., Kaipainen, E., Kivimäki, J., Niemi, H., Leissla, M., and Palosaari, S. (1990). *Biotechnology*, 547.
47. Lorber, B., Jenner, G., and Giegé, R. (1996). *J. Cryst. Growth*, **158**, 103.
48. Lorber, B. and Giegé, R. (1996). *J. Cryst. Growth*, **168**, 204.
49. Chayen, N. E. (1996). *Protein Eng.*, **9**, 927.
50. Salemne, F. R. (1972). *Arch. Biochem. Biophys.*, **151**, 533.
51. Zuk, W. M. and Ward, K. B. (1991). *J. Cryst. Growth*, **110**, 148.
52. Koeppe, R. E., Stroud, R. M., Pena, V. A., and Santi, D. V. (1975). *J. Mol. Biol.*, **98**, 155.
53. Chin, C.-C., Dence, J. B., and Warren, J. C. (1976). *J. Biol. Chem.*, **251**, 3700.
54. Rhim, W.-K. and Chung, S. K. (1990). *Methods: a companion to methods in enzymology*, **1**, 118.
55. Barynin, V. V. and Melik-Adamyan, V. R. (1982). *Sov. Phys. Crystallogr.*, **27**, 588.
56. Lenhoff, A. M., Pjura, P. E., Dilmore, J. G., and Godlewski, T. S., Jr. (1997). *J. Cryst. Growth*, **180**, 113.
57. Sazaki, G., Yoshida, E., Komatsu, H., Nakada, T., Miyashita, S., and Watanabe, K. (1997). *J. Cryst. Growth*, **173**, 231.
58. Ataka, M., Katoh, E., and Wakayama, N. I. (1997). *J. Cryst. Growth*, **173**, 592.
59. Taleb, M., Didierjean, C., Jelsch, C., Mangeot, J.-P., Capelle, B., and Aubry, A. (1998). *J. Cryst. Growth*,
60. Feigelson, R. S. (1988). *J. Cryst. Growth*, **90**, 1.
61. Lorber, B. and Giegé, R. (1992). *J. Cryst. Growth*, **122**, 168.
62. DeMattei, R. C. and Feigelson, R. S. (1993). *J. Cryst. Growth*, **128**, 1225.
63. Stoddard, B. L., Strong, R. K., and Farber, G. K. (1988). *J. Cryst. Growth*, **110**, 312.
64. Bosch, R., Lautenschlager, P., Potthast, L., and Stapelmann, J. (1992). *J. Cryst. Growth*, **122**, 310.
65. McPherson, A. (1996). *Crystallogr. Rev.*, **6**, 157.

6

Crystallization in gels and related methods

M.-C. ROBERT, O. VIDAL, J.-M. GARCIA-RUIZ, and F. OTALORA

1. Introduction

From the first studies showing the feasibility of macromolecular crystal growth in gels (1), an increasing attention has been paid to applications of gel techniques to the domain of biological macromolecules. Confidence in these techniques is such that kits of crystallization in gels are now commercially available (Hampton Research, Laguna Hills, CA, USA).

Basically, the protein crystallization process consists of two consecutive steps:

- first, the transport of growth units towards the surface of the crystals
- second, the incorporation of the growth units into a crystal surface position of high bond strength.

The whole growth process is dominated by the slowest of these two steps and is either transport controlled or surface controlled. Avoiding convection in the growth environment will increase the possibility of growing the crystal under slow diffusive mass transport providing that the surface interaction kinetics are faster than the characteristic diffusive flow of macromolecules (in the range of 10^{-6} cm^2/sec for proteins). The ratio between transport to surface kinetics, which can be tuned by either enhancing or reducing transport processes in the solution, has been shown (2) to control the amplitude of growth rate fluctuations (which is thought to reduce crystal quality). These are the main reasons why gels (as well as capillaries and microgravity conducted experiments), if correctly designed, are expected to enhance the quality of crystals. This quality enhancement (3), as well as the possibility of getting crystals when conventional solution techniques failed (4), have been experimentally demonstrated. However, up to now, gel methods have been used on a rather empirical basis, as a simple transposition of solution techniques, and recent fundamental studies of nucleation and growth in gels show that the situation is not as simple as first expected (5, 6).

After summarizing the main characteristics of crystal growth in gels, we will examine what are the best conditions using a gel method. Recipes for the preparation of different gel growth experiments will be given. Considering gel growth as a possible simulation of experiments under reduced gravity, recent results of space experiments will be reviewed. Mention will also be made to growth under hypergravity conditions.

2. General considerations

Gels used for crystal growth are hydrogels with a growth solution soaking a polymeric network. For physical gels like gelatin or agarose, sol-gel transition is obtained by decreasing the temperature (physical parameter variation). Polymerization corresponds to the formation of weak bonds and this process is reversible with some hysteresis (~ 50 °C for agarose). For chemical gels, such as polyacrylamide, polymerization corresponds to strong bonding and is not reversible. Although formation of silica gels also results from a chemical reaction, it rather corresponds to an intermediate case between chemical and physical gels: as a matter of fact, the chemical reaction leads to the formation of dense beads which further aggregate by weak bonding (7). As far as we know, only agarose and silica gels have been successfully used for macromolecule crystal growth.

2.1 Formation and structure of gels

2.1.1 Agarose gels

Agarose is a polysaccharide extracted from seaweed. Its basic repeat unit is agarobiose (*Figure 1a*) with different substituents like *O*-methyl or *O*-sulfate groups which vary according to the agarose origin and subsequent chemical treatment. The sol-gel transition temperature varies according to the nature and content of these different substituents so that one can find commercially available (e.g. Sigma) agaroses in the 15–40 °C gelling temperature range.

The gelling process is not yet fully understood; it has been thought for a long time that polysaccharide chains first associate to form double helices, then aggregate to form fibres. Recent studies (8) contradict the existence of double helices but suggest that agarose strands form ordered lateral associations interconnected through disordered junction zones. (*Figure 1b*). In any case, such associations leave large voids through which very large molecules can migrate. That is, for 0.6% (w/v) agarose gels, the pore size is around 7.000 Å, and for gels less than 0.3% (w/v), their average size is larger than 10.000 Å. Indeed, gel media are largely used for electrophoresis of biological macromolecules.

2.1.2 Silica gels

Silica gels result from the polycondensation of silicic acid which occurs either by neutralization of sodium metasilicate:

$$Na_2SiO_3 + H_2O + 2\,HCl \rightarrow Si(OH)_4 + 2\,NaCl \qquad [1]$$

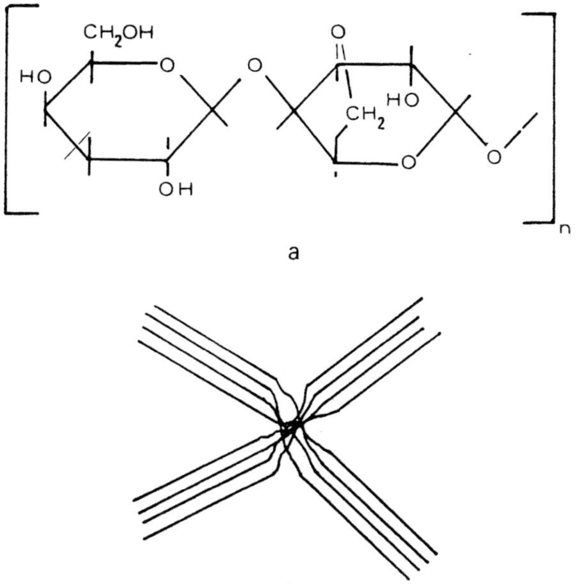

a

b

Figure 1. (a) Basic repeat unit of agarose chains: the agarobiose. (b) Schematic structure of the agarose gel showing the disordered junction zones.

or by hydrolysis of a siloxane, like tetramethoxysilane (TMOS) (or tetra-ethoxysilane, TEOS):

$$(CH_3O)_4Si + 4H_2O \rightarrow Si(OH)_4 + 4CH_3OH \quad [2]$$

The waste products of the reactions are, in the first case a sodium salt and in the second case an alcohol (if necessary, the waste products can be washed off, after gel setting).

The silicic acid tends to polymerize according to the reaction:

$$\text{-Si-OH} + \text{HO-Si} \rightarrow \text{Si-O-Si} + H_2O$$

Polymerization proceeds via formation of rings to which monomers add to form dense particles (silica beads) leaving OH groups outside (*Figure 2a*). The size and charge of these particles are pH-dependent. Above the iso-electric point (at pH 2.0), the particles are negatively charged. Gelation easily occurs in the pH 2–7 range; according to the pH value and the salt content, the silica beads either grow (*Figure 2b*) or aggregate (*Figure 2c* and *2d*) to form branched chains and then a 3D network. This process is accelerated by increasing the temperature. A detailed description of the whole process is given in ref. 7.

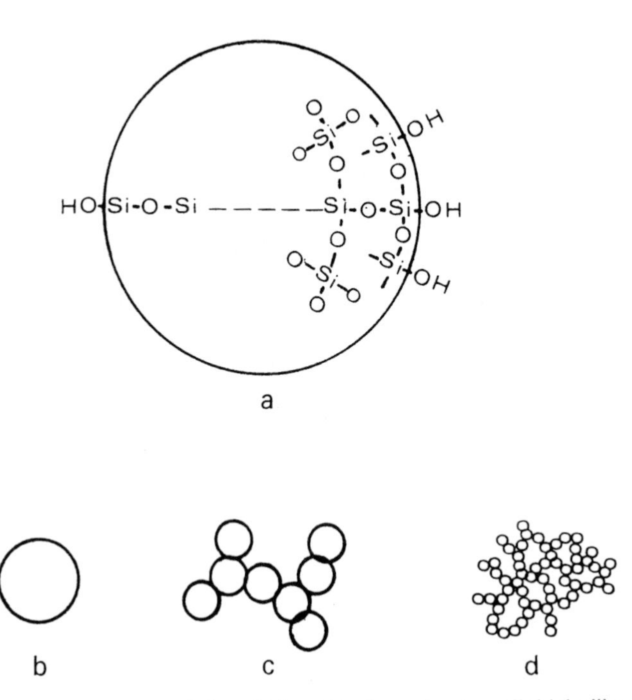

Figure 2. (a) The polymerization of the silicic acid gives rise to colloidal silica beads whose internal structure is due to siloxane bonding, leaving silanol groups on the surface. (b) The growth of beads depends on the chemical conditions: (c) network obtained at high pH, (d) network obtained at low pH.

2.2 Gel properties related to crystal growth

2.2.1 Diffusion of species

Entrapping a growth solution by a gel network prevents the onset of convection which is unavoidable on earth in ungelled solutions: a growing crystal is surrounded by a depletion zone which is less dense than the bulk so that density driven convection shifts the lighter layers upwards. Such movements are not possible in a gel network where mass transfer only proceeds by diffusion. The diffusion coefficients of macromolecules like lysozyme in light gels (i.e. 0.2% (w/v) agarose or 0.4% (w/v) silica) are not significantly changed with respect to diffusion in a gel-free medium. This is not the case for species the dimensions of which come close to the pore size of the gel network. Furthermore an adsorption process can be superimposed to the diffusion process when the diffusing species interacts with the gel.

The diffusion properties of the different gel media are illustrated on interferograms taken during growth of lysozyme crystals in an agarose gel (*Figure 3a*) or a silica gel (*Figure 3b*) (9). Mass transfer proceeds by diffusion from the solution to the crystals and the fringe patterns allow the solute

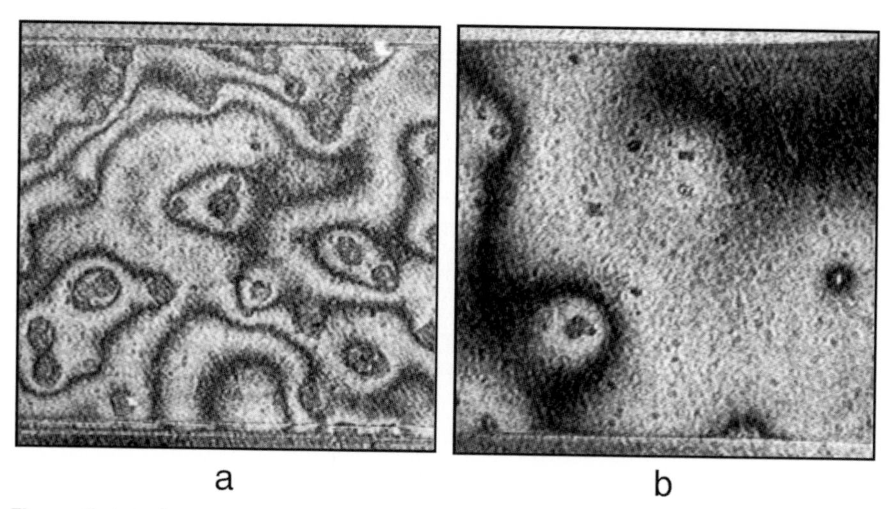

Figure 3. Interferograms taken during growth of lysozyme tetragonal crystals (450 microns) (a) after 24 h in a 0.15% (w/v) agarose gel and (b) after 48 h in a 0.81% (w/v) silica gel. The conditions are the following: 25 mg/ml of HEWL with 0.4 M NaCl in 0.1 M acetate buffer solution (pH 4.5) at T = 17 °C for the agarose gel case and T = 11 °C for the silica gel case.

concentration profile around each crystal to be calculated. In silica gels, there are fewer fringes: growth of crystals of similar size does not cause the same concentration decrease. Indeed, by decreasing supersaturation, molecules fixed on the silica gel progressively desorb which limits the solution depletion. Application of diffusion properties are developed in Section 3.2.

2.2.2 Suspension of crystals

Crystals growing in gel do not sediment as they do in free solution; they develop at the nucleation site, sustained by the gel network. For small molecule crystals grown in silica gel, the gel often fissures and forms cusp-like cavities around crystals (10), and a thin liquid film, that reduces contamination risk, separates the crystal from the gel. Such cavities have not been seen in macromolecule crystals.

Recent studies have even shown that silica gel can be incorporated in the crystal network (11) almost without disturbing the crystal lattice (*Figure 4*). Such crystals, that still diffract to high resolution, are mechanically reinforced (they can be manipulated with tweezers or even with the fingers) and are more resistant to dehydration, because the silica gel framework embedded in the crystal lattice slows down water loss due to its hygroscopic properties. This can be beneficial for X-ray diffraction and crystal growth experiments (*Figure 5*) as well as for the future use of protein crystals as a material for technological applications.

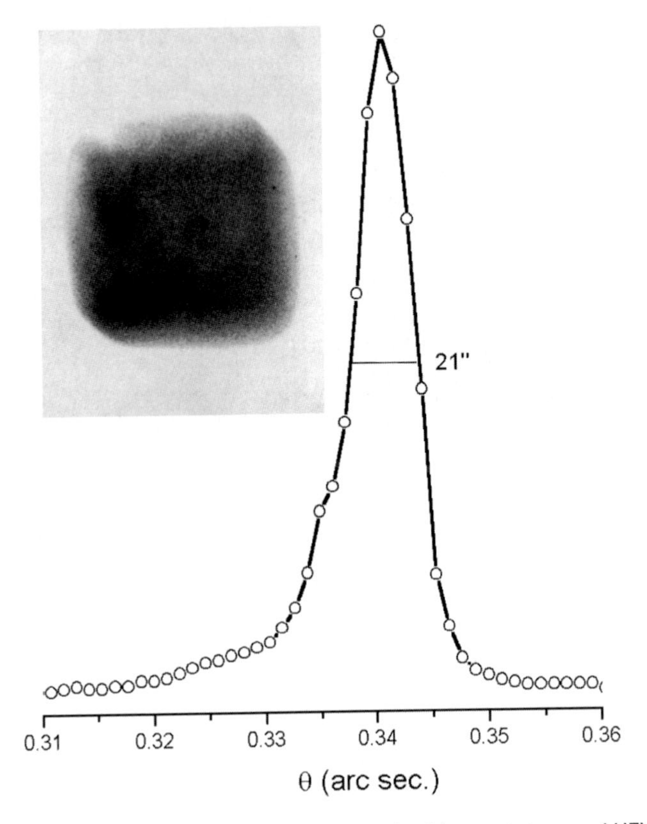

Figure 4. Rocking curve and topography (inset) acquired from a tetragonal HEW lysozyme crystal grown in high concentration gel. Although the gel network is embedded into the crystal lattice, the crystal quality is preserved.

High resolution diffraction has been observed for crystals grown from rather firm agarose gels (12). However the use of light gels is advisable, except for the special cases discussed above, first to make removal of crystals out of the gel matrix easier, and secondly to minimize the gel contamination.

2.2.3 Nucleation inside the gel

Although seeding can be used, it appears that most of the gel-grown crystals are obtained by spontaneous nucleation inside a macroscopically homogeneous gel. When the gel well adheres to the walls of the container (without intercalated liquid film), no nucleation occurs on the cell walls, neither on dust or fibres which have been embedded in the gel. So heterogeneous nucleation is strongly reduced, if not suppressed. Another type of nucleation, namely secondary nucleation, is due to attrition of a previous crystal by the solution flux. It is quite clear that, in gels, this type of nucleation is prevented. One can

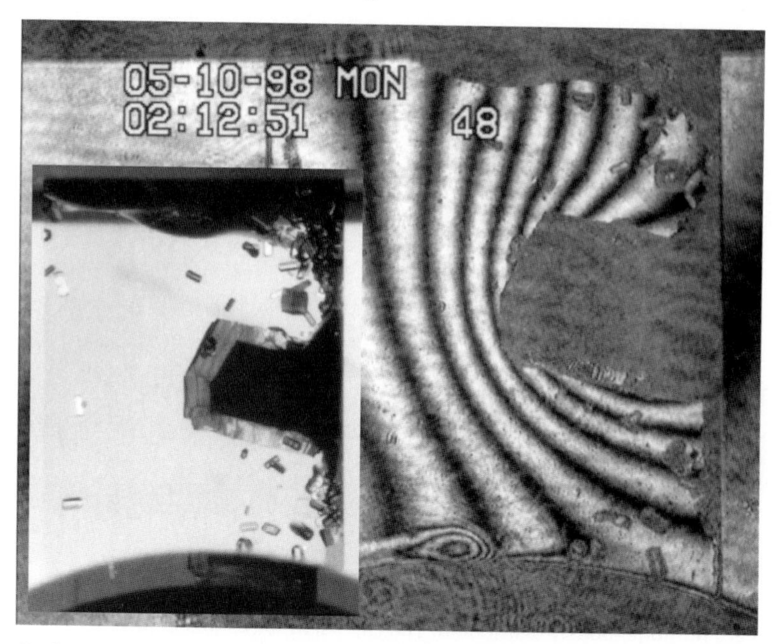

Figure 5. Growth of a tetragonal HEW lysozyme followed by interferometry. This experiment, performed inside a FID APCF reactor (protein chamber is 5 mm wide), was intended to study the depletion zone around a growing crystal. In this experiment, the presence of overlapping depletion zones around several small crystals degrades the data quality, so we implemented a seeding set-up using a large reinforced crystal grown in high concentration silica gel as seed. The seed crystal was glued to one of the walls of the reactor, benefiting from their improved mechanical properties. This image corresponds to a preliminary on-ground experiment during the preparation of the mission. The depletion zone around the crystal in on-ground experiments was not observed because of the homogenization of the solutions inside the reactor by buoyancy convection. To solve this problem, low concentration agarose gel was set inside the growth chamber after gluing the seed. Using this assortment of gel techniques we were finally able to follow on-ground the evolution of a large diffusive depletion zone around an isolated large protein crystal. The inset at left shows a picture of the experiment in which the seed (dark part inside the crystal) and the overgrown volume (clear part around the seed) can be clearly seen.

even take advantage of the absence of convection to apply feeding techniques (13): a simple one consists of putting on the gelled droplet containing a growing crystal a small droplet of protein solution at a concentration higher than inside the droplet. In solution, such a procedure gives rise to a shower of secondary nuclei which is not the case in gels where a unique crystal keeps growing.

When nucleation occurs inside the gel, one observes that all the crystals appear at the same time and consequently have about the same size; they are homogeneously distributed in the whole volume (*Figure 6*). This is not the

Figure 6. Example of a nucleation of tetragonal lysozyme crystals in agarose gel; the crystals are homogeneously distributed. The capillary diameter is 2 mm.

case in gel-free solutions. One generally observes that, in silica gels, the number of nucleated crystals decreases by increasing the gel content. This has been shown with a large variety of macromolecules (14). In agarose gels, on the contrary, the number of crystals increases by increasing the gel content (15, 16).

The influence of the gel media on the nucleation rate cannot be explained by changes of solubility values as far as a solubility curve relates equilibrium between a crystalline phase and a solution. However, supersaturation can be lowered as observed in silica gels where protein molecules can adsorb on the gel surface. Small angle neutron scattering (SANS) spectra of proteins, differing markedly from those corresponding to the gel-free solution, account for this effect. So, with HEW lysozyme solutions at pH 4.5, part of the protein molecules are adsorbed on the gel through electrostatic and H-bond interactions, which reduces the content of free molecules remaining in solution. The concentration of protein adsorbed increases by increasing the total protein content (until binding sites on the gel surface are covered) and by increasing the supersaturation (through protein–protein interaction) (see e.g. *Figure 5* in ref. 6). The latter process is reversible. Thus, reduction of the nucleation rates is simply explained by a reduction of the actual supersaturation. One can counterbalance this effect by using mixed silica gels (TMOS and MeTEOS) (Section 3.1.1), which increases the nucleation rate.

In agarose gels, the initial free protein concentration is the same as in gel-free solution. Differences in the SANS signals of gelled and gel-free HEW lysozyme solution are only visible in the very small scattering vector range. These signals are related to the presence of aggregates (in the 100 nm range): when the protein solution is trapped in agarose gel the signal is enhanced (5). For macromolecular crystals, it is assumed that nucleation could occur via restructuring of amorphous aggregates (17, 18). Here the concomitant observations of enhanced aggregation and enhanced nucleation, both increasing with the gel content, supports this hypothesis. As large aggregates cannot sediment in gel, they are maintained in the whole bulk as shown by refractive index measurements as a function of time during the nucleation process (19). An opposite effect is visible in solution under normal gravity or hypergravity conditions (Section 5.2).

From a practical point of view, use of either silica or agarose gels is interesting because situations exist where, in solution, nucleation is either too abundant or too scarce.

2.2.4 Parameters influenced by the presence of a gel

Taking into account the above considerations, one can select from the list of parameters influencing biological macromolecule crystal growth (see Chapter 1, *Table 1*) those which are influenced by the presence of a gel structure. They are parameters either related to the supply of reactants or related to the mechanical behaviour of the solid or liquid phases.

A priori, gels are not expected to improve crystal growth with regards to biochemical parameters like purity or degree of denaturation of the macromolecule. However, this assertion is refuted by a comparative study (in gel and in solution) on the effect of contamination by a parent molecule (20, 21). Indeed, it was shown that crystals of good quality can be obtained, even when contamination levels are much higher in the gel than in free solution. In current crystallization conditions, impurities are either rejected or incorporated in the crystal. Consequently, the impurity concentration nearby the interface increases (or decreases) with respect to the concentration in the bulk. In free solution, on earth, convections provoke fluctuations of the impurity content at the interface resulting in time-dependent incorporation of impurity in the crystal (growth striations). In gels, due to the supply of solute by diffusion, such fluctuations are damped; furthermore, with a slow growth, foreign molecules or molecules having a distorted conformation can be rejected from the growth interface instead of being buried in the crystal network. In silica gels, the adsorption–desorption process could also act as a purification process, assuming that ill-folded molecules are more strongly bound to the gel surface.

3. Practical consideration

Gel growth is a particular case of solution growth so that it must always be considered downstream with respect to classical solution growth. It results that the same chemical components would be chosen among those which have given the best results in solution. Then using gels, one tries to offset the drawbacks encountered during these first trials. A first possibility is to nucleate and grow the crystals inside the gel. This implies that the protein solution is either gelified or brought into a gel previously set. The different procedures are detailed in Section 3.2.1. A second possibility is to use the gel as a diffusion medium to monitor the supply of reactants, the crystal growing outside the gel. This technique, known as gel acupuncture method, will be described in Section 3.3.1.

3.1 Gel preparation

3.1.1 Silica gel

i. Preparation of silica sols

Silica sols can be prepared either from hydrolysis of a siloxane or by neutralization of sodium metasilicate.

Protocol 1. Preparation of 2% (w/v) silica sol

Equipment and reagents

- Tetramethoxysilane (TMOS)
- Methyltriethoxysilane (MeTEOS)
- 0.5 M sodium metasilicate solution
- 2 M HCl solution
- pH meter
- Magnetic stirrer

A. *Hydrolysis of siloxane*

1. Tetramethoxysilane or tetraethoxysilane are liquids very soluble in alcohol but not much in water. Add drop by drop 1 ml TMOS to 20 ml buffer solution (e.g. 0.1 M acetate buffer solution). Dissolution occurs through a vigorous stirring of siloxane droplets in water (*Figure 7a*); this emulsifying provides a large contact surface, which allows the dissolution of a small amount of siloxane. The reaction proceeds according to *Equation 2* (Section 2.1.2) and methanol is progressively released which makes easy the complete siloxane dissolution. This process consumes 12.5 ml water and releases 54.5 ml methanol per litre of solution. This must be taken into account to know the final growth solution composition.
2. The homogenization step must be achieved as quickly as possible because it competes with the polymerization reaction step, which begins as soon as monomers are available in the medium. It is possible to delay the polymerization step by keeping the mixture in a water/ice bath (*Figure 7b*).
3. The lower the pH, the more rapid is homogenization.
4. The mixture first looks like an oily emulsion in water, then, when no parasitic reaction occurs, it becomes clear and homogeneous.

5. One can reduce the interactions between protein and silica gel by adding some amount (~ 30% in weight) of methyltriethoxysilane, MeTEOS, $(C_2H_5O)_3Si\text{-}CH_3$ (for which the hydrolysis reaction is similar) to the TMOS so that $Si\text{-}CH_3$ groups are substituted to Si-OH groups on the gel surface (6).

Siloxanes are corrosive liquids. Careful protection of skin and eyes are recommended. All vessels in contact with them must be thoroughly rinsed with alcohol prior to water cleaning.

B. *Neutralization of sodium metasilicate*

1. Fill a burette with 0.5 M sodium metasilicate set over a container containing 2 M HCl. Add drop by drop the metasilicate up to the desired pH (control the pH with a pH meter or with some coloured indicator on a test sample). A careful and rapid cleaning of the pH electrode is recommended to avoid plugging by silica.
2. Under these conditions, the neutralization reaction releases NaCl at a concentration of 0.66 M.
3. If other Na salts are preferred as crystallizing agents, e.g. nitrate, the corresponding acid must be used for neutralization in the following step.

ii. Polymerization

After having added an aliquot of the silica sol to the solution at the required composition, the preparation is thoroughly mixed (*Figure 7c*). It must remain homogeneous (no flocculation). The mixture is poured in clean, dried crystallization containers and allowed to gelify without mechanical disturbances. The gel must stick to the container cell walls. It can look somewhat opalescent, but without macroscopic heterogeneities such as fissures. Dehydration of the gel surfaces must be avoided, either by sealing them with a minimum air volume enclosed or by closing them in a vessel containing a reservoir of solution giving the suitable vapour pressure (*Figure 7d*).

Gelation time depends on many parameters such as concentration of gelling agent, nature and concentration of species in solution, pH, and temperature. So, at room temperature it can vary from a few minutes at pH 7.0 to several hours at pH 4.0. With thermally stable solutions, one can shorten this time by increasing temperature (typically 40 °C for further use at room temperature). The gel can be considered as set when it resists pouring, though it undergoes some further evolution as shown by light scattering techniques.

3.1.2 Agarose gel

Commercially available agaroses are powders which can be dissolved in water as described in *Protocol 2*. Homogeneous preparations at concentrations higher than 2% (w/v) are difficult to achieve.

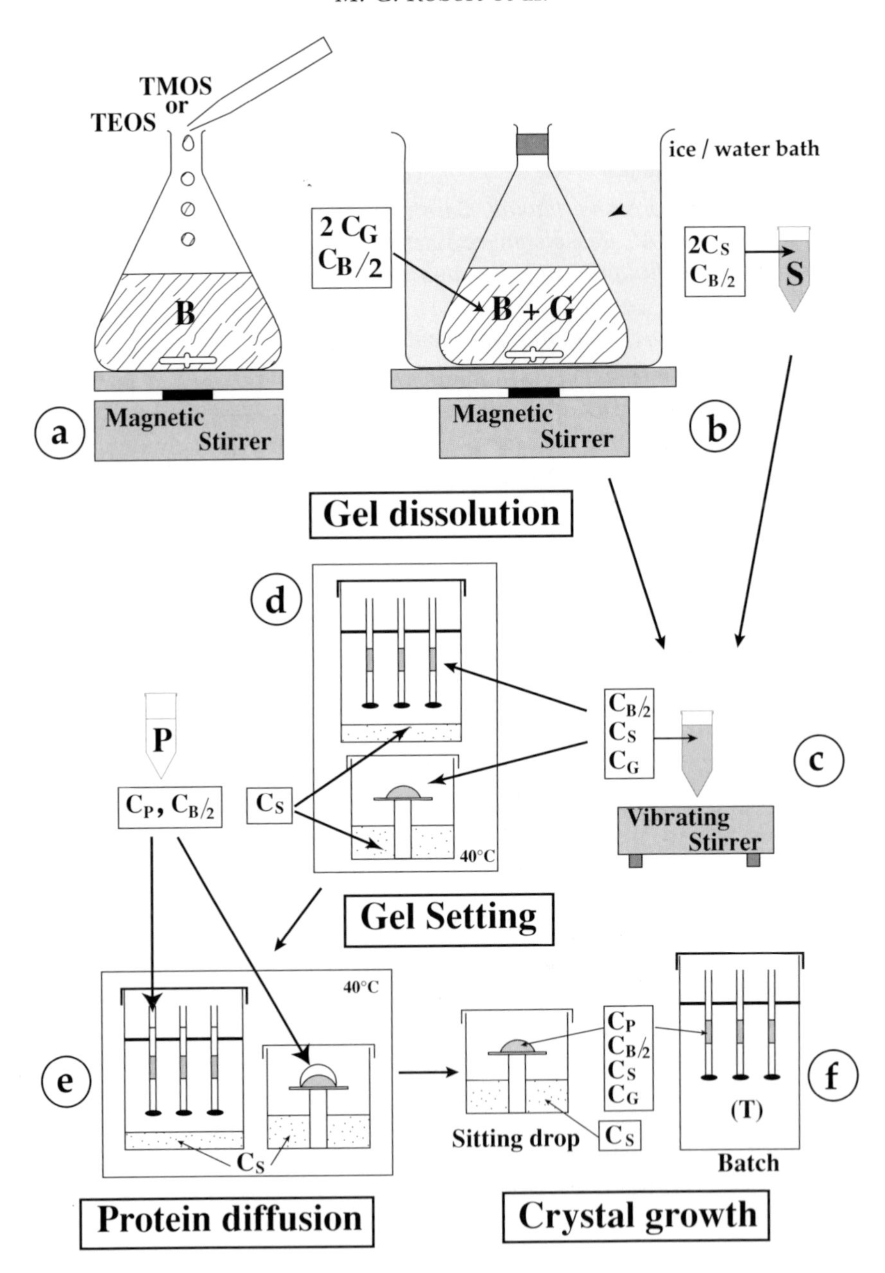

Figure 7. Two-step procedure for crystal growth in a C_G silica gel. After setting the silica gel at T = 40°C, the protein solution diffuses into the gel. Then, the crystal growth occurs under the final concentrations of protein C_P, salt C_S, and buffer $C_{B/2}$.

Protocol 2. Preparation of a 1% (w/v) agarose sol

Equipment and reagents

- Agarose powder
- Magnetic stirrer
- Water-bath

Method

1. Add progressively 0.1 g agarose to 10 ml water at room temperature (not the reverse which could leads to agglomerates) with a slow stirring (*Figure 8a*).
2. Keep stirring for a couple of hours at ambient temperature.
3. Raise the temperature to 100 °C (possibly in a water-bath to avoid overheating) and maintain the slow stirring. The mixture must rapidly become as clear and transparent as water (*Figure 8b*).
4. Keep it on the hot plate (80 °C) until use.

3.2 Gel methods

3.2.1 Crystallization inside the gel: batch method

In order to gelify a protein solution one can either add the protein solution to the agarose (or silica) sol and then carry out the sol-gel processing (*one-step procedure*) or gelify a protein-free sol and then bring the protein into the gel by diffusion (*two-step procedure*). The two procedures are equivalent with agarose gels, because there are no significant interactions between protein and agarose as shown by small angle neutron scattering (SANS) (5). Here, the presence of protein does not perturb the gelation process. This is not the case for silica gels, because protein molecules may adsorb on the gel surface (Section 2.2.3). In that case, all components (protein, salt) and the pH of the growth solution influence the gelation process, and therefore the two-step procedure is advised.

i. Case of silica gels

Due to the strong interactions between protein and silica gel in the presence of salt, the protein solution will preferably be brought into a gel already set.

Protocol 3. Preparation of growth solution with silica

Equipment and reagents

- Glass capillaries
- Silica sol (see *Protocol 1*)
- Water-bath

Protocol 3. *Continued*

Method

1. The silica sol is prepared as explained in *Protocol 1* in a buffer at concentration C_B. It is kept in a water/ice bath to avoid a premature gelation. The different growth settings are presented *Figure 7*.
2. Prepare a growth solution with a protein concentrations C_P, salt concentration C_S, gel concentration C_G.
3. Mix equal volumes of stock salt solution (at concentration 2 C_S) and silica sol (at concentration 2 C_G) (*Figure 7c*).
4. Suck a few microlitres of this preparation in capillaries (or prepare droplets of this preparation on coverslips). Place the capillaries opened at one end (or the coverslip) in a closed vessel containing a reservoir of salt at concentration C_S to avoid dehydration during gelling. Gelling can be accelerated by setting the closed vessel in an incubator at 40°C (*Figure 7d*).
5. (a) For hanging drops. When the gel is set, pour carefully a volume V of protein solution at concentration C_P on the gel surface taking care not to touch it with the pipette (*Figure 7e*).

   (b) For capillaries. Set again the capillaries (or droplets) in the closed vessel. Diffusion of protein in the gel matrix and dehydration of the liquid droplet occurs simultaneously so that the final result is a gelled droplet with buffer, salt, and protein at concentrations C_B, C_S, and C_P respectively.
6. Use the droplets as in usual hanging drop or sitting drop techniques. Seal the capillaries and proceed as for a batch technique, i.e. by keeping them at constant temperature. One can also apply a regulated temperature variation to increase supersaturation.

ii. Case of agarose gel

In the protocol presented on *Figure 8*, the protein is added to the agarose sol before gelation; so, the different components of the sol are kept at a temperature above the gelling point T_G before use. To prepare a growth solution containing gel, protein, salt, and buffer at final concentrations, respectively C_G, C_P, C_S, C_B, one starts from (warm) stock solutions (*Figure 8c*) whose compositions are:

gel:	10 C_G
protein:	2 C_P and C_B
salt:	4 C_S
buffer:	5 C_B

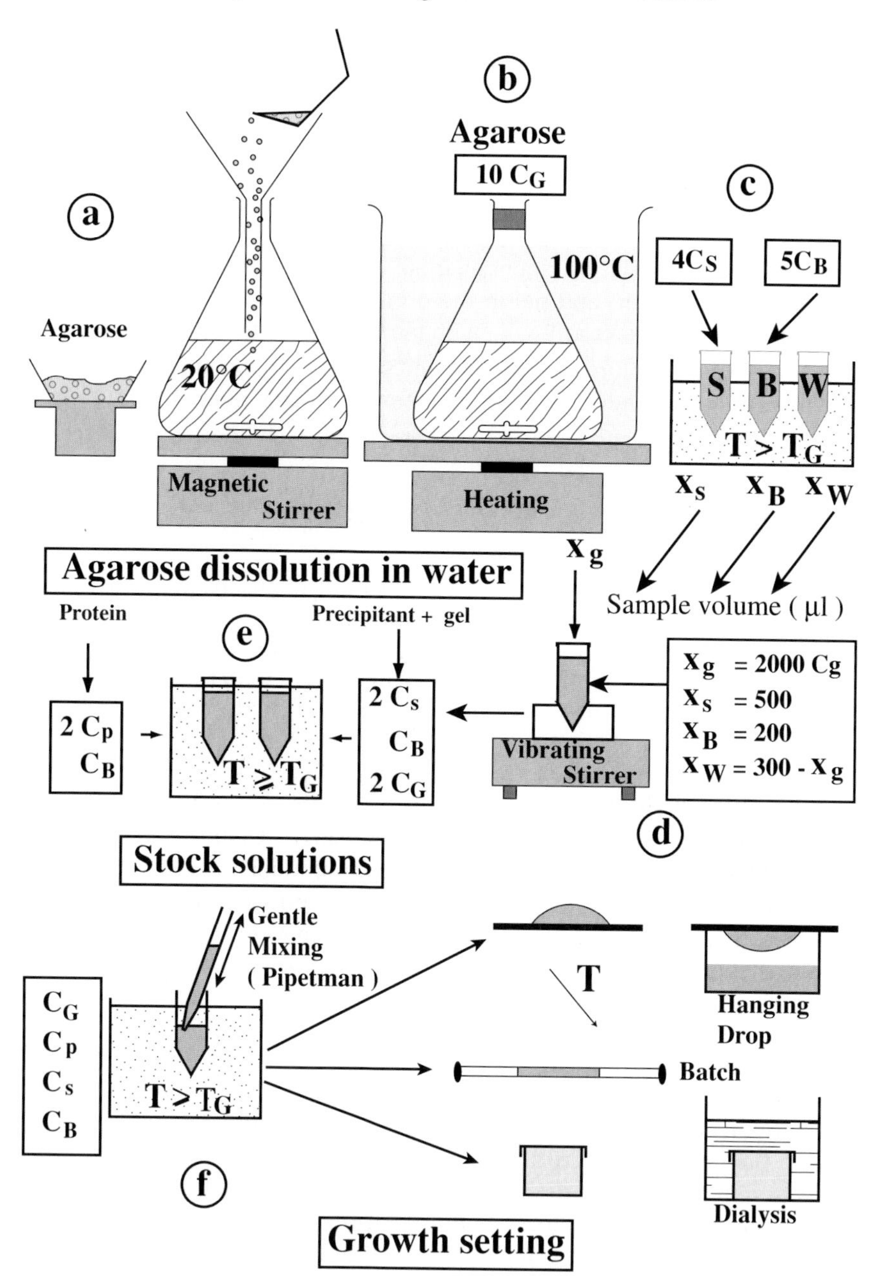

Figure 8. One-step procedure for crystal growth in a C_G agarose gel. The volume proportions are given in order to have final concentrations of protein C_P, salt C_S, and buffer C_B. For calculating X_g, take C_g in %. Agarose sol uptake needs reverse pipetting.

Protocol 4. Preparation of growth solution with agarose

Equipment and reagents

- Thermostated water-bath
- Agarose sol

Method

1. Prepare a crystallizing agent/agarose solution by thoroughly mixing salt, buffer, water, and gel in the proportions indicated in *Figure 8d*. This solution is kept at $T > T_G$ with the protein solution (*Figure 8e*).
2. Sample equal volumes of protein and crystallizing agent/agarose solutions in an Eppendorf tube and gently mix with a Pipetman (*Figure 8f*).
3. Use this preparation as you do for classical solution growth techniques (see Chapter 4).
4. Decrease the temperature under T_G to allow the gel to set.

To avoid a temperature-induced denaturation of the protein, one has to select an agarose of medium gelling point. Due to the 50 °C hysteresis associated with the gel-sol process, once the gel has set, one can use it at a temperature higher than T_G without altering the gel structure. For example, one can grow orthorhombic HEW lysozyme crystals at 40 °C with an agarose of 36 °C gelling point.

3.2.2 Counter-diffusion

All techniques currently used to grow protein single crystals (see Chapter 5) can be implemented using gels instead of free solutions in order to minimize convective flow and to avoid movements of the growing crystals, including sedimentation. Among these techniques, those in which there is a continuous change of supersaturation in space (through the growth reactor) and time (during the experiment) are of interest. This is particularly the case when they are forced to work out of equilibrium allowing the self-search for the best crystallization conditions. Because of the geometry and mass transport involved, they are termed counter-diffusion techniques.

In counter-diffusion techniques, the interacting solutions are placed one in front of the other either in direct contact (free interface diffusion) or separated by a membrane (dialysis) or by an intermediate chamber working as a physical buffer. By definition, counter-diffusion techniques require avoiding convection. In addition to using gels, two other ways are known to reduce convection: to perform the experiments under microgravity conditions in space and/or to perform the experiment inside narrow volumes, for instance glass capillaries. As shown in *Figure 9*, all the three implementations share the same geometry. Note that the term 'buffer chamber' for the intermediate chamber has a physical meaning, i.e. a chamber that slows down the transport

1) Under microgravity

Free interface diffusion

Precipitant | Buffer | Protein solution

Dialysis method

Dialysis membrane

Precipitant | Protein solution

2) On Earth

a) In gelled media

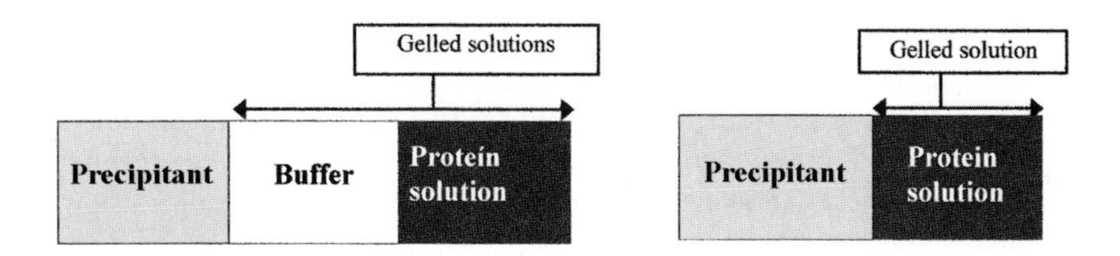

b) Into capillaries

for instance: Gel acupuncture method

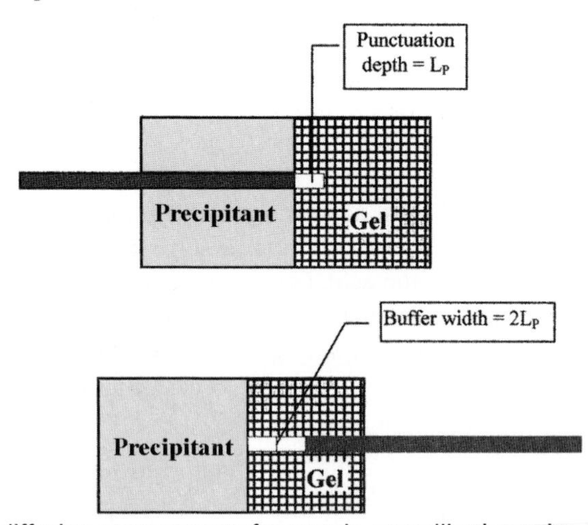

Figure 9. Counter-diffusion arrangements for protein crystallization using three different implementations sharing the same geometry. In the gel acupuncture method the physical buffer has a length two times the punctuation depth as this is the length of the diffusive path separating salt and protein reservoirs.

process, being filled either with a chemical buffer, with a porous network or with any other fluid such as water. Batch crystallization into capillary volumes was used by Feher *et al.* (22) to illustrate diffusional transport in protein crystallization. The use of capillaries for counter-diffusion methods dates back to the work of Zeppezauer *et al.* (23), and others reviewed by Phillips (24) (see also Chapter 5). It seems that these methods using capillaries were designed with the aim of reaching the critical supersaturation for nucleation very slowly, looking for a single nucleation event. Attempts to make use of Otswald ripening processes were also considered (25). Unlike these, recent studies have tried, starting from conditions far enough from equilibrium, to search for multiple nucleation events under conditions progressively approaching equilibrium. To illustrate the method we can use a simple technical implementation, the gel acupuncture technique, stressing that our discussion also applies for counter-diffusion arrangements using gelled protein solutions and microgravity experiments.

3.2.3 Crystallization outside the gel: gel acupuncture method

The gel acupuncture method is based upon the properties of gels, which are used to act as the mass transport medium for the precipitating agent and also to hold capillaries containing the ungelled protein solution. The experimental set-up is as simple as shown in *Protocol 5* (see ref. 26 for specific recipes to crystallize several proteins).

Protocol 5. A recipe to grow crystals of lysozyme by the gel acupuncture technique

Equipment and reagents

- Sodium silicate
- Agarose
- 1 M acetic acid

Method

1. Mix the sodium silicate as commercially supplied with four parts of water. Mix slowly under continuous stirring 12.5 ml of this solution with 10 ml of 1 M acetic acid. For this step, you can use a 50 ml vessel. In a few hours the silica gel is set with a pH about 5.8–6.
2. Prepare the following solutions while the gel sets down:
   (a) 20% (w/v) sodium chloride: pour 4.5 g of salt into 22.5 ml of water and stir until complete dissolution.
   (b) 100 mg/ml protein solution: weigh 100 mg of lysozyme into a small tube (e.g. an Eppendorf tube) and pour 1 ml of bidistilled water. Stir gently until complete dissolution.
3. Fill the capillaries with the protein solution. Introduce one of the ends of the capillary into the protein solution. You will see that the solution

flows up by capillarity. Once it reaches a level 1 cm from the other end of the capillary, remove it from the solution. The solution holds inside the capillary. Then seal the upper end of the capillary with a small piece of Plasticine. The next step is to punch the capillary in the gel layer (but be sure that the gel is set!). Insert the capillary into the gel about 0.5–1 cm, enough to maintain it straight.

4. Pour the solution of salt (22.5 ml) onto the gel layer and cover the experiment with a large vessel turned upside down.

In a typical experiment, the crystallizing agent and the protein start to counter-diffuse through the porous gel network used to hold the capillary (*Figure 10*). As the diffusion constant of macromolecules is one or two orders of magnitude smaller than that of the small molecules of the crystallizing agent (1), the latter reaches the open side of the capillary and starts to diffuse up into it while almost all the protein molecules are still in the capillary. This creates a set of supersaturation conditions $s(x, t)$ that changes in time t for every location in the capillary x. Eventually, at a given (x, t), the critical supersaturation value s^* for nucleation is achieved and protein starts to precipitate inside the capillary. The development of supersaturation in time and space is rather complex and needs to be studied by numerical methods (see ref. 27 for a computer simulation of the problem), but the basic behaviour can be briefly outlined.

As the experiment evolves (in time), the salt diffuses up through the capillary. Then the system experiences precipitation phenomena which take place at different values of both supersaturation and rate of supersaturation testing a wide range of plausible precipitation conditions along the capillary. The nucleation events take place under conditions very far from equilibrium (amorphous precipitation) at the entry of the capillary and then under conditions slightly closer to equilibrium (polycrystalline precipitation) further along the capillary because of a lower gradient of the reactants. Thus, as time advances, the precipitation system slowly approaches equilibrium iteratively, experiencing nucleation events, which yield successively fewer and larger crystals. In the middle and upper part of the capillary, few large and well-faceted single crystals form and, under optimum starting conditions, a single crystal completely filling the capillary diameter can be obtained (28) (*Figure 11*). In short, as soon as the protein molecules start to precipitate, the system behaves like a saturation ($\beta = C/C_s$, see Chapter 11) wave leading to complex precipitation events in space and time. In fact, the typical output of any counter-diffusion arrangement is a patterned precipitation in space and time, the archetype pattern being the so-called Liesegang rings precipitation (10).

For practical purposes, the technique can be used to grow protein crystals with most of the classical crystallizing agents (including PEG) (29) and it

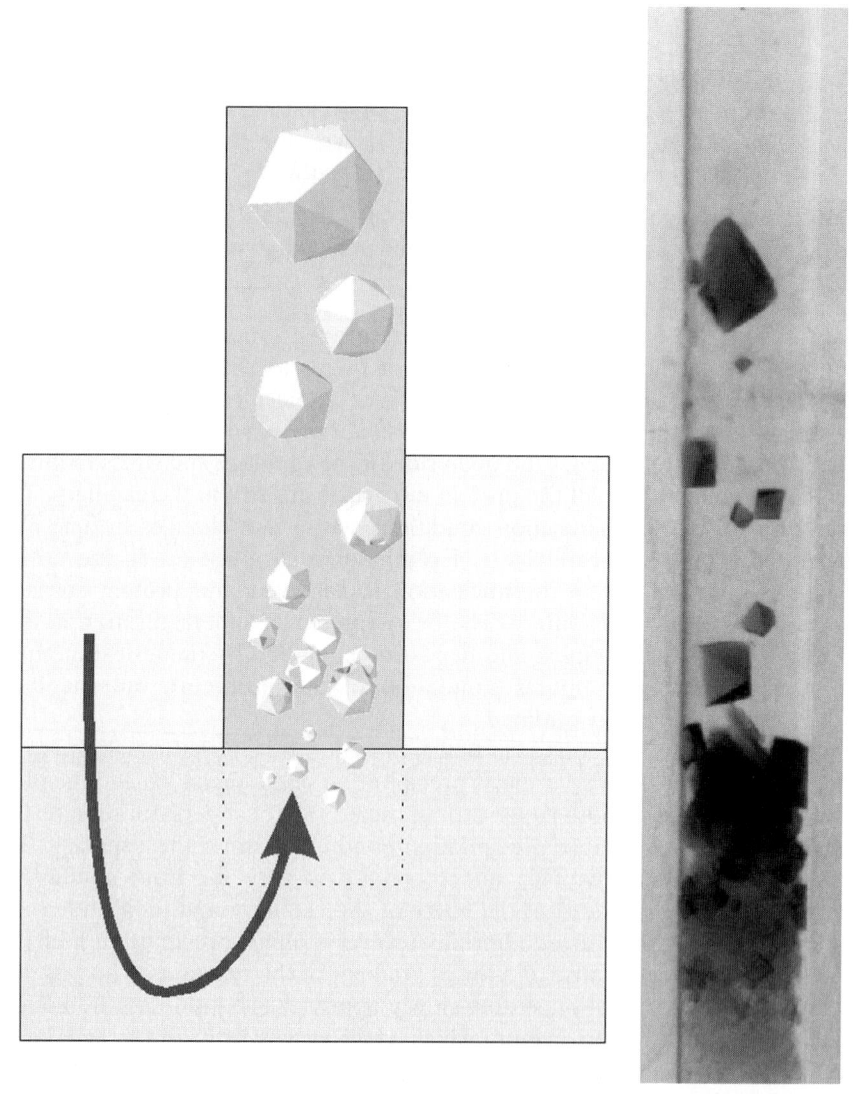

Figure 10. Protein crystallization by the gel acupuncture method. The diffusive path of the salt towards the capillary filled with protein solution is illustrated as well as the crystal size distribution obtained along the capillary. Picture at right shows an actual ferritin crystallization experiment using the gel acupuncture method.

consumes reasonable amounts of protein (5–50 mg). The main variables affecting the crystallization behaviour in the gel acupuncture method are:

- punctuation depth
- initial protein concentration
- initial crystallizing agent concentration.

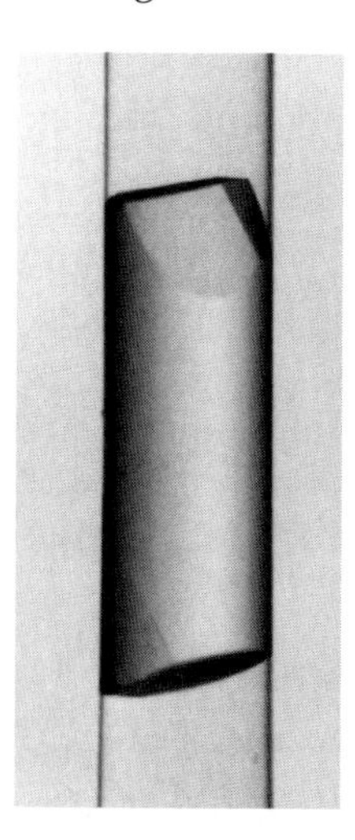

Figure 11. Rod-shaped crystal of tetragonal HEW lysozyme obtained by gel acupuncture method. The isolated crystal was grown until it completely filled the capillary and then growth continued at both ends of the cylinder. Capillary diameter 0.3 mm. Protein concentration 100 mg/ml. Salt concentration 10% (w/v) pH 4.5.

In the search for single crystals by this method, large protein concentrations are recommended to start with. From the point of view of classical crystallization methods, this can be surprising. However, it should be realized that in the gel acupuncture method the precipitation system itself searches for the best crystallization conditions. Thus the idea, when using large protein concentrations, is to trigger the nucleation of an initial amorphous precipitate in the lower part of the capillary and then leave the system to search for optimal growth conditions which typically occurs in a time scale of days. The crystallizing agent concentration is another important variable. A very high initial concentration will exhaust the protein in the capillary with amorphous precipitation, while a very low initial concentration will produce a batch-type precipitation behaviour due to the small salt gradients inside the capillary. Thus, an intermediate concentration, in the range known to precipitate the protein, is recommended. Finally, the punctuation depth affects the waiting time and the overall salt gradient inside the capillary. The suggested value of 8 mm has been experimentally found to be a good compromise between waiting time (longer for higher punctuation depth values) and mechanical stability of the capillary.

The gel acupuncture technique has been demonstrated for proteins of different types and diverse molecular weights. To work properly, the technique must avoid sedimentation of the crystals, as well as the buoyancy driven convection created as soon as the protein concentration falls in the lower part of the capillary. It is recommended to use, as the reservoir to be filled with gel, rectangular boxes made of two glass plates separated by a rubber frame hold with clips (30, 31). With this simple arrangement, the capillaries can be orientated perpendicular to the gravity field. Nevertheless, very large (up to 10 mm) crystals have already been obtained with this technique, showing a high diffraction resolution limit and very low mosaicity (1.2 Å and 9 arc second, respectively, for tetragonal HEW lysozyme crystals) (32, 33). These results are expected to be enhanced in future with *in situ* measurements during the growth process, ensuring mechanical stability (34). Finally, use of the technique for preparation of heavy-atom derivatives and search of crystallization conditions by pH variation (a method advocated by McPherson) (35) also seems promising.

4. Crystal preparation and characterization

Crystals grown in capillaries are ready for use but, to reduce diffuse scattering, it is recommended to pump out the gel surrounding the selected crystals using micropipettes or some filter paper wick. This is especially easy with light gels, which behave as viscous media.

For crystals grown in droplets or in dialysis buttons, the same procedure can be followed after introducing a capillary in the gel to suck up the selected crystal together with its surroundings. This avoids any direct contact of the crystal with the capillary.

The crystalline quality of different biological macromolecules crystals grown in agarose and silica gels has been characterized by measuring their resolution limit and mosaic spread. These studies concern lysozyme crystals (3, 11, 12, 36), but also crystals of higher molecular weight substances (37).

As to resolution, it appears that gel-grown crystals are, on average, better than solution-grown ones. In a few cases, comparison was made with space-grown crystals and plots of average I/s(I) values versus resolution show that the best diffraction data collected from gel-grown crystals lie in between those of earth-grown and space-grown crystals (3, 36). As to mosaic spread, differences have also been evidenced. In particular, misorientations between the different domains of a same crystal are generally less important in gel-grown crystals. Up to now, the best results have been obtained with silica gel.

5. Related methods

5.1 Microgravity

Space experiments share with growth in gels the ability to reduce buoyancy-driven convection, to reduce impurity concentration on the crystal surface, and to avoid sedimentation of crystals as well as the secondary nucleation of 3D protein clusters. In addition, the microgravity scenario removes the plausible chemical interaction of the gel with the reactants used in the chemical protocol, including the protein itself. In short, microgravity conducted experiments may be considered as 'clean' gel experiments.

As with gels, all techniques used in protein crystallization can be implemented under microgravity. A diverse range of facilities, covering vapour diffusion and liquid counter-diffusion techniques, are currently offered by several Space Agencies to grow single protein crystals under microgravity conditions (see refs 38–40 for a full description of facilities).

Because of the youth of microgravity science and the limited number of opportunities to fly experiments, space crystallization of biological macromolecules still faces a number of technical and conceptual problems, which need to be solved.

After ten years of microgravity-conducted experiments, some improvements of crystal size and crystal quality have been reported (41–43, see ref. 38 for a review), and in a few cases reduced mosaicity was reported. Often, however, the quality of the space-grown crystals, as evaluated by Wilson-type plots or mosaicity measurements, does not show a dramatic increment of quality, especially in terms of limit of resolution.

Current evaluation of space crystallization is basically performed by comparing space-grown crystals and crystals grown on earth using the same type of reactor and the same crystallization conditions. Considering that the typical dimensions of the reactors are large enough to allow density-driven and thermal convection on earth, it is evident that space-grown crystals would be expected to be of higher quality. In any case, what needs to be explained is why in some cases the reverse result was found. In the future, space crystallization has to face comparison with on-ground techniques emulating microgravity conditions, that is crystal growth within gels and/or capillary volumes.

The similarity between the geometry and mass transport properties of the gel acupuncture technique and the microgravity facilities is strong enough (*Figure 9*) to permit an extrapolation of the above discussion on supersaturation spatiotemporal evolution (see Section 3.2.3). The immediate advantage of microgravity conducted experiments on capillary methods is that, for the same path length, larger volumes of protein solutions can be used, which could yield larger crystals. Unfortunately, the typical linear dimensions in the direction of the diffusion path in the facilities currently available are too short (15 mm for PCF or 8 mm for APCF) to exploit the advantage of convection free counter-diffusion in large volumes. Therefore the typical and useful spatial heterogeneity of the counter-diffusion techniques is lost and only a limited set of the wide range of crystallization conditions that could be reached is tested in practice. This enforces the use of lower concentrations in all the experiments performed to date, converting the free interface diffusion and dialysis experiments into batch experiments because the characteristic time for the nucleation and transport processes are similar (44, 45). It is clear therefore that the use of longer protein chambers will permit the exploration of a larger set of local growth conditions.

An early criticism of space crystallization (46) was that the limited number of opportunities to fly experiments makes it impossible to employ the 'trial and error' methodology used so far on earth in the search for crystallization conditions. However, this restriction has, in fact, been a major driving force for the development of the current trend to rationalize protein crystal growth, as the advance in the understanding of fundamental aspects of protein crystallization (including the knowledge of the growth environment and crystal quality evaluation) directly derived from space-induced research has been substantial. Today, it seems evident that such a rationale will be obtained only by coupling on-ground and space crystallization research. In particular, the numerical simulation of mass transport and precipitation phenomena (27, 44–50), the development of growth techniques emulating microgravity conditions on-ground, the characterization of nucleation phenomena (51, 52), growth mechanisms, and surface kinetics (53) will be required in the future. In addition, the search for a relationship between growth history and crystal quality requires the improvement of the existing X-ray characterization tools (54, 55).

Finally, to evaluate properly the future of protein crystallization in space, it should be considered that beside producing crystals of improved perfection (with reduced mosaicity), another current interest is basically motivated by the need for high quality crystals larger than the size (tenths of millimetre) required for structural studies. The crystallization of biological macromolecules will face in the future the optimization of the final crystal size for purposes not only related to structural biology (e.g. neutron diffraction measurements) but also the characterization of their physical properties because their use for technological application is still an unexplored and exciting field. To grow large macromolecular single crystals, microgravity offers the best scenario without the limitations of capillary volumes, but first we need to learn how to properly control impurity distribution and its effect on the cessation of growth.

5.2 Hypergravity

Contrary to microgravity experimentation, kilogravity experimentation in centrifuges is a rather unexplored field, although successful results were reported as early as 1936 for tobacco mosaic virus (56).

In a review on this subject (57), Schlichta emphasized that gravity might be regarded as a variable in crystal growth and material processing. Besides convection and sedimentation which are drastically altered by increasing gravity forces, one has also to consider the increased compressive and shear stresses, as well as hydrostatic pressure (which can induce solubility variations) (58).

All these factors influence directly or indirectly the growth parameters. For example, due to forced sedimentation effect, supersaturated zones appear in an initially undersaturated solution, leading to crystal nucleation and growth (59). However, the role played by the gravity field may promote additional effects:

(a) Centrifugation can separate the molecules having different molecular weights (foreign macromolecule, impurities) or different conformer.
(b) Centrifugation will progressively modify the spatial distribution of the different populations of oligomeric species or aggregates or interacting monomers contained in a supersaturated macromolecular solution. Thus, one could set some local solution compositions suitable for nucleation.

The devices needed for hypergravity experiments are not necessarily complicated: the crystal growth presented in ref. 59 required a centrifuge currently available in any biochemistry laboratory.

More fundamental studies need ultracentrifuges equipped with observation set-ups such as Schlieren optics. As a matter of fact, the development of centrifugal crystal growth is related to the development of basic studies. A current feeling is that crystal grown under kilogravity would suffer from plastic deformation and would not diffract at very high resolution. This does not seem valid, at least considering the few examples found in the literature (57, 60–61).

References

1. Robert, M.-C. and Lefaucheux, F. (1988). *J. Cryst. Growth*, **90**, 358.
2. Vekilov, P. G., Alexander, J. I. D., and Rosenberger, F. (1996). *Phys. Rev. E*, **54**, 6650.
3. Miller, T. Y. and Carter, D. C. (1992). *J. Cryst. Growth*, **122**, 306.
4. Sica, F., Demasi, D., Mazzarella, L., Zagari, A., Capasso, S., Pearl, L. H., *et al.* (1994). *Acta Cryst.*, **D50**, 508.
5. Vidal, O., Robert, M.-C., and Boué, F. (1998). *J. Cryst. Growth*, **192**, 257.
6. Vidal, O., Robert, M.-C., and Boué, F. (1998). *J. Cryst. Growth*, **192**, 271.
7. Iler, R. K. (1979). *The chemistry of silica*. Wiley. Interscience, New York.
8. Ramzi, M. (1996). PhD Thesis, Université de Strasbourg.
9. Vidal, O. (1997). PhD Thesis, Université de Paris.
10. Henisch, H. K. (1988). *Crystal growth in gels and Liesegang rings*. Cambridge University Press.
11. García-Ruiz, J.-M., Gavira, J. A., Otalora, F., Guasch, A., and Coll, M. (1998). *Mat. Res. Bull.*, **33**, 1593.
12. Vidal, O., Robert, M.-C., Arnoux, B., and Capelle, B. (1999). *J. Cryst. Growth*, **196**, 559.
13. Bernard, Y., Degoy, S., Lefaucheux, F., and Robert, M.-C. (1994). *Acta Cryst.*, **D50**, 504.
14. Cudney, B., Patel, S., and McPherson, A. (1994). *Acta Cryst.*, **D50**, 479.
15. Provost, K. and Robert, M.-C. (1991). *J. Cryst. Growth*, **110**, 258.
16. Thiessen, K. J. (1994). *Acta Cryst.*, **D50**, 491.

17. McPherson, A., Malkin, A. J., and Kuznetsov, Y. G. (1995). *Structure*, **3**, 759.
18. Georgalis, Y., Umbach, P., Raptis, J., Saenger, W. (1997). *Acta Cryst.*, **D53**, 691 and 703.
19. Vidal, O., Bernard, Y., Robert, M.-C., and Lefaucheux, F. (1996). *J. Cryst. Growth*, **168**, 40.
20. Provost, K. and Robert, M.-C. (1995). *J. Cryst. Growth*, **156**, 112.
21. Hirschler, J. and Fontecilla-Camps, J. C. (1996). *Acta Cryst.*, **D52**, 806.
22. Feher, G. and Kam, Z. (1985). In *Methods in enzymology* (eds H. W. Wyckoff, C. H. W. Hirs, and S. N. Timasheff), Academic Press, Inc., London, Vol. 114, p. 77.
23. Zeppezauer, M., Eklund, H., and Zeppezauer, E. S. (1968). *Arch. Biochem. Anal.*, **126**, 564.
24. Phillips, G. N. (1985). In *Methods in enzymology* (eds. H. W. Wyckoff, C. H. W. Hirs, and S. N. Timasheff) Academic Press, Inc., London. Vol. 114, p. 128.
25. Weber, P. (1991). *Adv. Protein Chem.*, **41**, 16.
26. García-Ruiz, J.-M., Moreno, A., Otálora, F., Vieoma, C., Rondon, D., and Zautscher, F. (1998). *J. Chem. Educ.*, **75**, 442.
27. Otálora, F. and García-Ruiz, J.-M. (1996). *J. Cryst. Growth*, **169**, 361.
28. Moreno, A., Rondón, D., and García-Ruiz, J.-M. (1996). *J. Cryst. Growth*, **166**, 919.
29. García-Ruiz, J.-M. and Moreno, A. (1994). *Acta Cryst.*, **D50**, 484.
30. García-Ruiz, J.-M., López Martínez, C., and Martín-Vivaldi Caballero, J. L. (1985). *Cryst. Res. Techn.*, **20**, 1615.
31. García-Ruiz, J.-M. and Moreno, A. (1997). *J. Cryst. Growth*, **178**, 393.
32. Otálora, F., García-Ruiz, J.-M., and Moreno, A. (1996). *J. Cryst. Growth*, **168**, 93.
33. Otálora, F., Capelle, B., Ducruix, A., and García-Ruiz, J.-M. (1999). *Acta Cryst.*, **D55**, 644.
34. Otálora, F., Gavira, J. A., Capelle, B., and García-Ruiz, J. M. (1999). *Acta Cryst.*, **D55**, 650.
35. McPherson, A. (1985). In *Methods in enzymology* (eds. H. W. Wyckoff, C. H. W. Hirs, and S. N. Timasheff), Academic Press, Inc., London. Vol. 114, 125.
36. DeLucas, L. J., Long, M. M., Moor, K. M., Rosenblum, W. M., Bray, T. L., Smith, C., *et al.* (1994). *J. Cryst. Growth*, **135**, 183.
37. Lorber, B., Sauter, C., Ng, J. D., Zhu, D. W., Giege, R., Vidal, O., Robert, M.-C., Capelle, B., *et al.* (1999). *J. Cryst. Growth*, in press.
38. Giegé, R., Drenth, J., Ducruix, A., McPherson, A., and Saenger, W. (1995). *Prog. Cryst. Growth Charact.*, **30**, 237.
39. McPherson, A. (1996). *Cryst. Rev.*, **6**, 157.
40. Pletser, V., Stapelmann, J., Potthast, L., and Bosch, R. (1999). *J. Cryst. Growth*, **196**, 638.
41. Koszelak, S., Day, J., Leja, C., Cudney, R., and McPherson, A. (1995). *Biophys. J.*, **69**, 13.
42. Ng, J. D., Lorber, B., Giegé, R., Koszelak, S., Day, J., Greenwood, A., *et al.* (1997). *Acta Cryst.*, **D53**, 724.
43. Snell, E. H., Weisgerber, S., and Helliwell, J. R. (1995). *Acta Cryst.*, **D51**, 1099.
44. Otálora, F. and García-Ruiz, J.-M. (1997). *J. Cryst. Growth*, **182**, 141.
45. García-Ruiz, J.-M. and Otálora, F. (1997). *J. Cryst. Growth*, **182**, 155.
46. Leberman, R. (1985). *Science*, **230**, 370.

47. Wagner, G. and Linhardt, R. (1991). *J. Cryst. Growth*, **110**, 114.
48. Fowlis, W. W., DeLucas, L. J., Twigg, P. J., Howard, S. B., Meehan, E. J., and Baird, J. K. (1988). *J. Cryst. Growth*, **90**, 117.
49. Huo, C., Ge, P., Xu, Z., and Zhu, Z. (1991). *J. Cryst. Growth*, **114**, 486.
50. Savino, R. and Monti, R. (1996). *J. Cryst. Growth*, **165**, 308.
51. Malkin, A. J., Cheung, J., and McPherson, A. (1993). *J. Cryst. Growth*, **126**, 544.
52. Georgalis, J., Schüler, P., Frank, M., Soumpasis, D., and Saenger, W. (1995). *Adv. Colloid Interface Sci.*, **58**, 57.
53. Land, T. A., Malkin, A. J., Kuznetsov, Yu. G., McPherson, A., and Yoreo, J. J. (1995). *Phys. Rev. Lett.*, **75**, 2774.
54. Helliwell, J. R. (1988). *J. Cryst. Growth*, **90**, 259.
55. Chayen, N. E., Boggon, T. J., Casseta, A., Deacon, A. Gleichmann, T., Habash, J. *et al.* (1996). *Q. Rev. Biophys.*, **29**, 227.
56. Wyckoff, R. W. G. and Corey, R. B. (1936). *Science*, **84**, 513.
57. Schlichta, P. J. (1992). *J. Cryst. Growth*, **119**, 1.
58. Lorber, B., Jenner, G., and Giegé, R. (1996). *J. Cryst. Growth*, **158**, 103.
59. Pitts, J. E. (1992). *Nature*, **355**, 117.
60. Behlke, J. and Knespel, A. (1996). *J. Cryst. Growth*, **158**, 388.
61. Barynin, V. V. and Melik-Adamyan, V. R. (1982). *Sov. Phys. Crystallogr.*, **27**, 588.

7

Seeding techniques

E. A. STURA

1. Introduction

A seed provides a template for the assembly of molecules to form a crystal with the same characteristics as the crystal from which it originated. Seeding has often been used as a method of last resort, rather than a standard practice. Recently, these techniques have gained popularity, in particular, macro-seeding, used to enlarge the size of crystals. Seeding has many more applications, and the use of seeding in crystallization can simplify the task of the crystallographer even when crystals can be obtained without it. We will explore the various seeding techniques, and their applications, in the growth of large single crystals and the methods by which we may attempt to obtain crystals that diffract to higher resolution.

Crystallogenesis can be divided into two separate phases. The first being the screening of crystallization conditions to obtain the first crystals, the second consisting of the optimization of these conditions to improve crystal size and quality. Seeding can be used advantageously in both these situations. The first stage in crystallogenesis consists of the discovery of initial crystals, crystalline aggregates, or microcrystalline precipitate. This may result from a standardized screening method (1, 2), a systematic method (3), an incomplete factorial search (see Chapter 4 and refs 4 and 5), or by extensive screening of many conditions. This may be bypassed by starting with seeds from crystals of a related molecule that has been previously crystallized. Molecules that have been obtained by genetic or molecular engineering of a previously crystallized macromolecule fall in this category. This method is termed cross-seeding. It has been used to obtain crystals of pig aspartate aminotransferase starting with crystal from the chicken enzyme (6) and between native and complexed Fab molecules (7).

Whatever the method used to obtain the initial crystals, seeding may provide a fast and effective way to facilitate the optimization of growth conditions without the uncertainty which is intrinsic in the process of spontaneous nucleation. The streak seeding technique can be used to carry out a search quickly and efficiently over a wide range of growth conditions. Later the use of macroseeding and microseeding methods can be used to grow large crystals with a high degree of reproducibility.

2. Seeding

2.1 Supersaturation and nucleation

Details on the use of precipitants, together with general and theoretical considerations, and practical methods for macromolecule crystallization, are to be found in other chapters and in other publications (8–12). Here we will consider some of the aspects of crystallization that directly affect the application of seeding techniques. It is useful to separate the events leading to the spontaneous formation of a crystal nucleus and those conditions that allow a crystal or nucleus to grow. While both events depend on the degree of supersaturation of the protein, the physical processes involved are very different, and the degree of supersaturation required for nucleation is generally higher than that required for growth onto an existing crystal plane. In the case of spontaneous nucleation a new seed must be generated while other events are taking place and is driven by the requirement to lower the free energy of the supersaturated state. These other events involve the aggregation of molecules into various phases. During aggregation, reversible and irreversible processes are at work simultaneously. The formation of ordered nuclei may be competing for protein with irreversible processes that produce amorphous aggregates (such as precipitates and protein skins) and hence constantly lower the degree of supersaturation of the macromolecule. As supersaturation is decreased the chance of forming a stable nucleus is reduced. Since the occurrence of spontaneous nucleation depends on the relative rates at which these various competing events take place, crystals might never form even under conditions which might otherwise support crystal growth. The principle that there is a lower energy requirement in adding to an existing crystal surface than in creating a new nucleus has important consequences. If the inverse were true, the protein would partition into a very large number of small nuclei and large crystals would never grow. Instead, in many cases it is possible to grow large crystals in the absence of seeding. Since spontaneous nucleation is a statistical phenomenon, whose probability increases with increasing degree of supersaturation, the nucleation and growth of crystals is a process with negative feedback. As a nucleus is formed its growth reduces the degree of supersaturation of the solution, and hence decreases the probability that other nuclei will form. To take advantage of this, supersaturation must be achieved slowly. The degree of supersaturation should be just sufficient to obtain a small number of nucleation centres. With the proper choice of crystallization conditions and good control over the environmental factors, it is often possible to fulfil all of the above conditions. Determining the appropriate conditions can require many crystallization trials and is consequently time-consuming involving the use of many milligrams of macromolecule and other materials. Seeding can be used efficiently and effectively during the crystallization trials to minimize the quantity of protein required for this analysis.

2.2 Crystal growth

When crystal seeds are added to an equilibrated protein solution, the protein partitioned between the soluble phase and irregular aggregate phases will redistribute. The final equilibrium will be achieved only when crystal growth has ceased. From this we may understand why seeding can be used not only in situations where the protein remains in the soluble phase, but also in situations where most of the protein has precipitated. In fact, as crystals grow the degree of supersaturation is reduced and protein may be transferred from the various (reversible) amorphous phases to the soluble phase, and from this phase it can accrete onto crystal surfaces. Several factors affect the quality of the crystals obtained, including the rate of growth, the internal order of the initial nucleus, and the purity of the sample. For some macromolecules, the initial growth following nucleation may be too fast, because of the high degree of supersaturation, resulting in the incorporation of crystal defects, which may eventually lead to the premature termination of crystal growth and to poorly formed crystals. In certain cases, the nuclei which are generated spontaneously may be polycrystalline in nature. Growth from such nuclei may result in the formation of crystal clusters rather than individual single crystals. Recent studies have shown that lysozyme crystals have impurity rich cores of the order of 20–30 μm and that seeding may provide a method to avoid such problem (13, 14). By decoupling crystal growth from nucleation seeding provides a means by which growth conditions may be tailored for crystal growth rather than nucleation and lead towards the production of large, regular crystals. Seeding allows the experimenter to control not only the number of seeds but also reduce the supersaturation level of the protein and to decrease the incorporation of defects detrimental to crystal quality. Seeding provides a preformed, regular crystal surface onto which further molecules may aggregate in an orderly fashion. The seeds to be used in the seeding experiments can be selected from the best crystals previously grown, this will lead to the best results.

2.3 Seeding techniques

Seeding consists of three stages: a pre-seeding stage, an analytical stage to refine crystal growth conditions, and the final production stage using the refined conditions to produce large single crystals. The pre-seeding stage is essential for seeding to work in a consistent and reproducible manner. It can be separated into four important aspects:

- the environment, and the associated precautions necessary for seeding
- pre-equilibration of the protein solution to be seeded

- the determination of the appropriate supersaturation level for seeding
- choice of the crystals from which to obtain seeds and their preparation.

The design of the experiment is important since during seeding supersaturation may increase resulting in spontaneous nucleation leading to a shower of small crystals.

Seeding methods can be separated into:

(a) Microseeding which involves the transfer of microscopic crystals from a seed source to a non-nucleated protein solution.

(b) Macroseeding in which pre-grown crystals are washed and introduced individually into a pre-equilibrated protein solution. This method has been widely applied to tackle the problem of enlarging small crystals into crystals of a suitable size for X-ray diffraction studies (15, 16).

These methods are common to both homogeneous or heterogeneous seeding. Homogeneous seeding involves duplicating the three-dimensional lattice of the crystal using an identical macromolecular solution. Heterogeneous seeding is somewhat more complex and can be divided into:

(a) Cross-seeding, a form of seeding in which seeds come from crystals of a protein or macromolecular complex that is different from that being crystallized. A closely related crystal form is generally expected.

(b) In epitaxial nucleation, a regular surface rather than a three-dimensional lattice is the template for the growth of new crystals. It is this type of seeding that is exploited to grow more tightly-packed crystals that in some cases diffract to higher resolution than the original crystals. The nucleation of protein crystals on cellulose fibre impurities, which often end up in protein solutions, is another case of epitaxial nucleation.

In the latter of these methods the lattice dimensions of the crystals obtained is different from those of the seeds from which they grew. Heterogeneous seeding should be followed by homogeneous seeding as the core is likely to maintain some of the characteristics of the original seed.

3. Crystallization procedures

Crystallization by vapour diffusion is a relatively simple technique (see Chapter 5). This section deals with the establishment of crystallization procedures which are suited to the application of seeding.

3.1 Pre-seeding: sitting drop vapour diffusion

The crystallization procedures used in conjunction with seeding techniques may be different from those that would be used otherwise. The design of the crystallization experiment must allow for the introduction of seeds at some stage in the equilibration phase. This must be done with the minimum disruption to the crystallization environment. A compact variation of the sitting drop vapour diffusion method has been used successfully in the seeding of many proteins (17–20).

Vapour diffusion provides a controlled and relatively slow method of equilibration by the transfer of vapour between the protein and precipitant solutions.

i. Temperature

To achieve success in crystallization and seeding it is important to control the overall environment of the set-up. This must include temperature. Temperature regulation can be achieved by the combined use of a sitting drop vapour diffusion set-up using a glass pedestal and a constant temperature incubator. The sitting drop environment provides better temperature control than its hanging drop counterpart because of heat conduction between the reservoir solution and the protein solution in the inverted glass pot (see *Figure 1*). The problem of condensation on the glass coverslip, caused by temperature gradients and convection currents in the sealed set-up, are less likely to affect a sitting drop experiment where the protein drop is situated close to the surface of the reservoir. In contrast, in the hanging drop environment, the drop is effectively in thermal contact with the outside air. The thin coverglass absorbs the heat of condensation and dissipates it to the outside. Short-lived changes in temperature, such as opening the door of the constant temperature incubator containing the experiment, will rapidly vary the temperature of the hanging drop but not that of the reservoir because of its higher heat capacity. During a rise in temperature, vapour will distil away from the drop, increasing the degree of supersaturation, which may result in a shower of microcrystals. This will be more common for crystals that are grown at high salt concentrations. When the temperature decreases more vapour condenses onto the drop diluting the protein solution. It is not uncommon in low salt, or in crystallization trials using hanging drop vapour diffusion under low PEG concentrations, to observe an increase rather than a decrease in the volume of the protein-precipitant drop. The use of the sitting drop method reduces these problems.

ii. Multiwell sitting drop vapour diffusion plate

The set-up shown in *Figure 1* has been designed and fabricated in our laboratory to shield against short-lived temperature fluctuations, other similarly compact set-ups are now in use in other laboratories. Its compact size makes efficient use of incubator space. The sitting drop method gives easier access to the protein drop and the reservoir than hanging drop set-ups. This allows the precipitant concentration to be varied, when required, during the course of the experiment. By lifting the coverglass the protein-precipitant drop can be seeded, and the plate design enables minimum disturbance to the

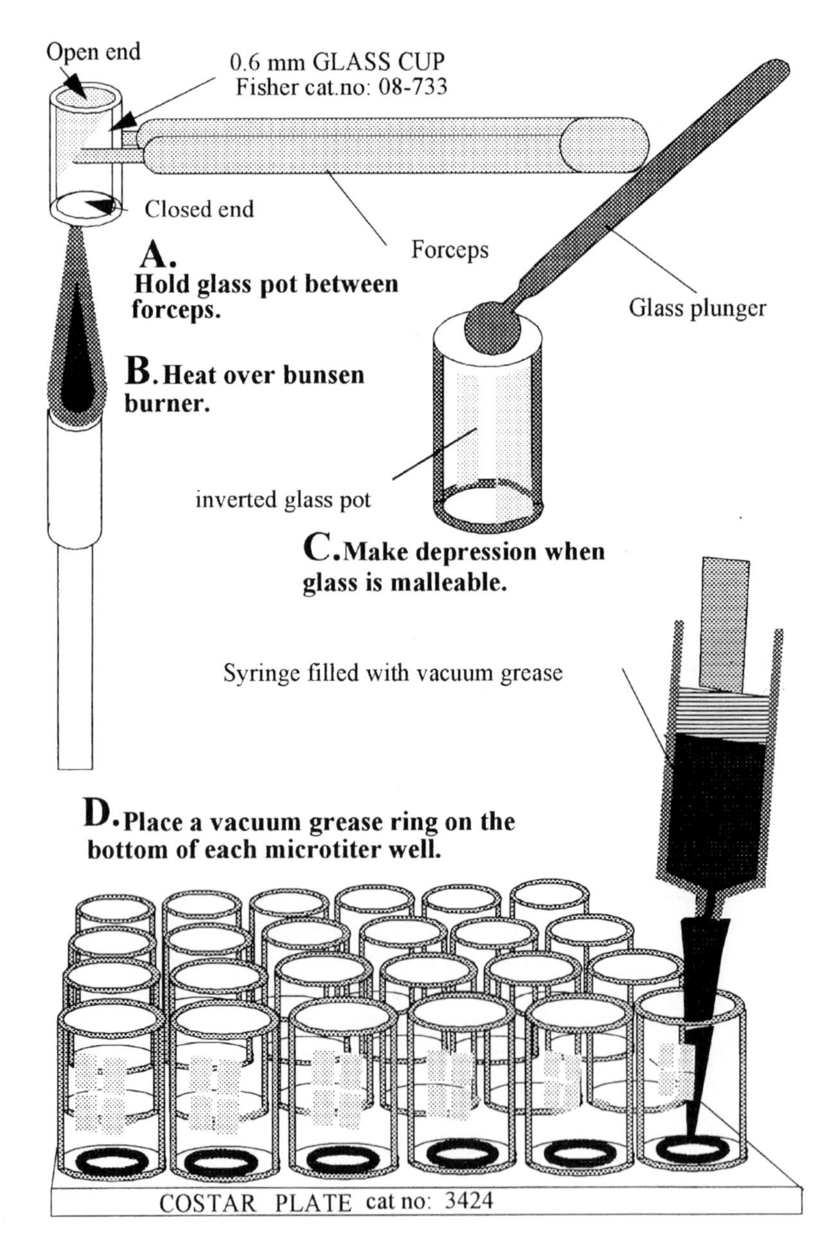

Figure 1. Schematic illustration of the sitting drop vapour diffusion plates. (A) A glass cup is held between long sized forceps over a Bunsen burner with the closed end down towards the flame. (B) The cup is heated until the glass becomes soft. (C) When the bottom is malleable the pot is inverted and a depression is made in it with a glass plunger. (D) A ring of silicone vacuum grease is placed in the bottom of each of the wells of the microtitre plate. Pots are placed onto top of the ring and pressed down. (E) Each depression is

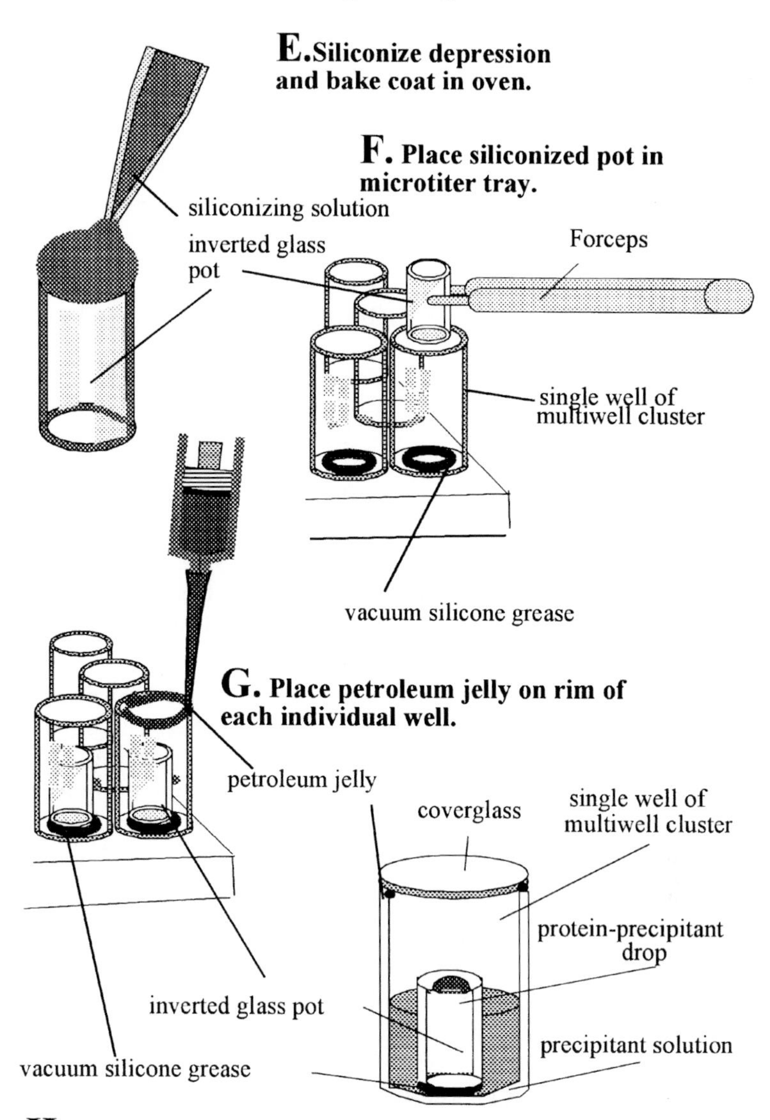

siliconized, washed repeatedly with distilled water, and the coat baked in an oven. The siliconized pots can now be placed in each of the wells in the multiwell cluster on top of the silicone grease ring which holds them in position. (G) The rim of the individual well is smeared with petroleum jelly to seal the well once the coverglass is placed on top. (H) The precipitant solution is placed around the inverted pot, the protein in the depression and the desired amount of precipitant mixed with it before sealing the experiment.

seeding environment. Such features can greatly enhance growth of quality single crystals. The tray consists of a 24-well Costar tissue culture plate from Hampton Research combined with 0.6 ml glass cups from Fisher Scientific. The wells are significantly smaller than those in the more commonly used Linbro plates, which use 22 mm circular coverglass slips instead of the 18 mm required for the Costar plates. The plates are made as described in *Protocol 1* and illustrated in *Figure 1*.

Protocol 1. Making multiwell sitting drop plates

Equipment and reagents

- 0.6 mm cups
- Bunsen burner
- Silicone grease
- Glass plunger
- Plastic syringe
- Forceps
- Pipette
- 18 mm round coverglass

Method

1. Make a depression in the bottom of the 0.6 ml cups by heating over a Bunsen flame and pressing down on the cylindrical base of each of the inverted cups with a rounded-end glass plunger.
2. Hold the resulting cup in place at the centre of each well of the tissue culture plate with Corning silicone vacuum grease. The open end of the cup sits on the bottom and the depression in the cup is at the volumetric centre of each individual well (sitting drop rods with depressions, made by Perpetual System Corporation, and micro-bridges are available from Hampton Research).
3. Siliconize the cavity created by the depression. Up to 100 μl of protein-precipitant mixture can be used with this set-up. The reservoir solution, typically 1 ml, occupies an annulus around the inverted cup.
4. Smear the edges of each well in the Costar tissue culture plate (available from Hampton Research) with petroleum jelly. Silicone grease can be used instead of petroleum jelly to give a better seal, although this will make covers harder to remove for seeding.
5. Place a 18 mm round microscope coverglass on top of each well **ensuring** that an airtight seal is achieved.

Since this method is similar in many ways to the hanging drop method, crystallization conditions determined for hanging drop experiments require little modification for implementation with sitting drop vapour diffusion. Another added advantage is that larger volumes can be used. Other sitting drop methods provide similar advantages although glass pedestals are recommended for temperature stability.

3.2 Analytical seeding

It is important to first determine under what conditions seeding will be effective. This is done by the use of an analytical seeding method such as streak seeding.

3.2.1 Streak seeding technique

i. Making the probe

A probe for analytical seeding is easily made with an animal whisker mounted with wax to the end of a pipette tip. The tip is then mounted on a wooden rod. Since the cross-section of the whisker varies along its length it is possible to obtain several probes from the same whisker, by repeatedly cutting the whisker to lengths of 5–20 mm from the end of the wax. The probes so obtained will be of different strength and thickness.

ii. Cleaning the probe

To clean the probes prior to their use the fibres are degreased using ethanol or methanol, and then washed in distilled water and wiped dry. Probes can be used several times by cleaning them with distilled water and tissue paper in between experiments. After a period of time the whiskers need to be trimmed as the end becomes frayed. A good probe should be able to transfer seeds to six to twelve drops consecutively without being dipped in seed solution. A good probe is best for titrating the number of seeds; old whiskers deposit many seeds in the first two drops and virtually none in further drops. Cat whiskers are generally used for making probes, although other animal whiskers have also been found suitable.

iii. Seeding

The end of the fibre is then used to touch an existing crystal and dislodge seeds from it (*Figure 2A*). Gentle friction against the crystal is normally sufficient. Some of the dislodged seeds remain attached to the fibre. The probe in now used to introduce seeds into a pre-equilibrated drop by rapidly running the fibre in a straight line across the middle of the protein-precipitant drop (*Figure 2B*).

Sitting drop set-ups are preferable since hanging drops tend to dry out when exposed to the ambient air, even in the short time interval between collecting the seeds on the fibre and streaking the new drop, typically 5–30 seconds. The precipitant collected from the first drop increases in concentration as it travels to the next drop though the air, and this can affect the conditions in the seeded drop. Therefore, the distance the whisker has to travel should be kept to a minimum, in most circumstances, less than 10 cm. Both the source well and the receiving well are resealed immediately after the transfer. Seed nucleated crystals grow along the streak line; any self-nucleated seeds will occur elsewhere in the drop. The pre-incubation time and the range

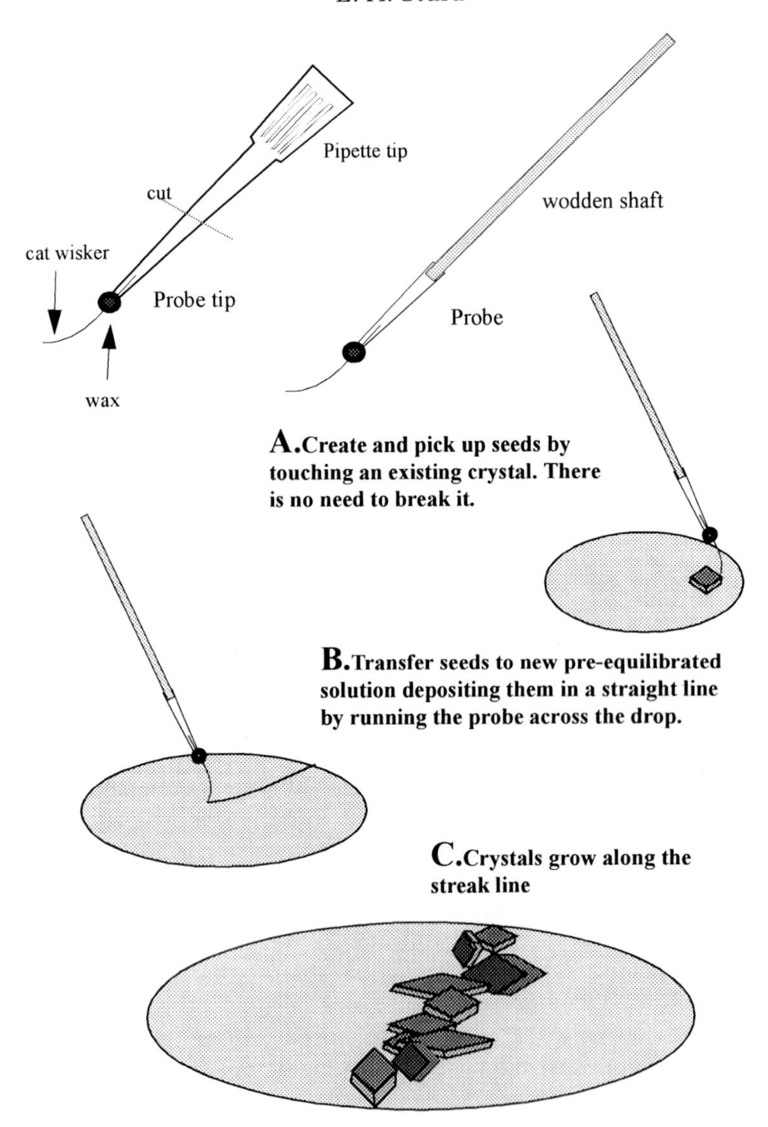

Figure 2. Schematic drawing of the stages in the application of the streak seeding technique for analytical seeding. (A) A probe made from a cut pipette tip mounted on a wooden shaft. On the end of the tip a short segment of an animal whisker is attached with molten wax. The probe so constructed is used to pick up seeds from an existing crystal, precipitate, or other ordered aggregate by simply touching it and displacing seeds from it. (B) The seeds remain attached to the whisker and can be transferred to a pre-equilibrated drop by running the end of the probe across the drop. Some seeds are deposited along the path, where they will either grow into large crystals or dissolve into the solution. (C) Growth of crystals along the streak line indicates that the conditions may be suitable for the application of other techniques such as micro- or macroseeding. Self-nucleated crystals will appear away from the streak line.

of supersaturation which allows for sufficient crystal growth without self-nucleation is determined experimentally by observing the growth (or lack of growth) of seeded crystals along the streak line (*Figure 2C*).

3.2.2 Protein pre-equilibration

In batch crystallization the precipitant concentration is slowly increased until the protein solution turns cloudy; further solvent is then added until it is clear again. The protein solution is continuously stirred until seeds are added. When seeding small volumes, it is best to avoid producing spontaneously nucleated seeds, as could be produced by exceeding the solubility threshold, since it is difficult to control their number, or ensure that they will later dissolve. By introducing the seeds well before sufficient supersaturation is reached, additional nuclei which might form spontaneously are prevented. However, if the protein is not suitably pre-equilibrated, the seeds will dissolve. Streak seeding conditions should be optimized by repeating the procedure after different pre-equilibration times, using different drops, although if seeds dissolve, the same drop can be seeded again.

3.2.3 Determining the degree of supersaturation for seeding

Initially, during the screening phase of a crystallization, we want to obtain results quickly in order to determine the many parameters that control the growth and morphology of the crystals. In the production phase, when conditions are optimized, we want to slow down the growth rate. As long as protein is not being lost to amorphous phases, drops are generally allowed to equilibrate fully before seeding, unless conditions for crystal growth and nucleation are tightly coupled. To optimize the conditions, drops are set up under conditions which vary only slightly from those previously determined in the fast growth experiments. Drop size, which was kept to a minimum to save material, should be increased at this stage, while simultaneously reducing the precipitant-protein ratio, and the precipitant concentration in the reservoir. This will have the effect of slowing down the rate of equilibration and the desired state of supersaturation will be approached more slowly. The final conditions will be established at reduced precipitant concentration but at a compensating, higher protein concentration. The precise protein-precipitant ratio and precipitant concentration are determined experimentally by allowing the drops to equilibrate, usually three to five days, and then streak seeded. The streak will not appear in drops where the concentration of the precipitant in the reservoir is too low, and where the precipitant concentration is too high crystals will initially appear along the streak line, followed later by the formation of others away from the streak line. Caution should be used when making this determination, as some seeds may drift away from the line along which they were deposited and sink down to the bottom of the drop, crystals that appear at the drop/air/glass interface are generally an indication that the well was left open too long during seeding rather than an indication that the seeding conditions were unsuitable. The experimental conditions where crystals grow only along the streak line determine the precipitant concentration range for production seeding. This method is also well suited for testing minor changes to growth conditions, such as adding a co-precipitant, testing new additives, or simply finely analysing the pH range. Small changes at this stage can result in significant improvements in the quality of the crystals obtained and can be essential for growing suitable crystals for high resolution X-ray structure analysis.

3.2.4 Assaying for microcrystallinity

Microcrystalline precipitates are often indistinguishable from their amorphous counterparts. Streak seeding can be used to distinguish between these two possibilities by using particles from an uncharacterized precipitate as a source of seeds. For example, the initial crystallization trials of a complex between the Fab' fragment from an anti-peptide murine antibody B13I2 with its 19 amino acid peptide antigen (myohaemerythrin residues 69–87) gave a precipitate composed of round or oval particles of roughly equal size (*Figure 5A*) (19). No indication of microcrystallinity could be deduced from microscopic observations of the precipitate since it was not appreciably birefringent. Three adjacent drops in which protein had not precipitated were streaked with this precipitate. Hexagonal-shaped crystals appeared in one of the drops as a result of the streak seeding experiment (*Figure 5B*). These crystals were then used for macroseeding onto other drops. After an adjustment in the crystallization conditions from 1.5 M sodium citrate pH 6.0, to 1.6–2.0 M mixed sodium and potassium phosphate pH 5.0–6.0 it was possible to grow crystals (*Figure 5C*) that diffracted 2.6 Å resolution (19). The presence of the peptide in the solution is essential to obtain these crystals and cross-seeding from the Fab'–peptide complex onto native Fab' solutions does not produce crystals or precipitate.

4. Production seeding methods

4.1 Microseeding

In microseeding microscopic crystal fragments are introduced into a prepared protein solution. By using an analytical seeding technique (Section 3.2) the supersaturation threshold can be accurately determined by scanning a range of precipitant concentrations. The streak seeding technique was initially developed for this purpose. Microseeding is composed of three stages:

- preparation of the seed stock
- repeated dilution of the seeds
- the seeding itself.

4.1.1 Preparatory steps

The two first stages in microseeding are described in *Protocol 2*.

Protocol 2. Microseeding

Equipment and reagents

- Tissue homogenizer
- Test tube
- Precipitant solution
- Small crystals
- Vortexer
- Seeding probe
- Pipette
- Microtest tubes for diluted seeds

A. *Preparation of the seed stock*

1. To produce seed stock wash three or four small crystals in a slightly dissolving solution to remove defects or amorphous precipitate from the crystal surfaces.
2. Stabilize the washed crystals in an appropriate precipitant solution and transfer them to a glass tissue homogenizer in which they are crushed (*Figure 3*).
3. Wash the crushed seeds from the sides of the homogenizer into the bottom by adding further solution. Transfer the solution from the homogenizer to a test-tube. This is the seed stock, which for most proteins can be stored in a constant temperature incubator for future use.

B. *Repeated dilution of the seeds*

The seed stock is normally diluted with further solution as it contains too many nuclei to be useful in the nucleation of only a small number of crystals.

1. Dilute the seeds, typically in the range 10^{-3} to 10^{-7}, vortexing the tube containing the microseeds between dilutions to evenly distribute the seeds. The reservoir solution is normally well suited for the initial dilutions as some of the smaller microseeds dissolve to provide a residual protein background. Further dilutions are likely to require extra precipitant, usually 10% or more above that of the reservoir to prevent the seeds from dissolving. Maintain the temperature constant as seeds may dissolve either on heating or cooling. One of the diluted seed solutions should be suitable to supply a small number of seeds into each drop.
2. Test the seed stock produced by streaking or by adding a measured amount of seed solution to several drops. The results should be visible in one to two days. If there is a sudden drop off in the number of

Protocol 2. *Continued*

crystals obtained, inconsistent with the a tenfold dilution, extra precipitant should be added to the stabilizing solution. If nucleation is independent of the dilutions, buffer should be added to the seed solution to reduce the precipitant concentration. Also test the drops by streak seeding from a crystal (*Protocol 3*) to ensure that the reservoir solution is appropriate for seeding.

4.1.2 Streak seeding as a microseeding technique

Since the seeds that are transferred from crystal to drop in the use of the streak seeding technique are microscopic, the technique is technically a microseeding method. But, while in the analytical streak seeding technique the deposition of many seeds along the path of the whisker is essential for the subsequent visualization, when growing large crystals for X-ray structure analysis only a few seeds should be deposited, as seeds compete with each other for the available protein. When changing the use of streak seeding from an analytical to a production seeding mode we must find a way of diluting the seeds quantitatively in a reproducible manner. The same probe is thus used in each repeat experiment to ensure that the volume of liquid and the number and size of the seeds which are transferred from one drop to another, remains constant. Thicker whiskers transfer more liquid containing microseeds and can potentially carry larger size seeds.

Protocol 3. Streak seeding

Equipment and reagents

- Seeding probe
- Seeds
- Preequilibrated protein drops
- Microscope
- Vacuum grease
- Cover glass
- Crystallization tray forceps

Method

1. By changing the angle at which the whisker is drawn out of the we can affect the size and number of seeds loaded onto the fibre. This should be kept constant for reproducibility. The whisker is lifted vertically upwards, maintaining it perpendicular to the drop's surface to minimize seed retention, seeds are scooped up to maximize size and the number of seeds picked up.
2. Pre-equilibrate the drops before seeding under conditions previously determined analytically by streak seeding.
3. Streak seed subsequent drops without loading the probe with new seeds to achieve seed dilution. To obtain greater dilutions the probe

can be dipped in and out of the reservoir in between streaks, to allow some seeds to drop into the precipitant solution.

4. Reduce the time the probe spends in the air by opening all the chambers to be seeded just before picking up the seeds. Streak the drops sequentially, as speed is important to prevent drying of the solution on the fibre.
5. Cover all the chambers without delay to reduce evaporation from the seeded drops.

The seed stock and seed dilutions prepared as previously described, can be used in streak seeding reliably by dipping the whisker into each of the diluted solutions including the seed stock and applying the seeds to new drops (*Figure 3*). It is common to start by dipping the probe in the most dilute solution first, streaking one drop, then progressing up the dilution series to the seed stock. Typically the results are analysed two days to one week after streaking.

4.2 Macroseeding

In macroseeding a single crystal is introduced into a suitably pre-equilibrated solution. Single prismatic crystals, which are free from twinning or any other crystallites, are most suitable for this technique. As in other seeding protocols it is important to take steps to maintain constant conditions, as even a slight dehydration of the drop being seeded could temporarily change the state of supersaturation and induce unwanted nucleation. Performing the experiment in a very humid environment and by using large drops can reduce dehydration. A beaker with a filter paper cylinder soaked in distilled water is such an environment. However it is more practical to use sitting drop multiwell trays, which have been used with high success in our laboratory. Macroseeding is done under a dissecting microscope where the small amount of heat generated from the microscope stage light bulb may actually be slightly beneficial in increasing the humidity level around the drop. The heat raises the temperature of the reservoir faster than that of the drop increasing the rate of evaporation from the reservoir, and counteracting evaporation into the room. Seeds are washed in a slightly dissolving solution to remove the top layer of protein, which contains possible defects, from the surface of the seed without causing excessive etching or cracking. They are then transferred to a stabilizing solution to re-equilibrate the crystals (*Figure 4*). Older seeds benefit the most from this treatment, whereas freshly grown crystals may be put directly through a series of washes in stabilizing solution. From the final wash solution each seed is then transferred to the protein-precipitant drop to be seeded (*Figure 4*).

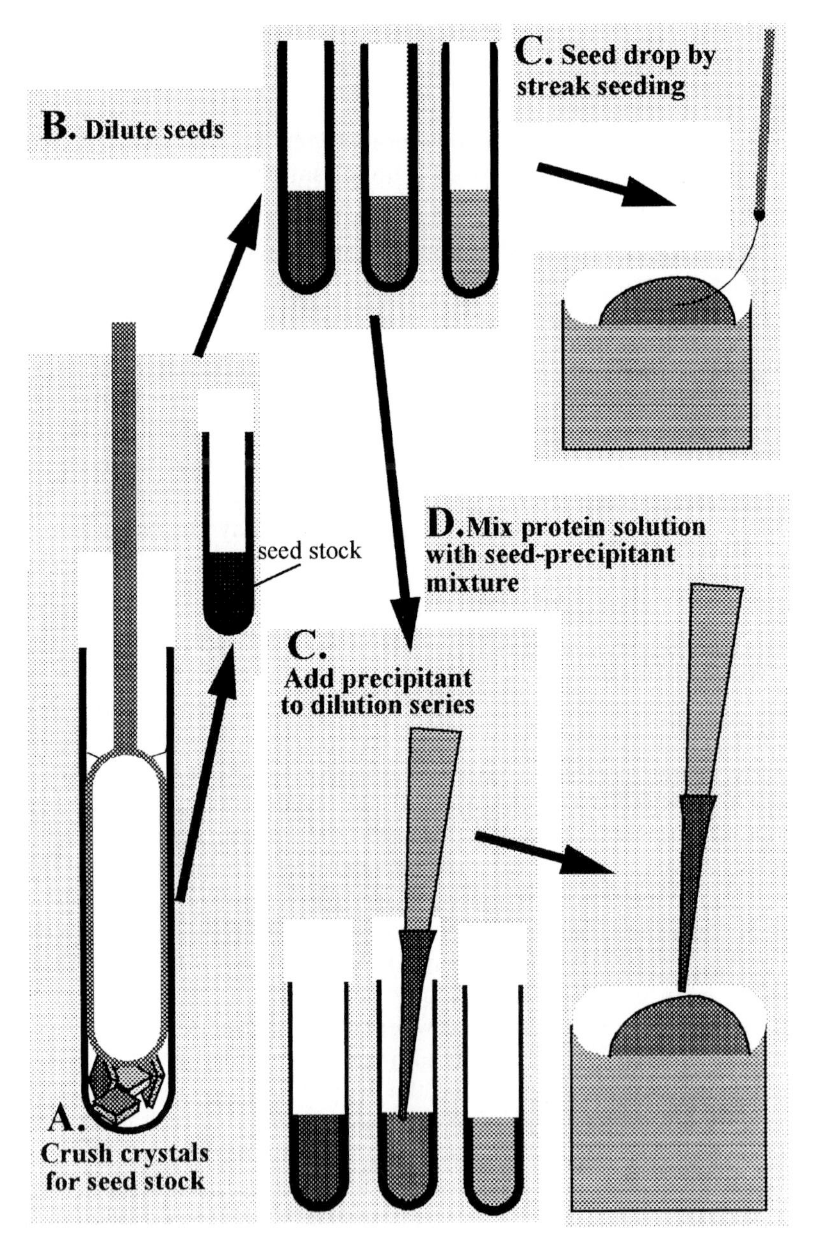

Figure 3. Diagrammatic illustration of the steps involved in microseeding. (A) Crystals of good morphology are crushed in a glass tissue homogenizer. The resulting seeds are washed into the bottom of the tube and stored in a test-tube. (B) The seed stock is diluted to produce a dilution series. (C) Seeds can be picked up from the diluted solutions by using a probe, or precipitant can be added to these, so that they may be mixed with protein solution (D). The wells are sealed by replacing the coverglass and the seeds are allowed to grow for several days.

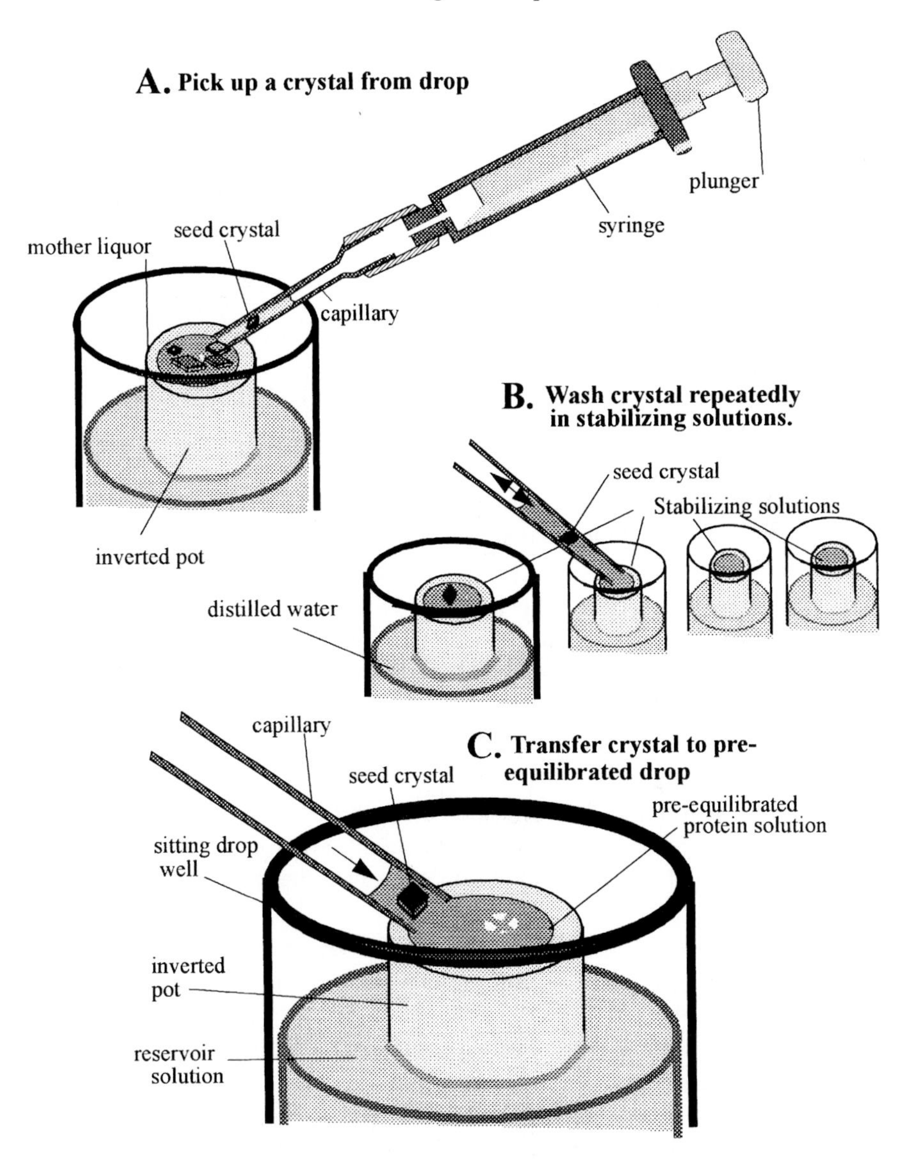

Figure 4. Illustration of the steps involved in macroseeding. (A) A single crystal is picked up from a drop. Crystals should be of good morphology and free from defects. (B) A series of washes is performed by repeatedly transferring the crystal from one depression to another, taking care not to damage the seed. (C) The seed is finally transferred to a pre-equilibrated drop for further enlargement.

4.2.1 Details of crystal handling in macroseeding

The handling of crystals is especially important to avoid the generation of microseeds or unwanted nuclei in the transfer to the drop being seeded.

Protocol 4. Handling of crystals

Equipment and reagents

- Glass syringe
- C-flex tubing
- X-ray capillary
- Microscope
- Forceps
- Seeding tray
- Seeding probe

Method

1. Connect a glass or quartz capillary to a 1 ml glass syringe with a short piece of rubber tubing such as c-flex (Fisher, 14-169-5c) which gives an excellent seal.
2. Snap open the end of the capillary with tweezers or scissors, and siliconize if the experimental situation can benefit from diminished adhesion of the solution to the glass capillary. After siliconizing it should be extensively washed.
3. Pick up the crystals under a dissecting microscope, using a magnification of × 10 to × 100.

   Crystals from hanging drops should first be washed into a secondary vessel; in a sitting drop vapour diffusion set-up it is possible to do this directly from the depression in which the crystals have been growing.

   After the coverglass sealing the vapour diffusion chamber is lifted off, the tip of the capillary is inserted into the drop and a crystal is drawn into the capillary.

   If the crystals are adhering to the well, by withdrawing liquid from the drop and gently ejecting it onto a chosen crystal, it is often possible to dislodge the crystal. Unfortunately some crystals have severe adhesion problems and cannot be dislodged without breaking them. To reduce this problem, the depressions in glass pots should be coated with a thin film of Corning vacuum silicone grease before the protein-precipitant drop is added. The seeds will remain suspended on top of the grease, and the final crystals are mounted for X-ray diffraction use without the recurrence of this problem.
4. Once in the capillary the crystal is brought to the middle, and then allowed to sink and adhere sufficiently to the inside wall of the capillary so that the liquid can be moved over the crystal.

   (a) For crystals that fail to sink to the bottom and adhere to the capillary wall, a long hair or whisker may be wedged against the crystal to stop movement while liquid is drawn out.
   (b) For soft crystals there is the danger that during this procedure microseeds may be dislodged from them by the hair with obvious consequences. A series of washes will minimize the number of microseeds that will be transferred but not eliminate the risk that one or more will still be present in the solution which is transferred together with the crystal.

5. Once the crystal has been separated from the bulk of the mother liquor, the hair withdrawn and the mother liquor ejected from the capillary and returned to the original drop, the crystal should remain in a small pool of liquid inside the capillary. Removing more of the remaining solution from around the crystal may help diminish the number of microseeds and aggregated material, and since the solution may now be at a higher precipitant concentration due to evaporation during the handling, transferring less of this solution may avoid creating conditions which are unsuited to the seeding.

The crystal can be repeatedly washed in a stabilizing solution (typically the reservoir solution is used) prior to transferring it to the new drop as described in *Protocol 4*.

Protocol 5. Washing crystals

Equipment and reagents

- Multiwell sitting drop plate
- Distilled water
- Stabilizing solution
- Microscope
- Tray with source crystals
- Tray to be seeded

Method

1. Fill four depressions of a multiwell sitting drop plate with about 100 μl of solution from the reservoir where the seeds originated (or prepare a solution identical to this).

2. Fill the reservoirs around these drops with 1.5 ml of distilled water to maintain a high degree of moisture around the solution (*Figure 4*).

3. The crystal is repeatedly transferred and picked up from each of these stabilizing solution drops until finally, it is picked up into the capillary. Because the addition of stabilizing solution to the new drop would unnecessarily modify the equilibrium, or dilute the equilibrated protein-precipitant solution, it is best to minimize the amount of liquid that remains around the crystal.

Protocol 5. *Continued*

4. Remove the excess liquid with a small thin strip of filter paper or a very thin capillary.
5. Resuspend the crystal in new mother liquor drawn in from the drop to be seeded and return it to the well for equilibration and further growth. Alternatively, after the series of washes, the crystal is allowed to sink towards the open end of the capillary, so that when the capillary touches the solution of the drop being seeded, the crystal falls directly into this solution with little transfer of wash solution.

4.2.2 Macroseeding of needles

Macroseeding using needles as seeds is more complicated since needles have a tendency to bend while being transferred from the original growth solution to the new solution. The stresses created in the crystals during this process may result in defects at each stress point, and each of these points may act as a nucleation site for growth of new needles. By breaking the needles with a sharp object (glass or metal) into smaller segments, the resulting fragments can be used for macroseeding. The sharper the instrument the less damage will be done to the seeds. A small number of these needle fragments are then transferred to a stabilizing solution. From this point the procedure is essentially the same as for prismatic crystals, as each is then transferred from this solution to another container with the same stabilizing solution, and then to a third and a fourth, to wash away any microseeds that may be transferred to the pre-equilibrated growth medium.

5. Heterogeneous seeding

The principle that there is a lower energy requirement in adding to an existing surface than in creating a new nucleus (Section 1.2) holds for many surfaces. Such aggregation onto surfaces may be considered more of a problem than an advantage. However, regular surfaces may offer a charge distribution pattern which is complementary to a possible protein layer and could provide a suitable starting point for the nucleation of new crystals. The work of McPherson (21) with various inorganic minerals provides strong support for the idea that regular planes are able to *catalyse* the nucleation of crystals of macromolecules, even if the lattice dimensions of the crystalline minerals differ from those of the resulting protein crystals. In these studies, nucleation occurred preferentially on the mineral substrate at a lower degree of supersaturation than was required for the same crystals to nucleate in the absence of the minerals. Crystals of related macromolecules can also be used to induce nucleation of proteins; the resulting crystals may maintain some, but not all, of the lattice dimensions or symmetry axes of the initial seeds. In such cases, where the protein in the crystals from which the seeds are obtained is related to the protein in the solution being seeded, the operation is termed cross-seeding. When crystals of the same macromolecule are used to induce a related crystal form under different crystallization conditions, an epitaxial jump (by analogy with quantum jumps) has been achieved.

5.1 Cross-seeding

5.1.1 Cross-seeding between Fab–peptide complexes

In λ-type light chain dimers, the dimers pack so as to form an *infinite β-sheet* maintaining one cell dimension in common, 72.4 (± 0.2) Å, along one of the 2_1 axes (22). Such packing in preferred planes for certain classes of protein molecules, may indeed provide suitable surfaces for nucleation for other members of that class. A similar observation was made in the course of our work with different anti-haemagglutinin monoclonal Fab–peptide complex crystals, where it was noticed that these have a common crystal lattice plane with cell dimensions 73.0 (± 1.0) Å along the 2_1 axis and 66.4 (± 2.5) Å along one of the other axes (19, 23, 24).

Within the description of the three-dimensional structure of the complex of Fab 26/9 that recognizes the same six residue epitope of an immunogenic peptide from influenza virus haemagglutinin (HA1 75–110) as Fab 17/9, it was possible to understand the hierarchy in the crystal contacts responsible for the differences and similarities between these crystal forms (25). In brief, 26/9 and 17/9 antibodies are very similar, but their interaction with the peptide are slightly different. Structural and sequence analysis suggests that amino acid differences near the peptide binding site are responsible for altering slightly the specificity of 26/9 for three peptide residues. Since the peptide is essential for one of the crystal interactions, we can understand the influence of peptide length on the crystallization and the similarity in crystallization between these antibodies. Cross-seeding, using the streak seeding method can bridge the gap between the various peptide complexes of 26/9. Initial crystals were obtained by spontaneous nucleation with a nine residue peptide (HA1 100–108). The quality of these crystals was improved by using the streak seeding technique as a microseeding method. Seeds from these crystals were used to search for growth conditions of complex crystals of Fab 26/9 with longer peptides, for which no conditions for spontaneous nucleation had been found. Both the 13-mer (HA1 98–110) and the 23-mer (HA1 88–110) peptide–Fab mixtures responded positively to the seeding. Seeds obtained from this cross-seeding were used to seed repeat experiments to dilute out the effect of the heterogeneous seeds and optimize the crystallization conditions for the new complexes. The crystals obtained from this second seeding diffract to 2.5 Å resolution.

A third anti-peptide antibody 21/8 also belonging to this same panel but their heavy chains belong to different classes; Fab 21/8 is derived from an

IgG_{2b}, while 26/9 is cleaved from an IgG_{2a}. Nevertheless, seeds from Fab 26/9–13-mer complex crystals have been used to induce crystallization in solutions of Fab 21/8–13-mer mixtures, under identical crystallization conditions. The Fab 21/8–13-mer crystals first obtained by the streak seeding experiment (*Figure 5D*) were very thin needles and although optimization of the conditions resulted in better crystals, the real breakthrough was achieved from the refined conditions which yielded large crystals of an unrelated form by spontaneous nucleation.

Cross-seeding does not need to be carried out by streak seeding as, for example, large crystals from the chicken mitochondrial aspartate aminotransferase were used as seeds in the cross-seeding of the pig enzyme (6) (initial cross-seeded crystals of the chicken enzyme were badly twinned but were improved to X-ray quality in a second cycle of macroseeding). However streak seeding can substantially increase the speed with which crystals can be obtained.

5.1.2 Cross-seeding from native Fab to Fab–peptide complex

Anti-peptide Fab 50.1, that recognizes an epitope of the gp120 surface glycoprotein of HIV-1, is an example where native crystals were used to seed the Fab–peptide complexes. The native crystals can be grown in three different morphologies. Spontaneous nucleation has not been observed for any of the peptide complexes tested. The peptide lengths vary from 13–40 residues. Crystals of the Fab–13-mer complex have now been obtained by streak seeding with the native Fab crystals in 12–24% PEG 10 000 pH 5–8 (26). In the Fab–peptide solution most of the protein is found partitioned in a gel phase covering the bottom of the drop. On addition of native seeds, crystals grow by acquiring protein from the gel phase surrounding the seeds. The morphology of the Fab–peptide complex crystals differs substantially from that of native crystals (*Figure 5E*). It is also worth noting that while the crystals of the native Fab are very mosaic, data have been collected on well

Figure 5. Photomicrographs of the results of crystallization experiments mentioned in the text. (A) The initial precipitate obtained in the crystallization of the complex of anti-peptide Fab′ B13I2 with the 19 residue peptide corresponding to the C-helix of myohaemerythrin. (B) When this precipitate was streaked onto drops under similar conditions to those that yielded the precipitate crystal were obtained. (C) These crystals were used in macroseeding experiments to yield X-ray quality crystals. (D) Cross-seeding between anti-haemagglutinin Fab 26/9 and Fab 21/8. (E) Crystals obtained by spontaneous nucleation from 21/8 under the optimized conditions for the cross-seeded crystals. (F) Crystals of anti-HIV-1 Fab 50.1 complexed with a 13 residue peptide from the glycoprotein gp120 sequence of the MN isolate. The crystals were obtained by cross-seeding from native crystals (G) that grow spontaneously under similar conditions. (H) Epitaxial nucleation on a cellulose fibre. The first crystals for this Fab nucleated from a drop without fibres after a period of six months. The seeded crystals were obtained in less than a week. Notice the number of crystals nucleated on the fibre compared to the number nucleated separately.

ordered crystals of the seeded complex that diffract to better than 2.8 Å (*Figure 5F*) sufficient for X-ray structure determination (27).

5.1.3 Cross-seeding chemically modified proteins

Because solubility changes as a result of modifications, proteins which have been altered either by chemical means or by site-directed mutagenesis, may not yield crystals spontaneously in crystallization trials. Selenolsubtilisin, in

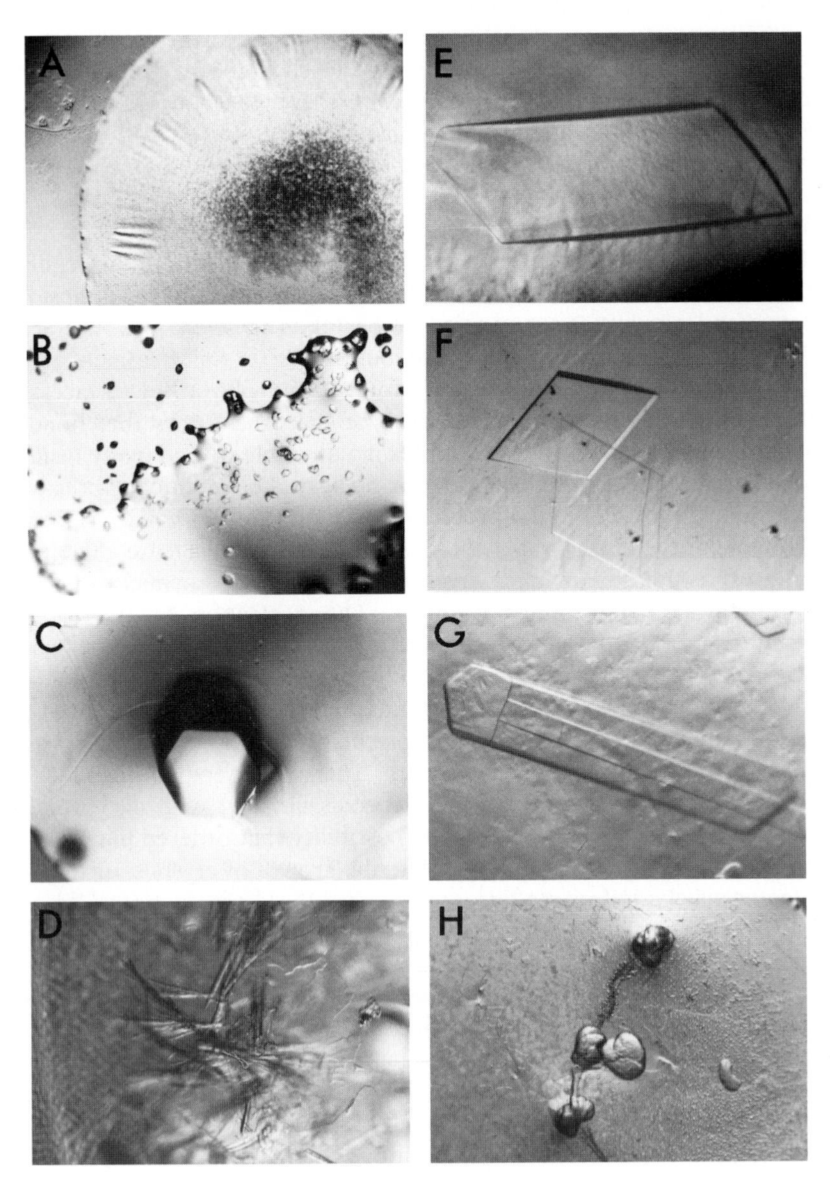

which serine 221, the active serine, of the bacterial protease (28–30), is converted into a selenolcysteine (31) is one such example. Even after extensive efforts to better purify the engineered enzyme (32) crystals could only be obtained by cross-seeding from crystals of the commercially available native subtilisin. Although the quality of these seed crystals, obtained from commercial grade enzyme was rather poor, the crystals of the selenolsubtilisin obtained from the cross-seeding experiment were of good morphology and size. The crystallization conditions under which the cross-seeding was done were similar to those for the native subtilisin. After further optimization it was possible to obtain good quality crystals, which were used for the structure determination of the modified enzyme (32). Under the optimized conditions some preparations nucleated spontaneously.

5.2 Epitaxial nucleation

Epitaxial nucleation is a particular instance of adhesion where the regularity of the surface facilitates nucleation. Many substrates mediate adhesion of proteins, and precautions may need to be taken to avoid this interaction. Glass surfaces are generally siliconized, but even after this treatment crystals are still found to be preferentially attached to the siliconized surfaces. The strength of the interaction can be stronger than the forces that bond the crystalline lattice. In crystallization trials it is possible to find many instances in which crystals or microcrystals can be nucleated on cellulose fibres which are accidentally present in the protein-precipitant drop (*Figure 5H*). Because of the regularity of the fibres this can be considered a case of epitaxial nucleation. In most cases microcrystals are also observed nucleating spontaneously away from foreign particles. The nucleation of crystals from aggregates and oils may also be due to epitaxial nucleation. Here ordered surfaces may be present within the random aggregation of macromolecules which come out of solution as oils and precipitates. These surfaces may provide platforms suitable for macromolecular nucleation, and may possibly support three-dimensional crystal growth. The streak seeding technique can be used not only with small unusable crystals but also with any promising aggregate or precipitate, to test for the possibility that ordered planes within such aggregates may be able to stimulate the growth of crystals, or that such aggregates may be polycrystalline.

Seeds can be used to seed supersaturated solutions equilibrated under conditions quite dissimilar from those from which the seed crystals originated. The resultant crystals may result in a lattice similar but not identical to that of the seeds. Epitaxial jumps can be induced by increasing the precipitant conditions of the reservoir, without any manual seeding.

Crystals of the tissue factor–factor VII–5L15 complex (33) were obtained using a combination of *Protocol 5B* and *A*. The epitaxial jump was accomplished by streak seeding a drop, equilibrated at the same precipitant concentration as that from which the seeds were grown but at lower protein concentration. Subsequently, the precipitant concentration of the well was increased by adding NaCl to the reservoir solution causing a rise in the precipitant and protein concentrations in the drop. Initially there was minimal growth of form 1 crystals (because of the low protein concentration) later as the precipitant concentration was raised, a more compact form nucleated off the crystals of the first form. The crystals so obtained were thicker and larger than the initial form 1 crystals, but similar to form 1 crystals grown under high protein concentration. In this case the first crystal form (P321; a = b = 127.2 Å, c = 110.7 Å; V_m = 3.4 $Å^3$/Da) diffracted to 7 Å and could not be distinguished from the second form (P321; a = b = 67.2 Å, c = 314.8 Å; V_m = 2.7 $Å^3$/Da) that diffracted to 3.2 Å by its morphology.

Protocol 6. Epitaxial jumps

Equipment and reagents

- Forceps
- Seeding probe
- Precipitant
- Seeding syringe
- Pipette
- Microscope

A. *Jumps without manual seeding*

1. Set up drops under the *usual* crystallization conditions. Wait for crystals to nucleate and grow. After growth is completed the level of supersaturation of the drop should be substantially lower.
2. Increase the concentration of the reservoir. This may consist of a series of gradual additions of precipitant to the reservoir or one large jump in concentration. Observe the crystals after each addition of precipitant for new crystals. These are likely to nucleate off the original crystals and grow in a different morphology and direction from the original crystals. An epitaxial jump may have occurred. Use streak seeding to make use and propagate the new crystals.

B. *Jumps by streak seeding*

1. Set up drops under any desired supersaturated conditions. Different protein and precipitant concentrations should be tried. To try to obtain more tightly-packed crystals, the protein concentration should be halved and new supersaturated conditions established at higher precipitant concentrations.
2. Streak seed drops after equilibration with any crystal form obtained for the same macromolecule. The seeds may not be able to enlarge under the new conditions, typically characterized by a higher precipitant, lower protein concentrations, but may be able to provide a template to enable a switch to a more compact crystal form.

Protocol 6. *Continued*

3. Check for the development of a line. If a line develops use the crystals obtained to seed other drops. If growth is slow, increase the precipitant concentration.

C. *Propagation of epitaxially grown crystals*

1. Use seeds from either part A or part B to seed new drops. Once crystals can be grown large enough for X-ray diffraction, these should be checked for increased diffraction limit. Crystals which have been grown in this manner and have a lower solvent content are termed 'squeezed' crystals.

Crystals of class I deoxyribose-5-phosphate aldolase from *Escherichia coli* were originally obtained for the unliganded enzyme and in complex with its substrate, 2-deoxyribose-5-phosphate $P2_12_12_1$ with cell dimensions a = 183.1 Å, b = 61.4 Å, c = 49.3 Å and a = 179.2 Å, b = 60.5 Å, c = 49.1 Å, respectively (34). In one instance, after these crystals were mounted the coverglass was replaced and the chamber sealed as it is routinely done. During the subsequent months the reservoir solution slowly increased because of evaporation through the petroleum jelly seal. When the experiment was viewed months later, new small crystals had appeared. This is often the case, except that the new crystals were morphologically different from those which had been previously obtained. Seeds from these crystals were streak seeded into pre-equilibrated drops and then macroseeded to obtain crystals large enough for X-ray studies. The resulting crystals were even more stable in the X-ray beam than the original crystal form and diffracted to 2.1 Å surpassing those of the first form which diffracted to 2.6 Å. The space group for the new form is P1, a = 49.2 Å, b = 51.3 Å, c = 54.0 Å, α = 77.23°, β = 78.4°, γ = 77.8°. A third example is the anti-testosterone Fab AN (35). In this case an epitaxial jump was achieved by streak seeding from crystals grown in PEG 600 to drops equilibrated in the same concentration of PEG 4000. The results were obvious (*Figure 5G*); the first form consisting of thin needles ($P2_1$; a = 55.1 Å, b = 70.6 Å, c = 66.3 Å, β = 106.3°) the second of more prismatic crystals ($P3_121$; a = b = 72.0 Å, c = 156.0 Å).

6. Crystallization of complexes

6.1 Considerations in the crystallization of complexes

When crystallizing complexes, such as a receptor–ligand, an enzyme–inhibitor, or an Fab–antigen complexes, it is important to consider the resultant heterogeneity of the system. Both members of the complex will be somewhat heterogeneous, and the resulting mixture will be composed of complexed and uncomplexed molecules in different ratios depending on the molar ratio of the two molecules in the solution and the dissociation constant. To reduce the heterogeneity it is important to optimize the number of complexed versus the uncomplexed molecules.

For protein complexes with small ligands it is possible to increase the number of complexed protein molecules by adding an excess of ligand. Theoretically, the excess ligand that is necessary to achieve the desired ratio of bound to unbound may be calculated from the dissociation constant if known. In practice it is best to set up experiments at different protein:ligand ratios, typically 1:1 to 1:20. Larger excesses are generally unnecessary and may even inhibit crystal growth.

Co-crystallization of enzyme substrate complexes presents a different level of complexity. Catalysis of the substrate into product will result in a mixture of free enzyme, enzyme–product complex, and enzyme–substrate complex. Such experiments are best attempted by using non-productive substrate analogues, inhibitors that mimic the transition state, and undissociable end-products. Triphosphate nucleotides, such as ATP, are easily hydrolysed, hence, non-hydrolysable analogues such AMP-PNP, adenosine diphosphate γ-S (ADPγS), and their analogous guanosine derivatives are now commonly used for crystallization instead of ATP or GTP. Vanadate, molybdate, and tungstate are commonly used as phosphate mimics (36).

When the complex consists of two or more macromolecules of comparable size the addition of an excess of one increases rather than decreases the heterogeneity of the system. In such cases, if the affinity between the macromolecules is 10^8 or better, the complex can be purified, otherwise it is best to mix the macromolecules in the appropriate stoichiometric ratio. When the stoichiometry of the system is not known it is best to set up the crystallization of the complex at different receptor:ligand ratios. Uncomplexed molecules are likely to adhere to the lattice of the complex crystal, and interfere with the growth of such crystals. In order for the crystal to continue its growth without defects the unbound molecule must either become complexed while still maintaining its lattice contacts, else it must break all lattice bonds and diffuse away from the crystal surface to be replaced by a complexed molecule. The energy involved in each of the lattice interactions that must be broken will determine the inhibitory effect of the uncomplexed molecule with respect to the growth of the complex crystal. Assuming that the number of lattice bonds is proportional to the surface area of the molecule, we expect that the inhibiting effect of an uncomplexed molecule will be roughly proportional to molecular weight. Hence a small excess of the smaller molecule is expected to be less damaging to the crystallization of the complex than an excess of the larger molecule. When the crystallization proceeds slowly, the crystal is less likely to incorporate unbound molecules as defects. If we consider the deleterious effect of incorporating a defect into the lattice the absence of the larger molecule will carry a greater energy penalty, and hence it will be less likely to occur. Therefore, it is better to have a slight excess of the smaller macromolecule when growing complex crystals, except during the initial search, when suitable growth conditions have not been established, and the most ordered nucleus is likely to occur with a slight excess of the larger macromolecule.

When the affinity between the molecules in the complex is not high, and the off rate is substantial, we must screen for possible crystallization conditions where the relative solubility of the complex is lower than that of the uncomplexed molecules. It is also important to use high concentrations of both macromolecules to push the equilibrium in favour of the complex, although the use of lower concentrations will allow for a larger number of trials. The number of trials can be minimized by using a screening approach as described in ref. 3. When crystals are obtained, streak seeding can be used to determine whether the complex or either macromolecule has crystallized.

6.2 Use of streak seeding in protein complex crystallization

The optimal ratio of the two molecules in the complex for the nucleation of crystals is not necessarily the optimal ratio for crystal growth. By using streak seeding, it is easy to determine the optimal ratio and concentrations of macromolecules experimentally from the response along the streak line. Initial trials should be started as soon as possible and success in obtaining microcrystals, microcrystalline aggregates, or even crystals can often be achieved with less than absolutely pure macromolecules, as determined by SDS–PAGE and isoelectric focusing (IEF) gels. In later stages, it is important, as in all crystallizations, to attempt to obtain higher sample purity in order to grow X-ray quality crystals (see Chapters 2 and 3). The use of affinity columns ensures that the only macromolecules that can form complexes are present in the crystallization trials. Such method can be advantageous as long as the subsequent elution conditions do not affect adversely the macromolecules being crystallized. The ability of the unliganded macromolecules to crystallize generally correlates well with purity and their ability to also crystallize complexes.

6.3 Analytical techniques for determination of crystal content

6.3.1 Protein–protein complexes

SDS–PAGE (37) can be used to determine the composition of crystals of a putative complex. Because oligomerization through the formation of disulfides may occur, reduced and non-reduced gels should be used.

Protocol 7. Preparing crystals for SDS–PAGE analysis

Equipment and reagents

- Multiwell sitting drop plate
- Seeding probe
- Syringe with capillary
- Cryo-loop
- Microscope
- Washing solution
- SDS-PAGE apparatus

Method

1. Separate several crystals from the mother liquor from which they have been growing, and wash them to remove residual mother liquor. The procedure described for the handling of crystals for macroseeding may be used here. Alternatively, the crystals can be lifted from the drop using the same probe used for steak seeding or a loop as for cryo-crystallography (Chapter 13).
2. If a skin has formed first remove the skin by running the probe around the drop. Skins are often correlated with oligomerization through the formation of disulfides (it is advisable to analyse the sample under both reducing and non-reducing conditions). Place the probe under the crystal and lift it out from the solution. Repeated attempts are generally needed.
3. Dissolve the crystals in distilled water typically with a final volume of a few microlitres. If a probe is being used, just touch the end of the probe with the crystal on top and the crystal will drop into the water and dissolve. More crystals can then be picked up if the crystals are small.
4. Check under a microscope that the crystals have dissolved. Crystals that do not dissolve under these conditions may have to be dissolved under more acidic, more basic, or higher salt conditions, or by adding urea or SDS directly to the crystals. When using urea or high salt to dissolve the crystals, the urea or the salt will have to be dialysed out before running the polyacrylamide gel. Silver staining of the gel may be needed if the crystals are small.
5. Run the solutions used for crystallization in one of the lanes. Some oligomerization will occur as a result of boiling in SDS under non-reducing conditions. This should be more accentuated for the control, which will be at higher protein concentration than the dissolved crystals.

The comparison between the control and the dissolved crystals should be able to determine the macromolecular content of the crystals.

6.3.2 Complexes with small ligands

For small charged ligands a native polyacrylamide gel electrophoresis (native PAGE) (38, 39) using a PhastGel Gradient 8–25 (Pharmacia, Piscataway, NJ), or other gel system, can be run with separate lanes for the complexed and uncomplexed macromolecules, and for the dissolved crystals (*Protocol 6*). The comparison may allow for the determination of the crystal content. This has been successful for Fab–peptide complexes. The bound peptide modifies the mobility of the Fab, and usually shows as an identifiable shift in the position and distributions of the protein bands.

6.3.3 Chemical reactions in the crystal

The channels in protein crystals are typically large enough to allow for the diffusion of many small molecules throughout the lattice. Heavy-atom derivatization relies on this fact (Chapter 13). If a chemical reaction can be done on the compounds which are presumed bound to the protein in the complex, so that colour or fluorescence is developed, the reaction can be tried on the crystals. For example, in the crystallization of steroid complexes of the anti-progesterone Fab DB3 (40), the presence of the steroids was verified before collecting X-ray data. Since progesterone has a free ketone group at position 3 on the A-ring, it reacts with 2,4-dinitrophenyl hydrazine. The reagent was diffused into the crystals, and for 5 min dilute HCl was added. As the outside of the crystals dissolved, a brownish red precipitate developed, demonstrating the presence of the steroid. As controls, the same reaction was repeated with crystals of 'uncomplexed' Fab DB3 and of an unrelated Fab. Those of the unrelated Fab remained clear. Those of the 'uncomplexed' Fab DB3, did not dissolve, but developed a light yellow colour. Subsequently it was determined that the Fab preparations contained 10–20% progesterone. Crystals of *truly* uncomplexed DB3 could be grown from steroid-free preparations, where the antibody was produced in cell culture rather than ascites.

7. Concluding remarks

The application of seeding methods in macromolecular crystallization has proven invaluable for obtaining high resolution X-ray quality crystals when conventional methods have failed. It provides a means of analysing many conditions without requiring large amounts of protein solution. Streak seeding is particularly valuable as it provides a fast method of analytical seeding with easy visualization of the results. Cross-seeding is a powerful tool to crystallize a given protein with seeds from a related protein. The application of micro- and macroseeding methods can result in the production of large single crystals for X-ray structure determination. Without such methods many projects would not have been viable.

Acknowledgements

The contribution of Dr Ian Wilson to the editing of the previous version as well as the support provided through his grants by National Institutes of Health Grants AI-23498, GM-38794, and GM-38419 is here acknowledged. The French Atomic Energy Commission (CEA) has provided support for the revisions.

References

1. Stura, E. A., Nemerow, G. R., and Wilson, I. A. (1992). *J. Cryst. Growth*, **122**, 273.
2. Jancarik, J. and Kim, S.-H. (1991). *J. Appl. Cryst.*, **24**, 409.
3. Stura, E. A., Satterthwait, A. C., Calvo, J. C., Kaslow, D. C., and Wilson, I. A. (1994). *Acta Cryst.*, **D50**, 408.
4. Carter, C. W., Jr. and Carter, C. W. (1979). *J. Biol. Chem.*, **254**, 12219.
5. Betts, L., Frick, L., Wolfenden, R., and Carter, C. W., Jr. (1989). *J. Biol. Chem.*, **264**, 6737.
6. Eichele, G., Ford, G. C., and Jansonius, J. N. (1979). *J. Mol. Biol.*, **135**, 513.
7. Wilson, I. A., Rini, J. M., Fremont, D. H., and Stura, E. A. (1990). In *Methods in enzymology* (ed. Langone, J. J.), Academic Press, London. Vol. 203, p. 270.
8. McPherson, A. (1982). *Preparation and analysis of protein crystals*. Wiley, New York.
9. Blundell, T. L. and Johnson, L. N. (1976). *Protein crystallography*. Academic Press, New York.
10. Arakawa, T. and Timasheff, S. N. (1985). *Methods in enzymology*, (eds Wychoff, H. W., Hirs, C. H. W., and Timasheff, S. N.), Academic Press, London, Vol. 114, p. 49.
11. Feher, G. and Kam, Z. (1985). In *Methods in enzymology*, ibid ref. 10. Vol. 114, p. 77.
12. McPherson, A. (1985). In *Methods in enzymology*, ibid ref. 10. Vol. 114, p. 112.
13. Vekilov, P. G., Monaco, L. A., Thomas, B. R., Stojanoff, V., and Rosenberger, F. (1996). *Acta Cryst.*, **D52**, 785.
14. Rosenberger, F., Vekilov, P. G., Muschol, M., and Thomas, B. R. (1996). *J. Cryst. Growth*, **168**, 1.
15. Thaller, C., Weaver, L. H., Eichele, G., Karlsson, R., and Jansonius, J. (1981). *J. Mol. Biol.*, **147**, 465.
16. Thaller, C., Eichele, G., Weaver, L. H., Wilson, E., Karlsson, R., and Jansonius, J. N. (1985). In *Methods in enzymology*, ibid. ref 10, Vol. 114, p. 132.
17. Stura, E. A. and Wilson, I. A. (1991). *J. Cryst. Growth*, **110**, 270.
18. Stura, E. A. and Wilson I. A. (1990). *Methods*, **1**, 38.
19. Stura, E. A., Stanfield, R. L., Fieser, T. M., Balderas, R. S., Smith, L. R., Lerner, R. A., *et al.* (1989). *J. Biol. Chem.*, **264**, 15721.
20. Stura, E. A., Johnson, D. L., Inglese, J., Smith, J. M., Benkovic, S. J., and Wilson, I. A. (1989). *J. Biol. Chem.*, **264**, 9703.
21. McPherson, A. (1989). *Sci. Am.*, **260**, **3**, 62.
22. Schiffer, M., Chang, C.-H., and Stevens, F. J. (1985). *J. Mol. Biol.*, **186**, 475.

23. Schulze-Gahmen, U., Rini, J. M., Arevalo, J. H., Stura, E. A., Kenten, J. H., and Wilson, I. A. (1988). *J. Biol. Chem.*, **263**, 17100.
24. Wilson, I. A., Bergmann, K. F., and Stura, E. A. (1986). In *Vaccines '86* (ed. R. M. Channock, R. A. Lerner, and F. Brown), pp. 33–7. Cold Spring Harbor Laboratory, Cold Spring Harbor, NY.
25. Churchill, M. E., Stura, E. A., Pinilla, C., Appel, J. R., Houghten, R. A., Kono, D. H., *et al.* (1994). *J. Mol. Biol.*, **241**, 534.
26. Stura, E. A., Stanfield, R. L., Rini, J. M., Fieser, G. G., Silver, S., Roguska, M., *et al.* (1992). *Proteins*, **14**, 499.
27. Rini, J. M., Stanfield, R. L., Stura, E. A., Salinas, P. A., Profy, A. T., and Wilson, I. A. (1993). *Proc. Natl. Acad. Sci. USA*, **90**, 6325.
28. Markland, F. S., Jr. and Smith, E. (1971). In *The enzymes*, 3rd edn (ed. P. D. Boyer), Vol. III, pp. 561–608. Academic Press, New York.
29. Kraut, J. (1971). In *The enzymes*, 3rd edn (ed. P. D. Boyer), Vol. III, pp. 547–60. Academic Press, New York.
30. Neidhart, D. J. and Petsko, G. A. (1988). *Protein Eng.*, **2**, 271.
31. Wu, Z.-P. and Hilvert, D. (1989). *J. Am. Chem. Soc.*, **111**, 4513.
32. Syed, R., Wu, Z. P., Hogle, J. M., and Hilvert, D. (1993). *Biochemistry*, **32**, 6157.
33. Stura, E. A., Ruf, W., and Wilson, I. A. (1996). *J. Cryst. Growth*, **168**, 260.
34. Stura, E. A., Ghosh, S., Garcia-Junceda, E., Chen, L., Wong, C.-H., and Wilson, I. A. (1995). *Proteins*, **22**, 67.
35. Stura, E. A., Charbonnier, J. B., and Taussig, M. J. (1999). *J. Cryst. Growth*, **196**, 250.
36. Arvai, A. S., Bourne, Y., Williams, D., Reed, S. I., and Tainer, J. A. (1995). *Proteins*, **21**, 70.
37. Laemmli, U. K. (1970). *Nature*, **227**, 680.
38. Andrews, A. T. (1981). In *Electrophoresis. Theory, techniques and biochemical and clinical applications* (ed. A. R. Peacocke and W. F. Harrington), p. 63. Clarendon Press, Oxford.
39. Hames, B. D. (1981). In *Gel electrophoresis: a practical approach* (ed. B. D. Hames and D. Rickwood). IRL Press Limited, London, Washington DC 4.
40. Stura, E. A., Arevalo, J. H., Feinstein, A., Heap, R. B., Taussig, M. J., and Wilson, I. A. (1987). *Immunology*, **62**, 511.

8

Nucleic acids and their complexes

A.-C. DOCK-BREGEON, D. MORAS, and R. GIEGÉ

1. Introduction

At first glance crystallizing nucleic acids poses the same problems as crystallizing proteins since most of the variables to investigate are alike. It is thus astonishing that crystallization data banks (1) that describe so many successful protein crystallizations are so poor in information on nucleic acids. This relies on the physico-chemical and biochemical characteristics of nucleic acids distinguishing them from proteins. The aim of this chapter is to underline features explaining the difficulties often encountered in nucleic acid crystallization and to discuss strategies that could help to crystallize them more readily, either as free molecules or as complexes with proteins. Other general principles, in particular for RNA crystallization, are discussed in ref. 2.

Among natural nucleic acids only the smaller ones provide good candidates for successful crystallizations. Large DNAs or RNAs can a priori be excluded because of their flexibility that generates conformational heterogeneity not compatible with crystallization. Thus the smaller RNAs with more compact structures (with 75–120 nt), especially transfer RNAs (tRNAs), but also 5S RNA, were the first natural nucleic acids to be crystallized (3, 4). At present attempts are being made with other RNA systems, such as ribozymes and introns, fragments of mRNA, viroids, viral and other tRNA-like RNAs, SELEX-evolved RNAs, and crystallization successes leading to X-ray structure determinations were reported for RNA domains of up to 160 nt long, with the resolution of the P4-P6 domain of the self-splicing *Tetrahymena* intron (5).

The recent excitement in nucleic acid crystallography, and particularly in RNA crystallography, have partly been due to technological improvements in the preparation methods of the molecules. Advances in oligonucleotide chemical synthesis provide opportunity for making large amounts of pure desoxyribo- and more recently of ribo-oligomers of any desired sequence. This led to the crystallization of a number of DNA and RNA fragments and was followed by the co-crystallization of complexes between proteins and such synthetic fragments. Transcription methods of RNAs from synthetic DNA templates were also essential for rejuvenating the structural biology of

RNAs. In the case of complexes of proteins with RNAs, the main difficulty was to purify large quantities of homogeneous biological material with well defined physico-chemical properties. The problem has now been overcome in many cases and problems of larger complexity are now addressed such as improved crystallizations of ribosome. Virus crystallization will also be briefly discussed.

2. Preparation of nucleic acids

General aspects on the characterization of nucleic acids and advice for the preparation of homogeneous samples aimed at crystallogenesis are also outlined in Chapter 2.

2.1 Synthetic nucleic acid fragments for crystallogenesis

2.1.1 Design of DNA fragments

Because of their large size and plasticity, genomic DNA molecules cannot be crystallized as such. Thus the first problem concerns the choice of the appropriate DNA fragments amenable to crystallization. It can be dictated by the biological significance of the fragment, methodological aspects related with its preparation, or considerations on its crystallizability potency.

Critical evaluation of results on DNA oligonucleotide crystallizations have shown that these molecules crystallize in a limited number of crystal packing families (6), the molecules adapting their conformation according to sequence and crystallization conditions. For example, in the B-form of the d(**CGC**G-AATTC**GCG**)$_2$ dodecamer duplex (7), the terminal **CGC**xxxxxx**GCG** boxes led to crystallization in the orthorhombic $P2_12_12_1$ space group by favouring formation of specific hydrogen bonds between the minor grooves of the staggered duplexes, whereas the presence of C residues at particular positions (i.e. positions 3 and 6) would allow a major groove–backbone interaction and crystallization in the trigonal R3 space group (8). The practical consequence is the possibility to design B-DNA molecules containing packing driving boxes that will guide crystallization in well defined crystal lattices (6).

The biological relevance of the sequence and the stability of the nucleic acid moiety (and for DNA:protein complexes, the stability of the complex) have to be taken into account. Also the size of the oligonucleotide and the way it is terminated, blunt or extended end, can both have advantages according to packing or stability. The best bet is to first try the sequence that is closest to the biological one. That can be obtained from sequence comparisons (9) and from structural probing data in solution (10) designed to find the most stable structures, closing helices by, e.g. C–G rather A–U base pairs. When crystallization is not successful the design can be more difficult; try to cut out what seems to provide instability and try to add or eliminate symmetry.

2.1.2 Design of RNA fragments

Except in the few cases where small RNAs have rather globular and compact structures (e.g. tRNAs, hammerhead ribozymes), RNAs are large, flexible, and multidomain molecules. Such characteristics are detrimental for crystallization and make RNA crystallization challenging. Therefore use of biological knowledge gained for instance by comparative computer analysis of RNA sequences (9) and/or structural solution studies of these molecules with chemical or enzymatic probes (11, 12) followed by computer modelling (13) is mandatory to define compact domains which in a second step can be prepared by chemical or *in vitro* transcription methods. RNA domains with pre-defined function can also be generated by combinatorial SELEX-type methods (14). Such RNA domains often contain motives, as internal loops, bulges, non-Watson–Crick pairs, pseudo-knots, which may be prepared *per se* for crystallization purposes. Examples include a synthetic RNA 12-mer that folds in a duplex structure containing two G(anti)–A(anti) base pairs (15), a 34-mer ribozyme with a 13-mer DNA inhibitor (16), an RNA 12-mer containing the *Escherichia coli* Shine–Dalgarno sequence (17), an RNA helix incorporating an internal loop with G–A and A–A base pairing (18), and a duplex RNA mimicking the amino acid acceptor stem of *E. coli* $tRNA^{Ala}$ (19). A promising approach consists of engineering crystal contacts by introducing appropriate structural modules in RNA domains. This was done by using the GAAA tetraloop:tetraloop receptor interaction and allowed crystallization of group II introns and hepatitis delta virus ribozyme RNA constructs. The method led to crystals diffracting at 3.5 Å resolution (20). Knowledge-based design of RNA motifs was also essential for the crystallization of RNA:protein complexes (21).

2.1.3 Chemical synthesis of DNA fragments

Synthesis of DNA has been automated and DNA synthesizers are commercially available. At present it is straightforward to prepare DNA fragments (up to 100 nt) and most molecular biology institutes provide this facility. Due to technical advances, production costs of deoxyoligonucleotides is decreasing and now it is often advantageous to order the oligonucleotides from industrial producers. However, large scale synthesis of DNA fragments for structural studies remains expensive and requires 10 μmol solid support resins (i.e. ~ 20 mg of a 20-mer can be synthesized in such a way).

After cleavage from the resin by ammonia, the final product is generally purified by HPLC. The expected DNA is contaminated by shorter molecules that result from incomplete reaction within a cycle of synthesis when the previous product was not fully deprotected. Other contaminants result from side-reactions, like depurination that may occur during the acid treatment for the removal of the 5′-protecting group at the end of each cycle. Purification can be performed before or after the complete deblocking of the protecting groups by ammonia (22).

The purification procedure calls for the different properties of oligonucleotides which are electrolytes with some hydrophobic regions. The resins are generally adapted to HPLC. These are ion exchangers or reverse-phase chromatography (RPC) columns. They are generally made of silica particles modified with functional groups in C4, C8, or C18. They can be used in two different ways, according to the counterion choice. With ammonium ions, retention of oligonucleotides is mainly due to hydrophobicity and this method is used to separate oligonucleotides of similar length but of different sequences. With triethylammonium ions, the ion pairing phenomena comes into play; the oligonucleotides are adsorbed to the stationary phase via their counterions. In this case the strength of the interaction is dependent on the hydrophobicity of the counterion and also on the length of the oligonucleotide, since the strength of the interaction is proportional to the charge of the oligonucleotide (22). In both cases, the oligonucleotides are eluted by increasing concentrations of acetonitrile. An advantage of the second method is that triethylammonium salts are volatile, so that products are easy to recover by lyophilization.

2.1.4 *In vitro* transcription methods for RNA preparation

Another way of producing large amounts of RNA is by *in vitro* transcription with phage polymerases. The most used is T7 RNA polymerase, which is cloned and over-produced (23). The yield of *in vitro* transcription can reach several hundred moles of transcripts per mole of template. When used for RNA production, the template is made according to the strategy described in *Protocol 1*. Note that the procedure is simplified when short molecules (12–35 nt) are synthesized since in this case the template can be synthesized directly. There are, however, a number of drawbacks that can limit the use of *in vitro* RNA synthesis:

(a) The yield depends on the sequence of the 5′-end of the RNA product, i.e. the +1 to +6 region that is part of the promoter of the polymerase. Thus, the polymerase works more efficiently when residue +1 is a G (23, 24). For poorly transcribed sequences, transcription yield may be improved by adding synthetic polyamines in incubation mixtures (25) or by optimizing experimental conditions by statistical methods (26). An alternative method consists of transcribing precursor RNAs with strong promoters that are processed at the desired position by RNase H after hybridization of the precursor with an appropriate DNA oligonucleotide (27).

(b) *In vitro* transcription produces RNAs which lack modified nucleosides. In the case of tRNAs, most transcripts are correctly recognized by their cognate synthetases but have less stable tertiary structures (28).

(c) A major problem is transcription termination, since polymerase sometimes adds one or two nucleotides at the 3′-end.

For small RNAs resulting from *in vitro* transcription, the desired molecule can be separated from the template and the nucleotide monomers by gel filtration. Transcripts can be further purified and resolved in individual species by electrophoresis in polyacrylamide gels. HPLC methods also give good results. Other methods, of more general use for large scale preparations of virtually any homogeneous RNA sequence, are based on transcription of the RNA together with flanking ribozyme sequences, so that the desired RNA is yielded after self-cleavage, and the 5′-promoter sequence as well as the heterogeneous 3′-end eliminated (29).

Protocol 1. *In vitro* synthesis of RNA by a molecular biology strategy[a]

Equipment and reagents

- DNA synthesizer (DNA oligonucleotides may be of commercial origin)
- Gel electrophoresis and HPLC equipment
- Standard reagents for molecular biology of RNA
- T7 RNA polymerase

Method

1. Construct an insert ending with restriction sites, and containing the T7 promoter and the RNA sequence. This is made by ligation of synthetic DNA oligomers chosen to hybridize unambiguously in tandem so as to give the correct, double-stranded, sequence.
2. Insert this synthetic gene into a plasmid, digested with the appropriate restriction enzymes.
3. Amplify the plasmid by cell culture.
4. Extract the DNA.
5. Linearize the DNA template at the restriction site.
6. Transcribe the DNA template in an appropriate medium containing the polymerase and the nucleotide monomers.
7. Remove the RNA from the transcription medium, and if needed purify by gel electrophoresis or HPLC.

[a] Detailed experimental procedures are described in refs 23–25.

2.1.5 Chemical synthesis of RNA fragments

Chemical synthesis is the method of choice for the preparation of short RNAs. RNA, however, is more difficult to make than DNA, because of the need to protect the 2′-OH group of the ribose. Several protecting groups have been designed and the automated solid phase method has been adapted for RNA synthesis (e.g. ref. 30). Commercial DNA synthesizers can be used with RNA monomers with the same facilities as for DNA synthesis, but giving somewhat lower yields. The great advantage of chemical upon enzymatic synthesis is that modified nucleosides, or even deoxynucleotides, can be introduced at a specific position (30), and that any sequence can be designed (the disadvantage is the lack of synthons for many of the modified nucleotides one would like to insert in synthetic RNAs). Homogeneous preparations of milligram quantities of chemically synthesized RNA, devoid of false sequences and of incompletely deprotected material, and suitable for crystallogenesis, can be obtained by anion exchange HPLC (31).

Chemical synthesis of the tetradecamer $U(UA)_6A$ in the 10 mg range has led to the crystallization of this short RNA (32). Other examples concern crystallization of a mispaired RNA double helix (33) and that of 15-mer and 19-mer RNA sequences corresponding to a hairpin or a helix with an internal loop and to an RNA pseudo-knot (31).

2.2 Preparation of natural small RNAs

2.2.1 Sources

These preparations concern essentially tRNAs, the decoding molecules of the genetic message, and 5S RNAs, a ribosomal constituent. 5S RNA can be prepared from ribosomes by a phenol extraction and is separated from tRNAs and other RNAs by molecular sieving (34). Crystals have been obtained with 5S RNA from *Thermus thermophilus* (35) and *E. coli* (4).

Bulk tRNA from yeast *Saccharomyces cerevisiae* and *E. coli* is commercialized (e.g. by Boehringer or Sigma). Purification of a single species of a yeast tRNA is better done starting from brewer's yeast bulk tRNA because of the structural integrity of their -CCA at 3′-end, which it not the case for tRNAs from baker's yeast (36). Other sources need a phenol extraction from the cells (*Protocol 2*). Thermophilic bacteria provide RNAs of potential great interest for crystallization, since these are more stable at higher temperature (37). Halophilic organisms may represent another interesting alternative for preparation of RNAs or nucleic acid–protein particles intended for improved crystallization, as was exemplified in the case of ribosomes (38).

Protocol 2. Preparation of small RNAs by phenol extraction from cells

Equipment and reagents

- Vortex and centrifuge
- Phenol
- Ethanol and ether
- Extraction buffer: 0.1 M Tris–HCl pH 7.5, 20 mM $Mg(Ac)_2$, 1 mM EDTA

Method

1. Prepare phenol by adding 50% (w/w) water. When melted add a few drops of 1 M KOH to bring the pH of the supernatant around 7.0.
2. Suspend the cells (5 ml/g) in extraction buffer.

3. Add to the cell suspension the same volume of phenol. Shake vigorously for 30 min at room temperature and centrifuge 5 min at 3000 g (room temperature). Recover the upper phase.
4. Add 10–20% of the initial volume of extraction to the phenol phase, shake vigorously, and recover again the upper phase. Mix the aqueous phases.
5. Repeat steps 3 and 4 (with some cells, like *T. thermophilus*, it is advisable to add 0.1% SDS in the aqueous phase for the second extraction). Phenol may be removed from aqueous phase by ether extraction. Caution: ether is volatile and easily flammable.
6. Precipitate RNAs with ethanol (see *Protocol 3*).

Advances in genetics has made cloning a possibility when a specific tRNA is chosen. This procedure was first used for *E. coli* $tRNA^{Gln}$ which was crystallized with glutaminyl-tRNA synthetase (39). Cloning allows over-production of a single isoacceptor and then proper purification. However, this can lead to under-modification of the tRNA, and in turn to microheterogeneous tRNA samples, when modifying enzymes become limiting for processing of large quantities of overexpressed tRNAs.

2.2.2 Purification of transfer RNAs

With natural tRNAs the purification problem is complicated since bulk tRNA contains about 60 different species with similar structures. Also the quantities of purified tRNA species needed for crystallization projects have to be in the 5–50 mg range, which excludes purification by effective micromethods (e.g. 2D gel electrophoresis).

Countercurrent fractionation is a powerful first step in a tRNA purification procedure, but is no longer operating in most laboratories because of the complexity of the instrumentation. Selectivity relies upon differential solubility of individual tRNAs in two solvents and of their distribution between an aqueous and an organic phase. The method allows large quantities material to be handled, typically 5 g, and enrichments can be excellent (40).

Different chromatographic supports interacting with the negatively charged tRNAs have been used, such as DEAE–Sephadex (41) or hydroxyapatite (42). Ionic interactions take place between phosphates of the tRNA and positively charged groups of the DEAE matrix or calcium ions of the hydroxyapatite crystals. Additional weak interactions, sensitive to the presence of Mg^{2+} ions or urea, to pH and temperature, are tuned by the tRNA structures. The resolution of such columns is limited since electrostatic forces, related to the number of accessible phosphates are poor discriminators of tRNA species. BD–cellulose provides greater resolution; it is a DEAE–cellulose modified by the addition of benzoyl groups (43). The tRNAs are sorted by electrostatic interactions between phosphate and DEAE groups, and hydrophobic interactions between the accessible bases and the benzoyl moieties. Resolutive RPC systems consisting of an inert support coated with a quaternary ammonium of high molecular weight not miscible with water were also employed (44). Several of the classical ionic exchange methods were adapted to FPLC or HPLC systems (45). We obtained excellent results with monoQ columns (Pharmacia or equivalent) which can purify to homogeneity a tRNA species starting from 20% enriched preparations.

Hydrophobic interaction chromatography (HIC) is efficient for tRNA purification. It was first used with a Sepharose 4B support and a reverse gradient of $(NH_4)_2SO_4$ (46), so that tRNAs are separated according to their solubility. The advantages of HIC are its high resolution, the possibility to process large quantities of material (1 ml of Sepharose 4B can adsorb 8 mg of unfractionated tRNA) (46), and its adaptation to HPLC (HIC has improved with the *n*-alkylated silica supports available for medium and high pressure chromatography). A resolutive HPLC system using C4-bonded silica gel has been reported (47). A disadvantage of HIC is the presence of salt at high concentration in the enriched or pure tRNA fractions. Getting rid of $(NH_4)_2SO_4$ is necessary, especially if ethanol precipitation is the following step. It can be done either by dialysis or by buffer exchange in a concentration set-up (see *Protocol 4*).

Purification of a single tRNA species results from the combination of these different methods. For crystallization, the need of large quantities of pure material ($\geq$ 5 mg) directs the choice of the first steps to methods that can handle large quantities of material; e.g. fractionation on hydrophobic matrices with elution by reverse $(NH_4)_2SO_4$ gradients. The last steps will be the most resolutive (RPC, HIC).

2.2.3 General principles for RNA handling

Beside the problem of sensitivity towards RNases which requires work in sterile conditions, RNAs are sensitive to alkaline hydrolysis (48) and therefore alkaline pH should be strictly avoided. The cleavage of the polyribonucleotide chain is also favoured by some metal ions, of which the most effective is lead (49). A chelating agent, like EDTA, is generally introduced into buffers, at a concentration 0.1–0.5 mM, to complex the traces of heavy metals. At the end of the purification the RNA preparation must be checked for its integrity. The simplest method is electrophoresis in a denaturing polyacrylamide–urea gel.

Concentrations are obtained from optical density measurements at 260 nm with

$$A_{260\,\mathrm{nm}} = \varepsilon \times c \times l$$

where ε = 25 ml/mg (this ϵ value applies for most RNAs, but for exact measurements it may be necessary to determine it experimentally) (50), l is the optical path in cm, and c the concentration in mg/ml.

Two ways described in *Protocol 3* and *Protocol 4* can be used to concentrate RNAs. The method described in *Protocol 4* is convenient to change the solvent. When the RNA is concentrated to a small volume, dilute it in the new solvent and concentrate again. Repeat several times.

Protocol 3. Concentration of nucleic acids (20–120 nt) by ethanol precipitation

Equipment and reagents

- Refrigerated centrifuge
- Deep freezer
- Ethanol

Method

1. Prepare the solution. It should contain Mg^{2+} ions (≥ 2 mM) and a Na salt such as Na acetate (≥ 10 mM, generally at pH 6.0). For good recovery, the RNA (e.g. tRNA) solution should be ≥ 0.1 mg/ml. If not, raise the Na acetate concentration to ≥ 100 mM. Take care that the solution does not contain too much salt (i.e. after a chromatography, dialyse first in water).
2. Add two or three volumes of ethanol (best quality). The precipitate forms.
3. Leave to precipitate completely at –20 °C (2 h or more) or at –80 °C (20 min or more). For the shortest oligonucleotides or low concentrations, use the lowest temperature.
4. Centrifuge at the lowest possible temperature, 10 min at ≥ 5000 g should be sufficient.
5. Dry the pellet under vacuum in the presence of solid KOH.
6. Dissolve the pellet in the desired amount of buffer.

To gain more homogeneity, the RNA solution to be crystallized is dialysed thoroughly in a buffer at low concentration of Mg^{2+}. For tRNA crystallization one can use a 2 mM $MgCl_2$ and 10 mM Na cacodylate buffer at pH 6.0. The same result can be obtained by buffer exchange. The RNA samples can be stored frozen in such solution at –20 °C or –80 °C.

Protocol 4. Concentration of nucleic acids on a membrane

Equipment and reagents

- Amicon or Centricon-type dialysis concentrators
- Dialysis membranes
- Compressed nitrogen

Protocol 4. *Continued*

Method

1. Prepare a set-up of the type Amicon (for large volumes) or Centricon (for volumes of a few millilitres). Use membranes of correct cut-off (usually 10 000).
2. Concentrate by pushing the solvent through the membrane under nitrogen pressure (for the Amicon set-up) or by centrifugation (for Centricon). The RNA (or DNA oligonucleotide) concentrates on the membrane.
3. Recover the solution when the desired volume is obtained.

3. Crystallization of nucleic acids

Several examples (arbitrarily chosen) of oligonucleotide crystallizations are given in *Table 1*. Notice that some oligonucleotides have palindromic self-complementary sequences. Such sequences are favourable for crystallogenesis and were among the first to be crystallized. Additional data are in refs

Table 1. Some examples of crystallization conditions of oligonucleotides

Sequence Precipitant	Temp (°C)	Oligo	Concentration Buffer[a]	Spermine	Mg^{2+b}	Crystals	Ref.
DNA A-form							
GGCCGGCC MPD 30%	–	1.2 mM	25 mM pH 7.0	0.6 mM	3.0 mM	$P4_32_12$ 2.25 Å	54
CTCTAGAG MPD 7 *vs* 50%	18	1.2 mM	60 mM pH 6.8	1 mM	25 mM	$P4_12_12$ or $P4_32_12$ 2.15 Å	55
GTACGTAC MPD 5 *vs* 30%	20	2.0 mM	14 mM pH 6.0	8 mM	15 mM	$P4_32_12$	56
DNA B-form							
CGCATATATGCG MPD 10 *vs* 40%	–	0.5 mM	–	0.4 mM	22 mM $Mg(Ac)_2$	$P2_12_12_1$ 2.2 Å	57
CCAAGATTGG, with G:A mismatch MPD 45%	4	3.0 mM	–	None	0.7 M	C2 1.3 Å	58
5′-ACCGGCGCCACA TGGCCGCGGTGT-5′ MPD 40%	4	1.0 mM	50 mM pH 6.0	1.2 mM	18 mM $Mg(Ac)_2$	R3 2.8 Å	8
CCAGGC^{Me}CTGG MPD 40% microdialysis	4	2.0 mM	20 mM pH 7.5	0.0 mM	50 mM	P6 2.25 Å	59

Table 1. *Continued*

Sequence Precipitant	Temp (°C)	Oligo	Concentration Buffer[a]	Spermine	Mg^{2+b}	Crystals	Ref.
DNA Z-form							
CGCGCG Isopropanol 5%		2 mM	30 mM pH 7.0	10 mM	15 mM	$P2_12_12_1$ 0.9 Å	60
m^5CGTAm^5CG MPD 8 *vs* 50%	–	4 mM	30 mM pH 7.0	7 mM	10 mM	$P2_12_12_1$ 1.2 Å	61
$(^5BrCG)_3$ MPD 10 *vs* 60%	18 or 37	0.5 mM	20 mM pH 6.5	–	200 mM NaCl	$P2_12_12_1$ 1.4 Å	62
m^5CGUAm^5CG MPD 8.5 *vs* 30%	Room	4 mM	28 mM pH 7.0	–	15 mM	$P2_12_12_1$ 1.3 Å	63
Four-stranded intercalated DNA							
CCCC MPD 20%	–	2.7 mM	100 mM pH 5.5	–	–	I23 2.3 Å	64
RNA:DNA hybrid							
r(GCG)d(TATACGC) MPD 40%	–	1.5 mM	30 mM pH 6.0	8 mM	15 mM	$P2_12_12_1$ 1.9 Å	65
RNA							
$U(UA)_6A$ MPD 35%	35	4 mM	40 mM pH 6.5	None	0.4 M	$P2_12_12_1$ 2.25 Å	32
5′-GGCC(GAAA)GGCC-3′, with internal loop PEG 400 30%	Room	2 mM	50 mM Tris pH 7.5	–	5 mM $MnCl_2$ + 20 mM NaCl	$P6_522$ 2.3 Å	18
5′-GGGGCUA[c] CCUCGAU-5′ MPD 6 *vs* 35–45%	25	1 mM	12.5 mM pH 6.5	1 mM	50 mM $MgSO_4$	C2 1.7 Å	19
DNA:drug complexes							
CGCG + ditercalinium MPD 6 *vs* 30%	Room	0.7 mM + 0.2 mM drug	16.8 mM pH 6.0	0.3 mM	0.8 mM + 14 mM NH_4Ac	$P4_12_12$ 1.7 Å	66
CGCGAATTCGCG + berenil MPD 20 *vs* 50%	5	3 mM + 2 mM drug	10 mM pH 7.0	None	30 mM	$P2_12_12_1$ 2.5 Å	67

[a] When the crystallization medium is buffered, the buffer is always sodium cacodylate.
[b] Most often, $MgCl_2$; in other cases, the salt is specified.
[c] Amino acid accepting stem of $tRNA^{Ala}$ with G:U mismatch.

51 and 52. For tRNAs, a compilation of crystallization conditions is given in ref. 3. Other general ideas on RNA crystallization can be found in refs 2 and 53.

3.1 General features

3.1.1 Crystallizing agents and concentration of nucleic acids

The more widely used are alcohols, and especially methyl 2,2 pentane diol (MPD) which is not volatile and therefore easy to handle. It is used in the range of 10% (v/v) in the case of tRNAs and 30% in that of oligonucleotides. Isopropanol has also given good results with tRNAs, and especially with $tRNA^{Phe}$ (68). For tRNAs another successful crystallizing agent is $(NH_4)_2SO_4$. It gave good results with yeast $tRNA^{fMet}$ (69) and yeast $tRNA^{Asp}$ (70). Polyethylene glycol (PEG) precipitates tRNAs at concentrations of a few per cent for a medium sized PEG (M_r 4000–8000) and a different crystal form of yeast $tRNA^{Asp}$ could be obtained with PEG (3). Crystals of the Z-DNA hexamer $d(CG)_3$ were obtained with smaller sized PEG (61). A similar observation came from the crystallization of RNA oligomers (71). Mixtures of PEG or MPD with NaCl or NH_4Cl, or $(NH_4)_2SO_4$ are also interesting possibilities; the salt acts as an electrostatic shield and modulates the interaction between RNA and additives.

For tRNAs, the crystallization is generally tried in the order of 5–20 mg/ml, i.e. 0.2–0.8 mM. Higher molar concentrations are generally used for oligonucleotides (*Table 1*).

3.1.2 Temperature, pH, and buffers

The temperature stability of nucleic acids allows examination of a large range of temperatures, from 4 °C (usual cold room temperature) to 30 °C or 35 °C (in a bacteriologic incubator or an oven). The 35 °C assays bring new paths towards crystallization especially when mixed precipitants are tried since it modifies phase partitions.

The pH appears to play a smaller role in nucleic acid than in protein crystallization where the overall charge of a protein, and then its capacity of packing in a certain way, may be tuned through pH variations. The situation is quite different with nucleic acids, which are negatively charged polyelectrolytes. At pH 4.0–5.0 cytidines are protonated, and such a pH range can therefore promote crystallization, when there is an accessible cytidine, by introducing a potential additional interaction. A too low pH could, however, induce local structural artefacts. Taking pH into account is also of importance for mismatched oligonucleotides. In the case of RNA the problem of degradations forbids use of alkaline pH.

The buffer is often Na cacodylate (pH range 6.0–7.0) which pH is rather temperature-insensitive and has the additional advantage of preventing bacterial growth (a problem in PEG). In $(NH_4)_2SO_4$, the buffer concentration must be high enough (i.e. 100–300 mM) to maintain pH against variations due to ammoniac evaporation (72) (Chapter 5). In low ionic strength media (PEG or MPD) the buffer itself can introduce an electrostatic shield, and variations of its concentration may modulate the electrostatic interactions between nucleic acids and additives.

3.2 Specific features: additives

Nucleic acids are polyelectrolytes and therefore the counterions are important additives for crystallization. Two families of cations are generally used, polyamines and divalent cations. Their role in crystallization differs subtly and parallels their structural effects.

3.2.1 Polyamines

Polyamines are involved in many biological processes including DNA condensation and protein synthesis (73). Some examples of natural polyamines are given in *Table 2*. They often enter in crystallization media but their presence is not always needed (*Table 1*). Spermine is the most used. It is a linear molecule with four positive charges at neutral pH and became popular because of its key role in the production of the first crystals of a tRNA (74). After growth of high diffracting yeast $tRNA^{Phe}$ crystals (reviewed in ref. 3), spermine was systematically tried with nucleic acids, including oligonucleotides. Spermidine, which is an asymmetric molecule bearing three positive charges at neutral pH, was also reported to promote crystal growth, but with less success. Positive effects, including resolution improvements, of several synthetic cyclic polyamines on $tRNA^{Phe}$ crystallization were recently reported (75).

Spermine binds in the grooves of nucleic acids. The refinement of the structure of yeast $tRNA^{Phe}$ has identified two spermine molecules (76). One is coiled in the deep major groove of the anticodon arm of the tRNA, at the junction of D- and T-stems. It is H-bonded to four phosphates on both sides of the groove. The second spermine molecule interposes a string of positive charges between the extended polynucleotide chain of the variable region and the P9-P10 sharp turn. Spermine has also been identified in crystals of Z-DNA oligonucleotides (60, 77) but was not found yet in crystals of A-type oligonucleotides.

Table 2. Natural polyamines used in nucleic acid crystallization

Putrescine	$H_2N\text{-}(CH_2)_4\text{-}NH_2$
Cadaverine	$H_2N\text{-}(CH_2)_5\text{-}NH_2$
Spermidine	$H_2N\text{-}(CH_2)_3\text{-}NH\text{-}(CH_2)_4\text{-}NH_2$
Thermine	$H_2N\text{-}(CH_2)_3\text{-}NH\text{-}(CH_2)_3\text{-}NH\text{-}(CH_2)_3\text{-}NH_2$
Spermine	$H_2N\text{-}(CH_2)_3\text{-}NH\text{-}(CH_2)_4\text{-}NH\text{-}(CH_2)_3\text{-}NH_2$

3.2.2 Divalent cations

Divalent cations, and especially Mg^{2+} ions, are involved in the stabilization of nucleic acids structures and play an important role in their functions. Crystallography has given a first insight into the structural effect of Mg^{2+} on the conformation of tRNA. Preferential Mg^{2+} sites are located mostly in the non-helical regions of the tRNA molecule and appear to stabilize loops and bends of the tertiary structure. Some of these Mg^{2+} sites are of interest for crystallization, since they are bridging tRNA molecules and therefore seem to stabilize the crystal packing (e.g. one Mg^{2+} in the D-loop of the refined $tRNA^{Phe}$ structure) (78). Structures of oligonucleotides in the A- or B-helical forms, refined to better resolution than tRNAs, unfortunately have brought little additional information about the preferred co-ordination of Mg^{2+} ions. The Z-structures, on the contrary, give generally more details about ion binding, a consequence of their better resolution. Examples of Mg^{2+} binding to Z-DNA can be found in the structure of d(m^5CGTAm^5CG) (61). One Mg^{2+} ion is surrounded with six oxygen atoms, one of which is a phosphate oxygen of the backbone and the others are water molecules. Other examples are found in the structure of $d(CG)_3$ (79) or d(CGTACGTACG) (80) where intermolecular Mg^{2+} sites are described. The presence of such sites confers probably, with the H-bonding possibilities, an increased stability to the crystal packing, and may explain why Z-DNA crystals often diffract to higher resolution.

Other divalent cations can be used instead of Mg^{2+}, or in addition to it. For tRNA crystallization, different divalent cations have been tried, like manganese, calcium, cobalt, nickel, barium, mercury. Care must be taken, however, since some metal ions may induce hydrolysis of the phosphodiester bonds in RNA, especially lead (49). Crystallographic structures of mono- or dinucleotides give an insight on the mode of binding of several ions to nucleic acids, e.g. calcium binding to ApA (81). These ions sometimes provide stabilization of local structures or new packing possibilities. More complex ions can also be tried, of which cobalt hexamine is an interesting case. It stabilizes Z-DNA with an efficiency that is five orders of magnitude greater than Mg^{2+}. Cobalt hexamine favoured crystallization of $d(CG)_3$ (79) and d(CGTACGTACG) (80) in the Z-form. Cobalt hexamine was also identified as an helix-stabilizing agent in the case of $tRNA^{Phe}$ (82).

3.2.3 Monovalent ions

The example of the cluster of ions in d(m^5CGTAm^5CG) (61) has shown that Na^+ ions can also play a role in helix stabilization, and therefore can favour the crystallization of nucleic acids. Na^+ only, without Mg^{2+}, was used for the crystallization of d(Br^5CG)$_3$ in the Z-form (62), and the structure shows how Na^+ bridges two neighbouring molecules in the crystal. Compared to Mg^{2+}, the octahedral co-ordination of Na^+ is less precisely defined and in certain cases can be accommodated more easily.

3.2.4 Concentration of the counterions

A important parameter is the relative concentration of spermine and Mg^{2+}, as well as the ratio spermine or Mg^{2+}/nucleic acid molecule. For magnesium a 'rule of thumb' for first trials is 0.5–1.0 Mg^{2+} ion per phosphate (e.g. for a tRNA at 0.4 mM corresponding to a phosphate concentration of 30 mM, try Mg^{2+} concentrations of 15 mM and 30 mM). Smaller or larger concentrations may be tried if results are disappointing; e.g. in this range the tetradecamer $U(UA)_6A$ did crystallize readily, but crystals showed poor diffraction (with maximal resolution of $\sim$ 7 Å). The best crystals were obtained at a Mg^{2+} concentration of 400 mM (32). For spermine the 'rule of thumb' is one spermine molecule for 10–12 bp (e.g. for a tRNA at 0.4 mM the spermine concentration to try is 3 mM). Since spermine and Mg^{2+} act as counterions, the ionic strength of the medium has to be taken into account; in $(NH_4)_2SO_4$ solutions or when monovalent salts are added, the concentrations of Mg^{2+} and spermine have to be somewhat higher than in PEG or MPD. The relative concentration of spermine and Mg^{2+} is also to be considered; at higher Mg^{2+} concentration, higher spermine concentrations can be tested since Mg^{2+} brings its own shielding effect. An excess of spermine, especially at low ionic strength, often produces crystals which do not diffract. Some assays without spermine should also be tried (examples in *Table 1*).

3.3 Crystallization strategies

3.3.1 Design of crystallization conditions

As for proteins, parameters influencing crystallization of nucleic acids are numerous. As seen above a number of them are of a special type, like those mediated by polyamines or metal ions, and should therefore be always assayed. Nevertheless an extensive screen of possible parameters remains demanding in terms of macromolecular material needed. Therefore a factorial design of crystallization experiments is advised (Chapter 4). A more pragmatic approach, like in the protein field, is the use of crystallization condition sparse matrices. Several such matrices, primarily designed for RNA crystallization trials, have been described (83, 84). Commercial screening kits designed for crystallizing nucleic acids and their complexes are available (Nucleic Acid Mini Screen or Natrix™, Hampton Research, Laguna Hills, CA) and could be used as a start. However, design of new or refined sparse matrices, based on the increasing knowledge of experimenters, should not be forgotten.

3.3.2 Refinements of crystallization conditions

Formation of crystals of poor quality is a often encountered with nucleic acids. This is often the case of crystals obtained after rapid screening with sparse matrices. For small duplexes, this may be due to the geometry of the helices, which can pack easily despite rotational disorder. The answer is to play with additives, temperature, pH, trying to find a way of introducing structural change or to bind additional small molecules which could act as a lever promoting lattice building.

3.3.3 Engineering crystallization and heavy-atom derivatives

Sequence variations in nucleic acid oligomers may be a more powerful strategy for obtaining high quality crystals than variations in crystallization conditions. This mostly applies to structural characteristics of particular nucleic acid sequences that favour ordered assembly of the molecules. For instance overlapping sequences in DNA duplexes that allow H-bonding with the neighbouring molecules (85) or anticodon/anticodon and stacking interactions in tRNA (86) have been shown to trigger crystallization. Such effects have been rationalized for the crystallization of B-DNA oligomers (6) and were discussed for that of RNA domains (20, 87).

Thus in case of unsuccessful crystallizations it may be advantageous to engineer the nucleic acid sequence by introducing structural elements that favour packing interactions (20) or stabilize the nucleic acid structure. For RNAs it can be advised to remove CpA sequences that are preferential hydrolytic cleavage points (88) or to introduce stable tetraloops (20).

Preparation of heavy-atom derivatives often remains another bottleneck for structure determinations, especially in the nucleic acid field, since the polyanionic nature of nucleic acids offers too many possibilities of metal chelations. However, the synthetic procedures for nucleic acid preparations allow incorporation into DNA or RNA oligomers of brominated or iodinated nucleotides at well designed positions which in principle can serve for phase determinations. This engineering strategy was used to prepare heavy-atom derivatives of the large P4-P6 160 nt domain of *Tetrahymena* intron (89).

3.4 The special case of DNA:drug co-crystallizations

The pharmacological importance of DNA:drug complexes is obvious and explains the interest of structural biologists to crystallize them. As compared to free oligonucleotides, crystallization of drug complexes does not show any particular features (52). Two typical examples are displayed in *Table 1* and concern an intercalating (66) and a minor groove binder (67) drug.

4. Co-crystallization of nucleic acids and proteins

Many basic biological mechanisms and particularly those regarding storage and expression of the genetic message involve interactions between proteins and nucleic acids. This promoted a need for 3D structural knowledge. When the nucleic acid moiety of the complex of interest is a small ligand, like nucleotides or short oligonucleotides, it is sometimes possible to diffuse it into a 'receptor crystal' (e.g. dT_4 in Klenow fragment of DNA polymerase I from *E. coli*) (90). With larger substrates that cannot penetrate, or when much conformational changes occur, co-crystallization is a necessity.

A first practical advice for newcomers is to consider crystallization of a protein:nucleic acid complex as a new problem, different from the crystallization of the protein or nucleic acid alone. Nevertheless, knowledge on solubility and other behaviours of the free components can guide the search of crystallization conditions for the complex.

From the early 1980s to now more than 60 different DNA binding proteins have been co-crystallized with DNA oligonucleotides of biological significance (91), and ~ 20 RNA binding proteins with natural (tRNAs) or synthetic RNA substrates (see *Tables 3* and *4*). When looking at the successful attempts, one is impressed by the increasing number of high quality crystals of complexes with synthetic deoxyoligonucleotides with diffractions that can pass the limit of 2 Å resolution. Noticeable, in several cases crystallization of the complex was more straightforward than that of the protein alone. Similarly, RNA:protein complexes often crystallize more readily than the individual components. The reason for that is the conformational stabilization of the nucleic acid and/or protein components in the complex. After the first successes of crystallized DNA:protein complexes of prokaryotic origin, like restriction enzymes or phage repressors, the tremendous improvement of protein over-production and purification methods (Chapters 2 and 3) enabled structural biologists to tackle new challenges in the eukaryotic world. Thus complexes of DNA fragments with nuclear receptors or transcription factors could be crystallized. For RNA:protein complexes, a number of successes were obtained with components originating from thermophilic organisms. The higher stability of the proteins and RNAs from such organisms, certainly is the key factor explaining the improved crystallizability of the complexes.

4.1 General features of nucleic acid:protein co-crystallization

4.1.1 Co-crystallization or crystallization of pre-existent complexes

When dealing with complexes, a variety of situations can be encountered and the crystallization strategy must adapt to each particular problem. Some complexes can be isolated and purified from natural sources (e.g. viruses, ribosomes, nucleosome) while many others are transient and require independent purification of each component. Sometimes, a purification step of the reconstituted complex is advisable.

The heterologous nature of complexes introduces additional problems. Protein nucleic acid recognition involves specific interactions between macromolecules of different electrostatic properties, and for a given protein the binding areas are adapted to this complementarity. In order not to be a competitive site, the other part of the molecule will act as a repellent to the nucleic acid substrate. In many crystal structures of complexes the crystal packing is built upon contacts between macromolecules of the same type, i.e. protein:protein or nucleic acid:nucleic acid interactions.

4.1.2 Stability

Difficulties may arise regarding the stability of the complex under crystallization conditions (pH, ionic strength, and so on) and it is important to ascertain the physical existence of the particle in such conditions. A K_d value of 10^{-5} M is an upper limit of stability for a crystallizable complex. The time scale of experiments creates another problem when nucleic acids are substrates of enzymatic reactions. If RNA is the substrate it is now possible to chemically synthesize a mixed nucleic acid with the reactive ribo- being replaced by a deoxyribonucleotide (30). When DNA is the substrate various solutions have been found, like pH changes or removal of the cations necessary for the enzymatic reactions; e.g. omit Mg^{2+} in the co-crystallization of *Eco*RI (92) or add a chelating agent for Klenow fragment (90). In special cases, as with DNase I, the cleaved oligonucleotide was co-crystallized with the enzyme (93).

4.1.3 Homogeneity of samples, stoichiometry, and purity

A problem which is specific to natural samples containing large nucleic acids like nucleosome and ribosomes, is the heterogeneity of the nucleic acid part of the samples. Even if the size of the nucleic acid component can be defined with some accuracy, the random dispersion of the nucleotide sequence is a major problem. In the case of nucleosome core particle, this problem was one of the major limitation to the formation of high resolution diffracting crystals. Chemical synthesis or *in vitro* genetic engineering techniques, which enables a large scale preparation of long oligonucleotide sequences up to a hundred nucleotides, can provide solutions. Rendering DNA homogeneous in nucleosome particles enabled quality crystals to be produced and to solve their structure at 2.8 Å resolution (94).

A slight excess of substrate is a general trend of all experiments. For the complex between yeast $tRNA^{Asp}$ and aspartyl-tRNA synthetase, where stoichiometry was well analysed, a variation of the tRNA concentration around the stoichiometric value (2:1) was the main cause of polymorphism (95). In some cases, however, excess of DNA over protein concentration was used, as for the crystallization of a λ repressor fragment (1–92) with a 20-mer operator, and the correct stoichiometry, one DNA duplex per protein dimer, was found in the crystals (96).

Purity is of general concern (Chapter 2). Since we are dealing with two molecules the problem is more crucial here. An illustration of the importance of the nucleic acid purity is given with crystallization of the operator binding domain of the λ repressor with the λ operator site. Much better crystals were obtained when the synthetic operator was further purified with HPLC (96).

For the crystallization of yeast aspartyl-tRNA synthetase with $tRNA^{Asp}$, improvement of the protein purification protocol produced a new and better-diffracting crystal form (97).

4.1.4 Crystallizing agent and pH

High salt conditions were long believed to be unfavourable to the stability of nucleic acid protein complexes. This was the main reason for the success of crystallization attempts with alcohols, and among them MPD is the most popular (in the range 15–25%). However, many successful attempts were realized with $(NH_4)_2SO_4$ at high concentration (in the range of 2 M). This can be explained by a screening effect of ammonium and/or sulfate ions which hamper non-specific contact and prevent aggregation. It is an experimental fact that high salt concentrations are disruptive of complexes. However high concentrations of $(NH_4)_2SO_4$ or ammonium citrate do not have such disruptive effects and thus can be used in crystallization attempts. Despite their disruptive effects, salts such as NaCl can also sustain crystallization, as for phage λ Cro repressor complexed with its operator (98). We believe that the salts have stronger disruptive effects on non-specific than specific complexes. Thus choice of adequate amounts of salt may favour formation of homogeneous samples of specific complexes and hence their crystallization. Along these lines, mixtures of salts and PEGs are also of particular interest. Slightly acidic or neutral pH seem the best bet although attempts at slightly basic pH (7.5–8.0) are not uncommon.

4.1.5 Additives

For this part we enter in more specific problems linked to the nature of the systems investigated. $MgCl_2$ and $CaCl_2$ are the most common additives used. Phosphate salts have to be avoided for two reasons: they often lead to insoluble compounds and act as competitors for nucleic acid binding sites. When existing, cofactors or small substrates (like ATP, GTP, L-tryptophan) should be used as an important variable in crystallization screenings.

4.2 Complexes of synthetic oligodeoxynucleotides and proteins

Table 3 illustrates the diversity of complexes that were crystallized and the diversity in crystallization conditions employed. Crystals have been obtained with salts, alcohols, or PEGs as crystallizing agents. A tendency that emerges from a survey of recent crystallizations of DNA:protein complexes is the usefulness of PEGs of rather low molecular weight and of mixtures of medium sized PEGs with salts. Of frequent use is the addition of protein stabilizing agents, like glycerol or ethylene glycol (111). Interestingly, and in contrast to what observed for RNA:protein complexes, spermine is used as an additive in many cases.

Table 3. Sampling of crystallization conditions for DNA:protein complexes

Protein and DNA Precipitant[a]	Temp (°C)	Concentration Protein[b]	DNA	Buffer	Additives	Crystals	Ref.
Phage 434 repressor (fragment 1–69) + operator (14-mer, symmetric, blunt ends)							
$(NH_4)_2SO_4$ 1.3 M	4	0.5 mM (2:1)	0.25 mM	Na phosphate 5 mM pH 4.7		I422 3.2–4.5 Å	99
Phage 434 repressor (fragment 1–69) + operator (20-mer, asymmetric and complementary overhangs of 1 nt)							
PEG 3000 12–14%	4	2 mM	1 mM		NaCl 100 mM, $MgCl_2$ 120 mM, spermine 2 mM	$P2_12_12_1$ 2.5 Å	100
Phage λ repressor (fragment 1–92) + operator (20-mer, asymmetric, complementary overhangs of 1 nt)							
PEG 400 10 *vs* 20%	20	0.91 mM	0.91 mM	BTP 15 mM pH 7.0	NaN_3 1 mM	$P2_1$ 2.5 Å	96
Phage λ Cro repressor + operator (17-mer, asymmetric, blunt ends)							
NaCl 0.1 M *vs* 3.5–4.0 M or slow evaporation	–	2.5 mg/ml	2.5 mg/ml	Na cacodylate 20 mM pH 6.9		$P6_2$ ($P6_4$) 3.7 Å	98
E. coli trp repressor + Trp + operator (18-mer + overhanging 5′-T)							
MPD 20 *vs* 40%	20	0.4 mM	0.6–0.8 mM	Na cacodylate 10 mM pH 7.2	L-Trp 2 mM $CaCl_2$ 11 mM	$P2_1$ 2.5 Å	101
E. coli CAP protein + cAMP + DNA binding site (30-mer + overhanging 5′-G)							
PEG 3350 5–10%	–	4–6 mg/ml	1.5-fold molar excess	MES 50 mM pH 5.0–6.0	NaCl 0.2 M, $CaCl_2$ 0.1 M, cAMP 2 mM, spermine 2 mM, 0.02% NaN_3, DTT 2 mM, 0.3% *n*-octyl-glucoside	$C222_1$ 3.0 Å	102

E. coli Klenow fragment of Pol I + DNA substrate (8 bp + 3 bases, single-stranded 5′ overhang)							
$(NH_4)_2SO_4$ 38%	–		2-fold molar excess		EDTA 1 mM		90
E. coli restriction endonuclease *Eco*RI + DNA substrate (13-mer with overhanging 5′-T)							
PEG 400 8 *vs* 16%	4	2.7 mg/ml	2.8 mg/ml	BTP 40 mM pH 7.4	NH_4Ac 0.5 M, dioxane 15%	P321 2.6 Å	92
Bovine pancreatic DNase I + DNA substrate (8-mer, nicked)							
PEG 600	4				EDTA 15 mM	$C222_1$ 2.0 Å	93
GCN4 (leucine zipper protein) + DNA substrate (20-mer pseudopalindrome with complementary overhangs of 1 nt)							
PEG 400 12%	22	0.95 mM	0.57 mM	MES 25 mM pH 5.8	$MgCl_2$ 30 mM, spermine 1 mM, NaCl 0.15 M	$P2_12_12_1$ 2.9 Å	103
GLI (Zn finger protein) + DNA substrate (21-mer with complementary overhangs of 1 nt)							
PEG 400 20–25%	22	0.5 mM	0.6 mM	BTP pH 7.0	$CoCl_2$ 1 mM, $MgCl_2$ 60–100 mM	$P2_12_12_1$ 2.6Å	104
HNF-3/fork head + DNA substrate (13 bp with blunt end)							
NH_4 acetate 550 mM salting-in	4	Complex 1 mM		K acetate 20 mM pH 5.5	KCl 100 mM, $MgCl_2$ 2 mM, DTT 20 mM	$P3_1$ 2.5 Å	105
γδ Resolvase + DNA substrate (34 bp)							
PEG 3350 15 *vs* 30% + ethylene glycol 2.5 *vs* 5%	Room	0.36 mM	0.72 mM	MES 50 mM pH 6.0 + Tris 10 mM pH 8.0	EDTA 0.5 mM, $(NH_4)_2SO_4$ 0.2 M	$P2_12_12_1$ 3.0 Å	106
Oestrogen receptor DNA binding domain + DNA substrate (17 bp with overhanging 1 nt)							
MPD 10%	20	Complex 70 μM		MES 20 mM pH 6.0	Spermine 1.8 mM, $ZnCl_2$ 2 μM, $CaCl_2$ 2–8 mM, NaCl 30–80 mM	$P2_12_12_1$ 2.4 Å	107

Table 3. *Continued*

Protein and DNA Precipitant[a]	Temp (°C)	Protein[b]	DNA	Concentration Buffer	Additives	Crystals	Ref.
Yeast TATA binding protein + DNA substrate (29 nt:12 bp stem and 5 nt loop)							
PEG 8000 15 *vs* 30%	22	0.25 mM	0.5 mM	BTP 20 mM pH 7.5	NaCl or KCl 500 mM, glycerol 3.75%, ethylene glycol 1%	$P4_3$ 2.5 Å	108
Heterodimeric transcription factor c-Fos–c-Jun + DNA substrate (20 nt with complementary overhangs of 1 nt)							
PEG 400 5.5–7.5 *vs* 11–15%	20	0.5 mM	0.6–0.8 mM	bisTris 50 mM pH 6.7	NH_4acetate 100–200 mM, NaCl 150 mM, $MgCl_2$ 27.5 mM, spermine 1.0 mM, DTT 5–10 mM	$P2_12_12$ 3.0 Å	109
Ternary complex TBP:TFIIB:DNA substrate (16 bp blunt end)							
Salting-in from NH_4 acetate 300 mM	4	Complex 0.3–0.4 mM		Tris–HCl 40 mM pH 8.5	KCl 40 mM, $MgCl_2$ 5 mM, $CaCl_2$ 5 mM, DTT 10 mM, Zn acetate 10 μM, glycerol 10%, ethylene glycol 2%	$P2_12_12_1$ 2.7 Å	110

[a] The concentration in the reservoir is given, or initial concentrations in the form: C(drop) versus C(reservoir).
[b] The molar ratio of protein monomers versus DNA duplex is given in brackets.
[c] When several DNA duplexes were tried, the conditions indicated are those producing the best crystals.

The main problem with binary complexes is the choice of the best DNA sequence and of its optimal length. Two major constraints have to be taken into account: the biological relevance of the sequence and the stability of the duplex. An effect of the number of base pairs (which should have been a multiple of seven) was thought to be important after the co-crystallization of the DNA binding domain of phage 434 repressor and its operator (99). Later examples were no longer in this line. Some co-crystallizations were made with blunt-ended oligomers: e.g. phage λ Cro repressor with a 17-mer operator (98) or phage 434 repressor with a 14-mer operator (99). Others underline the importance of overhanging nucleotides. These could reinforce the end-to-end stacking of DNA duplexes which seem to be a common mode of packing. Clearly, there is no generally applicable rationale that specifies the optimal length and terminal structure of the oligonucleotides to be used in crystallizing protein:DNA complexes. The principal limitation of the choice seems to be the production of the oligonucleotides, especially if the required sequence is large. This problem has been nicely overcome in the crystallization of the CAP protein complexed with DNA (102). Ten oligonucleotides, up to 20 nt in length, were synthesized. These are able to self-hybridize and were mixed to generate 19 different double-stranded segments (of 28–36 bp) with symmetric overhangs of zero, one, or two bases. Crystallization conditions were examined with 26 different DNA segments, 28 or more bp in length, that explored a variety of sequences (symmetric or not), length, and extended 5′- or 3′-termini. Crystals of variable quality were produced, one of them diffracting to 3.0 Å resolution.

4.3 Complexes of RNAs and proteins

Complexes between tRNAs and their cognate aminoacyl-tRNA synthetases were the first examples of co-crystallization of proteins and RNAs. Their stability range is not very high (K_d values within 10^{-6} to 10^{-9} M). Well characterized crystals, which led to high resolution structure determinations, were first obtained in the *E. coli* glutamine (39) and the yeast aspartate (95, 97) systems. More recently, several other tRNA:synthetase complexes have been crystallized: e.g. the phenylalanine complex from *T. thermophilus* with a tetrameric synthetase (115), a serine complex with a long variable loop tRNA (116), the *E. coli* glutamine complex with unmodified $tRNA^{Gln}$ (113), the heterologous aspartate complex between the *E. coli* synthetase and yeast tRNA (112), and the lysine complex, either homologous with *T. thermophilus* partners but unmodified tRNA or heterologous with the thermophilic synthetase and *E. coli* $tRNA^{Lys}$ (mnm^5s^2UUU) (114), as well as other RNA: protein complexes, e.g. the RNA binding domain of the U1A spliceosomal protein with an RNA hairpin (118) or EF-Tu with $tRNA^{Phe}$ (117).

An initial limitation in the field has been the poor understanding of the conditions leading to complex formation in the presence of crystallizing agents.

Table 4. Sampling of crystallization conditions for RNA:protein complexes

Stoichiometry	Concentration RNA and protein	Temp (°C)	pH	Buffer	Precipitant	Additives	Ref.
Aminoacyl-tRNA synthetase:tRNA complexes							
AspRS:tRNA^{Asp} (yeast)							
1:2	[protein] 10 mg/ml (80 μM) [tRNA] 4.8 mg/ml (190 μM)	4	7.5	Tris–maleate 40 mM	$(NH_4)_2SO_4$ 1.0 *vs* 2.4 M	$MgCl_2$ 5 mM	97
AspRS (*E. coli*)**:tRNA^{Asp}** (yeast) heterologous complex							
1:2	[protein] 2 mg/ml [tRNA] 1 mg/ml seeding required	17	6.7	BTP 75–100 mM	$(NH_4)_2SO_4$ 1.5 *vs* 1.9 M	$MgCl_2$ 1 mM, AMP-PCP 0.5 mM, aspartic acid 1 mM	112
GlnRS:tRNA^{Gln} (*E.coli*)							
1:1	[complex] 10 mg/ml	17	6.8–7.0	Pipes 80 mM	Na citrate 44–64%	$MgCl_2$ 20 mM, ATP 4 mM, 2-mercaptoethanol 20 mM	39
or							
			7.0–7.5	Pipes 80 mM	$(NH_4)_2SO_4$ 1.8–2.0 M	$MgSO_4$ 20 mM, ATP 8 mM, NaN_3 0.02%	
GlnRS:tRNA^{Gln} (*E. coli*), but with unmodified tRNA transcript							
1:1	10–15 mg/ml	17	6.5–7.5	Pipes 80 mM	$(NH_4)_2SO_4$ 1.6–2.0 M	$MgSO_4$ 20 mM, ATP 8 mM, NaN_3 0.02%	113
LysRS (*T. thermophilus*)**:tRNA^{Lys}** (*E.coli*) + lysyl-adenylate analogue							
1:2	[protein] 4 mg/ml [tRNA] 2.5 mg/ml	–	7.6	Tris–maleate 50 mM	$(NH_4)_2SO_4$ 24–26%	$MgCl_2$ 10 mM, NaN_3 1 mM, Lys-AMS 325 μM	114

PheRS:tRNA^{Phe} (*T. thermophilus*)							
1:1	[protein] 5–7 mg/ml	15	7.2	Imidazole 20 mM	$(NH_4)_2SO_4$ 15 *vs* 25–30%	$MgCl_2$ 1mM	115
SerRS:tRNA^{Ser} (*T. thermophilus*)							
1:2	[protein] 5.6 mg/ml [tRNA] 2.6 mg/ml	20	7.2	Tris–maleate 25 mM	$(NH_4)_2SO_4$ 20 *vs* 32%	NaN_3 1 mM, $MgCl_2$ 2.5 mM	116
Other protein:RNA complexes							
EF-Tu (*T. aquaticus*)**:GDPNP:Phe-tRNA^{Phe}** (yeast)							
1:1	[complex] 15 mg/ml	4	6.7–7.0	Tris 20 mM 7.6 Mes 3 mM 2.7	$(NH_4)_2SO_4$ 35 *vs* 47–49%	NaN_3 0.5 mM, $MgCl_2$ 7 mM, DTT 0.5 mM, GDPNP 0.4 mM	117
RNA binding domain of U1A spliceosomal protein + RNA hairpin (21-mer)							
1:1 [RNA] 2.6mg/ml	[protein] 5.6 mg/ml	20	7.0 40 mM	Tris–HCl 1.8 M	$(NH_4)_2SO_4$	Spermine 5 mM	118

Because complex formation between proteins and nucleic acids involves electrostatic interactions (119), crystallizations were for long not tempted in the presence of salts. Only when it was realized that tRNA:synthetase complexes are stable and even active in the presence of $(NH_4)_2SO_4$ (120), crystals could be obtained in the presence of this salt. Another limitation has been the poor supply of biological material with reliable physico-chemical integrity. This prevented a good survey of crystallization conditions. In the case of the yeast aspartate system, the first attempts of large scale purifications of the synthetase and the tRNA were set up from wild-type yeast cells and commercial bulk tRNA (95). Preparation of the enzyme necessitated three weeks of work with rather poor yield of intact enzyme due to proteolysis. An improvement of the purification procedure reduced the time scale to three days with concomitant increase of the yield. Together with revised conditions of crystallization, this improvement resulted in much better crystals diffracting to 2.7 Å resolution (97). In the case of the *E. coli* glutamine system, the problem of sample quantities was overcome by cloning and construction of over-producing strains for both the enzyme and the tRNA (39). At present, all crystallized complexes are obtained from over-produced synthetases and tRNAs.

For crystallization, most of the problems with tRNA:synthetase systems are similar to those described in the general part. *Table 4* shows conditions leading to crystals of such complexes. Except in one case, crystallizations were realized in high $(NH_4)_2SO_4$ conditions. Spermine, which is an important additive for obtaining good diffracting crystals of free tRNA, is not necessary to obtain co-crystals of complexes. Interestingly, spermine is often used for co-crystallization of DNA:protein complexes *(Table 3*). On the other hand, Mg^{2+} is systematically included when crystallizing with RNA, which is not the case with DNA. This difference should be related to the different nature of the nucleic acids, the 2'-OH of riboses introducing the possibility of making new contacts with protein as well as with neighbouring RNA molecules.

4.4 Ribosomes and their subunits

Protein biosynthesis takes place on the large ribonucleoprotein particles called ribosomes. These organelles are made of two subunits which associate upon initiation of protein synthesis to form a full particle. Although early observations of crystalline material were made *in vivo* as part of a mechanism of hibernation in a variety of lizards, the only real successful attempts to grow large 3D crystals of ribosomes was achieved with bacterial particles. In bacteria the smallest subunit (30S) has a molecular weight of 700 kDa and contains ~ 20 proteins and one RNA chain (16S). The large subunit (50S) of 1600 kDa consists of ~ 35 different proteins and two RNA chains (23S and 5S). The large size of ribosomes, comparable to that of viruses, and the lack of internal symmetry combined with conformational heterogeneity transforms the problem in a formidable challenge.

Crystallization properties and biological activities of the particles are strongly correlated, i.e. inactive particles do not crystallize. Conversely, re-dissolved crystals are active. In the case of 70S ribosomes from *T. thermophilus*, the best crystals are grown from material obtained after dissolution of previously formed egg-like crystals (121). Interestingly, addition of two molecules of charged tRNA (phenylalanyl-$tRNA^{Phe}$) and of a piece of mRNA (35-mer poly U) improves crystal quality with diffraction resolution improved from 20 to 12 Å (122). The best 3D crystals with highest resolution (~ 3 Å), however, are obtained with the isolated 50S subunit from *Haloarculum marismortui* (122, 123); promising diffraction limits of 7.3 Å and 8.7 Å were also reported for crystals of the 30S and 70S particles of *T. thermophilus* (122).

Since high salt conditions are disruptive for most ribosomes, crystallization conditions were searched mostly with volatile organic solvents. Initially, the crystallization droplets contain no precipitant or a very small quantity of it. To reduce the rate of crystal growth and to avoid technical difficulties linked to the use of volatile solvents, crystallization assays are sometimes realized directly in X-rays capillaries (38). Crystallizing agents are often PEGs and MPD (38, 121–123). In contrast, halophilic ribosomes and their isolated subunits are stable at high salt concentrations and the high diffracting crystals of the 50S subunit of *H. marismortui* were indeed grown from $(NH_4)_2SO_4$ solutions (122). These growth conditions mimic to some extent the natural salt-rich environment within the halobacteria that contains KCl, NH_4Cl, and $MgCl_2$. Altogether, crystal quality depends on the procedure used for the preparation of the ribosomal material, the strain of a given bacterial species, and on the fine-tuning of conditions such as the balance between Mg^{2+} and monovalent ions.

A major problem of ribosome crystals is their poor limit of resolution and their extreme sensitivity to X-ray damage. Data collection at very low temperatures (–150 °C) increases crystal lifetime but does not allow improved resolution. Nevertheless, data sets-up to 10 Å could be collected and preliminary phasing at 7.9 Å could be achieved for crystals of the halophilic 50S particles (122). Further improvements led to a 9 Å resolution map (123). These promising crystallographic results became possible because of progress in crystallization methods, in data collection strategies, and in production of isomorphous crystals containing appropriate clusters of heavy atoms (122, 123).

4.5 Viruses

In viruses, proteins form the protecting shell which encapsidates the genetic material (RNA or DNA). The quaternary structure of viruses is dominated by the nature of protein–protein interactions within the external capsid. Therefore crystallization of viruses resembles that of proteins. Two main shapes are observed: helical rods as in tobacco mosaic virus and filamentous bacteriophages, and isometric capsids in spherical viruses. Viruses were among the

Table 5. Crystallization conditions of some viruses

Virus[a]	Concentration	Temp (°C)	Buffer pH	Precipitant	Additives	Ref.
RNA viruses						
Animal viruses (Picornaviruses)						
BEV	5 mg/ml	20	Na phosphate 0.1 M pH 7.6	NaCl 30%	NaN_3	125
Coxsackie B1	10 mg/ml	20	Na acetate 10 mM pH 5.0	$(NH_4)H_2PO_4$ 0.1 M	126	
HRV 14	5 mg/ml	20	Tris–HCl 10 mM pH 7.2	PEG 8000 0.25–0.5%	$CaCl_2$ 20 mM	127
Mengo	5 mg/ml	Room	Na phosphate 0.1 M pH 7.4	PEG 8000 2.8%		128
Plant viruses						
CCMV	20–50 mg/ml	Room	Succinate 0.3 M pH 3.3	PEG 8000 3.7–4.0%	NaN_3 1 mM, EDTA 1 mM	129
CpMV	35 mg/ml	20	K phosphate 50 mM pH 7.0	PEG 8000 2%	$(NH_4)SO_4$ 0.4 M	130
STMV	20 mg/ml	23	Cacodylate, Na phosphate, or Tris 40 mM pH 6, 6.5, or 7	$(NH_4)SO_4$ 10–18%	NaCl or $NaC_2H_4O_4$	131
STNV	10–12 or 7–8 mg/ml		Na phosphate 50 mM pH 6.5	PEG 6000 0.4%	Mg^{2+} 1 mM	132
TBSV	30 mg/ml	4	None	$(NH_4)SO_4$ 0.5 M		133
TYMV		25	MES 100 mM pH 3.7	$(NH_4)H_2PO_4$ 1.11–1.15 M		134

Insect and bacterial viruses						
BBV	8 mg/ml	20	Na phosphate 50 mM pH 6.9–7.2	$(NH_4)SO_4$ 13.5%		130
FHV	18 mg/ml	Room	bisTris 10 mM pH 6.0	PEG 8000 2.8%	$CaCl_2$ 20 mM	135
MS2	1%	37	Na phosphate 0.4 M pH 7.4	PEG 6000 1.5%	NaN_3 0.02%	136
DNA viruses						
CPV	10 mg/ml	Room	Tris 10 mM pH 7.5	PEG 8000 0.75%	$CaCl_2$ 6 mM	137
ΦX174	8 mg/ml	20, 4	bisTris methane 90–93 mM pH 6.8	PEG 8000 1.5–2.0%		138

[a] BBV, black beetle virus; BEV, bovine enterovirus; CCMV, cucumber chlorotic mottle virus; CpMV, cowpea mosaic virus; CPV, canine parvovirus; FHV, flock house virus; HRV, human rhinovirus; STMV, satellite tobacco mosaic virus; STNV, satellite tobacco necrosis virus; TBSV, tomato bushy stunt virus; TYMV, turnip yellow mosaïc virus; MS2 and ΦX174, two bacteriophages.

first crystallized biological materials (Chapter 1). Now many viruses have been crystallized and more than 20 structures of spherical viruses are determined (124). A list of typical representatives that yielded highly ordered crystals is given in *Table 5*. For the best, diffraction limit often exceeds 3.0 Å resolution, probably as a consequence of their symmetric and isometric structure.

The importance of the external capsid and hence the non-effect of RNA or DNA on crystal formation is nicely demonstrated with cowpea mosaic virus (CpMV). The genome of this virus consists of two RNA molecules, RNA1 (5.9 kb) and RNA2 (3.5 kb), which are encapsidated in separate particles. Empty capsids are also formed *in vivo*. All three components are of the same size and appear to have identical surfaces. Isomorphous crystals were obtained with each of the isolated components or with a mixture of the three components, and the same ratio of components was found in the crystals and in the crystallizing solution (130).

As for other macromolecular systems, a wide diversity of conditions led to crystal formation. Details on crystallization conditions can be found in *Table 5* and in ref. 124. PEGs (2–3%), alone or mixed with $(NH_4)_2SO_4$ in the 0.5 M range, are the most currently used crystallizing agents. Interestingly, their concentration range is low when compared to other systems. Crystallizations are usually done at room temperature (20°C). The pH range is larger than for nucleic acids and reaches the acidic domain (i.e. 3.3–7.5). It is only limited by the stability of the capsid. Thus, turnip yellow mosaic virus, an RNA spherical virus, was crystallized at pH 3.7 (134). Finally, attempts leading to precipitation should not be discarded since crystals of viruses can also grow from heavy precipitates by Ostwald ripening mechanisms as exemplified for tomato bushy stunt virus (TBSV) (139). However, under such circumstances duration of crystal growth can be long (several weeks and more).

References

1. Gilliland, G. L. and Ladner, J. E. (1996). *Curr. Opin. Struct. Biol.*, **6**, 595.
2. Lietzke, S. E., Barnes, C. L., and Kundrot, C. E. (1995). *Curr. Opin. Struct. Biol.*, **5**, 645.
3. Dock, A.-C., Lorber, B., Moras, D., Pixa, G., Thierry, J.-C., and Giegé, R. (1984). *Biochimie*, **66**, 179.
4. Abdel-Meguid, S. S., Moore, P. B., and Steitz, T. A. (1983). *J. Mol. Biol.*, **171**, 207.
5. Cate, J. H., Gooding, A. R., Podell, E., Zhou, K., Golden, B. L., Kundrot, C. E., *et al.* (1996). *Science*, **273**, 1678.
6. Timsit, Y. and Moras, D. (1992). In *Methods in enzymology* (ed. D. M. J. Lilley and J. E. Dahlberg), Vol. 211, pp. 409–29. Academic Press, London.
7. Wing, R., Drew, H., Takano, T., Broka, C., Tanaka, S., Itakura, K., *et al.* (1980). *Nature*, **287**, 755.
8. Timsit, Y., Westhof, E., Fuchs, R .P. P., and Moras, D. (1989). *Nature*, **341**, 459.
9. Westhof, E., Auffinger, P., and Gaspin, C. (1996). In *DNA and protein sequence*

analysis: a practical approach (ed. M. Bishop and C. Rawlings), pp. 255–78. IRL Press, Oxford.
10. Sigman, D. S., Mazumder, A., and Perrin, D. M. (1993). *Chem. Rev.*, **93**, 2295.
11. Ehresmann, C., Baudin, F., Mougel, M., Romby, P., Ebel, J.-P., and Ehresmann, B. (1987). *Nucleic Acids Res.*, **15**, 9109.
12. Kolchanov, N. A., Titov, I. I., Vlassova, I. E., and Vlassov, V. V. (1996). *Prog. Nucleic Acid Res. Mol. Biol.*, **53**, 131.
13. Westhof, E. and Michel, F. (1995). In *RNA-protein interactions* (ed. K. Nagai and I. W. Mattaj), pp. 25–51. IRL Press, Oxford.
14. Gold, L., Polisky, B., Uhlenbeck, O., and Yarus, M. (1995). *Annu. Rev. Biochem.*, **64**, 763.
15. Leonard, G. A., McAuley, H. K., Ebel, S., Lough, D. M., Brown, T., and Hunter, W. N. (1994). *Structure*, **2**, 483.
16. Pley, H. W., Flaherty, K. M., and McKay, D. M. (1994). *Nature*, **372**, 68.
17. Schindelin, H., Zhang, M., Bald, R., Fürste, J.-P., Erdmann, V. A., and Heinemann, U. (1995). *J. Mol. Biol.*, **249**, 595.
18. Baeyens, K. J., De Bondt, H. L., Pardi, A., and Holbrook, S. R. (1996). *Proc. Natl. Acad. Sci. USA*, **93**, 12851.
19. Ott, G., Dörfler, S., Sprinzl, M., Müller, U., and Heinemann, U. (1996). *Acta Cryst.*, **D52**, 871.
20. Ferre-D'Amare, A. R., Zhou, K., and Doudna, J. A. (1998). *J. Mol. Biol.*, **279**, 621.
21. Oubridge, C., Ito, T., Teo, C. H., Fearnley, I., and Nagai, K. (1995). *J. Mol. Biol.*, **249**, 409.
22. Pingoud, A., Fliess, A., and Pingoud, A. (1989). In *HPLC of macromolecules: a practical approach* (ed. R. W. A. Oliver), pp. 183–208. IRL Press, Oxford.
23. Davanloo, P., Rosenberg, A. H., Dunn, J. J., and Studier, W. (1984). *Biochemistry*, **81**, 2035.
24. Milligan, J. F., Groebe, D. R., Witherell, G., and Uhlenbeck, O. C. (1987). *Nucleic Acids Res.*, **15**, 8783.
25. Frugier, M., Florentz, C., Hosseini, M. W., Lehn, J.-M., and Giegé, R. (1994). *Nucleic Acids Res.*, **22**, 2784.
26. Yin, Y. and Carter, C. W., Jr (1996). *Nucleic Acids Res.*, **24**, 1279.
27. Lapham, J. and Crothers, D. M. (1996). *RNA*, **2**, 289.
28. Giegé, R., Sissler, M., and Florentz, C. (1998). *Nucleic Acids Res.*, **26**, 5017.
29. Price, S. R., Ito, N., Oubridge, C., Avis, J. M., and Nagai K. (1995). *J. Mol. Biol.*, **249**, 398.
30. Perreault, J. P., Wu, T., Cousineau, B., Ogilvie, K. K., and Cedergren, R. (1990). *Nature*, **344**, 565.
31. Anderson, A. C., Scaringe, S. A., Earp, B. E., and Frederick, C. R. (1996). *RNA*, **2**, 110.
32. Dock-Bregeon, A.-C., Chevrier, B., Podjarny, A., Johnson, J., de Bear, J. S., Gough, G. R., *et al.* (1989). *J. Mol. Biol.*, **209**, 459.
33. Cruse, W. B., Saludjian, P., Biala, E., Strazewski, P., Prangé, T., and Kennard, O. (1994). *Proc. Natl. Acad. Sci. USA*, **91**, 4160.
34. Monier, R. (1971). In *Procedure in nucleic acids research* (ed. G. L. Cantoni and D. R. Davies), Vol. 2, pp. 618–28. Harper and Row, New York.
35. Morikawa, K., Kawakami, M., and Takemura, S. (1982). *FEBS Lett.*, **145**, 194.
36. Giegé, R. and Ebel, J.-P. (1968). *Biochim. Biophys. Acta*, **161**, 125.

37. Kowalak, J. A., Dalluge, J. J., McCloskey, J. A., and Stetter, K. O. (1994). *Biochemistry*, **33**, 7869.
38. Yonath, A., Frolow, F., Shoham, M., Müssig, J., Makowski, I., Glotz, C., *et al.* (1988). *J. Cryst. Growth*, **90**, 231.
39. Perona, J. J., Swanson, R., Steitz, T. A., and Söll, D. (1988). *J. Mol. Biol.*, **202**, 121.
40. Dirheimer, G. and Ebel, J.-P. (1967). *Bull. Soc. Chim. Biol.*, **49**, 1679.
41. Nishimura, S. (1971). In *Procedure in nucleic acids research* (ed. G. L. Cantoni and D. R. Davies), Vol. 2, pp. 542–64. Harper and Row, New York.
42. Spencer, M., Neave, E. J., and Webb, N. L. (1978). *J. Chromatogr.*, **166**, 447.
43. Gillam, I., Millward, S., Blew, D., von Tigerstrom, M., Wimmer, E., and Tener, G. M. (1967). *Biochemistry*, **6**, 3043.
44. Singhal, R. P., Griffin, G. D., and Novelli, G. D. (1976). *Biochemistry*, **15**, 5083.
45. Nishimura, S., Shindo-Okada, N., and Crain, P. F. (1987). In *Methods in enzymology* (ed. R. Wu), Vol. 155, pp. 373–8. Academic Press, London.
46. Holmes, W. M., Hurd, R. E., Reid, B. R., Rimerman, R. A., and Hatfield, G. W. (1975). *Proc. Natl. Acad. Sci. USA*, **72**, 1068.
47. Dudock, B. S. (1987). In *Molecular biology of RNA. New perspectives* (ed. M. Inouye and B. S. Dudock), pp. 321–9. Academic Press Inc., New York and London.
48. Brown, D. M. (1974). In *Basic principles in nucleic acids chemistry* (ed. P. O. P. Ts'O), Vol. 2, pp. 1–90. Academic Press Inc., New York and London.
49. Werner, W., Krebs, W., Keith, G., and Dirheimer, G. (1976). *Biochim. Biophys. Acta*, **432**, 161.
50. Guéron, M. and Leroy, J.-L. (1978). *Anal. Biochem.*, **91**, 691.
51. Kennard, O. and Hunter, W. N. (1989). *Q. Rev. Biophys.*, **22**, 327.
52. Neidle, S. (1996). In *Methods in molecular biology* (ed. C. Jones, B. Mulloy, and M. Sanderson), Vol. 56, pp. 267–92. Humana Press Inc., Totowa, NJ.
53. Wahl, M. C., Ramakrishnan, B., Ban, C., Chen, X., and Sundaralingam, M. (1996). *Acta Cryst.*, **D52**, 668.
54. Wang, A. H. J., Fujii, S., van Boom, J. H., and Rich, A. (1982). *Proc. Natl. Acad. Sci. USA*, **79**, 3968.
55. Hunter, W. N., Langlois D'Estaintot, B., and Kennard, O. (1989). *Biochemistry*, **28**, 2444.
56. Langlois d'Estaintot, B., Dautant, A., Courseille, C., and Precigoux, G. (1993). *Eur. J. Biochem.*, **213**, 673.
57. Yoon, C., Privé, G. G., Goodsell, D. S., and Dickerson, R. (1988). *Proc. Natl. Acad. Sci. USA*, **85**, 6332.
58. Privé, G. G., Heinemann, U., Chandrasegaran, S., Kan, L. S., Kopka, M., and Dickerson, R. (1987). *Science*, **238**, 498.
59. Heinemann, U. and Alings, C. (1991). *EMBO J.*, **10**, 35.
60. Wang, A. H. J., Quigley, G. J., Kolpak, F. J., Crawford, J. L., van Boom, J. H., van der Marel, G., *et al.* (1979). *Nature*, **282**, 680.
61. Wang, A. H. J., Hakoshima, T., van der Marel, G., van Boom, J. H., and Rich, A. (1984). *Cell*, **37**, 321.
62. Chevrier, B., Dock, A.-C., Hartmann, B., Leng, M., Moras, D., Thuong, M. Y., *et al.* (1986). *J. Mol. Biol.*, **188**, 707.
63. Zhou, G. and Ho, P. S. (1990). *Biochemistry*, **29**, 7229.
64. Chen, L., Cai, L., Zhang, X., and Rich, A. (1994). *Biochemistry*, **33**, 13540.

65. Wang, A. H. J., Fujii, S., van Boom, J. H., van der Marel, G., van Boeckel, S. A. A., and Rich, A. (1982). *Nature*, **299**, 601.
66. Gao, Q., Williams, L. D., Egli, L., Rabinowitch, D., Chen, S.-L., Quigley, G. J., *et al.* (1991). *Proc. Natl. Acad. Sci. USA*, **88**, 2422.
67. Brown, D. G., Sanderson, M. R., Skelly, J. V., Jenkins, T. C., Brown, T., Garman, E., *et al.* (1990). *EMBO J.*, **9**, 1329.
68. Kim, S. H., Quigley, G., Suddath, F. L., McPherson, A., Sneden, D., Kim, J. J., *et al.* (1973). *J. Mol. Biol.*, **75**, 421.
69. Johnson, C. D., Adolph, K., Rosa, J. J., Hall, M. D., and Sigler, P. B. (1970). *Nature*, **226**, 1246.
70. Giegé, R., Moras, D., and Thierry, J.-C. (1977). *J. Mol. Biol.*, **115**, 91.
71. Baeyens, K. J., Jancarik, J., and Holbrook, S. R. (1994). *Acta Cryst.*, **D50**, 764.
72. Mikol, V., Rodeau, J.-L., and Giegé, R. (1989). *J. Appl. Cryst.*, **22**, 155.
73. Sakai, T. T. and Cohen, S. S. (1976). *Prog. Nucleic Acid Res. Mol. Biol.*, **17**, 15.
74. Young, J. D., Bock, R. M., Nishimura, S., Ishikura, H., Yamada, Y., Rajbandhary, U. L., *et al.* (1969). *Science*, **166**, 1527.
75. Sauter, C., Ng, J. D., Lorber, B., Keith, G., Brion, P., Hosseini, M.-W., *et al.* (1999). *J. Cryst. Growth*, **196**, 365.
76. Ladner, J. E., Finch, J. T., Klug, A., and Clark, B. F. C. (1972). *J. Mol. Biol.*, **72**, 99.
77. Bancroft, D., Williams, L. D., Rich, A., and Egli, M. (1994). *Biochemistry*, **33**, 1073.
78. Holbrook, S. R., Sussman, J. L., Warrant, W. R., and Kim, S. H. (1978). *J. Mol. Biol.*, **123**, 631.
79. Gessner, R. V., Quigley, G. J., Wang, A. H. J., van der Marel, G., van Boom, J. H., and Rich, A. (1985). *Biochemistry*, **24**, 237.
80. Brennan, R. G., Westhof, E., and Sundaralingam, M. (1986). *J. Biomol. Struct. Dyn.*, **3**, 649.
81. Einspahr, H., Cook, W. J., and Bugg, C. E. (1981). *Biochemistry*, **20**, 5788.
82. Hingerty, B. E., Brown, R. S., and Klug, A. (1982). *Biochim. Biophys. Acta*, **697**, 78.
83. Doudna, J. A., Grosshans, C., Gooding, A., and Kundrot, C. E. (1993). *Proc. Natl. Acad. Sci. USA*, **90**, 7829.
84. Scott, W. G., Finch, J. T., Grenfell, R., Fogg, J., Smith, T., Gait, M. J., *et al.* (1995). *J. Mol. Biol.*, **250**, 327.
85. Dickerson, R. E., Goodsell, D. S., Kopka, M. L., and Pjura, P. E. (1987). *J. Biomol. Struct. Dyn.*, **5**, 557.
86. Moras, D. and Bergdoll, M. (1988). *J. Cryst. Growth*, **90**, 283.
87. Anderson, A. C., Earp, B. E., and Frederick, C. A. (1996). *J. Mol. Biol.*, **259**, 696.
88. Dock-Bregeon, A.-C. and Moras, D. (1987). *Cold Spring Harbor Symp. Quant. Biol.*, **52**, 113.
89. Golden, B., Gooding, A. R., Podell, E. R., and Cech, T. R. (1996). *RNA*, **2**, 1295.
90. Freemont, P. S., Friedman, J. M., Beese, L., Sanderson, M. R., and Steitz, T. A. (1988). *Proc. Natl. Acad. Sci. USA*, **85**, 8924.
91. Brown, D. G. and Freemont, P. S. (1996). In *Methods in molecular biology* (ed. C. Jones, B. Mulloy, and M. Sanderson), Vol. 56, pp. 293–318. Humana Press Inc., Totowa, NJ.
92. Grable, J., Frederick, C. A., Samudzi, C., Jen-Jacobson, L., Lesser, D., Greene, P., *et al.* (1984). *J. Biomol. Struct. Dyn.*, **1**, 1149.

93. Suk, D., Lahm, A., and Oefner, C. (1988). *Nature*, **332**, 464.
94. Luger, K., Mader, A. W., Richmond, R. K., Sargent, D. F., and Richmond, T. J. (1997). *Nature*, **389**, 251.
95. Lorber, B., Giegé, R., Ebel, J.-P., Berthet, C., Thierry, J.-C., and Moras, D. (1983). *J. Biol. Chem.*, **258**, 8429.
96. Jordan, S. R., Whitcombe, T. V., Berg, J. M., and Pabo, C. O. (1985). *Science*, **230**, 1383.
97. Ruff, M., Cavarelli, J., Mikol, V., Lorber, B., Mitschler, A., Giegé, R., *et al.* (1988). *J. Mol. Biol.*, **201**, 235.
98. Brennan, R. G., Takeda, Y., Kim, J., Anderson, W. F., and Matthews, B. W. (1986). *J. Mol. Biol.*, **188**, 115.
99. Anderson, J., Ptashne, M., and Harrison, S. C. (1984). *Proc. Natl. Acad. Sci. USA*, **81**, 1307.
100. Aggarwal, A. K., Rodgers, D. W., Drottar, M., Ptashne, M., and Harrison, S. (1988). *Science*, **242**, 899.
101. Joachimiak, A., Marmorstein, R. Q., Schevitz, R. W., Mandecki, W., Fox, J. L., and Sigler, P. B. (1987). *J. Biol. Chem.*, **262**, 4917.
102. Schultz, S. C., Shields, G. C., and Steitz, T. A. (1990). *J. Mol. Biol.*, **213**, 159.
103. Ellenberger, T. E., Brandel, C. J., Struhl, K., and Harrison, S. C. (1992). *Cell*, **71**, 1223.
104. Pavletich, N. P. and Pabo, C. O. (1993). *Science*, **261**, 1701.
105. Clark, K. L., Halay, E. D., Lai, E., and Burley, S. K. (1993). *Nature*, **364**, 412.
106. Yang, W. and Steitz, T. A. (1995). *Cell*, **82**, 193.
107. Schwabe, J. W. R., Chapman, L., Finch, T., and Rhodes, D. (1993). *Cell*, **75**, 567.
108. Kim, Y., Geiger, J. H., Hahn, S., and Sigler, P. B. (1993). *Nature*, **365**, 512.
109. Glover, J. N. M. and Harrison, S. C. (1995). *Nature*, **373**, 257.
110. Nikolov, D. B., Chen, H., Halay, E. D., Usheva, A. A., Hisatake, K., Lee, D. K., *et al.* (1995). *Nature*, **377**, 119.
111. Sousa, R. (1995). *Acta Cryst.*, **D51**, 271.
112. Boeglin, M., Dock-Bregeon, A.-C., Eriani, G., Gangloff, J., Ruff, M., Poterzman, A., *et al.* (1996). *Acta Cryst.*, **D52**, 211.
113. Arnez, J. and Steitz, T. A. (1994). *Biochemistry*, **33**, 7567.
114. Cusack, S., Yaremchuk, A., and Tukalo, M. (1996). *EMBO J.*, **15**, 6321.
115. Reshetnikova, L., Khodyreva, S., Lavrik, O., Ankilova, V., Frolow, F., and Safro, M. (1993). *J. Mol. Biol.*, **231**, 927.
116. Yaremchuk, A. D., Tukalo, I., Krikliviy, N., Malchenko, N., Biou, V., Berthet-Colominas, C., *et al.* (1992). *FEBS Lett.*, **310**, 157.
117. Nissen, P., Reshetnikova, L., Siboska, G., Polekhina, G., Thirup, S., Kjeldgaard, M., *et al.* (1994). *FEBS Lett.*, **356**, 165.
118. Oubridge, C., Ito, N., Evans, P. R., Teo, C. H., and Nagai, K. (1994). *Nature*, **372**, 432.
119. Record, M. J. J., Anderson, C. F., and Lohman, T. M. (1978). *Rev. Biophys.*, **11**, 103.
120. Giegé, R., Lorber, B., Ebel, J.-P., Moras, D., Thierry, J.-C., Jacrot, B., *et al.* (1982). *Biochimie*, **64**, 357.
121. Trakhanov, S., Yusupov, M., Shirokov, V., Garber, M., Mitschler, A., Ruff, M., *et al.* (1989). *J. Mol. Biol.*, **209**, 327.

122. Thygesen, J., Krumholz, S., Levin, I., Zaytzev-Bashan, A., Harms, J., Bartels, H., *et al.* (1996). *J. Cryst. Growth*, **168**, 308.
123. Ban, N., Freeborn, B., Nissen, P., Penczek, P., Grassucci, R. A., Sweet, R., *et al.* (1998). *Cell*, **93**, 1105.
124. Fry, E., Logan, D., and Stuart, D. (1996). In *Methods in molecular biology* (ed. C. Jones, B. Mulloy, and M. Sanderson), Vol. 56, pp. 319–63. Humana Press Inc., Totowa, NJ.
125. Smyth, M., Fry, E., Stuart, D., Lyons, C., Hoey, E., and Martin, S. J. (1993). *J. Mol. Biol.*, **231**, 930.
126. Li, T., Zhang, A., Iuzuka, N., Nomoto, A., and Arnold, E. (1992). *J. Mol. Biol.*, **223**, 1171.
127. Arnold, E., Erickson, J. W., Fout, S. G., Frankenberger, E. A., Hecht, H. J., Luo, M., *et al.* (1984). *J. Mol. Biol.*, **177**, 417.
128. Luo, M., Vriend, G., Kamer, G., Minor, I., Arnold, E., Rossman, M. G., *et al.* (1987). *Science*, **235**, 182.
129. Speir, J. A., Munshi, S., Baker, T. S., and Johnson, J. E. (1993). *Virology*, **193**, 234.
130. Sehnke, P. C., Harrington, M., Hosur, M. V., Li, Y. R., Usha, R., Tucker, R. C., *et al.* (1988). *J. Cryst. Growth*, **90**, 222.
131. Koszelak, S., Dodds, J. A., and McPherson, A. (1989). *J. Mol. Biol.*, **209**, 323.
132. Jones, T. A. and Liljas, L. (1984). *J. Mol. Biol.*, **177**, 735.
133. Harrison, S. C. and Jack, A. (1975). *J. Mol. Biol.*, **97**, 173.
134. Canady, M. A., Larson, S. B., Day, J., and McPherson, A. (1996). *Nature Struct. Biol.*, **3**, 771.
135. Fisher, A. J., McKinney, B. R., Wery, J. P., and Johnson, J. E. (1992). *Acta Cryst.*, **48**, 515.
136. Valegård, K., Unge, T., Montelius, I., Strandberg, B., and Fiers, W. (1986). *J. Mol. Biol.*, **190**, 587.
137. Luo, M., Tsao, J., Rossmann, M. G., Basak, S., and Compans, R. W. (1988). *J. Mol. Biol.*, **200**, 209.
138. Willingmann, P., Krishnaswamy, S., McKenna, R., Smith, T. J., Olson, N. H., Rossmann, M. G., *et al.* (1990). *J. Mol. Biol.*, **212**, 345.
139. Ng, J. D., Lorber, B., Witz, J., Théobald-Dietrich, A., Kern, D., and Giegé, R. (1996). *J. Cryst. Growth*, **168**, 50.

9

Crystallization of membrane proteins

F. REISS-HUSSON and D. PICOT

1. Introduction

Crystallization of membrane proteins is one of the most recent developments in protein crystal growth; in 1980, for the first time, two membrane proteins were successfully crystallized, bacteriorhodopsin (1) and porin (2). Since then, a number of membrane proteins (about 30) yielded three-dimensional crystals. In several cases, the quality of the crystals was sufficient for X-ray diffraction studies. The first atomic structure of a membrane protein, a photosynthetic bacterial reaction centre, was described in 1985 (3), followed by the structure of about ten other membrane protein families. Crystallization of membrane proteins is now an actively growing field, and has been discussed in several recent reviews (4–8).

The major difficulty in the study of membrane proteins, which for years hampered their crystallization, comes from their peculiar solubility properties. These originate from their tight association with other membrane components, particularly lipids. Indeed integral membrane proteins contain hydrophobic surface regions buried in the lipid bilayer core, as well as hydrophilic regions with charged or polar residues more or less exposed at the external faces of the membrane. Disruption of the bilayer for isolating a membrane protein can be done in various ways: extraction with organic solvents, use of chaotropic agents, or solubilization by a detergent. The last method is the most frequently used, since it maintains the biological activity of the protein if a suitable detergent is found. This chapter will be restricted to specific aspects of three-dimensional crystallizations done in micellar solutions of detergent. In some cases, it is possible to separate soluble domains from the membrane protein either by limited proteolysis or by genetic engineering. Such protein fragments can then be treated as soluble proteins and so will not be discussed further in this chapter. We refer to Chapter 12 and the review by Kühlbrandt (9) for the methodology of two-dimensional crystallization used for electron diffraction.

2. Crystallization principles

The general principles discussed in this book for the crystallization of soluble biological macromolecules apply for membrane proteins; the protein solution must be brought to supersaturation by modifying its physical parameters (concentrations of constituents, ionic strength, and so on), so that nucleation may occur. The main differences from the behaviour of soluble proteins stem from the following two points:

(a) The entity which is going to crystallize is the protein–detergent complex, not the protein alone. Yet, most of the detergent found in the crystal is disordered. This has been demonstrated for three detergents (C_{10}DAO, C_{12}DAO, and C_8G) associated with two bacterial reaction centres (10, 11) and OmpF porin (12). Usually, only a few ordered detergent molecules are seen in the electron density maps. But the amount of disordered detergent in the crystals is fairly high; about 200 molecules of detergent are associated with one reaction centre protein, and form a ring around the hydrophobic transmembrane α helices. The detergent ring is interconnected with its neighbours by bridges. Thus ribbon-like detergent structures run throughout the crystal. These findings explain why the characteristics of the detergent molecules (such as their length) are so crucial in the crystallization process. Indeed they should fit around the hydrophobic regions of the protein without hindering the interprotein contacts.

(b) The solubility of the protein–detergent complex is governed not only by the protein properties, but also (and mainly) by those of the detergent micellar solution. Generally, as will be discussed below, this detergent is non-ionic; its micellar solution therefore only exists in a limited range of concentration and temperature, defined in a phase diagram (*Figure 1*). Outside of this range, the micellar solution may spontaneously break apart into two immiscible aqueous phases; one being enriched in detergent, the other one remaining essentially depleted in detergent. The temperatures and concentrations at which phase separation is observed define a curve, called the consolution boundary. Depending on the detergent and the crystallizing agent used, this boundary may be reached starting from the micellar solution either by increasing or by decreasing the temperature as shown in *Figure 1*. When phase separation takes place, the solubilized membrane protein generally partitions into the detergent-rich phase (but exceptions are known for glycoproteins) (13). Phase separation is a function of all constituents of the solution such as detergent, protein, nature and concentration of salt, concentration of a crystallizing agent like PEG. Phase separation seems to play a major role in the crystallization because it is quite often observed that crystallization takes place right before phase separation occurs. Choosing crystallization con-

Table 2. Properties of some detergents used for crystallization

Detergent	Molecular weight	CMC (mM)	Ref.	Monomers per micelle	Suppliers[a]
C_8G	292.4	23	18	78	Various
C_9G	306.4	6.5	18		C, F
$C_{10}M$	482.6	2.2	17		C, F
$C_{11}M$	496.6	0.59			A
$C_{12}M$	510.6	0.16	17	130	Various
Hecameg	335.4	20	57		C, V
$C_{10}DAO$	201.4	10.4	4		F
$C_{11}DAO$	215.4				O
$C_{12}DAO$	229.4	2.0	20	73	Various
C_8E_4	306	8.5	58	82	B, K
C_8E_5	350	9.2	58		B, K
$C_{10}E_8$	515.1	0.10	58		C
$C_{12}E_8$	518	0.071	58	120	C, K
$C_{12}E_9$	583	0.071	58		C, F
C_8HESO	206	29.9	27		B, O
MEGA-10	349.5	5	27		O

[a] A, Anatrace; B, Bachem; C, Calbiochem; F, Fluka; K, Kohyo; O, Oxyl; V, Vegatec.

Protocol 3. Purification of C_8G[a]

Equipment and reagents

- Chromatography column (10 × 150 mm)
- C_8G
- Ethanol
- Strong mixed-bed ion exchanger (e.g. Rexin I-300 or Bio-Rad AG501X8) in the H-OH form

Method

1. Pour the column with the ion exchanger resin.
2. Wash the column with 120 ml ethanol then with 600 ml water. Stop the flow when water is draining the gel surface.
3. Dissolve 5 g C_8G in 50 ml water and put the solution on the gel. Elute at a flow rate of 0.2 ml/min. Then wash with water at the same rate.
4. Collect the first 100 ml of eluate. Lyophilize and store at –20 °C.

[a] This protocol may be scaled down for smaller detergent quantities.

3.2 *n*-Alkyl-thioglucosides

Although these detergents have not been used very often in crystallizations (23, 24), they could be assayed instead of the glucosides. The C_6, C_7, C_8, and C_{10} compounds are commercially available.

Table 3. Crystallization conditions for some membrane proteins

Protein[a]	**Organism**	**Detergent**[b]	**Precipitant**[c]	**Additive**[d]	**Method**[e]	**Ref.**
Reaction centre	*Rhodopseudomonas viridis*	$C_{12}DAO$	AS	HT	VD	59
Reaction centre	*Rhodobacter sphaeroides* R26	C_8G	PEG/NaCl		VD	60
Reaction centre	*Rhodobacter sphaeroides* R26	$C_{12}DAO$	PEG/NaCl	HT	VD	61
Reaction centre	*Rhodobacter sphaeroides* 241	C_8G	PEG/NaCl	HT	VD	62
Reaction centre	*Rhodobacter sphaeroides* Y	C_8G	PEG/NaCl		MD	63
Reaction centre	*Rhodobacter sphaeroides* 241	C_8G	PEG	HT/BZ	SD	64
Reaction centre	*Rhodobacter sphaeroides* 241	$C_{12}DAO$	KPi	HT/1,4-dioxane	VD	65
Reaction centre	*Chromatium tepidum*	C_8G	PEG/NaCl		HD	66
Reaction centre	*Chloroflexus aurantiacus*	$C_{10}E_8$	PEG	GAPA	SD	67
LH B800-850	*Rhodopseudomonas acidophila*	C_8G	PO_4	BZ	VD	68
LH B800-850	*Rhodospirillum molischianum*	$C_{11}DAO$	AS	HT	SD	69
Porin OmpF	*Escherichia coli*	C_8HESO C_8POE	$PEG/MgCl_2$		MD	70
Porin OmpF	*Escherichia coli*	C_8G C_8POE	PEG/NaCl		VD, MD	27
Porin	*Rhodobacter capsulatus*	C8E4	PEG/LiCl			35
Porin	*Rhodopseudomonas blastica*	C_8E_4	PEG/LiCl			71

Porin PhoE	*Escherichia coli*	C_8G C_8E_4	PEG/NaCl			72
Porin LamB	*Escherichia coli*	$C_{10}M$ $C_{12}E_9$	PEG/$MgCl_2$		MD	73
Porin ScrY	*Staphylococcus typhimurium*	C_8G C_6DAO	PEG/LiCl/ $MgSO_4$		SD	41
Porin	*Paracoccus denitrificans*	C_8G	PEG/KCl		SD	74
PS I	*Synechocystis elongatus*	$C_{12}M$	$MgSO_4$		MD	42
Cytochrome oxidase	Bovine heart	$C_{10}M$	PEG			75
Cytochrome oxidase/antibody complex	*Paracoccus denitrificans*	$C_{12}M$	PEG-ME/ NH_4 acetate			37
Cytochrome *bc*₁	Bovine	MEGA-10 (or SPC)	PEG/KCl	Glycerol		50
Cytochrome *bc*₁	Bovine	HECAMEG	PEG			49
Prostaglandin H synthase I	Sheep	C_8G	PEG/NaCl		HD	6
Prostaglandin H synthase II	Human	C_8E_5	PEG/NaCl			76
Prostaglandin H synthase II	Murine	C_8G	PEG-ME		VD	46
α-Haemolysin	*Staphylococcus aureus*	C_8G	AS PEG-ME			77
Phospholipase Plda	*Escherichia coli*	C_8G	MPD/$CaCl_2$		HD	43

[a] LH, light harvesting; PS, photosystem.
[b] For detailed properties see *Table 1*.
[c] AS, ammonium sulfate; PEG-ME, polyethylene glycol monomethylether; MPD, 2-methyl-2,4-pentane diol.
[d] HT, heptanetriol; BZ, benzamidine; GAPA, *N-N'*-bis (gluconamidopropyl)-amine.
[e] VD, vapour diffusion; SD, sitting drop; HD, hanging drop; MD, microdialysis.

3.3 *n*-Alkyl-maltosides (C_nM)

The C_{12}, C_{11}, and C_{10} compounds have been used in crystallizations. $C_{12}M$ was also used for purification of membrane proteins and is considered to be superior to C_8G. Its CMC is fairly low, and the size of the micelles in pure water is larger than for C_8G. Dialysis of the detergent is an extremely slow process. It is likely to hydrolyse and should be stored frozen. Contamination by dodecanol decreases its solubility and may be detected by appearance of a white precipitate in solutions kept at 5°C (18). Contamination with the α isomer influences the crystallization process (P. Fromme, personal communication) but can be checked by reverse-phase HPLC of detergent batches.

3.4 *n*-Alkyl-dimethylamineoxides (C_nDAO)

No hydrolysis is observed for this class of detergents; solutions are stable when kept at 5°C. The C_{11} and C_{12} compounds have been used successfully for the crystallization of bacterial reaction centres and antenna (see *Table 3*). They are zwitterionic at $pH > 3$ and cationic at $pH < 3$. The micellar size of $C_{12}DAO$ is similar to C_8G, but its CMC is lower, resulting in a slower dialysis rate. It is available from various sources, either as the pure C_{12} species, or as a cheaper mixture containing primarily the C_{12} species but also other chain lengths. Traces of H_2O_2 (left over from synthesis) may contaminate some batches. They can be removed by adding 10 μg of catalase per ml of a 30% stock solution of $C_{12}DAO$ in water. Shorter chain analogues (C_6–C_{10}) are available. Some of them have been used for crystallization added to C_{12} one.

3.5 *n*-Alkyl-oligoethylene glycol-monoethers (C_nE_m)

Besides pure $C_{12}E_8$ which has been used for biochemistry, pure compounds of shorter hydrocarbon chain (C_5–C_8) are commercially available with a defined number of ethylene glycol units ranging from one to five. The C_8E_4, C_8E_5, and $C_{12}E_9$ detergents have been used in crystallizations, either pure or mixed with C_8G. A cheaper polydisperse mixture, C_8E_m, has also been used mixed with $C_{12}M$ in some cases.

Aqueous solutions of these detergents are not stable; peroxides and aldehydes are formed on storage under air, particularly in the light. The purification protocol (*Protocol 4*) involves treatment with a reducing agent ($SnCl_2$ or Na_2SO_3) followed by solvent extraction (25).

Besides these five main classes, other detergents have been used for crystallization such as the *n*-alkyl-glucamides (MEGA-9, MEGA-10) (26), *n*-heptylcarbamoyl-methyl-α-D-glucoside (HECAMEG), *n*-methyl-*n*-decanoyl-maltosylamine, but the relatively poor solubility of the MEGA and HECAMEG detergents may however cause some problems (beware: they crystallize easily).

Protocol 4. Purification of *n*-alkyl-oligoethylene glycol detergents

Equipment and reagents

- 10% (w/v) solution of the detergent in distilled water
- $SnCl_2$
- 10% NaCl
- Dichloromethane
- 1% NaOH, 10% NaCl
- Anhydrous Na_2SO_4
- Flash evaporator (e.g. Büchi)

Method

1. Stir for 2 h a 10% (w/v) solution of the detergent in distilled water with $SnCl_2$ (0.5% (w/v) final concentration).
2. Add NaCl to 10% (w/v) final concentration, then add an equal volume of dichloromethane and mix thoroughly. Let stand until the two layers separate.
3. Discard the upper water layer and recover the lower organic layer which contains the detergent.
4. Extract the organic phase with an equal volume of 1% NaOH, 10% NaCl, then three times with 10% NaCl (the pH of the final NaCl layer must be 7). Each time discard the water layer.
5. Dry the organic phase for 24 h over anhydrous Na_2SO_4, then in a flash evaporator at 40 °C.
6. Store the purified detergent at –20 °C.

New detergents related to those described appear steadily on the market. Personal investigation of chemical catalogues may be fruitful.

4. Purification of membrane proteins before crystallization

General methods for solubilization and purification of membrane proteins have been given (14) and will not be detailed. This section only focuses on several specific points: purity requirement, procedures for detergent exchange and for sample concentration.

4.1 Purity requirements

As for the crystallization of soluble proteins, the starting protein solution should be as pure and homogeneous as possible and the same precautions should be taken to avoid denaturation, proteolysis, and microheterogeneities (see Chapter 2).

Protein purity is usually checked by SDS–PAGE. However one should be aware that contaminants may escape detection; e.g. lipopolysaccharides and lipoprotein contaminants of porin preparations are not stained by Coomassie Blue but only by silver staining (27).

Residual lipids represent another common source of impurity for membrane protein preparations. Indeed, they may withstand the solubilization by a mild detergent and remain associated with the detergent–protein complexes. Their non-specific, random binding prevented crystallization in several cases. Lipid content should therefore be checked by TLC (28) of organic solvent extracts (29) (see *Protocol 5*). Alternatively, phosphorus content of these extracts can be measured (30) as most residual lipids are phospholipids.

If present, lipids should be eliminated as much as possible; the ease of removal depends on the protein–detergent couple and there is no general recipe. Chromatographic techniques may be adequate even if they were not devised for this purpose. Ion exchange chromatography has been reported in several cases to lower the lipid content, probably because of the extensive detergent washes of the adsorbed protein. Another chromatographic step which has been used for purification of membrane proteins prior to crystallization is chromatofocusing in the presence of a detergent (31). Besides providing a homogeneous preparation of defined isoelectric point, it may also result in lowering the phospholipid content. Purity of the detergent may play a role in elution by IEF, as reported in the case of bacteriorhodopsin which seemed heterogeneous when impure C_8G was present (22).

Protocol 5. Analysis of lipids in membrane protein preparations

Equipment and reagents

- Protein sample (~ 1 mg/ml)
- Hexane:isopropanol (3:2, v/v)
- Nitrogen gas (or flash evaporator)
- Chloroform
- Reagents for TLC analysis (as described in *Protocol 1* except use chloroform:methanol: water (65:25:4, by vol.) as the solvent)

Method

1. Mix the protein solution (~ 1 mg/ml) with 20 vol. of hexane: isopropanol (3:2, v/v).
2. Shake well and then centrifuge at low speed (5000 *g*) for 20 min.
3. Recover the supernatant. Repeat steps 1 and 2 on the pellet.
4. Combine the supernatants and dry this under a stream of N_2 or with a flash evaporator.
5. Dissolve the residue in the minimal volume of chloroform.
6. Carry out TLC analysis of the lipid extract as described in *Protocol 1*, except use chloroform:methanol:water (65:25:4, by vol.) as the solvent. The neutral lipids run near the solvent front and the other lipids are fractionated into various classes. They are identified with reference to known standards and published R_f values, in addition to the use of specific stains (28) instead of iodine staining or H_2SO_4 charring.

Homogeneity of the preparation requires also its monodispersity, i.e. all the protein–detergent complexes should have the same composition. This is verified by gel filtration experiments, e.g. on FPLC columns (Superose gels, Pharmacia) or HPLC ones (such as the TSK-SW or TSK-PW gels, Toso Haas). From these experiments one can estimate the size of the whole complex, including detergent. On the other hand, the amount of bound detergent may be determined by several techniques (see ref. 32 for a review). Combining these results allows the aggregation state of the protein–detergent complex to be determined and controlled.

4.2 Detergent exchange

Purification of a membrane protein is often done in the presence of a detergent and crystallization performed with a different one. This may be because of the cost of one detergent, or because solubilization and purification require a particular detergent, or the variation of the detergent nature during crystallization trials. In all cases, exchange of detergent hereafter called detergent 1 and detergent 2 has to be performed; several methods may be used.

4.2.1 Dialysis

This is the simplest procedure but not applicable in all cases. The meaningful parameter of a dialysis membrane is its cut-off value. It must be low enough for retention of the protein–detergent complex. Thus for large complexes, highly permeable membranes can be used; e.g. for a complex of 100 kDa, a Spectrapor 7 membrane with a cut-off value of 50 kDa may be used. With such a pore size, exchange of detergents with CMC values higher than 1 mM is relatively rapid (a few days). On the other hand, for a small complex (e.g. 15 kDa) a Spectrapor 1 membrane (cut-off value 6 kDa) should be chosen and only detergents of high CMC (10 mM or so) will exchange at an acceptable rate. Since the diffusion rate between two detergents may differ by several orders of magnitude, care should be taken in order to avoid detergent depletion leading to irreversible aggregation or increase of detergent concentration causing inactivation of the protein.

In favourable cases dialysis may be performed in two steps; e.g. it is possible with the bacterial reaction centre, to exchange 0.1% $C_{12}DAO$ for 0.8% C_8G as follows:

(a) First dialyse for 48 h against a detergent-free buffer, with several changes of reservoir (Spectrapor semi-microtubing, cut-off 12 kDa). Removal of $C_{12}DAO$ results in increased turbidity of the sample.
(b) Transfer the bag in a C_8G containing buffer for another 24 h. Loss of turbidity indicates redissolution of the protein.

From a practical point of view, dialysis may be performed either in the familiar closed bags, or for small volumes (< 500 μl) in microdialysis cells, either home-built (see Chapter 5) or commercially available (Pierce, Amicon). Some dialysis membranes, particularly those stored wet (e.g. Spectrapor type 7, Amicon) are specially prone to fungi contamination. Before use, a good precaution is to boil membranes for 1 min in 1% (w/v) $NaHCO_3$, then to soak them three times in highly pure water (Milli Q grade), and to use them immediately afterwards.

4.2.2 Chromatography

The sample which contains detergent 1 is chromatographed on a column equilibrated with detergent 2, and eluted with detergent 2. This method is feasible with any type of detergent, with various chromatographic supports.

(a) Gel filtration is gentle and can be used with all detergents. However, it usually dilutes the sample appreciably.

(b) Ion exchange chromatography (with DEAE, CM exchangers, or sometimes hydroxyapatite) is restricted to non-ionic detergents; it has the advantage of concentrating the sample when elution is done by a steep salt increase, but one must check that the stability of the protein is not affected by the shift of the CMC, which is induced by the higher salt concentration.

If the presence of salt in the final sample is not wanted, a mixed column consisting of ion exchanger superposed on gel filtration matrix (Sephadex G25) will exchange the detergent and desalt the sample altogether (33). Microcolumns built from Pasteur pipettes are useful for such ion exchange procedure.

4.2.3 Precipitation

The protein in the presence of detergent 1 is first precipitated with cold ethanol, which is a solvent of detergent 1; the precipitate is washed to eliminate detergent 1, and redissolved in detergent 2. This method has been used only for porins as it requires a very sturdy protein. Salt or PEG precipitation may also be used in some cases but care should be taken to avoid phase separation. Furthermore, since precipitated protein still bound a large amount of detergent, several precipitation cycles are needed.

The efficiency of these procedures can be judged from the absence of detergent 1 in the final sample. Unfortunately, very few detergents exist in a labelled form and those are expensive. Colorimetric determination is possible for glucosides and maltosides with reagents specific for reducing sugars (34). In other cases, TLC of extracts is the only method.

4.3 Sample concentration

A concentrated stock solution (often 10 mg protein/ml or more) is required to prepare the samples for crystallization trials. As already mentioned (see

Section 4.2), concentration may be achieved during the detergent exchange by ion exchange chromatography. Another very useful method is ultrafiltration. Here again a large number of devices is available. For final volumes of 1 ml and up, stirred cells equipped with Amicon YM or XM membranes (with a wide choice of cut-off values) are convenient, and may be operated under nitrogen pressure. The major drawback is foaming during stirring. For smaller volumes, a number of devices allow concentration down to 100 μl or so with low speed centrifugation (Millipore, Amicon). Whatever the device, the cut-off value should be chosen as high as possible, taking into account the size of the protein–detergent complex; this avoids concentrating the detergent in solution in the same time as the protein–detergent complexes.

5. Crystallization protocols

For conducting crystallization trials with a membrane protein, similar strategies, like the incomplete factorial design (see Chapters 4 and 5), to those developed for the soluble proteins can be used (7). Several parameters have to be chosen (see Chapter 1, *Table 1*) to which should be added the nature and concentration of the detergent. Some of these parameters have been discussed earlier with reference to soluble proteins (see Chapter 5). One of these parameters, the purification of the protein, will influence the crystallization condition to such an extent that it may be as important to modify the purification as the crystallization protocol. This has been the case for porin (35) and prostaglandin H synthase (31), for which the change of the detergent used for solubilization was critical to obtain good crystals, even if an other detergent was then used for the crystallization.

Protein–protein, protein–detergent, and detergent–detergent interactions can be observed in membrane protein crystals. Depending on the nature of the protein and the detergent, the crystallization process will be more influenced by one or another type of interaction. Protein–protein interactions are more specific than detergent–detergent interactions and should thus yield better crystals. This may explain why crystals of the reaction centre from *Rps. viridis* diffract better than those from *Rb. sphaeroides* since the former is crystallized with an additional soluble subunit. Therefore, larger proteins, but with a larger polar domain, may be easier to crystallize than their more hydrophobic counterparts. An increase of the hydrophilic surface of cytochrome *c* oxidase of *Paracoccus denitrificans* has been obtained by forming a complex with a conformation specific engineered F_v fragment (36). This has allowed well diffracting crystals to be grown using the detergent $C_{12}M$, that is able to maintain the activity. Thus, the F_v fragment counterbalances the disadvantage of $C_{12}M$, i.e. to form large micelles (37). We will stress now a few specific points based on published crystallization protocols (*Table 3*).

5.1 Detergent

The choice of detergent is still empirical. The first criterion is to maintain the functional and structural integrity of the protein. Not only the type but also the detergent concentration are important. Furthermore, the behaviour and stability of the protein will be very different below and above the CMC as well as above and below the consolution boundaries. The optimal stability of a membrane protein is often observed around the CMC, which may therefore be a good starting detergent concentration for a crystallization experiment. A number of membrane proteins crystallize with a wide variety of detergents. In one case, a systematic search has been done over 23 detergents with Omp F, an *E. coli* porin (23). Among them, 16 non-ionic detergents (from classes described in *Table 1*) could be used successfully. Interestingly, OmpF crystallized also in micellar solutions of short chains lecithins (diC_6- or diC_7-glycerophosphatidylcholines) or lysolecithins ($monoC_{14}$- or $monoC_{16}$-glycerophosphatidylcholines) used as detergents. However, crystals could not be obtained with ionic detergents, nor with detergents derived from bile salts (cholate, CHAPS, or CHAPSO). A non-ionic, non-steroid polar group and a short alkyl chain seemed thus to be the only requirement, without narrow specificity.

The reverse situation may prevail, and strict requirements may exist for chain length or detergent type. For example, a light-harvesting chloroplast protein, LHCII, crystallizes reliably with C_9G and poorly with C_8G (4). Therefore, for an unknown protein, screening should be done using at least two homologues of each detergent class. However, before beginning an extensive screening, one could try the most popular C_8G and $C_{12}DAO$ detergents. Indeed they allowed a number of successful crystallizations such as bacterial reaction centres and light-harvesting complexes; *E. coli* porins gave crystals with one of them or both. From a certain point of view, these two detergents may be considered as equivalent: in two bacterial reaction centre crystals, the regions they occupied respectively around the hydrophobic α helices could be nearly superimposed (10, 11).

$C_{12}DAO$ should be tried alone and also in the presence of an additive (see Section 7) which was required in some cases. C_8G has been used either alone, or mixed with low amounts of other short chain detergents; these were however not essential for crystallization but improved crystal growth. Thus first trials may be done with pure C_8G only.

The initial concentration of detergent in the sample should be chosen only slightly higher than the CMC (see *Table 2*). Under this condition, the detergent is present either as monomers or as part of protein–detergent complexes, with very few pure detergent micelles. For porin crystallization, the optimal range of C_8G concentration is narrow: 8–9 mg/ml. Below the CMC (less than 7 mg/ml) or well above it (more than 10 mg/ml) growth rate and nucleation are excessive (38). The same optimal range was found for C_8G with bacterial reaction centres, and it does not seem to be very sensitive to the protein concentration.

5.2 Additives

Small molecules with amphiphilic character have been sometimes added to the crystallization media (see *Table 3*). Most often used is heptane-1,2,3-triol (high melting point isomer); hexane-1,6-diol, benzamidine, glycerol, and triethylamine phosphate were also used. Their effects are various.

(a) They may be absolutely required: best example is heptane-1,2,3-triol, essential for crystallizing *Rps. viridis* reaction centre with $C_{12}DAO$.
(b) They improve crystal growth and quality, but crystallization still takes place in their absence; this is the case of heptane-1,2,3-triol for *Rb. sphaeroides* reaction centre with $C_{12}DAO$.
(c) In other cases their presence has no effect whatsoever; it is the case of *E. coli* porin with C_8G.

These 'additives' have been usually used at quite high molarities, when compared to those of detergent; in the cases cited above heptane-1,2,3-triol was present at about 0.1 M. Their mode of action is still poorly understood. One hypothesis is that their small size and amphiphilic character allow them to localize between neighbouring protein molecules, in regions inaccessible to detergent, filling thus voids in the lattice (39). Another explanation, which has some experimental support, is they modify the micellar structure of the detergent by partitioning into the micelles (40); thus they change the consolution boundaries and could bring them in a favourable temperature range. Detergents with short aliphatic chains and a CMC too high to be used alone may also be used as additive (41). The addition of another detergent at low concentration (C_8E_4) may also have a similar effect to the other type of additive (31).

Whatever the case, such additives may be tried when all previous trials done in their absence have failed. It is better to test them beforehand on the protein in solution; indeed they may have a denaturing effect.

5.3 Crystallizing agent

In *Table 3*, one may notice that the conditions used so far to achieve supersaturation are not very diverse: either the presence of PEG at a suitable concentration, or 'salting-out' at high ionic strength. There are a few exceptions: the PSI complex has been crystallized under 'salting-in' conditions (42) and OMPLA from *E. coli* uses the organic solvent 2,4-methyl-pentane diol (43).

5.3.1 PEG

PEG, which is a classical crystallizing agent for soluble proteins, probably acts in the same way on detergent solutions by competing with the micelle polar groups for water molecules and by modifying the structure of the solvent. This destabilizes the micelles and the protein–detergent complexes. Thus, when PEG is continuously added to a membrane protein in detergent solution, the micellar solution is perturbed and one of two situations may occur: protein and detergent will precipitate, or the solution spontaneously separates in two immiscible liquid phases, with the protein and most of the detergent in one phase and PEG in the other (see *Figure 1*). At a given temperature, the solubility limit of the protein depends on all constituents of the solution (PEG, detergent, protein, and salts). Crystals will eventually form when the system is slowly approaching this limit by increasing PEG concentration.

PEG and the more recent PEG-ME (monomethyl ether) are available in a variety of polymeric ranges. The optimal range of PEG concentration for crystallization of a given protein depends on PEG molecular weight. It may be very narrow (about 1.5%) as observed for porin (38) and a bacterial reaction centre (44). Repurification is recommended before use (45) and is described in Chapters 2 and 5. Low molecular weight PEG and PEG-ME are also suitable with membrane protein (35, 46) and may be suitable for cryo-crystallography (46).

Crystallization of soluble proteins with PEG is usually performed at low ionic strength. On the contrary, for membrane proteins salt is generally required along with PEG (see *Table 3*). Again, the optimal range of salt concentration may be narrow. Thus the two meaningful parameters are the concentrations of PEG and of salt, for a given protein with a given detergent (pH and temperature being fixed). For C_8G at room temperature, these conditions are quite similar for unrelated proteins like *E. coli* OmpF porin, *Rb. sphaeroides* reaction centre, prostaglandin H synthase, and cytochrome bc_1 (47) (see *Table 3*), as if they were little influenced by the protein but mainly by the detergent. This observation, if it is generalized by further experiments, would greatly simplify the search for crystallization conditions with the system C_8G–PEG–salt.

No systematic comparison of the influence of various salts on the crystallization conditions of membrane proteins has been published. We have observed that in the case of crystallization of a bacterial reaction centre NaCl could be replaced by a number of other monovalent salts; their optimal concentrations were not identical and had to be optimized. However there was no significant influence of the salt nature on crystal growth or characteristics (Reiss-Husson, unpublished experiments).

5.3.2 'Salting-out'

Surprisingly, a few membrane proteins have been crystallized by high salt concentrations; ammonium sulfate and phosphate have been used in the presence of $C_{12}DAO$, $C_{11}DAO$, C_8G, or C_9G (*Table 3*). These salts decrease the solubility of a membrane protein–detergent complex by a mechanism probably more complicated than for the salting-out of soluble proteins (see

Chapter 9). Their presence may modify the interactions between water and the hydrophilic regions of the protein. More importantly, they induce a salting-out of the detergent itself, e.g. monovalent salts modify the upper consolution boundary of C_8E_5; the solubility shift is mainly determined by the anions and follows the Hofmeister series (48).

How to choose between these various precipitating agents? The choice may be restricted by considering the stability of the protein in their presence. For example LHCII, a light-harvesting chloroplast membrane protein, is denatured by PEG and by ammonium sulfate; it was therefore crystallized in the presence of high phosphate concentrations, which do not affect it. The biological activity of the protein should be checked in the presence of increasing amounts of these various precipitants before any crystallization trial. Then the useful range of precipitant concentrations is determined by measuring the lowest precipitant concentration leading to phase separation or precipitation, at given protein and detergent concentrations.

5.4 Optimization

Once crystals (most often microcrystals) have been observed in trial experiments, crystallization conditions have to be improved for crystal size and quality. The strategy is based on the same principles as for soluble proteins (see Chapter 4). Excessive nucleation, leading to a 'shower' of microcrystals, should be avoided; at the same time, growth rate should be kept low enough, as crystal defects are frequently observed when the rate is too high (hollow crystals may even be obtained). Practically, this implies repeating the trials with the different parameters (pH, concentrations, and so on) slightly modified around their initially positive values, over a fine grid. At this stage, use of an additive or of a small amount of a second detergent (compare Section 5.2) may be included as a further variation.

The crystal form of several membrane proteins has been shown to depend on several parameters: type of detergent, pH, nature of the buffer, and the ionic strength when PEG together with salt are present (38). By varying these parameters, it has been possible to select a form which grows better, or is more suitable for structure determination because of its symmetry or unit cell dimensions.

6. Experimental techniques

Crystallization of membrane proteins may be performed with all the experimental set-ups described in Chapter 5. Vapour diffusion and microdialysis have been more frequently used than batch crystallization and free liquid-liquid interface (see *Table 3*). Crystallization of cyt bc_1 in gel of agarose has also been described (26).

Because of the wetting properties of detergents, their drops tend to spread when formed on a planar glass slide, and to fall out when the slide is inverted. Therefore, vapour diffusion with hanging drops is restricted to drop volumes less than 10 μl. With sitting drops formed on depression slides, there is no restriction in volume.

Microdialysis is performed in capillaries or microtubes closed by dialysis membrane with sample volumes less than 150 μl, equilibrated against a reservoir. Choice of the cut-off value of the membrane should take into account the molecular weights of the components of the sample (see Section 4.2.1). Depending on this cut-off value, and also on the thickness of the membrane, dialysis rate, and diffusable species may be controlled; e.g. PEG 4000 and 6000 diffuse (but slowly) through a membrane of cut-off value 25 000 daltons, together with water and salts, but not if a cut-off value of 2000 daltons is used. We have observed that with these two membrane types, all other conditions being the same, crystallization of a bacterial reaction centre with PEG 4000 does not occur similarly (unpublished experiments).

Choosing between microdialysis and vapour diffusion is often a matter of personal preference. Cost of microdialysis is higher when the detergent is expensive, as detergent must be present in the microdialysis reservoir but may be omitted from the vapour diffusion reservoir. One of the advantages of the microdialysis method over vapour diffusion for screening experiments is the possibility of changing individual constituents of the mixture; furthermore the detergent concentration may be kept constant throughout the crystallization, by putting it at the same concentration in the sample and in the reservoir. Changing the dialysis reservoir is also very easy.

The main disadvantage of microdialysis is the rapid equilibration between sample and reservoir (much faster than through vapour diffusion), which may be troublesome if growth rate has to be slowed down. In that case, double dialysis (see Chapter 5) is recommended.

Finally, when crystals are obtained, it important to realize that they are extremely fragile and that their stabilization and manipulation are often difficult. For example, crystals of prostaglandin H synthase are stable in artificial mother liquor only with the detergent at its CMC; this value is critical enough that changes of the CMC due to the addition of salt or sucrose have to be taken into account (6). Recently, it has been shown that the cryo-crystallographic techniques used for soluble protein (see Chapter 13) may successfully be applied to membrane protein, provided that suitable stabilization conditions are found (46, 49, 50, 76).

7. Conclusion

The structures of several membrane proteins have been solved during the past few years, some of them to high resolution (51, 52). This has shown that the methodology originally developed for porin and bacteriorhodopsin has a more general validity. Furthermore, crystallization conditions worked out for one protein have been successfully used with other proteins, opening the way to the design of more systematic strategies. Various detergents are suitable. Their role is important: their properties and their phase diagrams influence the crystallization conditions; they are still associated with the protein in the crystal lattice. The methodology requires (as for soluble proteins) a systematic search over the different parameters, including the nature of the detergent. This adds one more factor to this empirical analysis. However, good quality crystals are still not easy to obtain; the difficulties encountered with bacteriorhodopsin provide the more vivid example. This has stimulated the search for alternative approaches; one of them takes advantage of the bicontinuous cubic phases of lipids where the lipid molecules are arranged in curved three-dimensional bilayers. Such a phase could incorporate bacteriorhodopsin and was used as a matrix for its crystallization (53). This allowed well-ordered crystals to be grown with an improved quality as compared to those previously grown in detergent solutions (54). It is to be hoped that this method could be of general use for other membrane proteins. On the other hand, the problem of maintaining a pure and active protein in solution has been recently tackled with the use of polymeric amphiphiles (55). Nevertheless, finding proper expression system to overexpress these types of protein is still a difficult task (56) that will have to be overcome before membrane protein structures could flood the Protein Data Bank.

References

1. Michel, H. and Oesterhelt, D. (1980). *Proc. Natl. Acad. Sci. USA*, **77**, 1283.
2. Garavito, R. M. and Rosenbusch, J. P. (1980). *J. Cell Biol.*, **86**, 327.
3. Deisenhofer, J., Epp, O., Miki, K., Huber, R., and Michel, H. (1985). *Nature*, **318**, 618.
4. Kühlbrandt, W. (1988). *Q. Rev. Biophys.*, **21**, 429.
5. Michel, H. (ed.) (1991). *Crystallization of membrane proteins*. CRC Press, Boca Raton.
6. Garavito, R. M., Picot, D., and Loll, P. J. (1996). *J. Bioeng. Biomemb.*, **28**, 13.
7. Song, L. and Gouaux, J. E. (1997). In *Methods in enzymology* (ed. C. W. Carter and R. M. Sweet), Academic Press, London. Vol. 276, p. 60.
8. Ostermeier, C. and Michel, H. (1997). *Curr. Opin. Struct. Biol.*, **7**, 697.
9. Kühlbrandt, W. (1992). *Q. Rev. Biophys.*, **25**, 1.
10. Roth, M., Lewit-Bentley, A., Michel, H., Deisenhofer, J., Huber, R., and Oesterhelt, D. (1989). *Nature*, **340**, 659.
11. Roth, M., Arnoux, B., Ducruix, A., and Reiss-Husson, F. (1991). *Biochemistry*, **30**, 9403.
12. Pebay-Peyroula, E., Garavito, R. M., Rosenbusch, J. P., Zulauf, M., and Timmins, P. A. (1995). *Structure*, **3**, 1051.
13. Bordier, C. (1981). *J. Biol. Chem.*, **256**, 1604.
14. Findlay, J. B. C. (1990). In *Protein purification applications: a practical approach* (ed. E. L. V. Harris and S. Angal), pp. 59–82. IRL Press, Oxford.
15. Zulauf, M., Fürstenberger, U., Grabo, M., Jäggi, P. M. R., and Rosenbusch, J. P.

(1989). In *Methods in enzymology*, (ed. S. Fleischer and B. Fleischer), Academic Press, London. Vol. 172, p. 528.
16. Rosenthal, K. S. and Koussale, F. (1983). *Anal. Chem.*, **55**, 1115.
17. De Vendittis, E., Palumbo, G., Parlato, G., and Bocchini, V. (1981). *Anal. Biochem.*, **115**, 278.
18. Van Aken, T., Foscall-Van Aken, S., Castleman, S., and Ferguson-Miller, J. (1986). In *Methods in enzymology*, ibid ref. 15. Vol. 125, p. 27.
19. Herrmann, K. W. (1966). *J. Coll. Int. Sci.*, **22**, 352.
20. De Grip, W. J. and Bovee-Geurts, R. (1979). *Chem. Phys. Lipids*, **23**, 321.
21. Lund, S., Orlowski, S., de Foresta, B., Champeil, P., le Maire, M., and Moller, J. V. (1989). *J. Biol. Chem.*, **264**, 4907.
22. Lorber, B., Bishop, J. B., and DeLucas, L. J. (1990). *Biochim. Biophys. Acta*, **1023**, 254.
23. Eiselé, J.-L. and Rosenbusch, J. P. (1989). *J. Mol. Biol.*, **206**, 209.
24. Soulimane, T., Gohlke, U., Huber, H., and Buse, G. (1995). *FEBS Lett.*, **368**, 132.
25. Ashani, Y. and Catravas, G. N. (1980). *Anal. Biochem.*, **109**, 55.
26. Yu, C.-A., Xia, D., Deisenhofer, J., and Yu, L. (1994). *J. Mol. Biol.*, **243**, 802.
27. Garavito, R. M., Markovic-Housley, Z., and Jenkins, J. (1986). *J. Cryst. Growth*, **76**, 701.
28. Kates, M. (1972). In *Techniques in lipidology* (ed. T. S. Work and E. Work), Vol. 3, Part II. North Holland, Amsterdam.
29. Radin, N. S. (1981). In *Methods in enzymology*, (ed. J. M. Lowenstein), Academic Press, London. Vol. 72, p. 5.
30. Bartlett, G. R. (1959). *J. Biol. Chem.*, **234**, 466.
31. Garavito, R. M. and Picot, D. (1990). *Methods: a companion to methods in enzymology*, **1**, 57.
32. Møller, J., Le Maire, M., and Andersen, J. P. (1986). In *Progress in protein lipid interactions* (ed. A. Watts and J. J. De Pont), Vol. 2, pp. 147–96. Elsevier, Amsterdam.
33. Rivas, E., Pasdeloup, N., and Le Maire, M. (1982). *Anal. Biochem.*, **123**, 194.
34. Rao, P. and Pattabiraman, T. N. (1989). *Anal. Biochem.*, **181**, 18.
35. Kreusch, A., Weiss, M. S., Welte, W., Weckesser, J., and Schulz, G. E. (1991). *J. Mol. Biol.*, **217**, 9.
36. Kleyman, G., Ostermeier, C., Ludwig, B., Skerra, A., and Michel, H. (1995). *Biotechnology*, **13**, 155.
37. Ostermeier, C., Iwata, S., Ludwig, B., and Michel, H. (1995). *Nature Struct. Biol.*, **2**, 842.
38. Garavito, R. M. and Rosenbusch, J. P. (1986). In *Methods in enzymology*, ibid ref. 15. Vol. 125, p. 309.
39. Michel, H. (1983). *Trends Biochem. Sci.*, **8**, 56.
40. Timmins, P. A., Hauk, J., Wacker, T., and Welte, W. (1991). *FEBS Lett.*, **280**, 115.
41. Forst, D., Schülein, K., Wacker, T., Diederichs, K., Kreutz, W., Benz, R., *et al.* (1993). *J. Mol. Biol.*, **229**, 258.
42. Krauss, N., Schubert, W.-D., Klukas, O., Fromme, P., Witt, H. T., and Saenger, W. (1996). *Nature Struct. Biol.*, **3**, 965.
43. Blaauw, M., Dekker, N., Verheij, H. M., Kalk, K. H., and Dijkstra, B. W. (1995). *FEBS Lett.*, **373**, 10.
44. Ducruix, A., Arnoux, B., and Reiss-Husson, F. (1988). In *The photosynthetic*

bacterial reaction center: structure and dynamics (ed. J. Breton and A. Vermeglio), Vol. 149, pp. 21–5. Plenum, New York.
45. Ray, W. J. and Puvathingal, J. M. (1985). *Anal. Biochem.*, **146**, 307.
46. Kurumbail, R., Stevens, A. M., Gierse, J. K., McDonald, J. J., Stegeman, R. A., Pak, J. Y., *et al.* (1996). *Nature*, **384**, 644.
47. Berry, E. A., Shulmeister, V. M., Huang, L.-S., and Kim, S.-H. (1995). *Acta Cryst.*, **D51**, 235.
48. Weckström, K. and Zulauf, M. (1985). *J. Chem. Soc. Faraday Trans. 1*, **81**, 2947.
49. Lee, J. W., Chan, M., Law, T. V., Kwon, H. J., and Jap, B. K. (1995). *J. Mol. Biol.*, **252**, 15.
50. Yu, C.-A., Xia, J.-Z., Kachurin, A. M., Yu, L., Xia, D., Kim, H., *et al.* (1996). *Biochim. Biophys. Acta*, **1275**, 47.
51. Weiss, M. S. and Schulz, G. E. (1992). *J. Mol. Biol.*, **227**, 493.
52. Song, L., Hobaugh, M. R., Shustak, C., Cheley, S., Bayley, H., and Gouaux, J. E. (1996). *Science*, **274**, 1859.
53. Landau, E. M. and Rosenbusch, J. P. (1996). *Proc. Natl. Acad. Sci. USA*, **93**, 14532.
54. Pebey-Peyroula, E., Rummel, G., Rosenbusch, J. P., and Landau, E. M. (1997). *Science*, **277**, 1676.
55. Tribet, C., Audebert, R., and Popot, J.-L. (1996). *Proc. Natl. Acad. Sci.USA*, **93**, 15047.
56. Grisshammer, R. and Tate, C. G. (1995). *Q. Rev. Biophys.*, **28**, 315.
57. Frindi, M., Michels, B., and Zana, R. (1992). *J. Phys. Chem.*, **96**, 8137.
58. Degiorgio, V. (1985). In *Physics of amphiphiles: micelles, vesicles and microemulsions* (ed. V. Degiorgio and M. Corti), pp. 303–35. North Holland, Amsterdam.
59. Michel, H. (1982). *J. Mol. Biol.*, **158**, 567.
60. Chang, C.-H., Schiffer, M., Tiede, D., Smith, U., and Norris, J. (1985). *J. Mol. Biol.*, **186**, 201.
61. Allen, J. P. and Feher, G. (1984). *Proc. Natl. Acad. Sci. USA*, **81**, 4795.
62. Yeates, T. O., Komiya, H., Rees, D. C., Allen, J. P., and Feher, G. (1987). *Proc. Natl. Acad. Sci. USA*, **84**, 6438.
63. Arnoux, B., Ducruix, A., Reiss-Husson, F., Lutz, M., Norris, J., Schiffer, M., *et al.* (1989). *FEBS Lett.*, **258**, 47.
64. Allen, J. P. (1994). *Proteins*, **20**, 283.
65. Buchanan, S. K., Fritzsch, G., Ermler, U., and Michel, H. (1993). *J. Mol.Biol.*, **230**, 1311.
66. Katayama, N., Kobayashi, M., Motojima, F., Inaka, K., Nozawa, T., and Miki, K. (1994). *FEBS Lett.*, **348**, 158.
67. Feiek, R., Ertlmaier, A., and Ermler, U. (1996). *FEBS Lett.*, **396**, 161.
68. Papiz, M. Z., Hawthornthwaite, A. M., Cogdell, R. J., Woolley, K. J., Wightman, P. A., Ferguson, L. A., *et al.* (1989). *J. Mol. Biol.*, **209**, 833.
69. Koepke, J., Hu, X., Muenke, C., Schulten, K., and Michel, H. (1996). *Structure*, **4**, 581.
70. Pauptit, R. A., Zhang, H., Rummel, G., Schirmer, T., Jansonius, J. N., and Rosenbusch, J. P. (1991). *J. Mol. Biol.*, **218**, 505.
71. Kreusch, A., Neubüser, A., Schiltz, E., Weckesser, J., and Schultz, G. E. (1994). *Protein Sci.*, **3**, 58.

72. Tucker, A. D., Jackman, S., Parker, M. W., and Tsernoglou, D. (1991). *J. Mol. Biol.*, **222**, 881.
73. Stauffer, K. A., Page, M. G. P., Hardmeyer, A., Keller, T. A., and Pauptit, R. A. (1990). *J. Mol. Biol.*, **211**, 297.
74. Hirsch, A., Wacker, T., Weckesser, J., Diederichs, K., and Welte, W. (1995). *Proteins*, **23**, 282.
75. Tsukihara, T., Aoyama, H., Yamashita, E., Tomizaki, T., Yamaguchi, H., Shinzawa-Itoh, K., *et al.* (1995). *Science*, **269**, 1069.
76. Luong, C., Miller, A., Barnett, J., Chow, J., Ramesha, C., and Browner, M. F. (1996). *Nature Struct. Biol.*, **3**, 927.
77. Gouaux, J. E., Braha, O., Hobaugh, M. R., Song, L., Cheley, S., Shustak, C., *et al.* (1994). *Proc. Natl. Acad. Sci. USA*, **91**, 12828.

10

From solution to crystals with a physico-chemical aspect

M. RIÈS-KAUTT and A. DUCRUIX

1. Introduction

Biological macromolecules follow the same thermodynamic rules as inorganic or organic small molecules concerning supersaturation, nucleation, and crystal growth (1). Nevertheless macromolecules present particularities, because the *intramolecular* interactions responsible of their tertiary structure, the *intermolecular* interactions involved in the crystal contacts, and the interactions necessary to solubilize them in a solvent are similar. Therefore these different interactions may become competitive with each other. In addition, the biological properties of biological macromolecules may be conserved although the physico-chemical properties, such as the net charge, may change depending on the crystallization conditions (pH, ionic strength, etc.). A *charged* biological macromolecule requires counterions to maintain the electroneutrality of the solution; therefore it should be considered as a *protein* (or *nucleic acid*) *salt* with its own physico-chemical properties, depending on the nature of the counterions.

To crystallize a biological macromolecule, its solution must have reached supersaturation which is the driving force for crystal growth. The understanding of the influence of the crystallization parameters on protein solubility of model proteins is necessary to guide the preparation of crystals of new proteins and their manipulation. Only the practical issues are developed in this chapter, and the reader should refer to recent reviews (2–4) for a description of the fundamental physical chemistry underlying crystallogenesis.

2. The concept of solubility and methods for solubility diagram determination

The solubilization of a solute (e.g. a biological macromolecule) in an efficient solvent requires solvent–solute interactions, which must be *similar* to the solvent–solvent interactions and to the solute–solute interactions of the compound to be dissolved. All of the compounds of a protein solution (protein, water, buffer, crystallizing agents, and others) interact with each other via various, often weak, types of interactions: monopole–monopole, monopole–dipole, dipole–dipole, Van der Waals hydrophobic interactions, and hydrogen bonds.

2.1 Solubility

Solubility is defined as the amount of solute dissolved in a solution in equilibrium with its crystal form at a given temperature. For example, crystalline ammonium sulfate dissolves at 25 °C until its concentration reaches 4.1 moles per litre of water, the excess remaining non-dissolved. More salt can be dissolved when raising the temperature, but if the temperature is brought back to 25 °C, the solution becomes *supersaturated*, and the excess of salt crystallizes until its concentration reaches again its *solubility* value at 25 °C (4.1 moles per litre of water).

In the case of biological macromolecules, the solubility is additionally defined by the characteristics of the solvent. Proteins are mostly solubilized in water which acts through hydrogen bonds. In some cases another protic solvent (an alcohol) or an aprotic solvent (e.g. acetone, DMSO, dioxane, . . .) is added at low concentration. In addition, the solvent solutions contain at least the ubiquitous buffer used to fix the pH of the solution and therefore the net charge of the protein. Salts are added not only to ensure an ionic strength but most often to reach supersaturation.

Throughout this chapter, protein *solubility* is defined as the concentration of soluble protein in *equilibrium* with the *crystalline* form at given temperature and pH values, and in the presence of a given concentration of solvent compounds others than the protein (i.e. water, buffer, crystallizing agents, stabilizers, additives). The solubility values depend on the physico-chemical characteristics of the protein itself (hydrophilicity, net charge, type of solvent exposed residues) and of the solvent (pH, dielectric constant, ionic strength, concentration, and nature of the additives).

Figure 1 illustrates the variability of protein solubilities, depending on the protein itself or on the protein salt (e.g. different lysozyme salts). The solubility of the three proteins: bovine pancreatic trypsin inhibitor (BPTI) (5) in ammonium sulfate, collagenase from *Hypoderma lineatum* (*Hl*) in ammonium sulfate (6), and hen egg white (HEW) lysozyme in NaCl (7), cover a very large range of both ionic strength and solubility values, in their standard crystallization conditions. Furthermore the solubility of a same protein, HEW lysozyme, can be changed drastically when changing the nature of the crystallizing salt, as shown by the solubility curves of lysozyme/KSCN, lysozyme/NaCl, and lysozyme/NH_4OAc.

In the literature other conventions of defining solubility are encountered; it may be the protein concentration measured *before* the actual equilibrium is reached, or it can be evaluated in the presence of *precipitate* instead of crystals (8). Their applications are discussed at the end of this section.

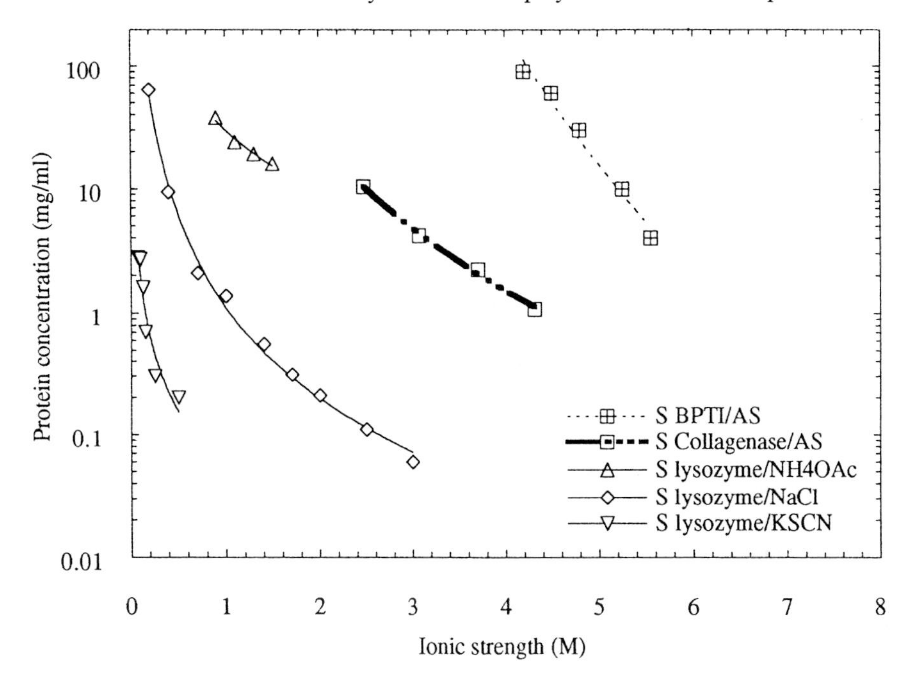

Figure 1. Solubility curves at 18°C of BPTI (5), collagenase from *Hypoderma lineatum* (6), and HEW lysozyme (7, 10). The crystallizing agents are AS (ammonium sulfate), NH_4OAc (ammonium acetate), NaCl (sodium chloride), and KSCN (potassium thiocyanate).

The zone of the solubility diagram where crystals appear (nucleation zone) depends on the *supersaturation*, which is the ratio, C_p/C_S, of the protein concentration over the solubility value, but also on the kinetics to reach these conditions.

The protein purity (see Chapter 2) must be checked before doing any screening experiments. In a mixture of proteins, the first crystals contain the most supersaturated protein and this may not be the most concentrated one. Furthermore the solubility of the major protein may be affected differently from the contaminant when changing a parameter.

2.2 Measurements of the solubility

Solubility measurements are necessary to understand the effect of crystallization parameters on the solubility of model proteins, which can then be transposed for the crystallization of a new protein.

2.2.1 Conventional methods

The solubility can be determined either by *crystallization* of a supersaturated solution (see *Protocol 1*) or by *dissolution* of crystals in an undersaturated solution. In both cases the protein concentration in the supernatant converges toward the same asymptotic value at equilibrium. For double checking of solubility values, both crystallization and dissolution methods can be run in parallel. For both crystallization and crystal dissolution, crystallization conditions must be previously defined.

During the course of the experiments, all parameters, except the one under investigation, (e.g. pH, temperature, salt and buffer concentrations, nature of salt and buffer) must be carefully kept constant and the stability of the biological macromolecule versus time and proteases must be checked.

Protocol 1. Solubility measurements by crystallization

Equipment and reagents

- ACA boxes
- 10 ml buffer
- Incubator
- 10 ml salt stock solution

Method

1. Define one parameter to vary (e.g. salt concentration), keeping all others strictly constant (e.g. pH, temperature, nature of salt and buffer).
2. Choose at least four values of the variable (different salt concentrations over a large range), because solubility curves usually do not fit with linear curves.
3. Set up the batch experiments ($\geq$ 10 μl) in duplicate at two or three different initial protein concentrations for a given parameter value.
4. Follow the decrease of the protein concentration of the supernatant, by withdrawing a crystal-free aliquot of the duplicate set-up, each week for optical density (OD) measurements. If microcrystals are present, filter or centrifuge the aliquot before the dilution for the OD measurement.

   Once crystallization has started, the protein concentration in the supernatant will converge to a constant value, solubility, with time. This value is identical for the different initial protein concentration at a same ionic strength.
5. When the protein concentrations remain constant for at least two weeks and are identical for the different initial protein concentration at the same ionic strength, confirm the measurement by testing the original undisturbed set of experiments.

Solubility measurements are often performed by *batch* methods (9, 10). Hanging, sitting, or sandwich drops systems can be used as long as the salt concentration is *identical* in both the drop and the reservoir all over the process. In this case the ratio of the salt concentration initially in the drop and in the reservoir is 1:1, whereas in a classical vapour diffusion experiment it is 1:2. The role of the reservoir here is only to keep vapour pressure constant during the experiment. The vapour diffusion technique is not less suited, because:

(a) Only the initial conditions at the beginning of the experiment are well known. The accurate ionic strength, once the drop/reservoir equilibration is achieved, is difficult to verify when working with small drop volumes.
(b) All components in the drop will concentrate: the crystallizing agent and the protein as expected, but also the buffer and additives (and impurities!). As a consequence more than one parameter may change during the experiment.

2.2.2 Alternative methods

More sophisticated methods for the solubility dependence with temperature have been described in the literature. They yield not only solubility values versus temperature, but also the crystallization enthalpies and sometimes the crystallization induction times.

i. Column method

A solution of either supersaturated or undersaturated protein solution is poured in a microcolumn filled with crystals. An aliquot of the solution is periodically withdrawn from the bottom of the column to follow the change of the protein concentration by optical density measurements. This method (11) is based on the maximization of the exchange between the available crystalline surface area and minimal free solution volume to reach equilibrium. It overcomes the problem of prolonged equilibration time, as equilibrium, i.e. the solubility value, is reached within one to five days.

Two microcolumns are run in parallel: one for crystallization (supersaturated state), one for dissolution (undersaturated state). This method is mostly appropriate for the study of the influence of temperature. To determine protein solubility at different concentrations of a given salt, crystals can be prepared from a same batch, but must then be equilibrated carefully at respective salt concentrations. When solubility is checked in different salts, a batch of crystals is prepared in each appropriate salt.

ii. Microscopic observation and OD measurements

About 0.5 ml crystallization solution containing crystals are placed in a glass vessel inserted in a thermoregulated cell. The temperature is monitored and controlled by a Peltier element ($\pm$ 0.1 °C). The whole set-up is placed under a microscope. Dissolution and growth are followed by microscopic observation in parallel with OD measurements of the crystallization solution (5, 12).

Equilibration is achieved within about 50 days but several cells can be run in parallel in the Peltier element.

iii. Scintillation

A thermoregulated cell (50–100 μl) is filled with the solution to crystallize. The bath temperature is changed until crystallites occur inducing scintillation which is detected by a photodiode signal. The temperature is changed backwards and forwards to define the solubility limit, defined by the appearance and disappearance of the crystallite detected by the scintillation signal. One solubility value is obtained within approximately 12–24 hours. In addition this technique (13) allows crystallization induction times to be measured which were shown to follow supersaturation.

iv. Temperature controlled static light scattering method

This method (14) is similar to the scintillation method described above, but using light scattering to follow the occurrence or dissolution of crystallites when changing the temperature. The crystallization solution (about 1 ml) is stirred by a Teflon coated magnet to maintain the particles in suspension. The solution is illuminated by a 1 mm^2 cross-section laser beam. The light scattered, normal to the incident beam, is focused on a photodiode whose signal is amplified and analysed by a phase-sensitive detector. The temperature is changed until faceted crystals nucleate and the scattered intensity reaches a plateau. Then the temperature is changed backwards until the crystals dissolve.

v. Michelson interferometry

A Michelson interferometer is used for the observation of concentration gradients around a crystal to determine whether the crystal is growing or dissolving when changing the temperature (15). The volume of the cell is about 70 μl. The equilibrium temperature is obtained within two hours.

vi. Calorimetry and OD measurements

The heat signal from a 1 ml crystallizing solution is recorded every two minutes over a period of two to three days, using a differential scanning calorimeter to follow the heat of crystallization. The final protein concentration, i.e. solubility, is obtained by removing an aliquot of crystal-free solution from the cell and measuring the absorbance at 280 nm (16).

2.2.3 Estimating the residual protein concentration

It is obvious that accurate solubility measurements are necessary in fundamental research to understand protein crystal growth. However when dealing with the crystallization of a new protein, one is probably not willing to invest much protein and time to define the solubility values in various conditions. Nevertheless it is very helpful to have at least an order of magnitude of the residual protein concentration, in contact with the crystals, to guide further experiments and handle the crystals (*Protocol 2*). The estimation of the residual protein concentration is easy to perform even with a small drop size.

Protocol 2. Measurement of the residual protein concentration in hanging drops[a]

Equipment and reagents

- Buffer solution
- U.V. spectrophotometer

Method

1. Open the coverslip with the drop in which crystals have grown for at least two weeks.
2. Withdraw 1 (preferably 2) μl of clear supernatant under the binocular. If too many microcrystals are present, centrifuge the drop and take the aliquot from the clear supernatant to avoid diluting crystals.
3. Dilute to the minimal volume required for an OD measurement at 280 nm. Eventually measure the protein concentration by the Bradford method using a coloured dye to increase the absorbance of the protein–dye complex.

[a] When using the dialysis technique, an aliquot can be withdrawn from the dialysis cell with a Hamilton syringe. The dialysis cell can of course no longer be immersed in the reservoir, but it can be rescued in a vessel with some reservoir solution around in order to avoid the solution drying.

Measuring the residual protein concentration is helpful for:

(a) Having an *estimate of the solubility*. Even though the drop has not reached equilibrium, this measurement gives an order of magnitude whether the solubility is low, medium, or high. It will be explained later in this chapter that low solubility conditions ($\leq$ 1 mg/ml) are difficult for the optimization of growing few and large crystals.

(b) Knowing the starting conditions, the ratio of initial over final protein concentrations tells an order of magnitude of the *supersaturation* where these crystals were obtained. This helps to choose the range of supersaturation, and therefore initial protein concentration, for further experiments.

(c) Defining the *amount* of protein which is available to grow crystals. For HEW lysozyme, 1, 8.5, 29, 68 μg of protein are necessary to grow respectively crystal of 0.1, 0.2, 0.3, 0.4 mm^3.

In *Figure 2* are illustrated two quite different conditions, A and B, for

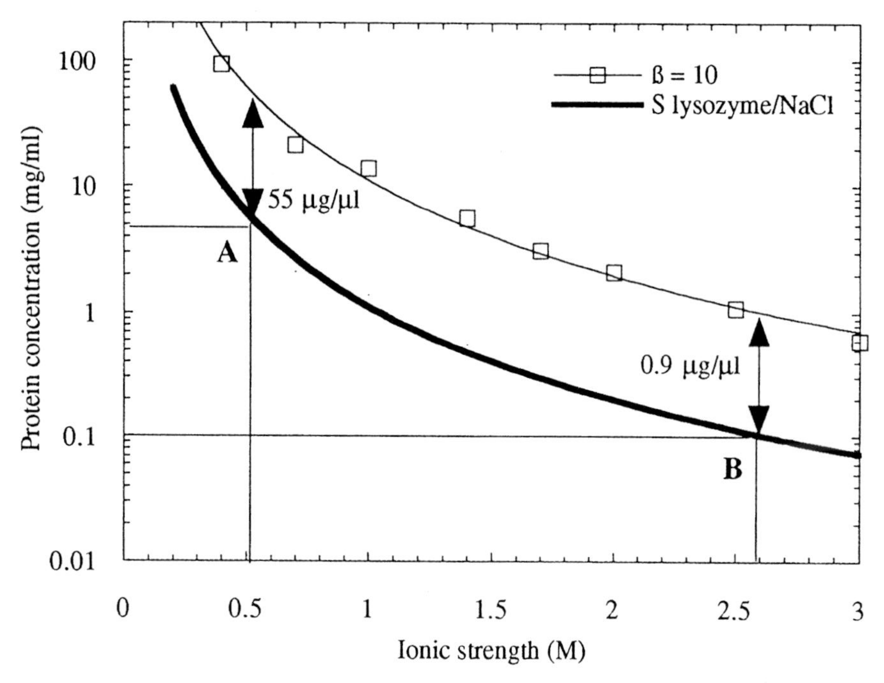

Figure 2. Schematic phase diagram showing the crystallization conditions at a supersaturation, β, of 10, and the solubility curve. Conditions A and B illustrate high and low solubility conditions, respectively.

which respectively 55 or 0.9 μg/μl of protein are in supersaturation, when starting from a tenfold supersaturated protein solution. A 0.3 mm^3 crystal can be grown from a drop of ≈ 0.5 μl in A but from 32 μl in B.

(d) Estimating the *slope* of the solubility curve, by measuring the residual protein concentration at three different values of the variable which can also be ionic strength, temperature, pH, etc. This helps for extrapolating the nucleation zone to lower or higher values of the variable depending on whether the solubility variation is steep or smooth.

(e) Guiding the preparation of *seeding* experiments. A classical vapour diffusion experiment, with a 1:2 salt concentration ratio between drop and reservoir, may evolve in two ways as shown on *Figure 3*.

   (i) The concentrations in the drop equilibrate from I (initial) to F (final), then nucleation occurs and the protein concentration drops from F to P2, the residual protein concentration, while the salt concentration remains constant in the drop and equals the one of the reservoir.

   (ii) The drop concentrations increase from I to an intermediate value B, at which nucleation starts already without ever reaching F. Then both protein and salt concentrations continue to change to reach P2.

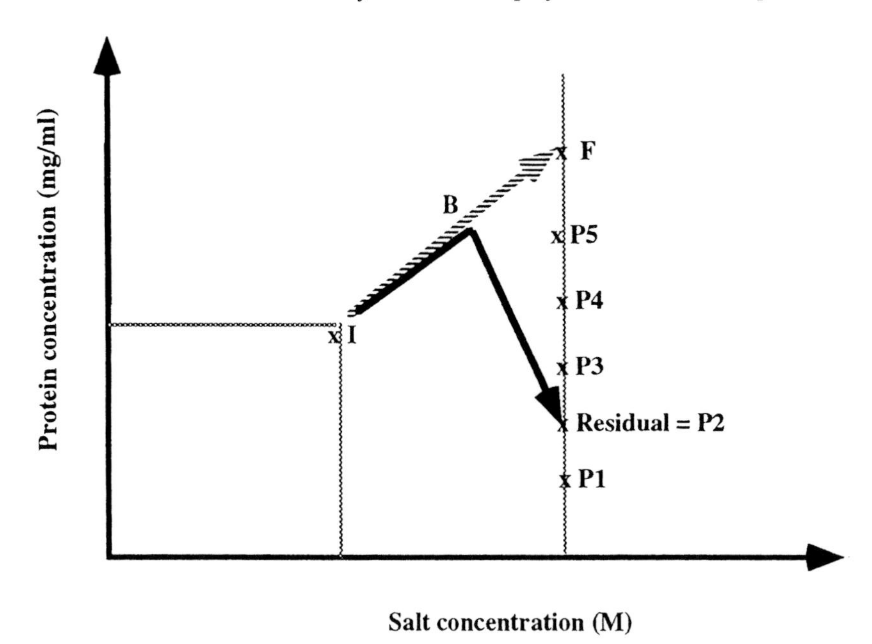

Figure 3. Residual protein concentration in a drop after a vapour diffusion process, and selection of the seeding conditions.

Once the residual protein concentration is measured, news drops are prepared at protein concentrations ranging from P1 to P5, as indicated on *Figure 3*, directly at the salt concentration of the reservoir (batch). The concentrations P3 to P5 aim at covering the *searched* metastable zone. The protein concentration P1 (≤ solubility) is added to have a better estimate of the lower limit of the metastable zone. Crystals dissolve in undersaturated drops (≤ P1) and remain unchanged in saturated drops (between P1 and P2). They grow in slightly supersaturated drops (P2 to P4, where you should seed) whereas new nucleation occurs in more supersaturated conditions (> P4–P5).

(f) *Mounting* the crystals. When a crystal is recovered from a drop, it is in a solution of a given protein and crystallizing agent concentrations as shown in *Figure 2*. Very often reservoir solution is used to transfer the crystal. In fact, this can be done safely only if the remaining protein concentration in the drop is lower than ≈ 0.5 mg/ml (e.g. B in *Figure 2*). If the solubility value is higher (e.g. A in *Figure 2*), then the crystal would start to dissolve as the reservoir contains no protein. Knowing the residual protein concentration in the mother liquor gives the amount of protein to introduce in additional mounting solutions.

Similarly protein should be added when soaking crystals in *cryo-protectant* solutions for cryo-crystallography *when crystals dissolve*. Cryo-protectants often change the protein solubility. Although the knowledge of the residual protein concentration of the crystallization drop is not representative of the solubility of the protein in the cryo-protectant solution, crystals should be soaked in a series of cryo-protectant solutions containing a protein concentration higher than the residual protein concentration.

2.3 Phase diagram

As the solubility of a biological macromolecule depends on various parameters (see Chapter 1, *Table 1*), a phase diagram is a useful representation of its solubility (mg/ml or mM biological macromolecule in solution) as a function of one parameter, all other parameters being kept constant.

The diagram, represented in *Figure 4*, comprises the following zones:

(a) The *solubility* curve delimits the under- and supersaturated zones. In an experiment where crystallizing agent and biological macromolecule concentrations correspond to solubility conditions, the *saturated* macromolecule solution is in *equilibrium* with the crystallized macromolecule. This corresponds to the situation at the end of the process of crystal growth: additional crystalline macromolecule does not dissolve, but adding

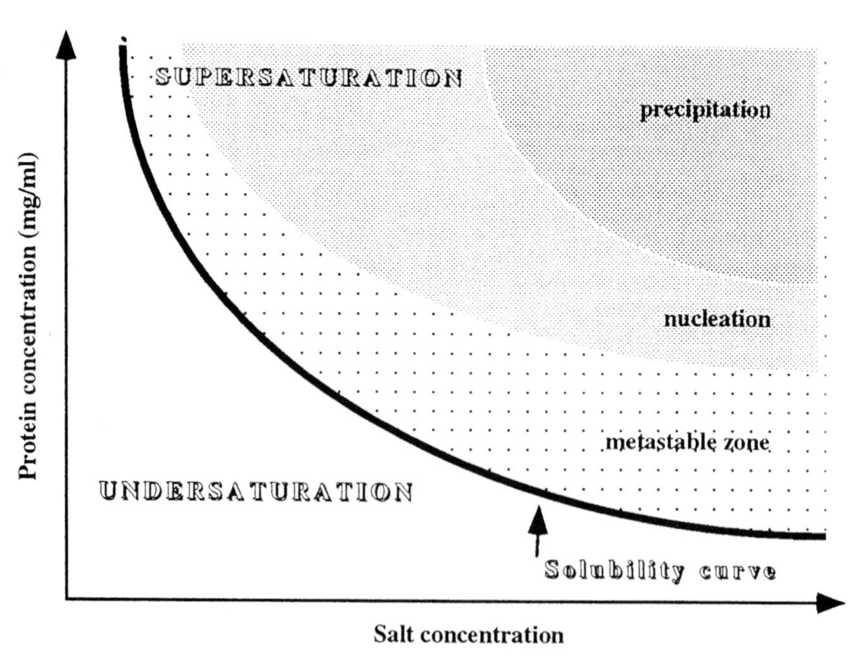

Figure 4. Schematic description of a two-dimensional solubility diagram showing the different zones of the supersaturation domain. Note that the metastable zone covers a larger range of supersaturation, when solubility is low (i.e. at high salt concentration) than when it is high (i.e. at low salt concentration). Conversely, the nucleation zone is larger for high solubility than for a low one.

reservoir solution without the macromolecule leads to the dissolution of the macromolecule crystals.

(b) Below the solubility curve the solution is *undersaturated*, the system is thermodynamically stable, and the biological macromolecule will never crystallize.

(c) Above the solubility curve, the concentration of the biological macromolecule is higher than the concentration at equilibrium. This corresponds to the *supersaturation* zone. A supersaturated macromolecule solution contains an excess of macromolecule which will appear as a solid phase until the macromolecule concentration reaches the solubility value in the solution (supernatant). In some cases the excess of macromolecule may concentrate in oily drops in a liquid–liquid separation. The *rate* of supersaturation is defined as the ratio of the biological macromolecule concentration over the solubility value. The higher the supersaturation rate, the faster this solid phase appears.

It is often difficult to understand how supersaturated macromolecule solutions are achievable. In terms of molarity, it must be remembered that macromolecule solubilities are very low (μM to mM) compared to small molecules or inorganic molecules (mM). This also corresponds to very low volume fractions of solute which allow macromolecule solutions to be prepared at supersaturations as high as 10 to 20 times the solubility. For small molecules supersaturation of only 1.1 to about 1.5 are achievable. Recently we observed crystallization of HEW lysozyme at supersaturations around 1.5 when working with 400 mg/ml (28 mM) protein which correspond to 30% volume fraction (17).

However the higher the supersaturation, the faster the solid phase appears in the solution, as described below. Contrary to macromolecule purification which implies precipitation, crystallization requires an accurate control of the level of supersaturation. This allows nucleation of crystals, i.e. a solid phase with a three-dimensional periodicity, by controlling the nucleation rate to yield few single crystals.

2.3.1 Precipitation zone

Precipitation occurs at very high supersaturation ($\approx$ 30 to 100 times the solubility value for HEW lysozyme). Insoluble macromolecules rapidly separate from the solution in an *amorphous* state. If the solution is centrifuged, the supernatant is in fact still supersaturated and crystallization may occur. To differentiate amorphous precipitate from microcrystals, fresh drops can be seeded (see Chapter 7) with this material; amorphous precipitate dissolves whereas microcrystals grow.

2.3.2 Nucleation zone

At a sufficient supersaturation, nucleation spontaneously occurs, once critical activation free energy is overcome. This is called *homogeneous* nucleation.

Crystallization often occurs at lower saturation when it is induced by vibrations or the presence of particles (dust, precipitate, irregularities of crystallization cell); it is then called *heterogeneous* nucleation. The latter is usually characterized by non-reproducibility, therefore it is recommended to filter all solutions and blow the coverslips with an air stream before setting up the hanging drops.

Nucleation requires a lower supersaturation than precipitation. To give an order of magnitude, the nucleation range for HEW lysozyme is ≈ 5 times the solubility for dialysis and batch crystallizations, and about 10 times the solubility for vapour diffusion. Crystals appear faster and in larger numbers with increasing supersaturation. High supersaturation may be useful to find the nucleation zone, but growing crystals for X-ray diffraction may benefit from a search of the optimal supersaturation where few but large crystals are grown.

The nucleation *rate*, defined as the number of nuclei formed per unit volume and unit time, is linked (1, 18) to:

(a) Supersaturation, as illustrated by the curve A of *Figure 5*. For supersaturations higher than β*, the critical supersaturation, nucleation occurs. When increasing the supersaturation, the number of crystals increases.

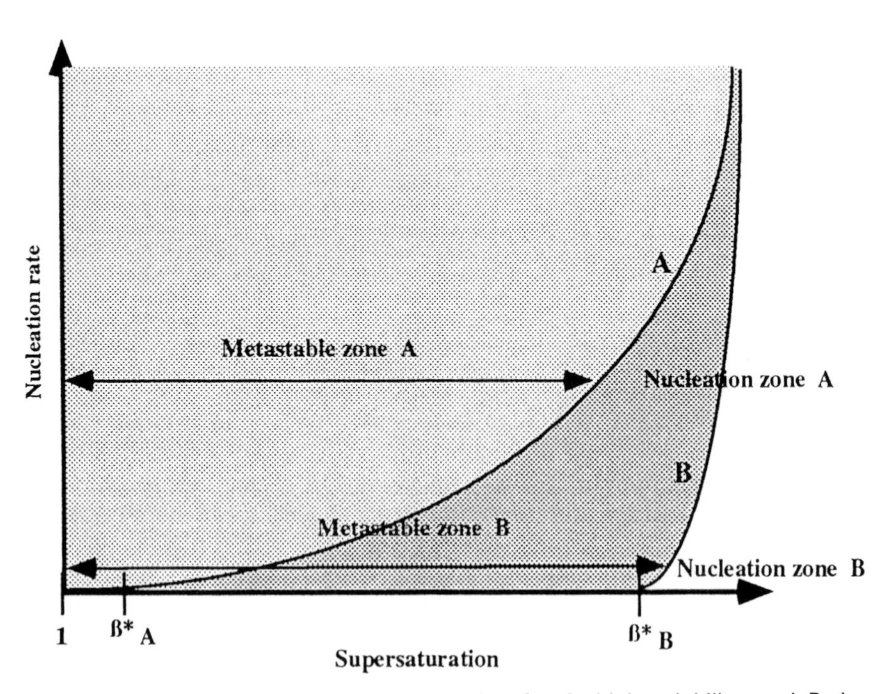

Figure 5. Nucleation rate versus supersaturation for A, high solubility, and B, low solubility. The curves A and B delimit the metastable zone from the nucleation one.

(b) The number of molecules per unit volume. When the solute is sparingly soluble, the solution remains in a metastable state over long periods. Nucleation requires much higher supersaturation to occur (*Figure 4*). Once β^* is reached, the nucleation rates becomes drastic, as illustrated by the curve B in *Figure 5*. The different curves A and B of *Figure 5* correspond respectively to the situations A and B of *Figure 2*.

Crystallization conditions for which the solubility is very low should be avoided. The precipitation curve is then very close to the solubility curve, the domain of crystallization becomes very narrow, which brings difficulties in defining the right conditions for growing large crystals. This is the case for the crystallization of HEW lysozyme/KSCN, where the crystallization zone is limited to a range of 100 mM KSCN. On the counterpart, HEW lysozyme/NaCl crystallizes over a broad range of 1400 mM. To enlarge the nucleation zone, solubility must be increased. This can be done by:

- decreasing the ionic strength while increasing the macromolecule concentration, if the slope of the solubility curve is smooth enough
- using another salt in which the solubility is higher (e.g. from HEW lysozyme/KSCN to HEW lysozyme/NaCl shown in *Figure 1*)
- changing the pH to increase the protein net charge
- changing the temperature.

2.3.3 Metastable zone

In the metastable zone, the critical supersaturation is not yet reached. Spontaneous nucleation does not occur, unless it is induced by vibrations or introduction of a particle which will promote heterogeneous nucleation.

As shown in *Figure 5*, the metastable zone is much larger when the solubility is very low. It becomes then extremely difficult to reduce the nucleation rate. One can use the metastable zone to seed crystals which will grow, fed by the amount of protein in supersaturation. When a low solubility system cannot be brought to higher solubility for technical reasons, seeding remains nearly the only way to grow large single crystals (Chapter 7).

2.4 Kinetic aspects

Dealing with a supersaturated protein solution implies a system which is thermodynamically out of equilibrium. Therefore nucleation and growth depend on various kinetics. In other words, the solubility is unchanged as long as the pH, the temperature, the ionic strength, and the nature of solvent constituents are constant, but the nucleation zone, as well as the precipitation zone, may be shifted depending on the crystallization technique, the geometry of the device, etc.

2.4.1 Time lag for nucleation

When the protein solution is directly prepared at a given supersaturation (using the batch method) or if the protein solution has completely equilibrated with its reservoir *before crystals have nucleated*, the crystals do not appear immediately, but after a time lag. This time lag is related with the supersaturation as illustrated for HEW lysozyme (18); the higher the supersaturation the faster crystals appear. It must be remembered that the number of crystals also increases with supersaturation.

2.4.2 Protein solution/reservoir equilibration

When using vapour diffusion or dialysis (see Chapter 5), the first kinetics to control are the equilibration of the protein solution with the reservoir.

These kinetics have been shown (19, and refs therein) to increase with the drop/reservoir distance, with the initial gradient of crystallizing agent between drop and reservoir, and when using PEGs instead of salts. These kinetics are obviously not relevant for the batch method because the protein solution is directly prepared at the protein and crystallizing agent concentrations to be tested.

Figure 6 shows the equilibration of a set of vapour diffusion experiments

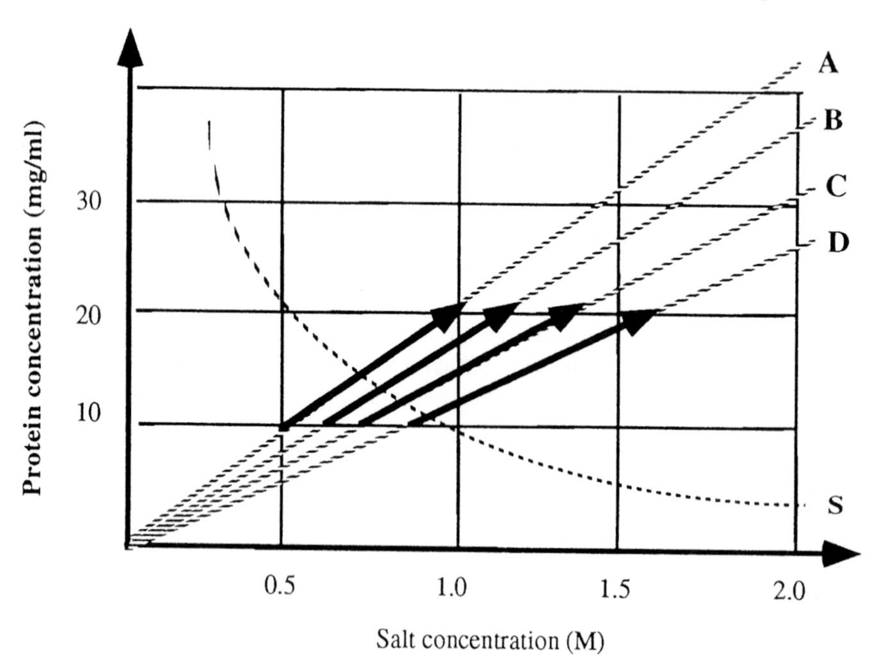

Figure 6. Variations of the protein and salt concentration in four drops A to D during equilibration with their reservoirs for a classical 1:2 ratio. The bold arrows start at the initial conditions in the drops, and end at the expected final conditions, if no crystallization occurs during the drop/reservoir equilibration.

with a typical 1:2 ratio for the salt concentration initially in the drop and in the reservoir. In condition A the salt gradient between the initial drop conditions and the reservoir is 0.5 M whereas it is 0.8 M in D. This implies a faster equilibration for A than for D. As a consequence the drop/reservoir equilibration may become much faster than the time lag for nucleation. Therefore nucleation occurs at a higher supersaturation, implying a higher nucleation rate. To help the system to crystallize before the drop/reservoir equilibration is achieved, the experiment may be run in two steps of lower gradient, or with a larger drop/reservoir distance, or in ACA instead of Linbro plates. Similarly for dialysis, capillaries may be preferred to Cambridge buttons, because the protein solution equilibrates more slowly with the reservoir solution, allowing nucleation to occur before equilibration is achieved.

2.4.3 Equilibration to reach the solubility value

Once crystallization has started, the protein concentration in the solution decreases until it reaches the solubility value. These kinetics depend on the *growth kinetics* of a given crystal form of a given biological macromolecule, the *number* of crystals growing, the *amount* of protein to crystallize, and *stirring* (or not) the solution.

For the photochemical reaction centre of *Rhodobacter sphaeroides* (20), the solubility value was reached within 12 days even though the unstirred batch method was used. In this case the crystals grew very quickly.

For tetragonal HEW lysozyme crystals, equilibration requires up to nine months if using unstirred batch methods. The delay can be reduced to two months by stirring (17) the crystallization vials. However this was not successful for very high protein concentrations which are very viscous. These experiments were kept for one month at low temperature to *accelerate* the crystallization before letting them equilibrate at 18 °C.

3. Proteins as polyions

Throughout the process of crystallization and of structure determination, a protein must be considered as a polyion of a given *net charge*, surrounded by *counterions*. The number of counterions is at least equal to the net charge to ensure the electrostatic compensation. Even though the biochemical activity may not be altered, the physico-chemical properties of a protein, and hence its behaviour in crystallization, may be changed significantly by variations of its net charge or by adsorption of small molecules or ions onto its surface. It is useful to begin a new crystallization project by first calculating the probable net charge of the protein as a function of pH, although it is an approximation.

Additives are well defined solvent constituents (chemical nature and concentration); otherwise they are *impurities*. To improve the reproducibility of crystallization experiments and the reliability during a structural investigation involving different protein batches, impurities should be eliminated whenever possible (see Chapter 2). This is usually done for biological contaminants, but seldom for small molecules or salts. After a routine purification step, aqueous protein solutions contain various additives, including at least a buffer, various salts from elution gradient, and additives to prevent oxidation of free SH groups, EDTA, NaN_3, stabilizers, etc. Some compounds or ions may be bound by the protein. Proteins from commercial sources often contain organic or inorganic compounds whose nature depends on the source, and the amount on the batch. Up to 14% (w/w) salt were observed for commercial HEW lysozyme (21), although the purity was otherwise excellent in terms of biological activity. All the uncontrolled organic and inorganic solutes should be eliminated or replaced by co-ions and counterions of known concentration and nature.

3.1 Estimation of the net charge

The net charge of a protein, Z_p, is the difference between the number of positive and negative charges arising from the deprotonation of acidic groups and the protonation of basic groups. On average, the pKa values of solvent-accessible charged residues are respectively: 3.5 (COOH-terminus), 4.5 (Asp, Glu), 6.2 (His), 7.6 (NH_2-terminus), 9.5 (Tyr), 9.3 (Cys), 10.4 (Lys), and 12.0 (Arg).

The estimated protein net charge, Z_p, for different pH values can be calculated with programs such as *Excell*, *Kaleidagraph*, according to:

$$Zp = \Sigma_{(+)}^{\alpha NH2,His,Lys,Arg} ni_{(+)} * \frac{[H+]}{K_{a(+)} + [H+]} - \Sigma_{(-)}^{\alpha COOHAsp,Glu,Tyr,CysSH} ni_{(-)} * \frac{K_{a(-)}}{K_{a(-)} + [H+]} \quad [1]$$

where $[H^+] = 10^{-pH}$; $K_a = 10^{-pKa}$ for the individual charged groups; n_i the number of each type of charged residue, and assuming that:

(a) Identical groups have the *same pKa* value. Of course actual local pH values may be different depending on their environment, but this is rarely known as long as the protein structure is not determined.

(b) All potentially charged groups are *accessible*, i.e. neither buried inside the protein, nor involved either in a salt bridge with another charged residue, or in ion complexation. However some ionizable side chains buried inside the protein may exist in their uncharged forms. Similarly, charged groups may complex ions (e.g. Zn, Fe, Mg, Ca, etc.). The contribution of such groups can be adjusted, a priori if they are known or, a posteriori, when comparing the estimated isoelectric points with the experimental ones or with titration data.

(c) The protein is *monomeric*. This requirement is important unless it is known which charged residues are accessible or buried in the interface of the di- or multimeric macromolecule.

Table 1. Calculation of protein charges versus pH[a]

pK =	α-COOH 3.5		Asp + Glu 4.5		Tyr 9.5		Cys-SH 9.3		α-NH_2 7.6		His 6.2		Lys 10.4		Arg 12		
n_i	$n_{\alpha\text{-COOH}}$ =		$n_{Asp + Glu}$ =		n_{Tyr} =		$n_{Cys\text{-}SH}$ =		$n_{\alpha\text{-}NH2}$ =		n_{His} =		n_{Lys} =		n_{Arg} =		
pH	z	z × n	z	z × n	z	z × n	z	z × n	z	z × n	z	z × n	z	z × n	z	z × n	Zp
2.00	−0.03		−0.003		−0.00		−0.00		+1.00		+1.00		+1.00		+1.00		
2.50	−0.09		−0.01		−0.00		−0.00		+1.00		+1.00		+1.00		+1.00		
3.00	−0.24		−0.03		−0.00		−0.00		+1.00		+1.00		+1.00		+1.00		
3.50	−0.50		−0.09		−0.00		−0.00		+1.00		+1.00		+1.00		+1.00		
4.00	−0.76		−0.24		−0.00		−0.00		+1.00		+1.00		+1.00		+1.00		
4.50	−0.91		−0.50		−0.00		−0.00		+1.00		+0.99		+1.00		+1.00		
5.00	−0.97		−0.76		−0.00		−0.00		+1.00		+0.97		+1.00		+1.00		
5.50	−0.99		−0.91		−0.00		−0.00		+0.99		+0.91		+1.00		+1.00		
6.00	−1.00		−0.97		−0.00		−0.00		+0.97		+0.76		+1.00		+1.00		
6.50	−1.00		−0.99		−0.00		−0.00		+0.91		+0.50		+1.00		+1.00		
7.00	−1.00		−1.00		−0.00		−0.00		+0.76		+0.24		+1.00		+1.00		
7.50	−1.00		−1.00		−0.01		−0.02		+0.50		+0.09		+1.00		+1.00		
8.00	−1.00		−1.00		−0.03		−0.05		+0.24		+0.03		+1.00		+1.00		
8.50	−1.00		−1.00		−0.09		−0.14		+0.09		+0.01		+0.99		+1.00		
9.00	−1.00		−1.00		−0.24		−0.33		+0.03		+0.00		+0.97		+1.00		
9.50	−1.00		−1.00		−0.50		−0.61		+0.01		+0.00		+0.91		+1.00		
10.0	−1.00		−1.00		−0.76		−0.83		+0.00		+0.00		+0.76		+0.99		
10.5	−1.00		−1.00		−0.91		−0.94		+0.00		+0.00		+0.50		+0.97		
11.0	−1.00		−1.00		−0.97		−0.98		+0.00		+0.00		+0.24		+0.91		
11.5	−1.00		−1.00		−0.99		−0.99		+0.00		+0.00		+0.09		+0.76		
12.0	−1.00		−1.00		−1.00		−1.00		+0.00		+0.00		+0.03		+0.50		
12.5	−1.00		−1.00		−1.00		−1.00		+0.00		+0.00		+0.01		+0.24		
13.0	−1.00		−1.00		−1.00		−1.00		+0.00		+0.00		+0.00		+0.09		

[a] The charge contribution z of the charged residue is given in each column for the indicated pH value. For a given protein, note: n_i, the number of the amino acid in each corresponding column. Multiply z_i by n_i of each column. Sum all charges for a given pH (row) to obtain the net charge at this pH.

In *Table 1* the contribution z_i of a given type of charged residue at a given pH value appears in each column. For a given protein, each column has to be multiplied by n, the number of a given type of amino acid. Summing all charges for a given pH (row) gives the net charge at this pH.

The estimation of a protein net charge is also accessible on the web at:

- `http://www-biol.univ-mrs.fr/d_abim/compo-p.html`
- `http://www.expasy.ch/sprot/protparam.html`
- `http://www.infobiogen.fr/service/deambulum`

The calculated pI (pH for a net charge of zero) should be supplemented by the electrophoretic measurement of the experimental pI. If a difference is observed between the estimated pI and the experimental one, it means that one of the conditions detailed for the estimation is not met or that some additives of the experimental conditions interact with the protein.

Water soluble proteins can be classified broadly, according to their pI (experimental or estimated from the content of charged residues), as:

(a) *Acidic* proteins, having a higher content of Asp and Glu, than His, Lys, and Arg. Their pI is lower than 6. This arbitrary value is linked to the pKa value of the basic group His.

(b) *Basic* proteins, with a higher content of His, Lys, and Arg, than Asp and Glu. Their pI is above 7.5–8.

(c) *Neutral* proteins containing roughly equal numbers of acidic and basic residues, and therefore presenting a pI near the neutrality.

In the pH range between pH 6–8, acidic proteins bear a *negative* net charge, basic proteins a *positive* one, and neutral proteins a net charge of *about zero*.

This classification does not include membrane proteins, which naturally occur in hydrophobic, lipid environments and which are known to be poorly soluble in water (see Chapter 9). Their solubilization requires detergents, and the physical chemistry of protein–detergent systems is very different from the discussion of this chapter. Nevertheless, the preceding discussion does apply to their water soluble surfaces, and should therefore also be considered in those studies.

3.2 Desalting of proteins

3.2.1 Dialysis against water

Dialysis removes most of the solvent compounds, except those tightly bound by the protein and the counterions necessary for electrostatic compensation (21). For example HEW lysozyme bears a net charge of about 10 at pH 5, thus the dialysed solution contains at least ten counterions leading to an anion concentration of 35 mM for a 50 mg/ml (3.5 mM) HEW lysozyme solution.

3.2.2 Mixed-bed resins

A more efficient procedure consists in passing the protein solution through strong cation and anion exchange resins in H^+ and OH^- form, where all cations and anions of the protein solutions are exchanged for H^+ and OH^-, except Li and F (4). The resins used for this purpose have a pore size of about 10^3 Da preventing the protein molecules from adsorbing in the exchange sites. The eluted protein solutions contain only the protein, water, H^+, and OH^-, and are isoionic by definition. Isoionic protein is perhaps the simplest possible system, free of the modulating effects of ligands and bound ions. So far we have tested the desalting procedure with six proteins having either an acidic or a basic pI, among which are HEW lysozyme, BPTI, and *Hl* collagenase. Two other proteins, which are very hydrophobic and show tendency to aggregate, were irreversibly adsorbed on the resins. Thus, it is recommended to first use a small amount in order to verify its stability in isoionic conditions. When using resins with dyes to indicate the saturation of the resins, we have observed by NMR the contamination of the eluted proteins with this dye.

Desalting can be performed in:

(a) One step by passing a previously dialysed protein solution through a mixed-bed resin. This is quicker than the two-step procedure, but the resins cannot be regenerated as they are mixed. It is recommended for small amounts of proteins, thus small quantities of resins, or when working with different proteins, as contamination is limited when changing the resins for each type of protein.

(b) Two steps by exchanging first the co-ions then the counterions (*Protocol 3*). Thus a solution of a protein presenting a basic pI is successively passed through a cation exchange resin and then a anion exchange resin, and in reverse sequence for proteins having an acidic pI. The advantage of a two-step desalting is the possibility to *regenerate* separately the cation exchange resins with HCl (1 M) and the anion exchange resins with NaOH (1 M).

Protocol 3. Preparation of up to 100 mg of isoionic basic protein

Equipment and reagents

- Two 5 ml syringes
- Bio-Rad AG 50W-X8 20–50 mesh, H^+ form (No. 142-1421)
- Bio-Rad AG 1-X8 20–50 mesh, OH^- form (No. 140-1422)

Method

1. Fill two 5 ml syringes with 1.5–3 ml of respectively Bio-Rad AG 50W-X8 20–50 mesh, H^+ form for the cation exchange, and Bio-Rad AG 1-X8 20–50 mesh, OH^- form for the anion exchange. Rinse five times with

Protocol 3. *Continued*

1 ml of pure water. Minimize the dead volume of water to avoid dilution of the protein sample.

2. Aspirate the dialysed protein solution (≤ 1 ml) in the syringe containing the cation exchange resin. Shake the syringe for ≈ 5 min, then remove the solution from the syringe through a 0.22 μm filter. The pH of the solution becomes more acidic (pH ≈ 3–4), depending on how extensive the dialysis was.
3. Aspirate the acidic protein solution in the syringe containing the anion exchange resin. Shake the syringe for ≈ 5 min, then remove the solution from the syringe through a 0.22 μm filter.
4. Aspirate 1 ml of pure water into the first syringe and shake for a few minutes to recover protein remaining in the dead volume. Remove this solution and rinse the second syringe. This step is repeated twice for a better recovery of the protein.
5. Isoionic protein solutions can be rapidly deep-frozen in liquid nitrogen and freeze-dried for storage. The freeze-dried protein is stored at −80 °C.
6. To prepare protein solutions, solubilize the isoionic protein powder in pure water, centrifuge the solution, and filter it to remove insoluble protein. Adjust to the required pH and add desired additives.

Alternatively the resin may be prepared in a small column, or on a 0.45 μm filter system equipped either with a vacuum system at the bottom, or using a nitrogen pressure on the top to accelerate the recovery of the solutions. This was necessary in the case of HEW lysozyme where a slight precipitation is observed, which may indicate some denaturation.

3.2.3 Exchange dialysis

Because the counterions of a charged protein cannot be eliminated by a dialysis step against water, and if the protein does not resist a treatment over a mixed resin, then the initial unknown counterions may be exchanged against the desired ions and buffer. This can be achieved by repeated dialysis steps or by repeated washing over concentration devices (e.g. Centricon or Microcon from Amicon, or Ultrafree with tangential flux from Millipore). The concentration must be higher than the counterions concentration (i.e. the protein concentration in mM multiplied by the net charge of the protein) and possibly follow the efficiency of ions described in Section 4.4.1.

3.3 Net charge and crystallization conditions

In June 1998, 3258 crystal forms of 2297 biological macromolecules were listed in the *Biological Macromolecule Crystallization Database* (22)

Table 2. Composition of three proteins having a basic pI

Protein	No. of amino acids	Experimental pI	% of neutral	% of polar	% of charged
BPTI	58	10–10.5	35	41	24
Erabutoxin	62	9.45	23	55	22
HEW lysozyme	129	11	35	44	21

(`http://ibm4.carb.nist.gov:4400/bmcd.html`). The data gathered in this database are limited to the information given by the authors of the published crystallization conditions, and very often the description is incomplete. The main question remains however of how to transpose the information of the previously crystallized protein to a new one.

Although the intrinsic solubility of a monomeric protein cannot yet be predicted, it is qualitatively related to the content of hydrophobic, hydrophilic, and charged amino acids. For the same molecular weight, a protein containing the highest content of charged amino acids will be more soluble. For example, the highly soluble cytochrome *c* contains 35% charged amino acids, whereas the poorly soluble elastase contains only 13%, and HEW lysozyme 21%. This can also be illustrated by three basic proteins, HEW lysozyme, BPTI, and erabutoxin (a snake venom) described in *Table 2*.

The information of the protein's pI and the net charge at the pH of crystallization is missing in the database, therefore render the search of 'related' proteins difficult. HEW lysozyme, BPTI, and erabutoxin are not biologically related, but have all three a basic pI. The estimated net charge for these proteins (*Figure 7*) at pH 4.5 is higher for BPTI than for the two others.

As to the nature and the concentration of the salt, the crystallization conditions of BPTI and erabutoxin at pH 4.5 could be found by starting from the knowledge of the crystallization conditions of HEW lysozyme at this pH. In all three cases, NaCl was much less effective for crystallizing the protein than KSCN (*Figure 8*).

Concerning the protein concentration, crystallization at pH 4.5 and 100–200 mM KSCN, requires a higher protein concentration for BPTI ($\approx$ 12 mM), than for erabutoxin ($\approx$ 6 mM), and for HEW lysozyme ($\approx$ 1.5 mM) (*Figure 8*). This may be linked to a difference of solubility due to their respective net charge. However for erabutoxin and HEW lysozyme, which bear approximately the same net charge at pH 4.5 (*Figure 7*), a difference of their respective content of polar and neutral amino acids may also been involved.

An estimation of the solubility change, and crystallization conditions, is also worth doing if the net charge of a protein is modified when preparing mutants or complexes of your protein.

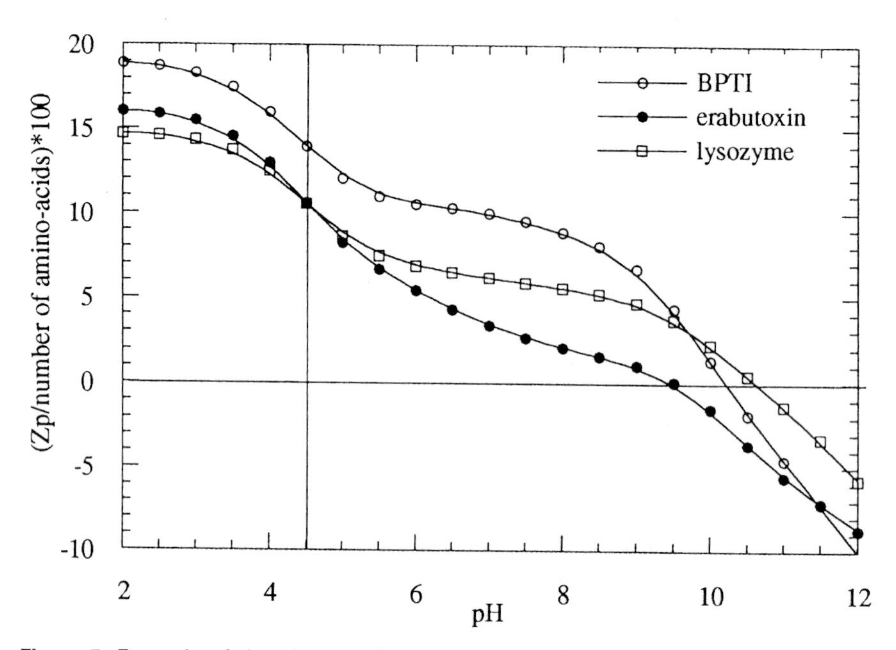

Figure 7. Example of the change of the protein net charge versus pH for three basic proteins. The pH for which the net charge is zero is the pI. At pH 4.5, the net charge of BPTI is higher than for the two other proteins.

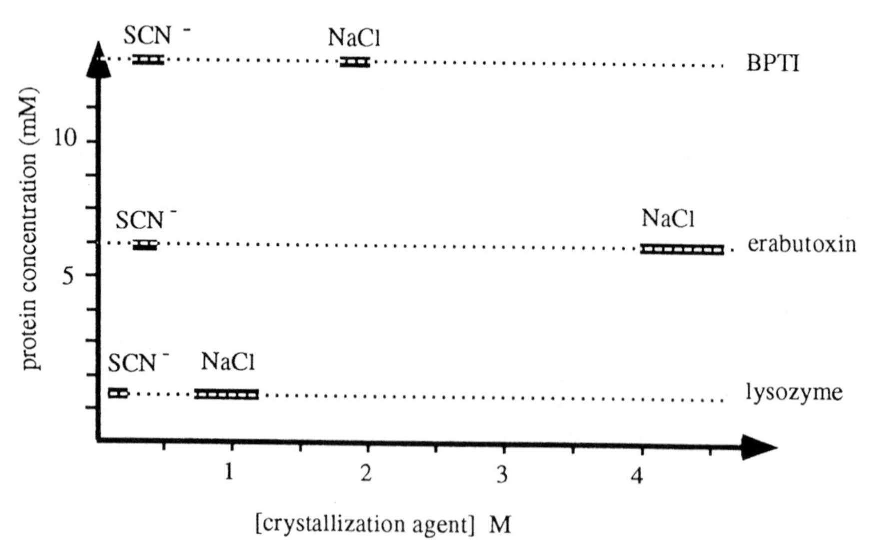

Figure 8. Crystallization conditions (bold segments) for three basic proteins, BPTI, erabutoxin, and HEW lysozyme, at pH 4.5 and 18°C.

4. Influence of physico-chemical parameter changes

Crystallogenesis depends on the chemical properties of the protein as well as on its interactions in a given solvent. Thus a physico-chemical characterization of crystallization conditions for a protein requires an exhaustive list of the chemical constituents of the solution, and the knowledge of how they change with the temperature and the pH of the solution.

The crystallization parameters which affect protein *solubility* are:

- temperature
- protein net charge, i.e. the pH and the pKa of the buffer
- the ionic strength, i.e. the concentration of the salts, the buffer, the counterions, and co-ions of the protein
- the nature of the crystallizing agent
- the dielectric constant, i.e. addition of organic solvents or heavy water.

These parameters affect the protein solubility but may also act on nucleation, protein crystal growth, and/or the crystallization kinetics.

Although exhaustive solubility diagrams are most often not achievable for a new protein available in small amounts, we describe here what general rules have been drawn from model proteins and how to adapt these to a particular case.

4.1 Interactions in a protein solution

A biological macromolecule is a polymer of amino acids or nucleotides, which is folded in a tertiary structure mainly by dipole–dipole interactions (H-bonds, e.g. C=O···H–N–, and Van der Waals interactions), by some covalent bonds (S–S bridges), and occasionally by salt bridges (e.g. $-COO^{-}\cdots{}^{+}H_3N-$) between charged residues.

'Water soluble' proteins contain mostly the hydrophobic side chains in the core, exposing the hydrophilic side chains on their surface. They are thus considered as polyions able to dissolve in water (protic polar solvent). The special case of membrane proteins (see Chapter 8) should be mentioned as they bear, at least on the surface embedded in the membrane, hydrophobic residues which interact in their natural medium with lipidic (apolar) compounds. In practice, detergents are added to the water solutions in order to induce hydrophobic interactions between the hydrophobic residues of the protein and the hydrophobic tail of the detergent. The hydrophilic head of the detergent allows then interactions with the solvent.

The balance of interactions controlling the solubility and/or the conformation of a macromolecule can be modified (23, 24) as summarized in *Table 3*.

The stability of biological macromolecules in solution relies on the competition of solvent–solute interactions with the intramolecular interactions which are necessary to maintain the tertiary structure.

Table 3. Effects of crystallizing parameters on the solvent and/or on the biological macromolecule

Parameter	Effect on solvent	Effect on the macromolecule	Examples
Temperature increase	Disorder of solvent molecules	Formation of conformations of higher total free energy	
pH	H^+ and OH^- concentrations	Protonation or deprotonation of charged groups	Arg, Lys, His, Asp, Glu, C- and N-terminus
Salts	Ionic strength	Chemical activity coefficient	
		Shielding of macromolecular electrostatic interactions	
		Monopole–monopole interactions with accessible charged residues	Anions with lysine or arginine side chains, or cations with glutamic or aspartic side chains
		Monopole–dipole interactions with dipolar groups of the macromolecule	Peptide bonds, amino, hydroxyl, or carboxyl groups, or amides
		Non-polar interactions between solvent exposed hydrophobic residues and the hydrophobic part of organic salts	Carboxylates, sulfonates, or ammonium salt Solubilization of solvent exposed hydrophobic residues by the hydrophobic tail of an ionic detergent
		Association with binding sites	Protein–ion interaction with a specific part, of well defined geometry, of the biological macromolecule
H-bond competitors		Competition at high concentration ($\geq$ 4 M) with H-bonds of water and the structural intramolecular H-bonds of the protein	Formamide, urea, guanidinium salts
Hydrophobic additives	Alteration of the solvent structure	Interaction with hydrophobic parts of the protein	Non-ionic detergents
Organic solvents	Modification of the dielectric constant	Interaction with hydrophobic or polar parts of the protein	Alcohols, DMSO, MPD

Two levels of protein–protein interactions can be distinguished:

(a) Long-range interactions (a few nm) are essentially governed by non-specific electrostatic interactions according to the Debye–Hückel theory. The individual macromolecules are then considered as spheres having a given net charge with randomly distributed charges. The Derjaguin–Landau–Verwey–Overbeck (DLVO) theory (25, 26) describes the interaction between two molecules as the *net* interaction resulting from:
   - an electrostatic repulsion and
   - a Van der Waals attraction (4).

   Considering a macromolecule as a particle with a given net charge Z_p, different from zero, the long-range electrostatic *protein–protein* interactions are repulsive at low ionic strength or in pure water. For electrostatic compensation, the solution provides at least Z counterions. Once the net charge is sufficiently screened, protein–protein interactions become less repulsive and finally attractive. The charges can be screened by increasing the salt concentration of the solution, because an increasing ionic strength lowers the repulsive contribution of the protein–protein interaction. Additionally, specific ions may be bound by the protein and therefore change its net charge. Thus binding of ions can also lead to less repulsive (or more attractive) protein–protein interactions. Attractive interactions are necessary, but not sufficient for protein crystallization. They may lead to amorphous precipitate, as well as to crystals.

(b) Specific short-range interactions, occurring at the intermolecular level, promote specific and periodic protein–protein contacts to build the crystal. The chemical equilibria become determinant when local charges on the protein surface interact specifically with another protein molecule, water, or a solvent component. Protein contacts in the crystal are due to hydrogen bonds, hydrophobic interactions/Van der Waals, and salt bridges (27). However, the balance of interactions finally leading to protein contacts in the crystal are difficult to predict. Moreover, the situation is complicated by the fact that both static and dynamic aspects should be considered.

4.2 pH

A change of pH implies a change of the protein net charge:

(a) Near the pKa of most numerous charged residues, solubility varies very rapidly.

(b) Outside the range of the pKa values of charged residues, the solubility changes smoothly.

(c) Solubility is *minimal at the pI* of the protein as shown in the case of insulin (28) (*Figure 9*), egg albumin (29), haemoglobin (30), and β-lactoglobulin (31). Conversely solubility is higher when the net charge increases (17).

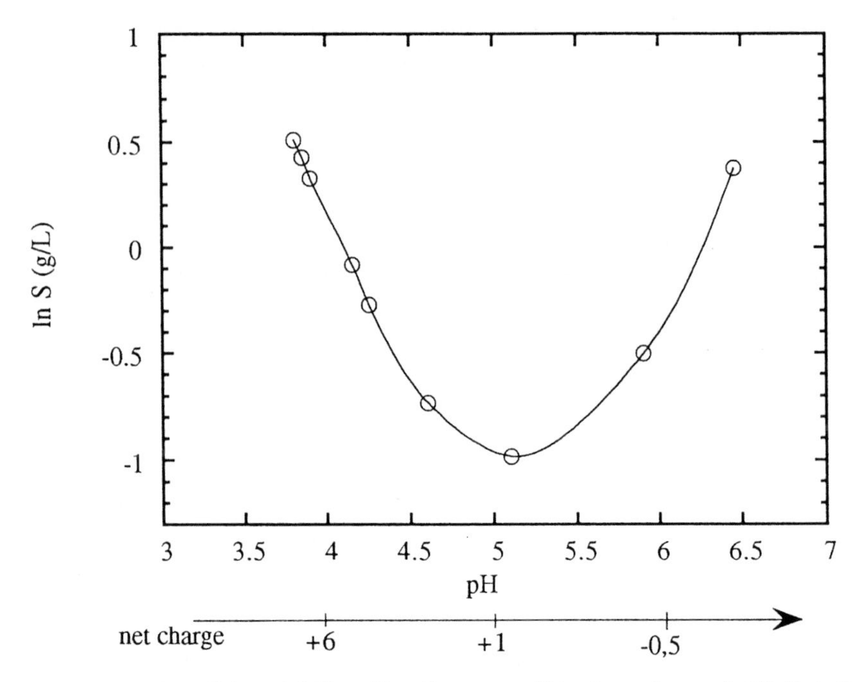

Figure 9. Variation of the solubility of insulin versus pH (redrawn from ref. 28). Note that decreasing the pH by 1 unit (i.e. from pH 5 to pH 4), the net charge changes from +1 to +5 and the solubility from ≈ 0.4 to ≈ 1.2 mg/ml. When increasing the pH by the same increment (+1 unit, but from pH 5 to pH 6), the net charge changes from +1 to only –0.5 and the solubility from ≈ 0.4 to only ≈ 0.6 mg/ml.

It is of practical interest to remember that the net charge is independent of the ionic strength at the pI (32). Consequently *solubility is also constant at the pI* whatever the ionic strength (17). A change of the protein net charge may induce polymorphism. The effect of pH on protein solubility is amplified at low ionic strength (17, 33).

The pH of a protein solution is set by the buffer, whose importance is often neglected. The buffering capacity of a weak acid, or base, is limited to a pH range from its pKa ± 1 pH unit. Some additives and crystallizing agents (phosphate, citrate, acetate) are themselves weak acids or bases. Their solutions have to be adjusted to the desired pH of crystallization. A polyacid changes its own charge depending on the pH; its efficiency can be different depending whether it is mono-, di-, or trivalent.

By definition buffers are soft acids or bases and may present preferential binding, even of low affinity, with the biological material. Even at exactly the same pH, the protein solubility can be different depending on the nature, and of course the concentration, of the buffer. Therefore two, otherwise identical, crystallization experiments may behave differently at a same pH when using different buffers.

To make use of the effect of pH on protein solubility, it is recommended to try crystallization conditions at the pI and on both sides of it (*Figure 9*). However the range of pH in which the protein is stable must first be checked. A change of pH is achievable with volatile acids or bases by vapour diffusion, otherwise by dialysis or batch method.

4.3 Ionic strength

The ionic strength, I, of a salt solution is due to the concentration C_i of this salt, but also to the valency Z_i of each ion. It is easily calculated by:

$$I = 1/2\ \Sigma C_i Z_i^2) \qquad [2]$$

Therefore the ionic strength of a 0.1 M salt solution of:

- a [1:1] electrolyte like NaCl is: $1/2\ (([Na^+] \times 1^2) + ([Cl^-] \times 1^2)) = 0.1$ M
- a [1:2] electrolyte like $(NH_4)_2SO_4$ is: $1/2\ ((2 \times [NH_4^+] \times 1^2) + ([SO_4^{2-}] \times 2^2)) = 0.3$ M
- a [2:2] electrolyte like $MgSO_4$ is: $1/2\ (([Mg^{2+}] \times 2^2) + ([SO_4^{2-}] \times 2^2)) = 0.4$ M

For salt concentrations above 0.2 M, the concentration should be corrected by the chemical activity coefficient which can be found in most handbooks. In a phase diagram, it is more convenient to express the salt concentration as ionic strength rather than molarity, especially for comparing the solubility in mono-, di-, or polyvalent ions.

If a salt of a weak acid or base is used, the actual concentration must be calculated depending on the pH of the solution in respect with the pKa, according to the same rules as detailed in Section 3.1 and *Table 1*. For a 0.1 M sodium acetate (pKa = 4.76) solution:

- at pH = 8 only acetate is present, $I = 1/2\ (([Na^+] \times 1^2) + ([AcO^-] \times 1^2)) = 0.1$ M
- at pH = pKa, 50% of the acetate is protonated and 50% is charged,
  $I = 1/2\ ((0.5[Na^+] \times 1^2) + (0.5[AcO^-] \times 1^2)) = 0.05$ M.

This shows the importance of how the buffer is prepared, if the pKa is reached by adding:

- acetic acid to a 0.1 M sodium acetate solution, I = 0.1 M
- sodium hydroxide to a 0.1 M acetic acid, I = 0.05 M.

As for cationic species, an increasing pH may also promote the formation of hydroxides, the cation then no longer acts as M^{n+}, but as $M(OH)^{(n-1)+}$, or more generally as $M(OH)_i^{(n-i)+}$.

The effects of salts on protein solubility are complex and rely on a balance between protein–water, protein–salt, and salt–water interactions. In addition, the variation of protein solubility over the whole range of salt concentration reflects the resultant effect of both electrostatic and hydrophobic interactions, the first being predominant at low salt concentration and the second at high salt concentration. The change of protein solubility at increasing salt concentrations was studied in term of *salting-in* and *salting-out* (34–36). However it was shown more recently that salting-in is not systematic at low ionic strength, but seems to be also correlated with the protein net charge (17).

4.3.1 Salting-in

Solubility data of carboxyhaemoglobin (*Figure 10a*) showed that protein solubility first increased (salting-in) and then decreased (salting-out) with increasing ionic strength (34). This phenomenon is explained by the decrease of the chemical activity of the protein when the ionic strength of its environment increases (36). It is worth emphasizing that the solubility variation of carboxyhaemoglobin at 25 °C and pH 6.6 corresponds to its minimal solubility, i.e. *near the pI*.

As for HEW lysozyme (*Figure 10b*) bearing a net charge different from zero, no salting-in could be evidenced (17). Here, the screening of the salt on the electrostatic protein–protein interactions seems to dominate the effect of the protein chemical activity. Furthermore, salting-in may be reduced or emphasized depending whether co-ions or counterions bind to the proteins, as will be discussed in Section 4.4.

4.3.2 Salting-out

Salting-out corresponds to a decrease of protein solubility at high ionic strength, where the protein behaves as a neutral dipole and solubility is mainly governed by hydrophobic effects. Theoretically a crystallizing agent added to the protein–water (solute–solvent) system can either bind to the protein (preferential binding) or be excluded (preferential exclusion) depending on preferential protein–additive or protein–water interactions (37). The net interaction of salting-out is preferential exclusion, even though molecules or additives *may bind* to the protein.

Protein solubility has been expressed (34) according to:

$$\log S = \beta - K_s m \qquad [3]$$

where S is the protein solubility (in mg/ml), m the molal salt concentration (g salt/1000 g water), and β the intercept at $m = 0$. β is a constant at high salt concentration and function of the net charge of the protein, thus strongly pH-dependent. Therefore it is minimal at the isoelectric point. The magnitude of β, as well as charge distribution, varies with temperature. K_s is the salting-out constant. It is independent of pH and temperature, but depends on the nature of the salt.

However, experimentally defined solubility curves rarely fit with a linear

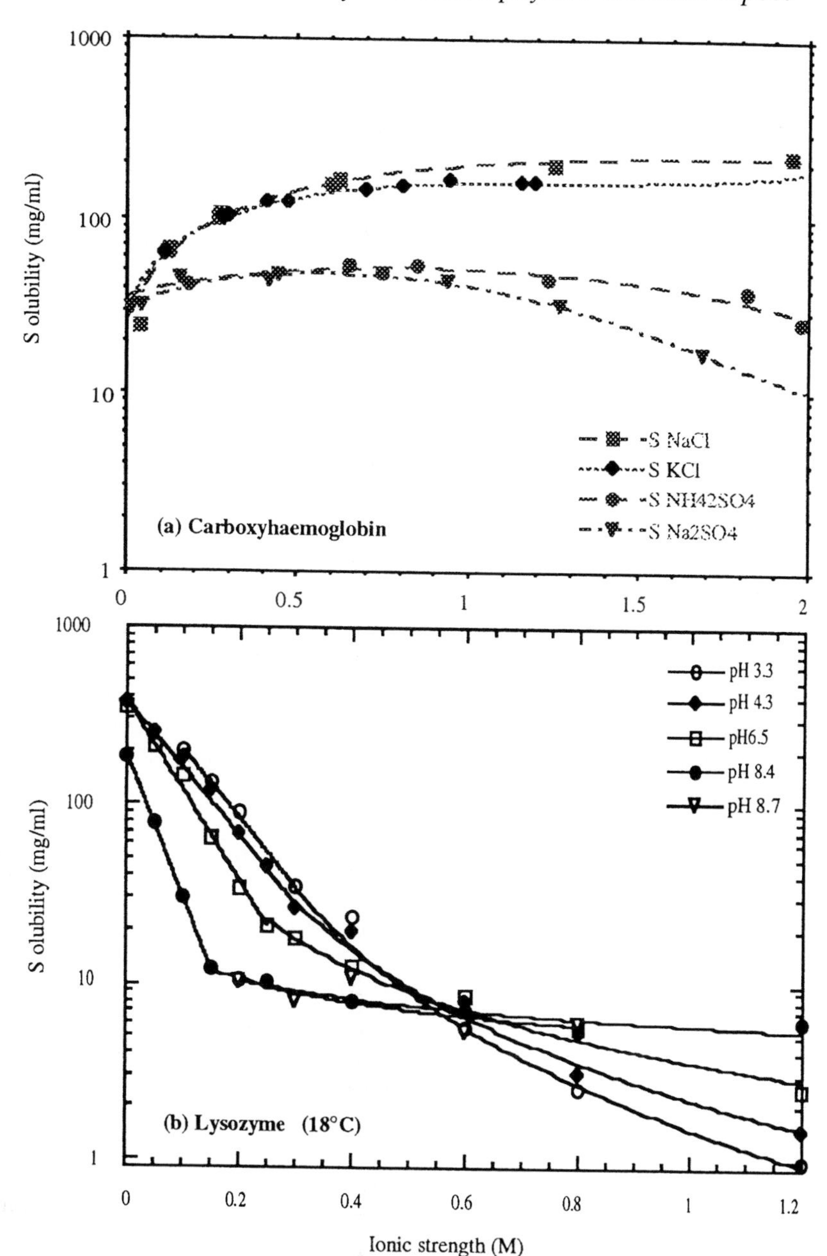

Figure 10. Variation of the solubility versus ionic strength. (a) Salting-in for carboxyhaemoglobin near its pI and in the presence of various salts (redrawn from ref. 34). (b) No salting-in for positively charged lysozyme in the presence of NaCl and at different pH values (17).

function (*Figure 1*). This non-linearity may be essentially due to the following reasons:

(a) The solubility tends to the solubility value of the protein in the buffer at low concentrations of the crystallizing agent, the efficiency of the crystallizing agent, and of the buffer becoming comparable. The curves of HEW lysozyme solubility in the presence of various salts (10) converge at low salt concentrations toward its solubility curve in the sodium acetate buffer.

(b) Higher amounts of protein are required for crystallization at low salt concentrations, so the solubility value can be affected by the presence of higher amounts of *protein related salts*; they can either be counterions, or salts which were not eliminated by a previous dialysis step.

(c) Protein binding of counterions can no more be neglected compared to preferential exclusion. The solubility is then affected by the change of both the net charge of the protein and the ionic strength.

4.4 Nature of salts

The way the nature of the salts acts on protein solubility is complex and not yet clearly understood. They affect the ionic strength depending on the concentration and the valency of their ions. Even though two ions bear an identical charge, their size and their polarizability are different, affecting therefore their own hydration, the interaction with charged residues of the protein, and potential binding sites of the protein.

4.4.1 Inversion of the Hofmeister series, depending on the protein net charge

A longstanding and apparently general observation is that ions differ greatly in their ability to salt-out protein solutions. In 1888, Hofmeister (38) ranked various ions toward their *precipitation* ability by adding increasing amount of salts to a mixture of hen egg white proteins. The Hofmeister series, sometimes called the lyotropic series, have since been associated with many biological phenomena and extensively reviewed (39, 40). It has been shown (24) that ions act on the structures of biological macromolecule structures according to the same series:

- cations: $Li^+ > Na^+ > K^+ > NH_4^+ > Mg^{2+}$
- anions: sulfate^{2-} > phosphate^{2-} > acetate$^-$ > citrate^{3-} > tartrate^{2-} > bicarbonate$^-$ > chromate^{2-} > chloride$^-$ > nitrate$^-$ >> chlorate$^-$ > thiocyanate$^-$

Ions such as sulfate reinforce the structures of water and biological macromolecules (they are called *lyotropic*), whereas ions such as chlorate and thiocyanate denature them (these are called *chaotropic*) (24).

i. Protein bearing a negative *net charge*

The solubility data of *Hl* collagenase (pI of 4.1) crystallized at pH 7.2 and at 18 °C, and in the presence of various ammonium salts (6) showed that the efficiency of anions to lower the solubility of *Hl* collagenase is consistent with the precipitation observations of Hofmeister:

$$\text{phosphate}^{2-}/\text{phosphate}^{-} > \text{sulfate}^{2-} > \text{citrate}^{3-}/\text{citrate}^{2-} >> \text{chloride}^{-}$$

Hl collagenase is moderately soluble in ammonium sulfate and becomes extremely soluble in ammonium chloride when it bears a negative net charge. This was used to increase the solubility of *Hl* collagenase and suppress growth of twinned crystals in ammonium sulfate by adding 200 mM NaCl.

ii. Protein bearing a positive *net charge*

Solubility measurements for basic proteins (10, 41) were undertaken at pH 4.5 (50 mM Na acetate) and 18 °C in the presence of a large variety of salts with HEW lysozyme (pI 11). The results of the solubility curves (some shown in *Figure 1*) show an inversion of the Hofmeister anion series, which becomes:

$$\text{thiocyanate}^{-} \sim \textit{para}\text{-toluene sulfonate (pTS}^{-}) > \text{nitrate}^{-} > \text{chloride}^{-} > \text{acetate}^{-} \sim \text{phosphate}^{-} > \text{citrate}^{2-}$$

whereas the efficiency of cations is weak and follows Hofmeister series,

$$K^{+} > Na^{+} > NH_4^{+} > Mg^{2+}$$

iii. Inversion of Hofmeister series

It is of considerable potential relevance to protein crystal growth, that the order of these effects on protein solubility depends on the net charge of the protein. This was evidenced by systematic solubility measurements for these two model proteins. It was since confirmed with a number of other proteins for which the solubility, as well as the nucleation zones and the precipitation zones are affected in the same order.

The anion series follows the order of Hofmeister in affecting the solubility of *Hl* collagenase, amylase (42), parvalbumin (43), the nucleation zone of Grb2 (44), and the precipitation of ovalbumin, the major protein in hen egg white tested by Hofmeister. All these proteins have an acidic pI and were tested at a higher pH where they bear a negative net charge.

The reverse order of the anion series is observed for the solubility of HEW lysozyme, BPTI (5), and the nucleation zone of toxins (45) (erabutoxin, fasciculin, and muscarinic toxin 2), and lysin from spermatozoa (46). These proteins have a basic pI and were tested at a lower pH where they bear a negative net charge.

Furthermore it was shown that the anion series is reversed depending whether the precipitation of a same protein is achieved below or above the pI in the case of the precipitation fibrinogen (47) and of insulin (28).

iv. Adsorption of anions by basic proteins

Thiocyanate, which is well known as a chaotropic agent at high concentration, appears to be very effective at crystallizing HEW lysozyme at low concentration. This is also observed with other proteins having a high isoelectric point: BPTI (48), toxins (45) (erabutoxin, fasciculin, and muscarinic toxin 2), and lysin from spermatozoa (46). A similar efficiency was observed with organic salts (sodium *p*-toluenesulfonate, benzenesulfonate, and benzoate) which successfully crystallized HEW lysozyme at low concentrations (48), typically 0.1–0.2 M. It appeared that the efficiency of carboxylates to crystallize HEW lysozyme was: $pTS^- \sim$ benzoate > propionate > acetate. The high efficiency of these anionic species was interpreted by the occurrence of anion binding to the protein, prior to the process of exclusion at high salt concentration. They were thought to interact with positively charged residues of the protein (48). The presence of one SCN ion could be demonstrated unequivocally in the electron densities of erabutoxin b (49) and turkey egg white lysozyme (50). Assuming a protein anion association constant of $\approx$ 0.1 M, a *protein salt* would be formed. This protein salt would have a lower net charge than the protein itself if a counterion is bound, or a higher net charge if a co-ion is bound. Therefore any solvent constituent, as well salts as buffers or other additives, may play an important role in the crystallization, if they interact with a protein.

4.4.2 Testing the Hofmeister series

The precipitation zone, the nucleation zone, and the solubility curve occur at different supersaturation in the phase diagram. However the relative efficiency of the salts can be detected on any of these zones. If no crystallization conditions are known, the Hofmeister series can be tested by precipitation tests. Precipitation is achievable with any crystallizing agent, when raising sufficiently either the protein or the salt concentration. An example of a rapid test is given in *Protocol 4*.

Protocol 4. Testing the Hofmeister series by dialysis

Equipment and reagents

- Five dialysis buttons
- Linbro box
- 2.0 M stock solution of KSCN, $NaNO_3$, NaCl, NaOAc, and NH_4SO_4

Method

1. Prepare five dialysis buttons with protein solution at a concentration as high as possible.
2. Prepare 2 ml reservoir solutions at 0.05, 0.1, 0.2, and 0.5 M of KSCN, and at 0.1, 0.5, 1.0, and 2.0 M ionic strength of $NaNO_3$, NaCl, NaOAc, and NH_4SO_4 in the same buffer as the protein.

3. Place a dialysis button in the reservoir at the lowest concentration of each salt. Close the well with a coverslip.
4. Transfer it after at least 2 h to the next concentration. Respect the same delay for each change.
5. As soon as precipitation is observed, place the dialysis button in the previous reservoir and prepare an intermediate concentration to refine the value of the precipitation limit.

The relative position of phosphate and citrate versus sulfate in the series may change, depending on the pH, since their ionic strength varies rapidly around their pKa values with Z_i^2 according to *Equation 1*.

4.4.3 Peculiar behaviours

The salt effect of the Hofmeister series was shown to be general, depending on the net charge of the protein. It was shown that formation of a *protein salt* may occur when an ion weakly binds to the protein surface, probably by the formation of an ion pair. This is different from binding to a specific site of the protein, because the binding is then stronger. It becomes typical of a given protein and no longer transposable to another one which has not this specific site. Apart from known binding sites of cations, like Zn or Ca, which intervene in the structure or the biological activity of a protein, binding may incidentally occur in given crystallization experiments. The two following examples illustrate such situations.

(a) HEW lysozyme crystallizes in similar concentrations when using Na benzoate, NapTS, or KSCN (48) as crystallizing agent. In the case of BPTI which also crystallizes at low concentration of KSCN, no crystallization could be achieved with NapTS, even when adding solid NapTS in the reservoir of a crystallization experiment where BPTI crystals had previously grown with NaCl. This led to the dissolution of the BPTI crystals inside the dialysis cell, although pTS had reached saturation and started to crystallize in the reservoir. This may possibly be due to an interaction of the *hydrophobic part* of pTS with a small protein like BPTI, thus acting like a solubilizing agent, instead of interacting through its sulfonic group.

(b) In the *Crystallization Database* (22), ammonium sulfate is more frequently cited than NaCl among the crystallizing agents (respectively about 800 citations versus 300). However HEW lysozyme is known to crystallize more easily with NaCl, and to resist crystallization with ammonium sulfate. Using the sulfate anion with diverse counterions (ammonium, sodium, or lithium) no crystallization occurred at pH 4.5. Crystals of HEW lysozyme sulfate could be grown at pH 8 instead of pH 4.5, from isoionic HEW lysozyme in the presence of sulfate ions up to 100 mM (51).

4.5 Temperature

The variation of protein solubility with temperature may be either *direct*, i.e. increasing with temperature, or *retrograde*. The behaviour of protein solubility with temperature cannot been foreseen, and is not characteristic, neither of the protein, nor of the crystallizing agent, but of the *protein salt* as illustrated with BPTI. This protein has a retrograde solubility change with temperature in the presence of ammonium sulfate (5) and sodium chloride (52), but direct with potassium thiocyanate (5) (*Figure 11*).

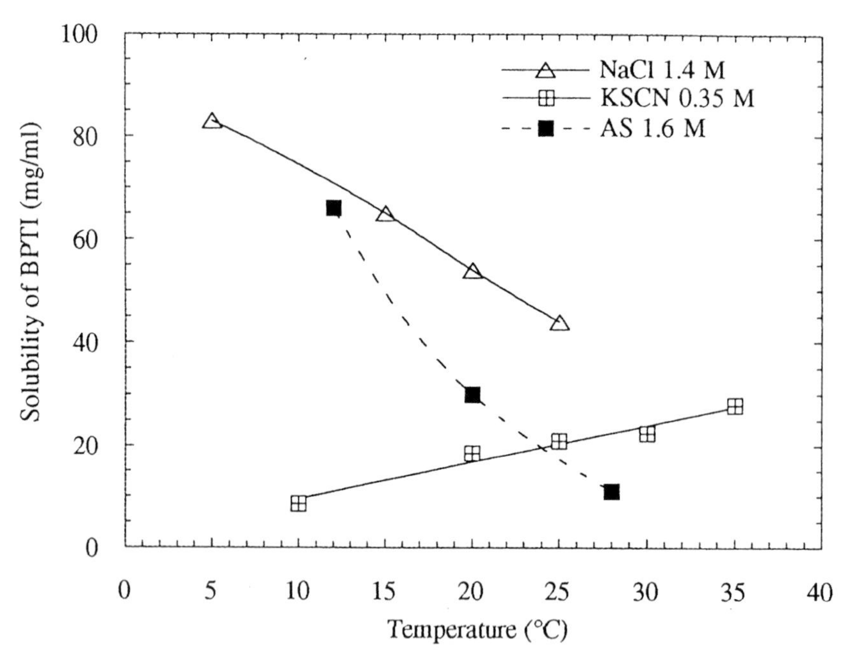

Figure 11. Solubility behaviour of different BPTI salts versus temperature. It is retrograde with ammonium sulfate (AS) and NaCl, but direct with KSCN. Redrawn from refs 5 and 52.

Protocol 5. Testing solubility changes versus temperature

Equipment and reagents

- Incubator at 4 °C
- Incubator at 37 °C

Method

1. Take a series of screening experiments (vapour diffusion, batch, or dialysis) covering the nucleation zone, i.e. from clear drops to ones containing slight precipitation at 18–20 °C.

2. Introduce the Linbro plate in a Styrofoam box and place these experiments in an incubator at 4°C.
3. On the next day, check whether the nucleation or the precipitation has increased or is reduced compared to the initial observations.
4. Bring the experiments again to 18–20°C for a day or two to check the reversibility.
5. Place these experiments in an incubator at 37°C for another day or two.
6. Check again whether the nucleation or the precipitation has increased or is reduced.
7. Analyse the results:
   (a) No change whatever the temperature. This may indicate that:
   - The sampling of ionic strength or pH is too large (and the nucleation zone very small). Repeat the experiments with drops differing by smaller steps of ionic strength or pH.
   - The ionic strength is too high. Repeat the experiments with drops at lower ionic strength and higher protein concentration.
   - The pH of crystallization is too close to the pI. Repeat the experiments with drops at lower or higher protein net charge.

   (b) The lower the temperature, the more intense the precipitation. Solubility is directly related to temperature. Verify nevertheless the reversibility of the precipitation by bringing the experiments to a higher temperature, to avoid confusion between precipitation which is reversible, with denaturation which may not be reversible.

   (c) The lower the temperature, the less intense the precipitation. Solubility is retrograde with temperature, at least with this salt combination. Repeat the experiments when changing the crystallizing agent.

Likely to the effect of pH, the variation of protein solubility with temperature is amplified at low ionic strength (33, 53). To benefit the effect of temperature this means working at rather low ionic strength, but conversely carrying out experiments at higher ionic strength for the transport of crystals with their crystallization solution to stabilize them (i.e. to avoid further nucleation or dissolution of the crystals).

4.6 H_2O versus D_2O

The measurement of the solubility of lysozyme in H_2O and D_2O (7) has shown that the solubility was decreased by about 30%. This has been explained by the difference of the density of the solvent (14).

4.7 Combined effects of crystallization variables

When testing one variable at a time toward crystallization or solubility, the other parameters are kept constant at a fixed value. The advantages of varying all variables at the same time are treated in Chapter 4. We illustrate here the combined effects of two variables on the solubility and on the nucleation zone.

4.7.1 Low ionic strength amplifies pH and temperature effects

It has been shown with different proteins and in different crystallization conditions that low ionic strength amplifies pH and temperature effects (33, 41, 53). The practical applications of this are the following:

(a) To make use of the pH or of the temperature effect, crystallization experiments should be carried out at low ionic strength.

(b) To prevent changes of solubility when the temperature is less controlled (e.g. transportation of the crystals), the samples should be gently soaked in high ionic strength solutions.

(c) It must be reminded that the solubility is constant at the pI whatever the ionic strength.

4.7.2 The nature of the salt and temperature

When combining the nature of the crystallizing salt and temperature on the crystallization of HEW lysozyme at pH 4.5 (41), a similar solubility of ~ 0.8 mg/ml was observed in the presence of NaSCN at 40 °C as well as of NaCl at 0 °C. In this case the temperature effect is direct for both protein salts. The effect of the very efficient NaSCN salt can be compensated at high temperature and become equivalent to a less efficient salt, NaCl, used at low temperature.

4.8 Stepwise replacement technique

When a known crystallization condition is unsatisfactory or if a crystallization parameter (new salt, other pH or temperature values) needs to be explored, the search to locate the nucleation zone in the new phase diagram can be done by testing the shift of the initial nucleation zone while replacing stepwise the parameters to be changed. This can be performed by hanging drop or dialysis method, the dialysis offering the advantage to work on the same sample while changing the nature of buffer or salt of the protein sample.

The stepwise replacement technique is very useful to qualitatively check the relative effectiveness of ions on protein solubility, as illustrated in *Protocol 6*.

Protocol 6. Stepwise replacement

Equipment and reagents

- Linbro box and silanized coverslips
- 25 ml of 3 M ammonium sulfate stock solution in buffer
- 25 ml of 4 M sodium chloride stock solution in buffer
- 100 ml buffer

Method

1. Choose a range of crystallizing agent concentration to cover the nucleation zone, i.e. from clear drops to those containing slight precipitation, at a constant protein concentration. As an example, these reference conditions may be 20 mg/ml protein and reservoirs from 0.8–1.3 M ammonium sulfate (with 0.1 M steps for the reservoirs).
2. Prepare a series of six drops in the reference conditions in the first row of a Linbro plate (A1 to A6).
3. Set up the second row at the same protein concentration (B1 to B6), but with reservoirs where 0.2 M ammonium sulfate is replaced by 0.3 M NaCl. It is recommended to take into account the ionic strength rather than molarity (see Section 4.3) when replacing crystallizing agents.
4. Observe whether crystallization occurs at higher or at lower ionic strength in the second row compared to the first one:
   (a) The nucleation zone is shifted by one column to lower ionic strength. NaCl is slightly more efficient than ammonium sulfate. Continue by setting up the third row while replacing 0.4 M ammonium sulfate by 1.2 M NaCl.
   (b) The nucleation zone starts already in B1. NaCl is much more efficient than ammonium sulfate. Set up the third row to centre again the nucleation zone by replacing 0.4 M ammonium sulfate by only 0.4 M NaCl and starting at lower ionic strength (from 1.2–2.7 M total ionic strength instead of 2.4–3.9 M total ionic strength).
5. Set up the last row at the same protein concentration (D1 to D6), but with reservoirs containing only NaCl at concentrations chosen depending on the previous results to centre the nucleation zone.

5. Crystallization

The aim of crystallization experiments is first to locate the nucleation zone, then to optimize the physico-chemical parameters and the kinetics to grow large single crystals. Testing a large number and combination of variables may yield different crystal forms. It is worth optimizing different polymorphs because the number of molecules in the asymmetric unit and the diffraction quality may be very different.

5.1 Crystallization strategies

Before the crystallization protocol is defined, the batch of protein has to be characterized as accurately as possible. This definitively helps the reproducibility when using news batches. The analyses should be done by different techniques: electrophoretic gel, IEF, mass spectrometry, UV spectrum, light scattering, etc. (see Chapter 2). All information about the stability of the protein should also be listed in order to select the parameters for which the protein is known to be stable with time, and eliminate those for which the protein denatures.

5.1.1 Solubilization in the buffer

When concentrating a protein solution, roughly three situations can be encountered:

(a) The protein concentration in the buffer solution is low ($\leq$ 5 mg/ml). Nucleation will be difficult to control, and little protein will be available to *feed* the crystals as described in Section 2.3.2. It is thus advisable to search for conditions where solubility is expected to be higher, e.g. at a different pH where the protein net charge is higher.

(b) The protein concentration in the buffer solution can reach at least 10–50 mg/ml. The screening tests can be performed.

(c) The protein concentration in the buffer solution is very high ($\geq$ 100 mg/ml). Nucleation will be easier to control and will probably occur at low supersaturation (1, 17). However if high protein concentration presents practical constraints, conditions of lower solubility can be sought, e.g. by approaching the pI.

5.1.2 Screening

The aim of the screening step is to locate the nucleation zone for each solubility diagram tested. This can be performed by the batch, the vapour diffusion, or the dialysis technique (*Protocol 7*). Dialysis is the most recommended as it includes exchange of the initial buffer and that of the unknown counterions. When working with hanging drops, in which the constituents are no longer exchanged (except water), a previous step of dialysis is recommended to exchange the chemicals coming from the protein purification steps.

In agreement with Section 2.3.2, the tests should benefit from working at high protein concentrations.

5.1.3 Refining the nucleation conditions

After the screening step, the list of crystallization parameters is examined again. Some conditions may have shown to be incompatible with the stability of the protein, others may be interesting for further investigations. For example if the carboxylate salt has given interesting results, other carboxylate salts with different cations may be worth testing. Additional parameters to test the crystallizability of a given protein salt can be added: mixing inorganic and organic crystallizing agents, effects of stabilizing agents, additives, glycerol, the temperature, etc. Depending on the number of combinations to test, it may be suitable to perform this step with an incomplete factorial design (compare Chapter 4).

Protocol 7. Search of the nucleation zone by dialysis

Equipment and reagents

- Three Linbro plates with silanized cover-slips
- 100 ml buffer at the three selected pH values
- 21–25 dialysis buttons
- 25 ml of stock solution of the six selected crystallizing agents in appropriate buffers

Method

1. According to the estimation of the variation of net charge with pH (see Section 3.1), select one pH close to the pI, and two on both sides of the pI so that the net charge is about the same value, but of opposite sign. When this is in conflict with the stability of the protein, select three other pH values for which the variation of the protein net charge is as large as possible. Taking the example of BPTI (*Figure 7*), a pH higher than the pI would not be well suited. Therefore one would select pH 10.5, 9, and 4.5 for which the net charge of this protein is 0, $\approx 7^+$, and $\approx 14^+$ respectively.
2. Prepare three Linbro plates one for each pH. Fill the six reservoirs of the first row (A) of each Linbro plate with 2 ml of buffer corresponding to the pH of each plate.
3. Prepare three dialysis buttons filled with the stock protein solution. Introduce one of them in A1 of each Linbro plate. Observe if the protein solution remains clear for a few days. If precipitation occurs:
   (a) It happens for the lower net charge. Prepare a dialysis button at the lower protein concentration until the drop remains clear for a week.
   (b) It seems not to be linked to the net charge. Transfer the sample in the original buffer to search for reversibility of the precipitation. Replace the buffer solution twice or more to ensure a good exchange of the buffer solution. If the precipitate remains, check for possible denaturation and choose another pH for the following steps.
4. Prepare 15 dialysis buttons filled with the stock protein solution.

Protocol 7. *Continued*

Introduce them in the remaining A2 to A6 reservoirs. Let them stand for one day. Eventually change the reservoir when using large protein volumes to ensure the buffer exchange.

5. Fill the six reservoirs of the row B with six different crystallizing agents (2 ml), each in the appropriate buffer of a given Linbro plate. The crystallizing agents should be chosen according to the protein net charge (see Section 4.4.1). They should preferably be of different chemical types; thiocyanate, halide (Cl^-, Br^-, I^-, or F^-), carboxylate (acetate, citrate, tartrate), sulfate (or phosphate), PEG, divalent cation (Mg^{2+} or Ca^{2+}). As a rule of thumb, the concentration of the first reservoir may be 0.1–0.5 M, or 5–10% PEG.
6. Transfer the dialysis buttons from row A to row B. After two to five days, observe the protein solutions:
   (a) Case 1: the solution is clear. Prepare the next reservoir C at twice the concentration of the one in B.
   (b) Case 2: the solution precipitates. Prepare the next reservoir C at half the concentration of B. Transfer the dialysis button, first back to row A to dissolve the precipitate, then to C.
   (c) Case 3: the solution B is neither clear, nor precipitated. Wait for another period of two to five days to decide whether the next reservoir concentration should be increased or decreased by only 10%.
7. Continue until the limits between clear solutions and precipitation (i.e. lower and upper limits of the nucleation zone) are defined.
8. Set up a new set of experiments to refine the concentration of each crystallizing agent at each pH with a small step in between the nucleation zone limits. At this stage, the vapour diffusion may be more suitable.

5.1.4 Optimization

Once the nucleation zone is defined, the optimal conditions to grow large single crystals must be sought. At this step the protein/reservoir equilibration kinetics should be included among the variables to be adjusted. The tools to perform optimization can also be found in Chapter 4.

5.2 Polymorphism

Table 4 illustrates the variety of crystal forms observed for lysozyme when changing the crystallization conditions (temperature, pH, nature of the crystallizing agent) or a combination of them. Apart from the crystal form, also the number of molecules in the asymmetric unit can change. Both can present an advantage for the crystallographer.

Table 4. Polymorphism of HEW lysozyme

Lattice	Space group	Parameters a, b, c in Å/ α, β, γ in °	z (mol/au)	Crystallizing agent	pH	T (°C)	Ref.
Tetragonal	$P4_32_12$	a = b = 79.2 c = 38.0	1	NaCl (0.3–1.5 M)	4.3–4.7[a]	18	10, 54–57
				or KCl (0.5–1.1 M)	4.3–4.7[a]	18	10
				or NH_4Cl (0.5–1.1 M)	4.3–4.7[a]	18	10
				or $MgCl_2$ (0.4–1.1 M)	4.3–4.7[a]	18	10
				or NH_4OAc (0.9–1.5 M)	4.3–4.7[a]	18	10
				or Na_2HPO_4 (1.1–1.2 M)	4.3–4.7[a]	18	10
				or Na *p*toluenesulfonate (0.08–0.25 M)	4.3–4.7[a]	18, 40	41
				or Na benzenesulfonate (0.21–0.24 M)	4.3–4.7[a]	18	48
				or Na benzoate (0.1–0.2 M)/benzoic acid	5[c]	18	48
		a = b = 78.1 c = 38.2	1	NaCl/HCl	4, 6, 8[c]	18	17
		a = b = 78.8 c = 38.3	1	NH_4citrate (0.5–1.2 M)	4.7[a]	18	10
		a = b = 78.9 c = 38.5	1	None/H_2SO_4	8	18	51
Orthorhombic	$P2_12_12_1$	a = 56.3 b = 65.2 c = 30.6	1	NaCl (0.88 M)	10		55–57
		a = 56.5 b = 73.9 c = 30.5	1	NaCl (0.88–1.37 M)	4.7[a]	37, 40	41, 54
		a = 58.6 b = 68.4 c = 54.3	2	Ethanol (55%) + NaCl	8.4[b]	18	58
Hexagonal	$P6_122$	a = b = 87.0 c = 70.4	1	Acetone (10%) + $NaNO_3$ (saturated)	8.4[b]	18	58
Monoclinic	$P2_1$	a = 28.1 b = 63.1 c = 60.6 β = 90.6	2	KSCN (0.075–0.2 M)	4.5[a]	18	10
				or NaSCN (0.1–1.0 M)	4.5[a]	18, 40	41
				or $NaNO_3$ (0.36 M)/HNO_3	4.5[c]	18	57
				or NaI (3%)	4.5[c]	18	57
		a = 28.6 b = 63.0 c = 60.6 β = 93.5	2	Na_2SO_4 (0.77 M) + NaAcO (0.5 M) /H_2SO_4	4.5	18	57
		a = 27.9 b = 63.0 c = 66.3 β = 114.2	2	KNO_3 (5%)	4.0	–	56
Triclinic	P1	a = 27.5 b = 32.1 c = 34.4 α = 88.3 β = 109.0 γ = 111.0	1	$NaNO_3$ (0.24 M)	4.5[a]	18	57
		a = 41.7 b = 58.4 c = 58.9 α = 116.5 β = 97.6 γ = 105.7	3	ND sulfo-betaine195 + $(NH_4)_2SO_4$	4.5[a]	18	59

[a] The pH is adjusted by 50 mM NaAcO buffer.
[b] The pH is adjusted either by $NaHCO_3$ buffer.
[c] The pH is adjusted only by the indicated acid.

By definition a *polymorph* is a variant of the crystal form for an *identical* molecule. Thus monoclinic HEW lysozyme thiocyanate is not a true polymorph of tetragonal HEW lysozyme chloride if we accept that the molecule which crystallizes is the *protein salt* and not the protein by itself. In the case of polymorphism of HEW lysozyme with pH, the importance of the initial desalting of the protein has also been shown to be important (60). Different crystal forms may appear during the process of crystallization, either because the crystallization conditions are at the borderline between two crystal forms, or because a parameter value has varied during the crystallization process (pH or temperature shift).

References

1. Boistelle, R. and Astier, J.-P. (1988). *J. Cryst. Growth*, **90**, 14.
2. Rosenberger, F., Vekilov, P. G., Muschol, M., and Thomas, B. R. (1996). *J. Cryst. Growth*, **168**, 1.
3. Chernov, A. A. (1997). *Phys. Rep.*, **288**, 61.
4. Riès-Kautt, M. and Ducuix, A. (1997). In *Methods in enzymology* (ed. C. Carter and R. Sweet), Academic Press, London. Vol. 276, Part A, Chap 3, p. 23.
5. Lafont, S., Veesler, S., Astier, J.-P., and Boistelle, R. (1997). *J. Cryst. Growth*, **173**, 132.
6. Carbonnaux, C., Riès-Kautt, M., and Ducruix, A. (1995). *Protein Sci.*, **4**, No. 10, 2123.
7. Broutin, I., Riès-Kautt, M., and Ducruix, A. (1995). *J. Appl. Crystallogr.*, **28**, 614.
8. Schein, C. H. (1990). *Biotechnology*, **8**, 308.
9. Ataka, M. and Tanaka, S. (1986). *Biopolymers*, **25**, 337.
10. Riès-Kautt, M. and Ducruix, A. F. (1989). *J. Biol. Chem.*, **264**, 745.
11. Cacioppo, E., Munson, S., and Pusey, M. L. (1991). *J. Cryst. Growth*, **110**, 66.
12. Boistelle, R., Astier, J.-P., Marcis-Mouren, G., Desseaux, V., and Haser, R. (1992). *J. Cryst. Growth*, **123**, 109.
13. Rosenberger, F., Howard, S. B., Sower, J. W., and Nyce, T. A. (1993). *J. Cryst. Growth*, **129**, 1.
14. Gripon, C., Legrand, L., Rosenman, I., Vidal, O., Robert, M.-C., and. Boué, F. (1997). *J. Cryst. Growth*, **177**, 238.
15. Sazaki, G., Kurihara, K., Nakada, T., Miyashita, S., and Komatsu, H. (1996). *J. Cryst. Growth*, **169**, 355.
16. Darcy, P. A. and Wiencek, J. M. (1998). *Acta Cryst.*, **D54**, 1387.
17. Retailleau, P., Riès-Kautt, M., and Ducruix, A. (1997). *Biophys. J.*, **73**, 2156.
18. Feher, G. and Kam, Z. (1985). *Methods in enzymology* (ed. S. P. Colowick and N. O. Kaplan), Vol. 114, pp. 77–112.
19. Luft, J. R. and DeTitta, G. T. (1997). *Methods in enzymology*, ibid ref. 4. Vol. 276, p. 110.
20. Gaucher, J. F., Riès-Kautt, M., Reiss-Husson, F., and Ducruix, A. (1997). *FEBS Lett.*, **401**, 113.
21. Jolivalt, C., Riès-Kautt, M., Chevallier, P., and Ducruix, A. (1997). *J. Synchrotron Rad.*, **4**, 28.

22. Gilliland, G. L., Tung, M., Blakeslee, D. M., and Ladner, J. E. (1994). *Acta Cyst.*, **D50**, 408.
23. von Hippel, P. H. and Schleich, T. (1969). *Acc. Chem. Res.*, **2**, 257.
24. von Hippel, P. H. and Schleich, T. (1969). In *Structure and stability of biological macromolecules* (ed. S. N. Timasheff and G. D. Fashman), pp. 417. Marcel Dekker, Inc., New York.
25. Israelachvili, J. (1992). In *Intermolecular and surface forces*. Academic Press, London, pp. 246.
26. Israelachvili, J. (1992). *Surface Sci. Rep.*, **14**, 113.
27. Salemme, F. R., Genieser, L., Finzel, B. C., Hilmer, R. M., and Wendoloski, J. J. (1988). *J. Cryst. Growth*, **90**, 273.
28. Fredericq, E. and Neurath, J. (1950). *Am. Chem. Soc.*, **72**, 2684.
29. Sørensen, S. P. L. and Høyrup, M. (1915–17). *Compt. Rend. Trav. Lab. Carlsberg*, **12**, 312.
30. Green, A. A. (1931). *J. Biol. Chem.*, **93**, 517.
31. Gronwall, A. (1942). *Compt. Rend. Trav. Lab. Carlsberg*, **24**, 8.
32. Tanford, C. and Wagner, M. (1954). *J. Am. Chem. Soc.*, **76**, 3331.
33. Mikol, V. and Giegé, R. (1989). *J. Cryst. Growth*, **97**, 324.
34. Green, A. A. (1932). *J. Biol. Chem.*, **95**, 47.
35. Melander, W. and Horvath, C. (1977). *Arch. Biochem. Biophys.*, **183**, 200.
36. Arakawa, T. and Timasheff, S. N. (1985). In *Methods in enzymology* (ed. S. P. Colowick and N. O. Kaplan), Academic Press, London. Vol. 114, p. 49.
37. Timasheff, S. N. and Arakawa, T. (1988). *J. Cryst. Growth*, **90**, 39.
38. Hofmeister, F. (1888). *Arch. Exp. Pathol. Pharmakol.*, (Leipzig) **24**, 247.
39. Collins, K. D. and Washabaugh, M. W. (1985). *Q. Rev. Biophys.*, **18**, 323.
40. Cacace, M. G., Landau, E. M., and Ramsden, J. J. (1997). *Q. Rev. Biophys.*, **30**, 241.
41. Guilloteau, J.-P., Riès-Kautt, M., and Ducruix, A. (1992). *J. Cryst. Growth*, **122**, 223.
42. Veesler, S., Lafont, S., Marcq, S., Astier, J. P., and Boistelle, R. (1996). *J. Cryst. Growth*, **168**, 124.
43. Lafont, S. (1996). Thèse de l'Université d'Aix-Marseille III.
44. Guilloteau, J. P., Fromage, N., Riès-Kautt, M., Reboul, S., Bocquet, D., Dubois, H., *et al.* (1996). *Proteins: Structure, Function, Genetics*, **25**, 112.
45. Ménez, R. and Ducruix, A. (1993). *J. Mol. Biol.*, **232**, 997.
46. Diller, T. C., Shaw, A., Stura, E. A., Vacquier, V. D., and Stout, C. D. (1994). *Acta Cryst.*, **D50**, 620.
47. Leavis, P. C. and Rothstein, F. (1974). *Arch. Biochem. Biophys.*, **161**, 671.
48. Riès-Kautt, M. and Ducruix, A. (1991). *J. Cryst. Growth*, **110**, 20.
49. Saludjian, P., Prangé, T., Navaza, J., Guilloteau, J.-P., Riès-Kautt, M., Ménez, R., *et al.* (1992). *Acta Cryst.*, **B48**, 520.
50. Howell, P. L. (1995). *Acta Cryst.*, **D51**, 654.
51. Riès-Kautt, M., Ducruix, A., and Van Dorsselaer, A. (1994). *Acta Cryst.*, **D50**, 366.
52. Lafont, S., Veesler, S., Astier, J.-P., and Boistelle, R. (1994). *J. Cryst. Growth*, **143**, 249.
53. Howard, S. B., Twigg, P. J., Baird, J. K., and Meehan, E. J. (1988). *J. Cryst. Growth*, **90**, 94.

54. Alderton, G. and Fevold, H. L. (1946). *J. Biol. Chem.*, **164**, 1.
55. Palmer, K. J. (1947). *Struct. Rep.*, **11**, 729.
56. Steinrauf, L. K. (1959). *Acta Cryst.*, **12**, 77.
57. Jollès, P. and Berthou, J. (1972). *FEBS Lett.*, **23**, 21.
58. Haas, D. J. (1967). *Acta Cryst.*, **23**, 666.
59. Vuillard, L., Rabilloud, T., Leberman, R., Berther-Colominas, C., and Cusack, S. (1994). *FEBS Lett.*, **353**, 294.
60. Elgersma, A. V., Ataka, M., and Katsura, T. (1992). *J. Cryst. Growth*, **122**, 31.

11

Diagnostic of pre-nucleation and nucleation by spectroscopic methods and background on the physics of crystal growth

S. VEESLER and R. BOISTELLE

1. Introduction

Unlike the crystallization of small inorganic molecules, the problem of protein crystallization was first approached by trial and error methods without any theoretical background. A physico-chemical approach was chosen because crystallographers and biochemists needed criteria to rationally select crystallization conditions. In fact, the problem of the production of homogeneous and structurally perfect protein crystals is set the same as the production of high-quality crystals for opto-electronic applications, because, in both cases, the crystal growth mechanisms are the same. Biological macromolecules and small organic molecules follow the same rules concerning crystallization even if each material exhibits specific characteristics.

This chapter introduces the fundamentals of crystallization: supersaturation, nucleation, and crystal growth mechanisms. Phase diagrams are presented in Chapter 10. Special attention will be paid to the behaviour of the macromolecules in solution and to the techniques used for their analysis: light scattering (LS), small angle X-ray scattering (SAXS), small angle neutron scattering (SANS), and osmotic pressure (OP).

2. Concentration and supersaturation

Before obtaining any nucleation or growth, it is necessary to dissolve the biological macromolecules under consideration in some good solvent. However, it may immediately be asked whether a good solvent is a solvent in which the material is highly soluble, or in which nucleation is easily controlled, or in which growth is fast, or solvent in which the crystals exhibit the appropriate morphology. In practice, the choice of the solvent often depends on the nature of the material to be dissolved, taking into account the well known rule which says that 'like dissolves like'. This means that, for dissolution to occur, it is necessary that the solute and the solvent exchange bonds: between an ion and a dipole, a dipole and another dipole, hydrogen bonds, and/or Van der Waals bonds. Therefore, the nature of the bonds depends on both the nature of the solute and the solvent which can be dipolar protic, dipolar aprotic, or completely apolar.

Once the material has dissolved, the solution must be supersaturated in order to observe nucleation or growth. The solution is supersaturated when the solute concentration exceeds its solubility. There are several ways to achieve supersaturation. The simplest is to partly evaporate the solvent, the drawback being that all species in the solution (salts, impurities) concentrate as well. This is the case for hanging drop, but, for a better control of growth, it is more advisable to cool or heat the solution depending on whether the solubility decreases with decreasing temperature or conversely. However this method is not recommended when the temperature dependence of solubility is too low. Besides, supersaturation can also be achieved by pH variation, chemical reaction, addition of a poor solvent or a precipitant, and so on. However, the evolution of the system is often more difficult to control with these latter methods.

Supersaturation is the driving force for nucleation and growth. From a thermodynamical point of view, it is the difference between the chemical potential of the solute molecules in the supersaturated (μ) and saturated (μ_s) states respectively. For one molecule which will crystallize one has:

$$\Delta\mu = \mu - \mu_s = k_B T \ln\beta \qquad [1]$$

where k_B is the Boltzmann constant, T the absolute temperature, and

$$\beta = C/C_s \qquad [2]$$

where C and C_s are the actual concentration and the saturation concentration, i.e. the solubility, respectively. This ratio is dimensionless but its value depends somewhat on the concentration units (g/litre, mol/litre, mol fraction, activities, and so on). For protein crystallization, the concentrations are mostly expressed as mg/ml, i.e. g/litre, which is the easiest way but probably not the best for explaining crystallization kinetics. Since activities of proteins cannot yet be calculated, molar fractions are the more appropriate units. Unfortunately, due to the complexity of the protein solutions, they are seldom used.

For the sake of simplicity, supersaturation is mostly defined as β, the ratio defined in *Equation 2*, or as another dimensionless ratio $\sigma = \beta - 1$:

$$\sigma = (C - C_s)/C_s \qquad [3]$$

As we will see in the sequel, several growth rate equations contain the term $\ln\beta$ included in *Equation 1*. Traditionally, for the growth of crystals made from small molecules, $\ln\beta$ is approximated by σ. This is permitted because supersaturations are often low or very low. For a supersaturation of 5%, i.e. $\beta = 1.05$, we have $\sigma = 0.05$ and $\ln\beta \sim 0.049$. However, proteins often nucleate and grow from highly supersaturated solutions, e.g. at supersaturation $\beta = 3$, we have $\ln\beta \sim 1.099$, whereas $\sigma = 2$. It is self-evident that in that case replacing $\ln\beta$ by σ should be avoided. In general, $\ln\beta$ is significantly different from σ as soon as β exceeds about 1.5 which is a very low value for protein crystallization.

It is also worth noting that supersaturation is sometimes defined as the difference $C - C_s$. In this case, its value drastically depends on the concentration units. The difference $C - C_s = 100$ mg/ml, for example, reduces to about 1×10^{-2} if the concentrations are expressed as mol fractions, for a molar weight of the solute $M_w = 10\,000$. In general, it is more suitable to use β or σ to solve the nucleation or growth rate equations. However this may conceal the specific influence of the concentration on crystallization. As an example, let us consider the case for which the solubility of the protein decreases when increasing the concentration of the crystallization agent, salt, or poor solvent. Thereby, in the case of BPTI in NaCl solutions (1), a supersaturation of twice the solubility, $\beta = 2$, can be achieved in the area of the solubility diagram where solubility is large (44 mg/ml in 1.4 M NaCl solutions at 25 °C) or low (3 mg/ml in 2.3 M NaCl solutions at 25 °C). In the former case the mass of solute which will be deposited is 44 mg/ml whereas in the latter case it is only 3 mg/ml. Despite the same β value, nucleation and growth will be favoured in the former case.

3. Nucleation

When a solution is supersaturated, the solid phase forms more or less rapidly depending on the conditions: concentration of solute, crystallization agent, pH, supersaturation, temperature, nature and concentration of impurities, stirring, presence of solid particles. Primary nucleation occurs in a solution that is clear, without crystals. It is called homogeneous nucleation if the nuclei form in the bulk of the solution. On the other hand, it is called heterogeneous if the nuclei preferentially form on substrates such as the wall of the crystallizer, the stirrer, or solid particles (dust particles, and so on). Secondary nucleation which is induced by the presence of already existing crystals is less frequent during protein crystallization because the crystallizers are rarely equipped with stirrers which generate attrition, shear at the crystal surface.

3.1 Nucleation rate

From a theoretical point of view (2–4) nucleation is considered as an addition of monomers to clusters made of a few molecules called i-mers. When the system is in a steady state, the rate of formation of an i-mer is equal to its rate of disappearance. If k^s_{i-1} is the rate constant for subtracting a monomer from a (i + 1)-mer and k^a_{i-1} is the rate constant for adding a monomer to a (i – 1)-mer, one has:

$$k^a_{i-1} N_{i-1}N_1 + k^s_{i+1}N_{i+1} = k^a_i N_i N_1 + k^s_i N_i \quad [4]$$

where N_1 and N_i are the concentrations of monomers and i-mer, respectively. The flux of clusters going from a lower class into an upper class is:

$$J = k^a_{i-1}N_{i-1}N_1 - k^s_i N_i = k^a_i N_i N_1 - k^s_{i+1}N_{i+1} \quad [5]$$

This is the steady state rate of nucleation, J being a number of clusters per unit time and unit volume of solution. It is further assumed that, at equilibrium between the size classes, the rates at which a monomer leaves or sticks on a cluster are equal, so that:

$$k^a_{i-1}N_{i-1}N_1 = k^s_i N_i \quad [6]$$

As it will be seen hereafter the small clusters turn into stable nuclei only if they contain a critical number of monomers. Accordingly, the nucleation rate J is mainly dependent on the class sizes around i*. It is the product of the nuclei concentration times the frequency at which they exceed the critical number i* by addition of a monomer. It expresses as:

$$J = k^a_{i*} N_1 Z \exp\left(-\frac{\Delta G^*}{k_B T}\right) \quad [7]$$

where ΔG^* is the activation free energy for forming a nucleus of critical size and Z the so-called Zeldovich factor:

$$Z = \left(\frac{\Delta G^*}{3\Pi k_B T^{*2}}\right)^{1/2} \quad [8]$$

In solution $Z \sim 1 \times 10^{-2}$ and the pre-exponential term typically ranges from 10^2 to 10^{20} nuclei per cubic centimetre and second. Usually *Equation 7* is rewritten as:

$$J = \vartheta N_1 \exp\left(-\frac{\Delta G^*}{k_B T}\right) \quad [9]$$

where ϑ is the frequency at which the critical size becomes supercritical allowing the nucleus to grow and turn into a crystal.

3.2 Activation free energy for homogeneous nucleation

In order to solve *Equation 9*, it is necessary to precisely calculate or estimate the activation free energy ΔG^*. Since the solute concentration is the same in the whole bulk, nucleation occurs if there are energy fluctuations, somewhere in the solution, around the mean value imposed by the supersaturation. To create a nucleus it is necessary to create a volume and a surface. Assuming that the nucleus is limited by only one type of face, the activation free energy for homogeneous nucleation is:

$$\Delta G = -ik_BT\ln\beta + A_1\gamma_1 \qquad [10]$$

where i is the number of molecules in the nucleus, A_1 the area of the nucleus, and γ_1 its interfacial free energy with respect to the solution. The first term represents the energy to create the volume whereas the second term is the excess energy to create the surface. To simplify the demonstration we can also suppose that the nucleus is a sphere so that:

$$\Delta G = -\frac{4}{3}\Pi r^3 k_BT\ln\beta + 4\Pi r^2\gamma_1 \qquad [11]$$

At equilibrium, when $\partial\Delta G/\partial r = 0$, the nucleus has the critical radius r*, as shown in *Figure 1*,

$$r^* = \frac{2\gamma_1 V}{k_BT\ln\beta} \qquad [12]$$

where V is the volume of a molecule.

Inserting *Equation 12* into *Equation 11* yields:

$$\Delta G^* = -\frac{16\Pi V^2\gamma_1^3}{3(k_BT\ln\beta)^2} \qquad [13]$$

which can also be written as:

$$\Delta G^* = \frac{1}{3}(4\Pi r^{*2}\gamma_1) \qquad [14]$$

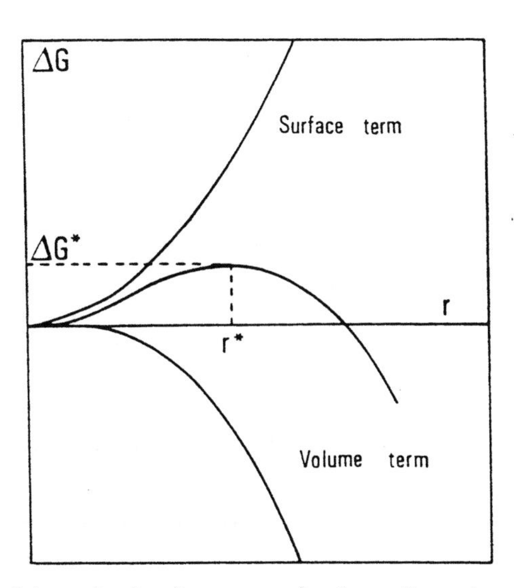

Figure 1. Variation of the activation free energy for three-dimensional nucleation versus nucleus size.

The critical activation free energy for creating the nucleus with critical radius r^* is one-third of the energy required for creating its surface. As shown in *Figure 1*, at the critical size r^*, the nucleus is in a very labile equilibrium. If it gains one molecule so that $r > r^*$ it grows. But if it loses one molecules so that $r < r^*$, then it spontaneously dissolves. In both cases there is a gain in energy. Inserting *Equation 13* or *Equation 14* into *Equation 9* allows for the calculation of the nucleation rate if nucleation is homogeneous.

3.3 Activation free energy for heterogeneous nucleation

Heterogeneous nucleation often occurs prior to homogeneous nucleation especially when superstauration is low. However, this implies that the solute molecules have some affinity for the substrate onto which they stick. Here also, it is convenient to consider that the nucleus is a sphere, actually cap-shaped, making the contact angle α with the substrate (*Figure 2*). Three surface free energies are involved in heterogeneous nucleation: γ_1 between the nucleus and the solution, γ_a between the nucleus and the substrate, and γ_0 between the substrate and the solution. They are related by Young's equation:

$$\gamma_0 = \gamma_a + \gamma_1 \cos \alpha \qquad [15]$$

If we name S_1 the area of the nucleus and S_a the area of the interface between the nucleus and the substrate, the activation free energy for heterogeneous nucleation is:

$$\Delta G_{het} = - ik_B T \ln\beta + S_1\gamma_1 + S_a\gamma_a - S_a\gamma_0 \qquad [16]$$

Taking *Equation 15* into account, ΔG_{het} becomes:

$$\Delta G_{het} = - \frac{4\Pi r^3}{3V} \frac{2 - 3\cos\alpha + \cos^3\alpha}{4} k_B T \ln\beta + 4\Pi r^2 \frac{1-\cos\alpha}{2} \gamma_1 - 4\Pi r^2 \frac{1-\cos^2\alpha}{4} \gamma_1 \cos\alpha \qquad [17]$$

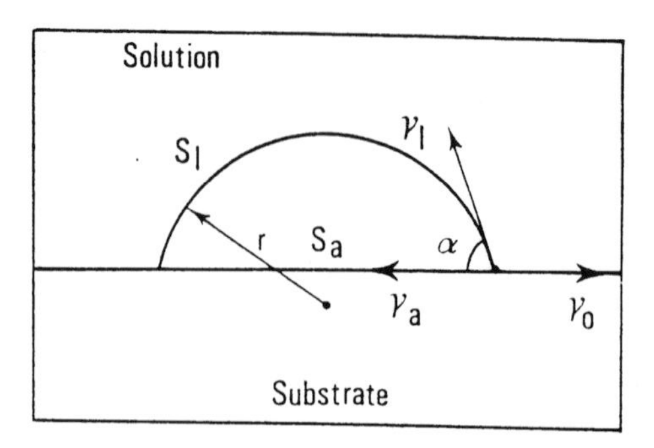

Figure 2. Cap-shaped nucleus forming by heterogeneous nucleation on a substrate.

At equilibrium, when $\delta\Delta G_{het}/\delta r = 0$, the radius of the critical nucleus is:

$$r^*_{het} = \frac{2\gamma_1 V}{k_B T \ln\beta} \qquad [18]$$

The critical radius of the nucleus formed by heterogeneous nucleation is the same as for homogeneous nucleation. However, the cap-shaped nucleus contains fewer molecules than does the full sphere. Inserting *Equation 18* into *Equation 17* yields:

$$\Delta G^*_{het} = \Delta G^*\left(\frac{1}{2} - \frac{3}{4}\cos\alpha + \frac{1}{4}\cos^3\alpha\right) \qquad [19]$$

which is the product of ΔG^* for homogeneous nucleation times a term depending on the contact angle. It is worth noting that for $\alpha = 180°$, $\Delta G^*_{het} = \Delta G^*$. The substrate does not have any effect on nucleation. For $\alpha = 90°$, $\Delta G^*_{het} = \Delta G^*/2$. If α tends toward zero, then ΔG^*_{het} tends also toward 0. That means that the substrate induces nucleation even at very low supersaturation since less and less energy is required to form the nucleus. The nucleation rate *Equation 9* drastically increases when the contact angle α decreases and subsequently the activation free energy for nucleation.

3.4 Examples

Let us first imagine a system for which nucleation of small molecules is homogeneous. To solve *Equation 9* we suppose that, in *Equation 13*, each molecule occupies a volume $V = (5 \times 5 \times 5) \times 10^{-24}$ cm^3. If the solubility is rather high (typically 10–50 g/litre) then the interfacial free energy, γ_1, is rather low, e.g. 10 erg cm^{-2} (10 mJ m^{-2}). Typically in *Equation 9* one has $\upsilon N_1 = 10^{20}$ cm^{-3} s^{-1}. Inserting T = 293 K and $k_B = 1.38 \times 10^{-16}$ erg/K yields:

$$J = 10^{20}\left(-\frac{3.96}{\ln^2\beta}\right) \qquad [20]$$

In order to have a nucleation rate J = 1 nucleus cm^{-3} s^{-1}, the supersaturation β has to be ~ 1.34, what is a rather low value. But, if β were only 1.2, then the nucleation rate would be catastrophically low (1.8×10^{-32} cm^{-3} s^{-1}). On the other hand, if β were 2.0, then the nucleation rate would be drastically high (2.6×10^{16} cm^3 s^{-1}). This demonstrates that nucleation is highly supersaturation-dependent. In the low supersaturation range the solution remains metastable over a long period of time, whereas in the high supersaturation range nucleation occurs spontaneously. That means, for protein crystallization, that a condition which leads to a large number of crystals can be improved by decreasing supersaturation.

The second parameter which greatly influences the nucleation rate is the surface free energy γ_1 of the nucleus. If the value of γ_1 is increased up to 20 or 50 erg cm^{-2}, all other parameters being unchanged in the previous example, then J = 1 nucleus cm^{-3} s^{-1} for β ~ 2.29 or 26.5, respectively. The metastable zone, where no nucleation occurs after a reasonable time lag, drastically widens out with increasing the interfacial free energy. In that case, the only way to avoid it is to change the solvent or the solution composition. The rule is 'the higher the solubility, the better the affinity of the solvent for the nucleus and the lower the surface free energy'. All other things being unchanged, a better solvent gives rise to higher solubilities and smaller surface free energies, both these facts contributing to increase the nucleation rate.

Considering again *Equations 9* and *13*, it can be seen that increasing the volume V of the molecule has the same effect on the nucleation rate as increasing the surface free energy. Accordingly, macromolecules should nucleate with a lower rate than small molecules if all other parameters are unchanged. Assuming that the volume of the macromolecule is $(50 \times 50 \times 50) \times 10^{-24}$ cm^3, instead of $(5 \times 5 \times 5) \times 10^{-24}$ cm^3, then *Equation 20* becomes:

$$J = 10^{20} \exp\left(-\frac{3690}{\ln^2\beta}\right) \qquad [21]$$

In that case a very large supersaturation is needed to obtain 1 nucleus cm^{-3} s^{-1}. In fact β equals 7718 which is completely unrealistic! However it is a good illustration of the difficulty often encountered for nucleating proteins. Furthermore, it is difficult to consider that the kinetic coefficient, taken as 10^{20} cm^{-3} s^{-1} in the previous example, would be greater for macromolecules than for small molecules. Hence, the only way to obtain reasonable values of J and of β is to assume that the surface free energy of a protein crystal is significantly lower than the energies usually encountered for crystals of small molecules. If to calculate *Equation 21*, $\gamma_1 = 1$ erg cm^{-2} is used instead of 10 erg cm^{-2} then the J values are given by *Equation 20* and subsequently the same low β values are obtained. Consequently lower surface free energies can compensate higher molecular volumes in the case of protein crystallization. Values of $\gamma_1 = 0.5$–0.7 erg cm^{-2} for thaumatin ($M_w = 22\,000$ Da) were observed by Malkin *et al.* (5).

As a concluding remark, it should be emphasized that there is no special reason that, a priori, all proteins should have a low or very low surface free energy. If the general rule holds for proteins, it might be, then for sparingly soluble proteins the surface free energies are relatively high. But, these proteins also nucleate, sometimes even after rather short induction periods. The only explanation would be that nucleation is heterogeneous. As a matter of fact, it cannot be homogeneous because the required supersaturation would be much too high. The existence of heterogeneous nucleation can be checked with a very well known model protein, i.e. with hen egg white lysozyme (HEWL). With β values of ~ 3, the solution deposits several tens of crystallites per cubic centimetre of solution within a few hours if the solution is not carefully filtered. On the other hand, it deposits sometimes only one or two crystals after one or two days, if most of the solid particles are removed by filtration.

Finally, in order to estimate the nucleation rate and the interfacial free energy γ_1, it is possible to measure the time lag for nucleation, or induction time t_i, as a function of different supersaturations. Assuming that after the time t_i $J = 1\ cm^{-3}\ s^{-1}$ on has:

$$Jt_i = \left\{\upsilon N_1 \exp\left[-\frac{16\Pi\upsilon^2\gamma_1^3}{3(k_B T)^3 \ln^2\beta}\right]\right\} t_i = 1 \qquad [22]$$

so that:

$$\ln \upsilon N_1 = \frac{16\Pi\upsilon^2\gamma_1^3}{3(k_B T)^3 \ln^2\beta} \qquad [23]$$

Plotting $\ln t_i$ versus $1/\ln^2\beta$ should give a straight line, the slope of which is proportional to γ_1^3 which is the only unknown in the term on the right side of *Equation 23*. This method was often used for determining the γ_1 values of crystals of small molecules. Due to the uncertainties of the measurements of t_i it only gives a good order of magnitude.

4. Pre-nucleation — investigation of the solution

The main questions addressed in this section are:

- what is the behaviour of the solution before nucleation occurs?
- why do some solutions lead to precipitate whereas others yield crystals?

From the classical nucleation theory, it is known that crystals form preferentially instead of precipitates when the first clusters (built up of several molecules) exhibit some crystalline arrangement. Therefore, the very first stage of nucleation, called pre-nucleation, is very important. Accordingly, the systematic investigation of the solutions, in both under- and supersaturated state, is essential to understand and control the crystallization process of proteins.

4.1 Methods

Depending on the techniques used, different information can be obtained on the solutions. Since we only give a brief survey of these techniques, the reader can refer to the different monographs where they are widely described. The references are given hereafter. In this section, solution scattering and osmotic pressure techniques will be presented. These techniques aim at obtaining information on molecules in solution: molecular weight, size, aggregation states, polydispersity, and interactions. As will be seen in Section 4.3 probing the protein interactions is very important in the field of protein crystallization. In dilute solution interactions include excluded volume term, repulsive electrostatic term, and attractive Van der Waals term, the last two terms are described in the so-called DLVO theory of colloidal stability (6).

4.1.1 Solution scattering techniques

For all scattering experiments in solution the principle is the same: a monochromatic beam of visible light, X-rays, or neutrons impinges on the protein molecules and induces an oscillating polarization of their electrons. The molecules then serve as a secondary source which is radiated and scattered. For neutrons the interaction with the matter is different, because neutrons are scattered by the atomic nuclei, the scattered intensity depending on the scattering length density.

i. Light scattering (7, 8)

Depending on the way the data are analysed, two types of experiments are possible: elastic or static light scattering (SLS) and quasi-elastic scattering or dynamic light scattering (DLS). The experimental set-up is shown in *Figure 3*.

Static light scattering (SLS)

The experiment consists in measuring the photons intensity scattered by the solution at different angles and for different concentrations. Let us consider two cases depending on the size of the protein with respect to the wavelength of the light.

(a) The light is scattered by spherical particles which are small compared to the wavelength of the light, $d < \lambda/10$, where d is the characteristic size of the protein and the wavelength of the light ($400 < \lambda < 650$ nm). In that case the particle is assumed to be a punctual source of light, and the intensity scattered is independent of the angle. The experiments are usually carried out at 90°, then:

$$K\left(\frac{dn}{dC}\right)^2 I_{ref}\frac{C}{\Delta I} = \frac{1}{M_w} + 2A_2C \qquad [24]$$

where K is a constant, dn/dC the increment of refractive index with protein concentration C, I_{ref} the intensity scattered by a reference at 90°, and

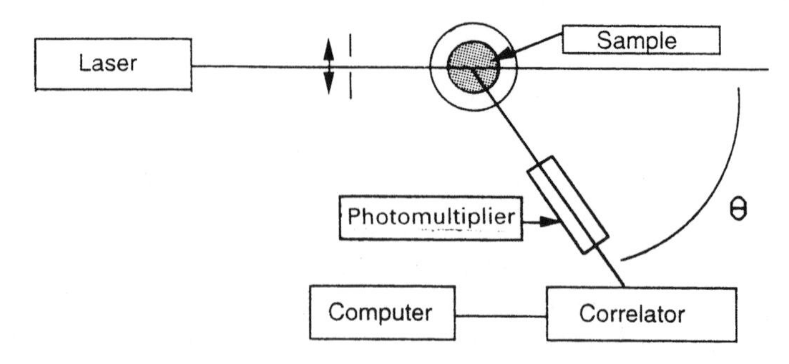

Figure 3. A schematic representation of the LS experiment.

ΔI the difference between the intensity scattered by the solvent and the solution at 90°.

Information obtained: M_w (the molecular weight of the particle) and A_2 (the second virial coefficient), also noted B_{22} in the literature. The sign of the second virial coefficient is indicative of the type of interactions: it is negative when the interactions between molecules are attractive and positive if the interactions are repulsive.

(b) The light is scattered by larger particles, $\lambda/10 < d < \lambda$; in that case the particle is assumed to be a multiple source of the light so that there is a phase difference between the light scattered by different portions of the particle at any time, the intensity scattered is angular dependent, and then:

$$K\left(\frac{dn}{dC}\right)^2 I_{ref} \frac{C}{\Delta I(\theta)\sin\theta} = \frac{1}{M_w}\left(1 + \frac{q^2R_G^2}{3}\right) + 2A_2C \qquad [25]$$

where $\Delta I(\theta)$ is the difference between the intensity scattered by the solvent and the solution at angle θ.

Information obtained: M_w, A_2, and R_G (the radius of gyration of the particle).

Dynamic light scattering (DLS)

The experiments consist in measuring the fluctuation of the scattered intensity at constant scattering angle but at different concentrations. These fluctuations are correlated to the diffusion (translational and rotational) of the particles in solution. In the following, it is assumed that the molecules are spherical and $d < \lambda/10$ so that rotational diffusion and internal motion of flexible macromolecules can be neglected.

Two cases must be considered:

(a) In the absence of interaction, there is no concentration dependence of the measured coefficient diffusion.
Information obtained: D_0 (the free particle diffusion coefficient), R_h (the hydrodynamic radius of the particle), and v (the quality factor or the polydispersity index which expresses the dispersion of the sizes of the particles in solution).

D_0 is related to the hydrodynamic radius by means of the Stokes–Einstein equation:

$$D_0 = k_BT/6\pi\eta_0R_h \qquad [26]$$

where k_B is the Boltzmann constant, T the absolute temperature, and η_0 the solution viscosity.

(b) In the presence of interactions, there is a concentration dependence of the measured diffusion coefficient.
Information obtained: D_{eff} (the effective diffusion coefficient of the particle) and K_D (the interaction parameter with $D_{eff} = D_0(1 + K_D\,C)$, C being the

protein concentration). K_D is negative when the interactions are attractive and positive when the interactions are repulsive.

ii. Small angle X-ray and neutrons scattering (9–12)

The experiments consist in obtaining the angular distribution of the X-ray or neutrons intensity scattered by the solution. There are also two cases depending on the interactions:

(a) In the absence of interactions, the intensity scattered is the sum of the scattering of the individual particles, namely, the form factor, and is given by the Guinier's law.
Information obtained: M_w, R_G, and the form factor.

(b) In the presence of interactions, the intensity scattered is the product of the form factor and the structure factor, the interference term related to particle distribution.
Information obtained: A_2 and the structure factor.

4.1.2 Osmotic pressure techniques (13, 14)

In an osmotic pressure experiment, the solvent and the protein solution are separated by a semi-permeable membrane. The excess pressure due to the difference in the chemical potentials of the two solutions creates a flux of solvent through the membrane. The osmotic pressure (Π) is therefore proportional to the number of particles in solution and is often expanded as a series of virial coefficients.

$$\Pi = CRT\left(\frac{1}{M_w} + 2A_2C + ...\right) \qquad [27]$$

where C is the protein concentration, R the molar gas constant, $8.31\ 10^7$ erg $mol^{-1}\ K^{-1}$, and T the absolute temperature.
Information obtained: M_w and A_2.

4.2 Practical recommendations

4.2.1 Generals

All the above methods (listed in Table 1) often require investigation of several solutions at different concentrations in order to precisely deduce the

Table 1. Comparison of the different methods

Method	Information	Domain–range	Wavelength
SLS	M_w, A_2	$d < 50$ nm	$\lambda \approx 500$ nm
SLS	M_w, A_2, and R_G	$50 \leq d < 500$ nm	$\lambda \approx 500$ nm
DLS	D_{eff}, D_0, R_h, v, and K_D	$1 < d < 1000$ nm	$\lambda \approx 500$ nm
SAXS	M_w, A_2, and R_G	$1 < d < 100$ nm	$0.1 < \lambda < 0.5$ nm
SANS	M_w, A_2, and R_G	$1 < d < 100$ nm	$0.2 < \lambda < 2$ nm
OP	M_w, A_2	$5000 < M_w < 10^6$ Da	

concentration dependence of the measured parameter. Furthermore, the experimenter has to check whether the particles are all the same in the whole concentration range under consideration.

4.2.2 LS

The volume of solution required for a measurement is about 100–300 μl. A typical experimental protocol is described in *Protocol 1*. Prior to the experiment it is necessary to control several points:

(a) Absence of fluctuation in the laser intensity.

(b) Absence of parasitic light due to reflections or refractions. These conditions are difficult to obtain in experiments carried out at angles below 30° with a commercial set-up.

(c) Absence of dust, air bubbles, glass particles, and other foreign tiny materials in solution.

(d) Measurement achieved at the proper angle.

(e) Good transparency of the solution is required in order to avoid the multiple diffusion, if needed a dilution must be done.

Therefore, before any experiment it is necessary to check the laser quality, the optical trajectory, and to filter and/or centrifuge the solution. To treat the signal it is also essential to know the refractive index and viscosity of the crystallization medium (buffer + crystallization agent). In addition, for SLS experiments, it is essential to know the increment of the refractive index of the solution as a function of the protein concentration.

4.2.3 SAXS-SANS

These experiments can only be carried out using a synchrotron radiation or a nuclear reactor. Runs are always allocated by a program committee to which the application must be submitted. Since the number of runs is limited, it is recommended to test the sample quality before using one of these techniques. For instance DLS is a good tool for checking whether the molecules are aggregated or not. If the polydispersity is high the SAXS and SANS will fail.

(a) SAXS: the volume of solution required for one measurement is ~ 100 μl. The electron density of the crystallization agent should be as low as possible, otherwise the signal due to the particles which is under investigation disappears in the background. Thus electron-rich buffer at high concentration (e.g. ammonium sulfate) should be avoided.

(b) SANS: the volume of solution required for one measurement is about 150 μl. In these experiments it is often necessary to dissolve the protein in D_2O solutions. Moreover, Broutin *et al.* (15) and Gripon *et al.* (16) recently showed a shift of the solubility of lysozyme when H_2O is replaced by D_2O. Furthermore, D_2O affects the interactions between the particles in solution, and care must be taken when working with materials which have a tendency to aggregate. Accordingly, the experimenter has to check the aggregation behaviour of the protein solution before any experiment.

4.2.4 OP

The volume of solution required for a measurement is about 120 μl. Actually, three measurements performed with 40 μl are necessary.

(a) The main point of this experiment concerns the equilibration on both sides of the membrane. The higher the salt concentration is, the more difficult the equilibration is. Practically the upper concentration limit is about 500 mM.

(b) High viscosities solutions can generate very long equilibrium time.

(c) It is important to check the temperature and the pH stability.

4.3 Examples

DLS is the most widely used method for the characterization of protein solutions and was first proposed as a diagnostic tool for protein crystallization by Kam *et al.* (17). Zulauf *et al.* (18) studied 15 proteins, in dilute solutions and in the absence of crystallization agent, and suggested that the detection of aggregates indicates that crystallization will not be successful. Ferré-D'Amaré *et al.* (19) have determined the crystallizability of three different RNAs by DLS with this criterion. In addition, more recent studies showed the absence of large molecular aggregates in supersaturated solutions for different proteins (1, 20, 21), contrary to Georgalis *et al.* (22, 23) who observed the formation of

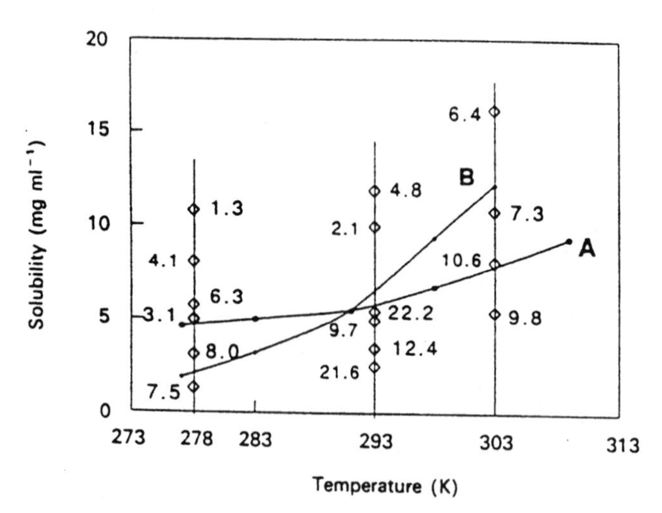

Figure 4. Polydispersities (%) measured by concentrating or diluting α-amylase solutions with respect to the solubility curves of the A and B polymorphs (15).

praggs (precipitating aggregates) or craggs (crystallizing aggregates) depending on the conditions.

Measurements of A_2 (or B_{22}) by SLS for different proteins in different solvents have shown that conditions which promote crystallization are grouped within a narrow range of A_2 values (24, 25). Moreover different studies have shown a systematic relationship between solubility and second virial coefficient (25). Recently Bonneté *et al.* (26) showed the complementary nature of DLS, SAXS, and OP in order to probe interaction in protein solution.

Rather than reviewing all the outputs for crystallogenesis of the above methods, we discuss here an example of the application of DLS with the associated experimental protocol, and another example dealing with an application of SAXS.

4.3.1 DLS study applied to porcine pancreatic α-amylase (20)

In this example special attention was paid to the polydispersity of under- and supersaturated solutions of α-amylase. The results are presented in *Figure 4*, and can be summarized as follows: polydispersity is very high ($v > 10\%$) when the protein concentration is much lower than solubility whereas it is very low ($v < 10\%$) when the protein concentration is nearly equal or even slightly higher than solubility. Even more important, monodispersity is a prerequisite for obtaining good crystals. Some polydispersity seems to be acceptable if there are no large aggregates in solution.

Protocol 1. DLS experiment

Equipment and reagents

- Protein in solution
- A solution containing the buffer plus the crystallization agent
- A high performance light scattering apparatus

Method

1. Prepare a solution of 300 μl of the protein at the desired concentration.
2. Filter three times the solution with the same LCR13 (Millipore) filter and pour the solution in a glass cell. Dust can also be removed by centrifugation (20 000–30 000 *g*, for 1–2 h).
3. Put the glass cell in the sample holder.
4. Switch on the laser and check the optical trajectory. Avoid continual illumination of the solution for several hours because of potential protein denaturation.
5. Run the analysis for 1–3 min; the correlator receives the signal from the detector (photomultiplier).
6. Analyse the data: the cumulant method (27) directly gives the diffusion coefficient and the polydispersity.

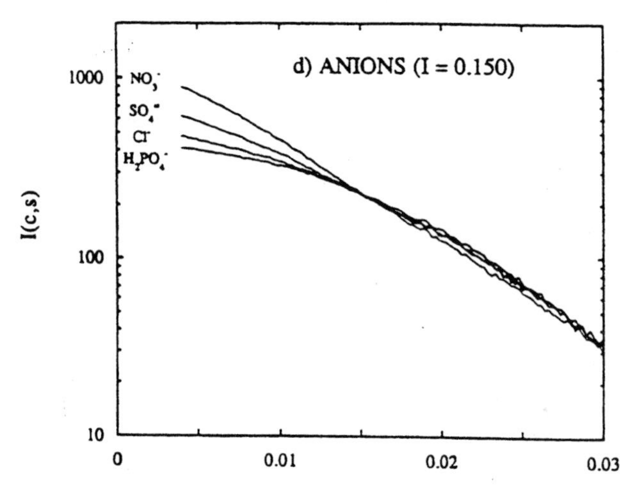

Figure 5. X-ray scattering curves recorded at 100 mg/ml lysozyme concentration in acetate buffer plus added salt. Anion series as indicated in the figure with Na^+ as counterion, ionic strength = 0.150 (17).

4.3.2 Lysozyme interactions as seen by SAXS (28)

In this example SAXS is used to characterize the influence of various salts on the protein–protein interactions in undersaturated lysozyme solutions at constant pH and temperature. Attractive and repulsive interactions respectively result in an increase or decrease of the structure factors at low scattering angles. Practically, as shown in *Figure 5*, when the scattering intensity at low angle is increased, the protein–protein interactions are moved toward attraction. Hence, the addition of different salts to lysozyme solutions results in changes from repulsion to attraction. Interestingly anions can be ordered according to their effectiveness to create attractive interactions and follow the reverse order of the Hofmeister series (see Chapter 10).

4.4 Practical considerations

The practical application of the studies on pre-nucleation is to diagnose whether a protein solution will deposit crystals or not. It is noteworthy that this ability should be studied under crystallization conditions, in both under- and supersaturated states. This implies that the phase diagram is known. As a general rule it seems that the solution monodispersity is a prerequisite to crystallization. Large molecular aggregates hinder growth especially if they do not dissociate into small aggregates or monomers once growth proceeds. Growth is also favoured by the occurrence of attractive interactions in the solution due to the addition of a crystallization agent. Finally there is a clear correlation between the occurrence of attractive protein interactions and the decrease of protein solubility.

From a practical point of view polydispersity of the protein solution can only be studied by DLS whereas the molecular interactions can be observed by LS, SAXS, SANS, and OP. Accordingly, LS is a powerful and interesting tool for knowing some important solution characteristics, and can be easily carried out at the laboratory scale.

5. Crystal growth

When a nucleus grows and transforms into a crystal, the different faces of the growing crystal exhibit growth mechanisms and rates that depend on external factors (supersaturation, impurities, temperature, and so on) and internal factors (structure, bonds, defects, and so on). According to the periodic bond chain (PBC) theory (29–32) there are three types of crystal faces (*Figure 6*).

- F (flat) faces: they contain at least two PBCs in the slice of thickness d_{hkl}, where d_{hkl} is the interplanar distance of the face (hkl).
- S (stepped) faces: they contain only one PBC in the slice d_{hkl}.
- K (kinked) faces: they do not contain any PBC in the slice d_{hkl}.

Let us just recall that a PBC is an interrupted chain of strong bonds running along a crystallographic direction in the crystal. Since all sites on the K faces are growth sites, more commonly called kinks, the K faces grow by direct incorporation of the growth units which hit them. The growth rate is high and, normally, these faces do not occur on the crystal morphology, because the growth form of the crystal is made up only of the faces which have the slowest growth rate.

Conversely, the F faces are poor in kinks. They grow by lateral spreading of

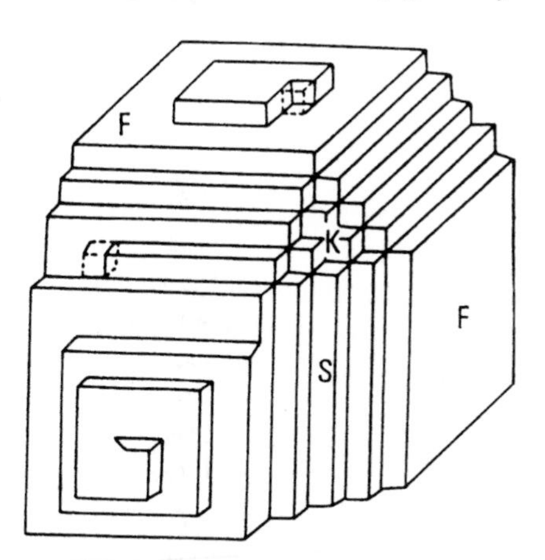

Figure 6. Schematic representation of a crystal exhibiting flat (F), stepped (S), and kinked (K) faces. The front face exhibits a polygonized growth spiral, whereas the top face exhibits a two-dimensional nucleus.

the growth layers. For being integrated into the crystal, the solute molecules must first adsorb on the surface, and later on diffuse toward the step of a growth layer along which they migrate toward a kink. Such faces grow either by a two-dimensional mechanism or a spiral growth mechanism (*Figure 6*). Since the number of kinks is low, the growth rates are low too.

At last, the S faces are in an intermediate situation. Their growth rate is lower than that of the K faces but higher than that of the F faces. Normally, the S faces do not appear on the crystal morphology, except when their growth rate is slowed down by adsorption of an impurity for example.

The growth mechanisms have been discussed in detail elsewhere (33). Hereafter we summarize the general trends.

5.1 Growth controlled by surface processes

5.1.1 Growth by two-dimensional nucleation

This growth mechanism occurs when the crystal face is perfect, without any defect. The molecules which adsorb, randomly diffuse on the surface, encounter, coalesce into a two-dimensional nucleus which spreads across the crystal face if its size exceeds a critical size. For a square-shaped 2D nucleus the critical size (number of molecules n^* in the nucleus) is:

$$n^* = \frac{2\lambda}{k_B T \ln\beta} \qquad [28]$$

where λ is the so-called edge free energy expressed here per molecule in the edge. In the mononuclear model, there is only one 2D nucleus which spreads across the surface so that the growth rate R of this face is:

$$R = B_2 dS \qquad [29]$$

where d is the height of the growth layer, S the area of the face, and B_2 the 2D nucleation rate, i.e. the number of nuclei forming per unit time and unit area ($cm^{-2}\ s^{-1}$). B_2 can be written:

$$B_2 = n_1 \upsilon \exp\left(-\frac{\Delta G_2^*}{k_B T}\right) \qquad [30]$$

where n_1 is the number of growth units adsorbed per unit area of the face and θ the frequency at which the 2D nucleus of critical size become supercritical and grows. ΔG_2^* is the activation free energy for 2D nucleation:

$$\Delta G_2^* = \frac{4\lambda^2}{k_B T \ln\beta} \qquad [31]$$

Inserting *Equation 31* into *Equations 30* and *29* shows that the growth rate is an exponential function of supersaturation. As in the case of three-dimensional nucleation, there is a critical supersaturation below which the growth rate is zero or nearly zero. A dead zone is observed at low supersaturation when measuring R as a function of β. Once this critical supersaturation is exceeded, the growth rate drastically increases with increasing supersaturation. Growth is difficult to control.

If several nuclei spread at the same time across the crystal face, growth is determined by a 2D polynuclear mechanism. In that case the expression for the growth rate is somewhat more complicated (33–35).

5.1.2 Growth by a spiral mechanism

When a screw dislocation emerges on a crystal face, it generates a growth spiral (*Figures 6* and *7*). Since the growth spiral is made of a parallel sequence of steps, growth can take place even at low supersaturation since the growth units which adsorb onto the crystal face easily find growth sites where they are incorporated into the crystal. It can be seen (*Figure 7*) that the growth rate of the face can schematically be written as:

$$R = \frac{vd}{y} \qquad [32]$$

where v is the lateral velocity of the steps, d their height, and y their equidistance. If the spiral is circular one has:

$$y = \frac{19\lambda a}{k_B T \ln\beta} \qquad [33]$$

where λ is also the edge free energy of the steps, a being the distance between two molecules in the step. Considering *Figure 7* and *Equation 33*, we see that an F face which exhibits a growth spiral is really flat only if the supersaturation is low (y large). Conversely, it takes a conical outline when the supersaturation is high (y small) due to the high step density.

The theories of the spiral growth mechanism were extensively discussed elsewhere (36–39). Here, we give only a few possibilities which can be derived from the general growth rate equation that is not commented here. Depending on the influence of the different parameters, growth depends on surface diffusion, kink integration kinetics, and so on. As an example, let us suppose

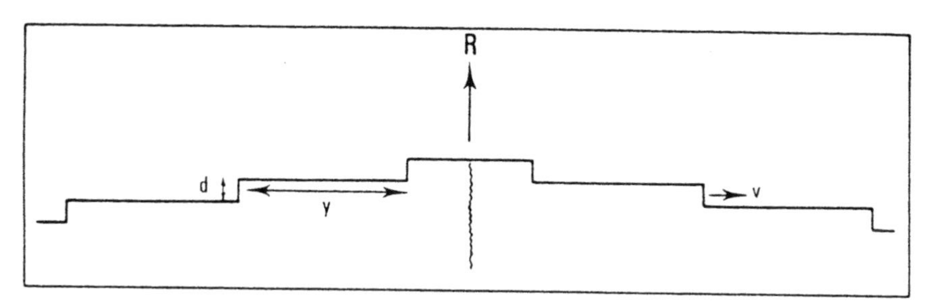

Figure 7. Profile of a face growing by a spiral growth mechanism.

that growth is controlled by surface diffusion of the growth units toward the steps. Then we have two possibilities.

At low supersaturation:

$$R = \frac{2n_1 D_s V}{x_s} \sigma \frac{1}{y} \qquad [34]$$

At high supersaturation, due to the high step density:

$$R = \frac{2n_1 D_s V}{x_s^2} \sigma \qquad [35]$$

In both equations, D_s is the surface diffusion coefficient and x_s the mean free path of diffusion; n_1 is again the number of adsorbed growth units per unit area, and V the volume of a molecule in the crystal.

Inserting *Equation 33* into *Equation 34* shows that R is proportional to $\sigma \times \ln\beta$. With the approximation $\ln\beta = \sigma$ the so-called primary quadratic growth rate law is obtained at low supersaturation:

$$R = k_1\sigma^2 \qquad [36]$$

whereas at high supersaturation the primary linear growth rate law is obtained:

$$R = k_1\sigma \qquad [37]$$

It is worth noting that *Equation 36* is valid only if the supersaturation is very low as discussed in Section 1. For protein crystals it is normally not allowed to replace $R = k_1\sigma\ln\beta$ by $R = k_1\sigma^2$ because of the high values of the supersaturation.

Among the different processes predicted by the spiral growth mechanism, there is one which directly depends on the kink integration kinetics. All volume diffusion and surface diffusion processes are supposed to be fast with respect to the kinetics at which the growth units enter into the kinks. The growth units reach the growth sites but need some time to find the proper conformation before being incorporated. This time is called relaxation time τ_k for entering into the kinks. It depends on the activation free energy for entering into the kinks:

$$\tau_k = \nu^{-1} \frac{\Delta G_k^*}{k_B T} \qquad [38]$$

where ν is a frequency often taken, in the vapour phase, as $k_B T/h \sim 6 \times 10^{12}\ s^{-1}$ where h is Planck's constant. In solution ν should have a lower value. If this mechanism is rate determining, one has:

$$R = \frac{a n_1 V \sigma}{\tau_k} \frac{1}{y} \qquad [39]$$

where a is the length of the elementary jump of the growth units which diffuse toward the kinks. Once more, if we insert *Equation 33* into *Equation 39*, and with the assumption that $\ln\beta = \sigma$ a secondary quadratic growth rate law is obtained:

$$R = k'_q \sigma^2 \quad [40]$$

Such a mechanism is likely to occur with large molecules such as proteins which have to reorient for being trapped into the growth sites. But the approximation $\ln\beta = \sigma$ should be avoided.

When the diffusion of the growth units towards the crystal is slow with respect to the surface processes, the theories (36–39) predict that growth is controlled by volume diffusion. This happens especially in stagnant systems where there is no solution flow. The basic equations describing the growth rates are rather cumbersome, but after some assumptions, some of them take a rather simple form. As an example, one has:

$$R = \frac{n_0 D_v V}{\delta} \sigma \quad [41]$$

where n_0 is the number of growth units per unit solution volume, D_v the volume diffusion coefficient, and δ the thickness of the boundary layer. This equation is similar to that derived from Fick's law (33, 40):

$$R = \frac{VD_v}{\delta} (C - C_s) \quad [42]$$

In the latter equation $C - C_s$ is expressed as a number of mol cm^{-3}, and V as $cm^3\ mol^{-1}$.

For theoretical reasons, it is sometimes interesting to know whether growth depends on volume diffusion or on surface kinetics. For doing this, the crystal is placed in a flow system and the growth rates of the faces are measured as a function of the flow velocity U of the solution. At least for crystals of small molecules, it is always observed that the growth rates first increase with increasing flow velocity up to a final value, a plateau, where they become independent of the flow velocity. Thus in the former case R is controlled by volume diffusion, whereas in the latter case it is controlled by surface processes. With growth rates, respectively as follows:

$$R = K_v\, U^{1/2} (C - C_i) \quad [43a]$$

$$R = K_s (C - C_s)^n \quad [43b]$$

where C, C_i, and C_s refer to the solute concentrations in the bulk of the solution, at the crystal–solution interface and at saturation, respectively. K_v and K_s are kinetic coefficients which depend on temperature, solvent, solubility, and so on. When both rates are equal:

$$\frac{R}{U^{1/2}} = K_v (C - Cs) \frac{K_v}{K_s^{1/n}} \quad R^{1/n} \quad [44]$$

Plotting $R/U^{1/2}$ versus $R^{1/n}$ for all curves obtained at different supersaturations, provides the highest possible growth rates (for $U = \infty$) by extrapolating the straight lines thus obtained to $R/U^{1/2} = 0$.

As concerns proteins crystallization, it is practically always carried out in stagnant systems. The reason for this is the missing of instrumentation, but perhaps also the fragility of proteins. Moreover, D_v of proteins are two order of magnitude smaller than the ones of small molecules. Accordingly, the growth rates are mainly controlled by volume diffusion (*Equations 39, 42*) or by the kink integration kinetics (*Equation 39*).

5.2 Kinetic measurements

Before leaving growth kinetics, we must emphasize that growth rate curves must be determined at constant temperature if one wants to speculate on the growth rate laws and growth processes. This means that supersaturation must be varied by changing the solute concentration. As a matter of fact the growth rate equations show that there are many parameters which depend on activation energies, i.e. on temperature (volume and surface diffusion, integration into the kinks, desolvation of the surface of the growth units, and so on). It is therefore not surprising that, at constant supersaturation, growth can drastically increase with increasing temperature. It is often not really important if there is a variation of $\pm$ 1°C around the mean crystallization temperature, but large gaps are prohibited. The second reason for which it is necessary to work at constant temperature is that misleading results and interpretations of the growth rate curves can be given. Changing the temperature is easier than changing the solution concentration in order to change supersaturation. Unfortunately, there are conflicting effects between increasing supersaturation and decreasing temperature. If, for instance, solubility decreases with decreasing temperature, cooling the solution results in an increase of supersaturation, but at the same time in a decrease of the diffusion coefficients, of the mean free paths for diffusion and in an increase of all relaxation times involved in the growth processes. This is the reason why the growth rate first increases with cooling down the solution, but later on passes through a maximum before drastically decreasing. A conclusion drawn from such a curve are therefore subject to criticism. The drastic effect of the temperature on the growth rate on protein crystal was pointed out earlier (41) for porcine pancreatic α-amylase crystallization. In *Figure 8* it clearly appears that the growth rate of the porcine pancreatic α-amylase crystals is extremely temperature-dependent above 18°C.

The most current observation on protein crystallization and growth measurements are made by optical microscopy. For instance, Boistelle *et al.* (41) measured the growth rate of the porcine pancreatic α-amylase as a function of supersaturation at 18°C. The principle of the experiment was the following. Once the solution was at the right temperature and supersaturation

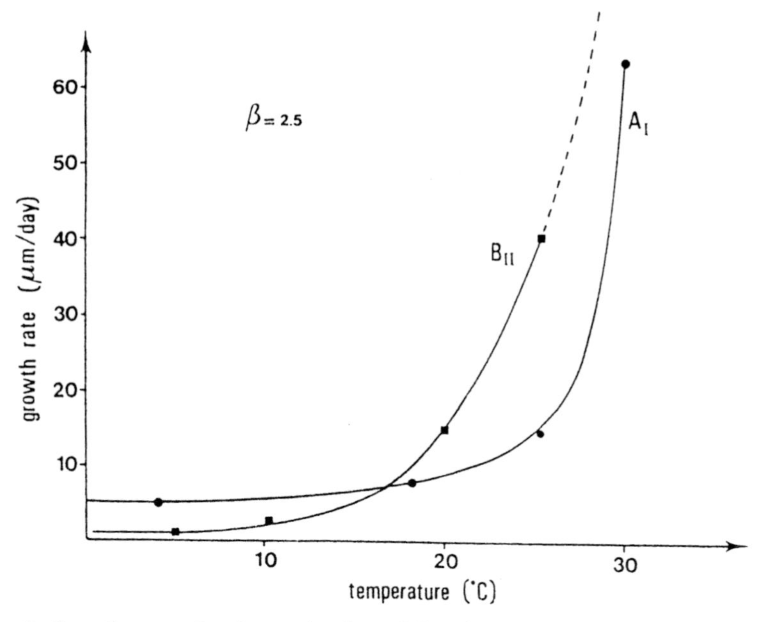

Figure 8. Growth rate of polymorphs A_I and B_{II} of porcine pancreatic α-amylase, at constant supersaturation, versus crystallization temperature (37).

a few seed crystals were introduced in the crystallization cell and the displacement of the faces was recorded as a function of time. A linear dependence with supersaturation was obtained and growth was interpreted as a process controlled by volume diffusion. On a more microscopic scale, direct measurements of the step velocity are possible using laser Michelson interferometry (42) or *in situ* atomic force microscopy (5). Such measurements are especially interesting if the step equidistance can be related to supersaturation in order to deduce the edge free energy of the step of the growth layer (*Equation 33*). Knowing the step velocity, also allows a better estimation of the parameters involved in the growth rate equations (mean free path for diffusion, surface diffusion coefficient, relaxation time for entering into the kinks, and so on).

6. Crystallization in the presence of impurities and additives

6.1 General trends

In the vocabulary of crystallization, the words impurities and additives play an important role. The latter word concerns foreign substances voluntarily added to the solution in order to obtain a special effect (inhibition of nucleation or growth, habit change of the crystal, and so on). On the contrary, the former word concerns foreign substances that exist in the solution, their source being either the material which is going to crystallize, the solvent, the crystallization agents, and so on. If impurities cannot always be avoided, it may be asked why additives are put into the solutions. Actually, there are several reasons. Some are biological in essence, for instance when adding a ligand to a protein, or any compound that will interact specifically with the protein. Other reasons, are of physico-chemical nature (Chapter 10). An additive properly chosen allows a better control of nucleation and growth. It can also induce a modification of the crystal morphology, more commonly called habit change. In some special cases, it induces phase transitions that are commonly observed with protein crystals. However, from a general standpoint, additives should also be considered as impurities as all other constituents of the solution.

The first impurities for the crystal are its own components. For small molecules this is easy to understand; calcium carbonates for instance can be grown in the presence of an excess of calcium or of carbonate. Nucleation, growth, and habits will be different from those obtained in a strictly stoichiometric solution. As concerns the crystallization of biological macromolecules, similar effects could be observed since the stoichiometry of the salts used as crystallization agents can be widely changed. Since these salts are inside the crystal structure, they can also be considered as constituents of the crystals. The solvent is also an impurity. Sodium chloride grows as cubes from aqueous solutions, but as octahedra from formamide solutions. Since organic solvents are often used to induce the crystallization of proteins by decreasing their solubility, they also influence the growth kinetics and the crystal habit. The pH variations affect both the interfaces between crystal and solution, and the nature and activities of the impurities. As an example glycine mainly exists as zwitterions $H_3^+N\text{–}CH_2\text{–}COO^-$ in the pH range 3.5–8.5, but as positively $NH_3^+\text{–}CH_2\text{–}COOH$ or negatively $NH_2\text{–}CH_2\text{–}COO^-$ charged ions outside this range. Adsorption of glycine on crystal faces, or interactions with other solute species will obviously be affected by the charges of the molecules. Many examples and interpretations concerning the different effects of impurities in the crystallization of small molecules are found in the literature (43–45).

The matter has also been discussed in the macromolecule field (Chapter 2). Here, the situation is even more complex, since salts (the most important additives used in protein crystallization) interact with the solvation shell of the macromolecules. In addition macromolecules can present conformational and sequence heterogeneities which will affect the crystallization process (46).

6.2 Additives, phases, and polymorphs

The polymorphs of a same chemical compound have all the same composition but different crystal structures, whereas different phases of a compound have both different compositions and crystal structures. *Stricto sensu*, calcium oxalates trihydrate, dihydrate, and monohydrate are three solid phases of calcium oxalate. They should not be called polymorphs. On the other hand, calcite, aragonite, and vaterite which are the rhombohedric, orthorhombic, and hexagonal varieties of calcium carbonate are real polymorphs.

From these definitions, the crystallization of a protein in different solutions, in the presence of different crystallization agents, gives rise not to real polymorphs but to different crystalline phases of the same protein. The crystallization agents, salts in general, belong to the crystal structure so that the phases of the same protein have different compositions. Despite these different compositions and for the sake of simplicity, it is accepted in the world of protein crystallization to call polymorphs crystal varieties which should be called phases of the same compound.

When the additive does not enter into the crystal structure, their main role is to stabilize metastable phases or polymorphs. Metastable phases form due to kinetic reasons and are favoured by high supersaturations. When several phases are possible in the same solution, each of them has its own solubility so that the solution can be supersaturated with respect to several phases at the same time. According to the Ostwald's rule of stages, the phase which first forms is not the most stable one, i.e. the less soluble one, but the phase lying nearest to the original state in free energy. In other words, nature prefers to follow a sequence of nucleations, growths, and phase transitions rather than using a high energy level to directly nucleate the most stable phase. The metastable phase later undergoes a phase transition as soon as nuclei of a more stable phase, i.e. a less soluble phase, occur. Several phases or polymorphs may temporarily coexist but all except one are subject to transformation. In most cases, the phase transition occurs by dissolution of the metastable phase and recrystallization into the stable one. It is called a solution-mediated phase transformation. If some impurities, or additives, strongly adsorbs on the crystals, the phase transition can be inhibited for a very long time. This explains the rather long metastability of some protein crystals.

6.3 Crystallization kinetics, impurities, and additives

Impurities adsorb on the terraces between the growth steps, along the steps or in the kinks, i.e. the growth sites. Depending on the energy of the bonds between impurity and adsorption sites, adsorption is more or less reversible. When growth proceeds, there is a competition between the kinetics of molecule incorporation and the kinetics of impurity adsorption and desorption. Accordingly, impurities hinder the crystallization processes so that nucleation and growth rates are sometimes drastically slowed down. Vekilov (47) has measured HEWL growth rates five to six times lower in the presence of impurity than in pure media at low supersaturation. When the impurity adsorption selectively takes place on a crystal face, the growth rate of this face is selectively reduced and its relative development rapidly increases at the expense of the development of the other faces. This induces the habit changes or influences the crystal quality. Lorber *et al.* (48) observed that the addition of ovalbumin or bovine serum albumin to pure HEWL is correlated with an increase of the proportion of twinned crystals. When the impurity adsorption takes place on all crystal faces, and is irreversible, i.e. without exchange with the surrounding solution, then growth is completely inhibited. Then the so-called growth cessation that often occurs with protein crystal is observed. The only way to overcome this difficulty is to drastically increase supersaturation; in some cases, new surface nuclei form and growth starts again. However, if the crystal surface is too energetically poisoned, three-dimensional nucleation becomes easier than growth. The solution deposits fresh crystals.

6.4 Impurity incorporation

For a long time, it was believed that the impurities which induce habit changes of crystals were incorporated into the crystals and that this incorporation was liable for the habit changes. Actually, impurity absorption can accompany the habit change but it is not its cause. This was demonstrated by growing negative crystals, i.e. holes, in solutions poisoned by impurities. A hole is first bored in a crystal. Then, all parts of the crystal but the hole, are coated by glue to prevent it from dissolution. The crystal is then immersed in a solution slightly undersaturated so that, by dissolution, the size of the hole increases while, at the same time, the hole becomes faceted. When the impurity provokes the occurrence of facets different from those which occur in pure solution, it is the proof that the habit change results from impurity adsorption and not from impurity incorporation. Impurities cannot be trapped in a crystal which is dissolving.

In the macromolecules field, impurity incorporation takes place especially when the molecule of the impurity resembles the molecule of the crystal. It was first observed in the case of small molecules (e.g. glutamic acid incorporated into asparagine monohydrate crystals) and later on in the case of biological macromolecules, contamination of turkey egg white lysozyme crystallizing solutions by HEWL (49). Pure materials are difficult to grow when impurity and crystal molecules are homologues.

References

1. Lafont, S., Veesler, S., Astier , J.-P., and Boistelle, R. (1994). *J. Cryst. Growth*, **143**, 249.
2. Abraham, F. F. (1974). *Homogeneous nucleation theory*. Academic Press, New York.
3. Zettlemoyer, A. C. (1969). *Nucleation*. Marcel Dekker, New York.
4. Katz, J. L. (1982). In *Interfacial aspects of phase transformations* (ed. B. Mutaftschiev), p. 261. D. Reidel, Dordrecht, The Netherlands.
5. Malkin, A. J., Kuznetsov Yu, G., Glantz, W., and McPherson, A. (1996). *J. Phys. Chem.*, **100**, 11736.

6. Israelachvili, J. N. (1991). *Intermolecular and surface forces*. Academic Press, London.
7. Tanford, C. (1961). In *Light scattering in physical chemistry of macromolecules*, p. 275. John Wiley & Sons, Inc., New York, London.
8. Pecora, R. (1985). *Dynamic light sacttering: application of photon correlation spectroscopy*. Plenum Press, New York and London.
9. Guinier, A. and Fournet, G. (1955). *Small-angle scattering of X-rays*. John Wiley, New York.
10. Tardieu, A. (1994). In *Neutron and synchrotron radiation for condenses matter studies* (ed. J. Baruchel, J. L. Hodeau, M. S. Lehmann, J. R. Regnard, and C. Schlenker), Vol. III, p. 145. Les Editions de Physique, Springer–Verlag.
11. Lindner, P. and Zemb, T. (1991). *Neutron X-ray and light scattering: introduction to an investigative tool for colloidal and polymeric systems*. North-Holland, Amsterdam.
12. Jacrot, B. and Zaccaï, G. (1981). *Biopolymers*, **20**, 2413.
13. Einsenberg, H. (1976). *Biological macromolecules and polyelectrolytes in solution*. Clarendon Press, Oxford.
14. Parsegian, V. A., Rand, R. P., Fuller, N. L., and Rau, D. C. (1986). In *Methods in enzymology*, (ed. L. Packer), Academic Press, London. Vol. 127, p. 400.
15. Broutin, I., Riès-Kautt, M., and Ducruix, A. (1995). *J. Appl. Cryst.*, **28**, 614.
16. Gripon, C., Legrand, L., Rosenman, I., Vidal, O., Robert , M.-C., and Boué, F. (1996). *C. R. Acad. Sci. Paris*, **323**, 215.
17. Kam, Z., Shore, H. B., and Feher, G. (1978). *J. Mol. Biol.*, **123**, 539.
18. Zulauf, M. and D'Arcy, A. (1992). *J. Cryst. Growth*, **122**, 102.
19. Ferré-d'Amaré, A. R., Zhou, K. Z., and Doudna, J. A. (1998). *J. Mol. Biol.*, **279**, 621.
20. Veesler, S., Marcq, S., Lafont, S., Astier, J.-P., and Boistelle, R. (1994). *Acta Cryst.*, **D50**, 355.
21. Muschol, M. and Rosenberger, F. (1996). *J. Cryst. Growth*, **167**, 738.
22. Georgalis, Y., Zouni, A., and Saenger, W. (1992). *J. Cryst. Growth*, **118**, 360.
23. Georgalis, Y., Zouni, A., Eberstein, W., and Saenger, W. (1993). *J. Cryst. Growth*, **126**, 245.
24. George, A. and Wilson, W. W. (1994). *Acta Cryst.*, **D50**, 361.
25. George, A., Chiang, Y., Guo, B., Arabshahi, A., Cai, Z., and Wilson, W. W. (1997). In *Methods in enzymology*, (ed. C. W. Carter and R. M. Sweet), Academic Press, London. Vol. 276, p. 100.
26. Bonneté, F., Malfois, M., Finet, J., Tardieu, A., Lafont, S., and Veesler, S. (1997). *Acta Cryst.*, **D53**, 438.
27. Koppel, D. E. (1972). *J. Chem. Phys.*, **57**, 4814.
28. Ducruix, A., Guilloteau, J.-P., Ries-kautt, M., and Tardieu, A. (1996). *J. Cryst. Growth*, **168**, 28.
29. Hartman, P. and Perdok, W. G. (1955). *Acta Cryst.*, **8**, 49.
30. Hartman, P. (1973). In *Crystal growth: an introduction* (ed. P. Hartman), p. 367. North-Holland, Amsterdam.
31. Hartman, P. (1982). *Geol. Mijnbouw*, **61**, 313.
32. Hartman, P. and Bennema, P. (1980). *J. Cryst. Growth*, **49**, 145.
33. Ohara, M. and Reid, R. C. (1973). *Modeling crystal growth rates from solution*. Prentice Hall, Englewood Cliffs, NJ.

34. Hillig, W. B. (1966). *Acta Metall.*, **14**, 1868.
35. Madsen, L. and Boistelle, R. (1979). *J. Cryst. Growth*, **46**, 681.
36. Burton, W. K., Cabrera, N., and Frank, F. C. (1951). *Phil. Trans. Roy. Soc.*, **243**, 299.
37. Chernov, A. A. (1961). *Sov. Phys. Usp.*, **4**, 116.
38. Gilmer, G. H., Ghez, R., and Cabrera, N. (1971). *J. Cryst. Growth*, **8**, 79.
39. Bennema, P. and Gilmer, G. H. (1973). In *Kinetics of crystal growth, an introduction* (ed. P. Hartman), p. 263. North-Holland, Amsterdam.
40. Nielsen, A. E. (1964). *Kinetics of precipitation*. Pergamon, Oxford.
41. Boistelle, R., Astier, J.-P., Marchis-Mouren, G., Desseaux, V., and Haser, R. (1992). *J. Cryst. Growth*, **123**, 109.
42. Vekilov, P. G., Ataka, M., and Katsura, T. (1993). *J. Cryst. Growth*, **130**, 317.
43. Boistelle, R. (1982). In *Interfacial aspects of phase transformation*, p. 621. Erice, Sicily.
44. Parker, R. L. (1970). In *Solid state physics*, Vol. 25, p. 151. Academic Press, New York.
45. Kern, R. (1968). *Bull. Soc. Fr. Mineral. Cristallogr.*, **91**, 247.
46. Giegé, R., Dock, A.-C., Kern, D., Lorber, B., Thierry, J.-C., and Moras, D. (1986). *J. Cryst. Growth*, **76**, 554.
47. Vekilov, P. G. (1993). *Prog. Cryst. Growth*, **26**, 25.
48. Lorber, B., Skouri, M., Munch, J.-P., and Giegé, R. (1993). *J. Cryst. Growth*, **128**, 1203.
49. Abergel, C., Nesa, P. M., and Fontecilla-Camps, J. C. (1991). *J. Cryst. Growth*, **110**, 11.

12

Two-dimensional crystallization of soluble proteins on planar lipid films

A. BRISSON, O. LAMBERT, and W. BERGSMA-SCHUTTER

1. Introduction

Electron crystallography of protein two-dimensional (2D) crystals constitutes a fast-expanding method for determining the structure of macromolecules at near-atomic resolution (1, 2). The main limitation in the application and generalization of this approach remains in obtaining highly ordered 2D crystals, as is the case of 3D crystals in X-ray crystallography.

Several methods of 2D crystallization are available which can be classified into two families, depending on the type of proteins under investigation, either membrane proteins (3, 4) or soluble proteins (5, 6). In both cases, 2D crystallization is a self-organization process which spontaneously occurs between macromolecules which are restricted to diffusing by translation and rotation in a 2D space, with a fixed orientation along the normal to this plane.

The scope of this chapter is restricted to the 2D crystallization of soluble proteins on planar lipid films, by the so-called 'lipid monlayer crystallization method' (5). Our aim is to present a step-by-step description of the experimental procedures involved in the application of this method.

2. Two-dimensional crystallization of soluble proteins on planar lipid films

The method of protein 2D crystallization on planar lipid films was introduced about 15 years ago (5) and has since been successfully applied to about 30 proteins (*Table 1*). Its principle is based on the specific interaction between soluble proteins and lipid ligands inserted in a lipid monolayer, at an air–water interface (*Figure 1*). In practice, a lipid monolayer is formed by spreading lipids dissolved in an organic solvent on a water surface. Proteins present in the aqueous subphase bind to their ligand of lipidic nature and spontaneously

Table 1. List of macromolecules crystallized on planar lipid layers

Protein	Ligand lipid	Ref.
	(a) Natural lipids	
Cholera toxin	G_{M1} ganglioside	1
Tetanus toxin	G_{T1} ganglioside	2
Botulinum toxin	G_{T1b} ganglioside	3
Staphylococcus α-toxin	platelets lipids	4
Annexin VI	14:0-PE / 18:1-PS	5
Annexin V	18:1-PS / brain extract	6
Coagulation factor Va	18:1-PS	7
Coagulation factor IX	18:1-PS	8
Protein kinase C	18:1-PS	9
	(b) Synthetic lipids made of a protein ligand coupled to a lipid molecule	
anti-DNP IgG	DNP-PE	10
Ribonucleotide reductase	dATP-PE	11
Streptavidin	biotin-PE	12
DNA gyrase B	novobiocin-PA	13
C-reactive protein	DS8PE	14
(His_6)-HIV-1 reverse transcriptase	Ni-NTA-18:1-PE	15
(His_6)-peptide-MHC	Ni-NTA-18:1-PE	16
(His_6)-HupR	Ni-NTA-DOGA	17
(His_6)-(M-MuLV) Nter-capsid protein	Ni-DHGN	18
RNA polymerase 1	Ni-NTA-DOGA	19
Streptavidin	Cu-DIODA	20
	(c) Charged lipids	
ferritin	eicosic-$N^+(CH_3)_3$	21
RNA polymerase (*E. coli*)	octadecylamine	22
RNA polymerase II, I (yeast)	octadecylamine / cetyl-$N^+(CH_3)_3$	23
α-actinin	DDMA	24
Mitochondrial creatine kinase	cardiolipin	25
50S ribosome	18:1-PS	26
Brush border myosin I	18:1-PS	27
Oestrogen receptor-LBD	14:0-PC	28
chaperonin		29

1) a) Ludwig *et al.* (1986) P.N.A.S., **83**, 8585; b) Mosser *et al.* (1992) *J. Mol. Biol.*, **226**, 23; 2) Robinson *et al.* (1998) *J. Mol. Biol.*, **200**, 367; 3) Schmid *et al.*, *Nature* (1993) **364**, 827; 4) Olofsson *et al.* (1990) *J. Mol. Biol.*, **214**, 299; 5) a) Newman *et al.* (1989) *J. Mol. Biol.*, **206**, 213; b) Benz *et al.* (1996) *J. Mol. Biol.*, **260**, 638; 6) Mosser *et al.* (1991) *J. Mol. Biol.*, **217**, 241; 7) Stoylova *et al.* (1994) *FEBS Lett.*, **351**, 330; 8) Stoylova *et al.*, *FEBS Lett.* (1998) **1383**, 175; 9) Owens *et al.* (1998) *J. Struct. Biol.*, **121**, 61; 10) a) Uzgiris, Kornberg (1983) *Nature*, **301**, 125; b) Uzgiris (1986) *Biochim. Biophys. Res. Comun.*, **134**, 819; 11) Ribi *et al.* (1987) *Biochemistry* **26**, 7974; 12) a) Blankenburg *et al.* (1989) *Biochemistry* **28**, 8214; b) Kubalek *et al.* (1991) *Ultramicroscopy*, **35**, 295; 13) Celia *et al.* (1994) *J. Mol. Biol.*, **236**, 618; 14) Sui *et al. FEBS Lett.* (1996) **388**, 103; 15) Kubalek *et al.*, *J. Struct. Biol.* (1994) **113**, 117; 16) Célia *et al.* (submitted); 17) Vénien-Bryan *et al.*, *J. Mol. Biol.* (1997) **274**, 687; 18) Barklis *et al.*, *EMBO J.* (1997).**16**, 1199; 19) Bischler *et al.* (1998) *Biophys. J.*, **74**, 1522; 20) Frey *et al.* (1998) *Biophys. J.*, **74**, 2674; 21) Fromhertz (1971) *Nature*, **231**, 267; 22) Darst *et al.* (1989) *Nature*, **340**, 730; 23) a) Darst *et al.* (1991) *Cell*, **66**, 1; b) Schultz *et al.* (1993) *EMBO J.*, **12**, 2601; 24) Taylor and Taylor (1993) *J. Mol. Biol.*, **230**, 196; 25) Schnyder *et al.*, *J. Struct. Biol.* (1994) **112**, 136; 26) Avila-Sakar *et al. J. Mol. Biol.* (1994) **239**, 689; 27) Célia *et al.*, *J. Struct. Biol.* (1996) **117**, 236; 28) 29) Ellis *et al.* (submitted).

Abbreviations: DDMA, didodecyldimethylammonium; DNP, dinitrophenyl; DHGN, dihexadecylglycero-; DOGA, dioleoylglyceroxyacetylamino; DS8PE, dioctadecanyl *N*-[*N'*-(aminoethyl phosphatoethyl) succinamido-*N*-yl]-aspartate inner salt; Ni-NTA, Ni-*N*-nitrilotriacetic acid-chelated nickel; PA, phosphatitic acid; PC, phosphatidylcholine; PE, phosphatidylethanolamine; PS, phosphatidylserine; 12:0, dilauroyl; 14:0, dimyristoyl; 18:1, dioleoyl.

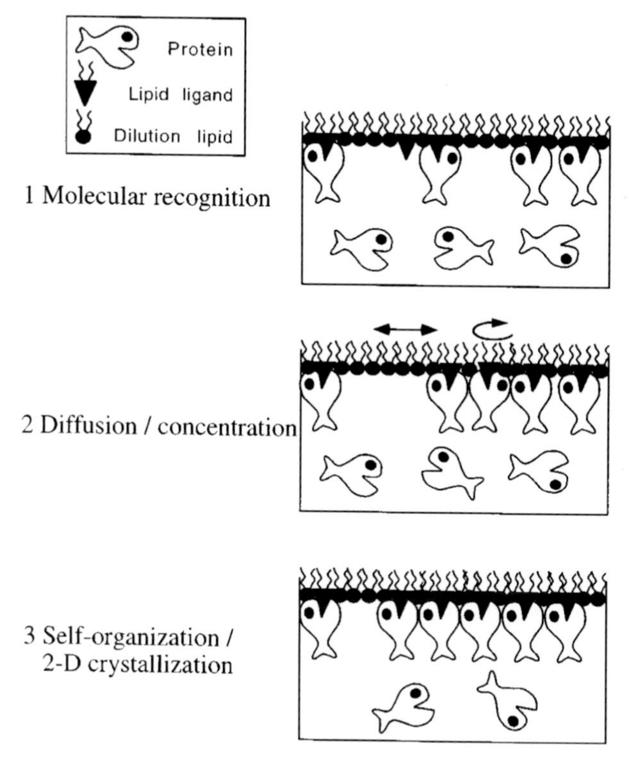

Figure 1. Scheme of the lipid-layer crystallization method. Lipids are deposited at an air–water interface and spread into a monomolecular layer. The monolayer presented here is made of two types of lipids, ligand lipids and dilution lipids. Proteins present in the subphase interact by molecular recognition with their ligand at the level of the planar layer (1). Protein–lipid complexes diffuse freely and concentrate in the plane of the monolayer (2). These complexes interact with each other and spontaneously assemble into 2D domains and 2D crystals, provided that favourable interactions are stabilized (3).

form 2D domains and, in favourable cases, 2D crystals. The process of 2D crystal formation relies on three successive steps:

(a) Molecular recognition between a protein and its ligand.
(b) Diffusion and concentration of the protein–lipid complexes in the plane of the lipid film.
(c) Self-organization of the proteins into 2D crystals.

As indicated in *Table 1*, three different types of systems can be distinguished, depending on the nature of the lipid ligand:

- natural lipids
- synthetic lipids made of a protein ligand coupled to a lipid molecule
- charged lipids.

Although most applications up to now have concerned proteins, this method has also been used to crystallize other types of macromolecules, as the 50S ribosomes on negatively charged lipids (7).

The main advantages of this method are:

(a) It is universal, as indicated by the wide diversity of proteins already crystallized.
(b) It is easy to apply and gives fast results (either positive or negative).
(c) Only small amounts of protein are required—about 1 μg per incubation.
(d) Crystallization conditions are gentle, as proteins can be maintained in a physiological buffer.
(e) 2D crystals can be ordered up to high resolution, around 3 Å (8–10), and thus amenable to high resolution structure determination by electron crystallography (11).

Several review articles have already been published on the lipid-layer crystallization method (12, 13). This chapter will focus on the experimental aspects involved in the application of this method, and will cover the following steps:

(a) Setting up a crystallization experiment.
(b) Transfer of protein–lipid films to an electron microscopy (EM) grid.
(c) Characterization of protein–lipid domains/crystals by EM. Only the first steps of the characterization will be described here, namely the preparation of negatively stained specimens and the analysis of electron micrographs by optical diffraction.

Examples taken from studies on annexin V and streptavidin will be used as illustrations.

3. Setting up a crystallization experiment

3.1 Lipid solutions

3.1.1 General considerations

Interfacial lipid films used for growing protein 2D crystals must fulfil both requirements of stability, as some experiments last for several days, and fluidity, allowing for the necessary diffusion of the protein–lipid complexes. Lipid molecules spread in monolayers at an air–water interface interact more or less strongly, depending on the chemical nature of the lipids, their concentration, surface pressure, temperature, or composition of the subphase. Three states are distinguished: gas, liquid expanded and condensed, or solid state, which reflect the strength of the intermolecular interactions (14, 15). Most, if not all, successful crystallization experiments reported until now have been performed with lipid layers in a liquid expanded or fluid state. This is the state of most unsaturated phospholipids, such as dioleoylphosphatidylcholine

(DOPC) or egg phosphatidylcholine (PC), at a temperature close to 20 °C. This is the reason why lipid solutions contain, in addition to the lipid ligands, a second lipid, often referred to as dilution lipid. DOPC or a mixture of unsaturated PC molecules are often used as dilution lipids. Mixtures with ligand lipid:dilution lipid molar ratios ranging from 1:3 to 1:6 are frequently used. It is highly recommended to characterize the physico-chemical properties of the monolayers formed by each new lipid mixture under study, by measuring surface pressure/area isotherms with a Langmuir balance or a Wilhelmy plate (15).

An illustrative example of the stabilizing role of dilution lipids is provided by the cholera toxin system (16). The monosialoganglioside G_{M1}, which is the natural lipid ligand of cholera toxin, has a tendency to form micelles in aqueous solutions and does not form stable monolayers. On the other hand, lipid films formed with G_{M1}/DOPC mixtures are stable and support the 2D crystallization of cholera toxin, over a wide range of G_{M1}/DOPC mixtures.

Most 2D crystallization experiments are performed in Teflon wells of small size (see Section 3.3). To saturate the non-specific lipid binding sites present at the Teflon surface, an excess of lipids is commonly used with respect to the amount needed to form one monolayer. Although there are no strict rules, a tenfold excess is often used in practice. As a first approximation, we can consider that lipid films formed with a lipid excess consist of a monolayer to which a reservoir of lipids is associated, probably in the form of small 3D aggregates adsorbed at Teflon edges. This hypothesis is supported by ellipsometry experiments which show no increase in film thickness after a monolayer is formed (17). Although it might be more correct to refer to planar lipid films instead of lipid monolayers, this is a rather semantic distinction.

3.1.2 Practical considerations

Particular attention must be given to the preparation and storage of lipid solutions. Most lipids can be dissolved in either chloroform or chloroform: methanol (2:1, v/v). Lipid mixtures used for spreading interfacial films often contain hexane in addition to these solvents, in a 1:1 (v/v) ratio. Lipid solutions must be stored in hermetic glass containers, such as Wheaton vials Z10 equipped with Teflon-faced rubber caps. Lipid solutions are stored at –20 °C. Pipettes or syringes made of either glass or metal must be used with lipid solutions. It is recommended to frequently control the level of the lipid solutions, as it is difficult to maintain tight seals and to prevent evaporation of organic solvents.

3.2 Protein solutions

The preparation of protein solutions is comparatively simple, as proteins can be used in their physiological buffer, as for other biochemical experiments.

Protein concentrations ranging from 50–200 μg/ml are used in practice. Concentrations as low as 10 μg/ml have been used with success (16).

Due to the known property of proteins to denature at air–water interfaces, it is often questioned whether it is better to inject proteins under a preformed lipid layer or to spread lipids over a protein solution. The few comparative studies performed until now have reported no significant differences between both procedures. For convenience, proteins are deposited first in the crystallization wells before spreading the lipids.

The affinity between proteins and lipid ligands is obviously a critical parameter for the success of crystallization. For several systems—streptavidin, cholera toxin, annexin V—apparent dissociation constants of the complexes are available, which in all three cases correspond to a very tight binding, with K_d ranging from 10^{-15} M to 10^{-9} M. However, high affinity is not sufficient to induce crystallization, as shown by the fact that only a few members of the annexin family form 2D crystals, while other members that bind negatively charged lipids with similar affinity ($K_d \sim 10^{-9}$ M) only form disordered 2D protein domains (Bergsma-Schutter and Brisson, unpublished results).

3.3 Preparation and cleaning of Teflon supports

Teflon supports are made of ordinary Teflon material. Discs of 60 mm diameter and 8 mm height are convenient in practice. Wells of 4 mm diameter and 1 mm depth are drilled at regular intervals in these discs; discs with 3 × 3 or 4 × 4 wells allow several crystallization experiments to be performed simultaneously. The diameter of each well—4 mm—is adapted to receive, for the transfer step, an EM grid with a standard diameter of 3 mm (see Section 4). Care must be given to the wells' edges, which must be devoid of irregularities, as they might constitute traps for lipids or affect the overall homogeneity of the planar lipid layer. To minimize possible problems due to uneven edges, it is recommended to select wells with sharp edges with a magnifying glass.

The state of hydrophobicity of the surface of Teflon wells plays an important role in crystallization experiments. The main goal of the cleaning step is to regenerate a highly hydrophobic Teflon surface, in order to prevent lipids overflowing over the Teflon surface.

Protocol 1. Cleaning of Teflon supports

Method

1. Handle Teflon discs with gloves.
2. Place a disc in a bath of 1% Hellmanex II for 2 h.
3. Rinse the disc extensively with warm tap-water for 1 h.
4. Rinse with several baths of deionized water.

5. Smash the disc against a piece of tissue paper lying on a table, to get rid of the water.[a] The disc is ready for use.[b]
6. For prolonged storage, keep the discs either in water or dry, in a box. Perform the treatment (steps 1–5) before use.

[a] A good indication of the state of hydrophobicity of a Teflon support is obtained by smashing it on a piece of tissue paper; the wells must look completely dry after two smashes. If some water remains at the bottom of the wells or at their edges, washing of the disc must be repeated. When discs have been used repeatedly, a second washing step might not be sufficient to obtain a hydrophobic surface. It is then recommended to brush the wells in order to eliminate lipids stuck onto their surface.
[b] Teflon surfaces tend to become electrostatic, which can lead to several surprising effects; a microscope grid may 'jump' upon deposition on the lipid-coated wells, or a lipid droplet approached from the Teflon surface may become deformed or even 'explode'. Although it is difficult to judge the influence of such a behaviour on crystallization, the best remedy is to discharge the surface with an antistatic device.

Protocol 2. Setting up a crystallization experiment, e.g. annexin V (see *Figure 2*)

Equipment and reagents

- A Hamilton syringe of 10 μl, with a 90° bevel, used for depositing lipids—rinse thoroughly with chloroform before and after use
- Protein solution: 100 μg/ml annexin V in 150 mM NaCl, 2 mM $CaCl_2$, 25 mM Hepes pH 7.5
- A Petri dish of 9 cm diameter, in which the Teflon disc is deposited, serves as a humid chamber—a small hole is made in the plastic lid to allow a gentle opening
- Lipid solution: 150 μM DOPS (dioleoylphosphatidylserine), 450 μM DOPC, in chloroform:hexane (1:1, v/v)

Method

1. Place a freshly conditioned Teflon disc in a Petri dish.
2. Poor water around the disc up to about mid-height.
3. Deposit in each well 17 μl of the protein solution.
4. Rinse the 10 μl syringe three times with chloroform.
5. Deposit 0.6 μl of the lipid solution[a] on top of each protein droplet.
6. Install the lid on the Petri dish and close the hole with a piece of tape, in order to limit evaporation.
7. Incubate.[b,c]

[a] Lipids must not flow over the edges. If this occurs, irreproducible results can be expected and it is wise to use another well or another Teflon disc.
[b] The incubation time required for crystal growth is variable and depends on the protein and lipid system. For example, one hour or less is sufficient to get 2D crystals of streptavidin or cholera toxin; in the case of annexin V, one hour is also sufficient to get one type of crystal, with p6 symmetry, while several days are required to get highly ordered p3 crystals (18).
[c] Most studies reported until now have been performed at ambient temperature, around 20 °C. It is of course important to use conditions in which the protein is stable.

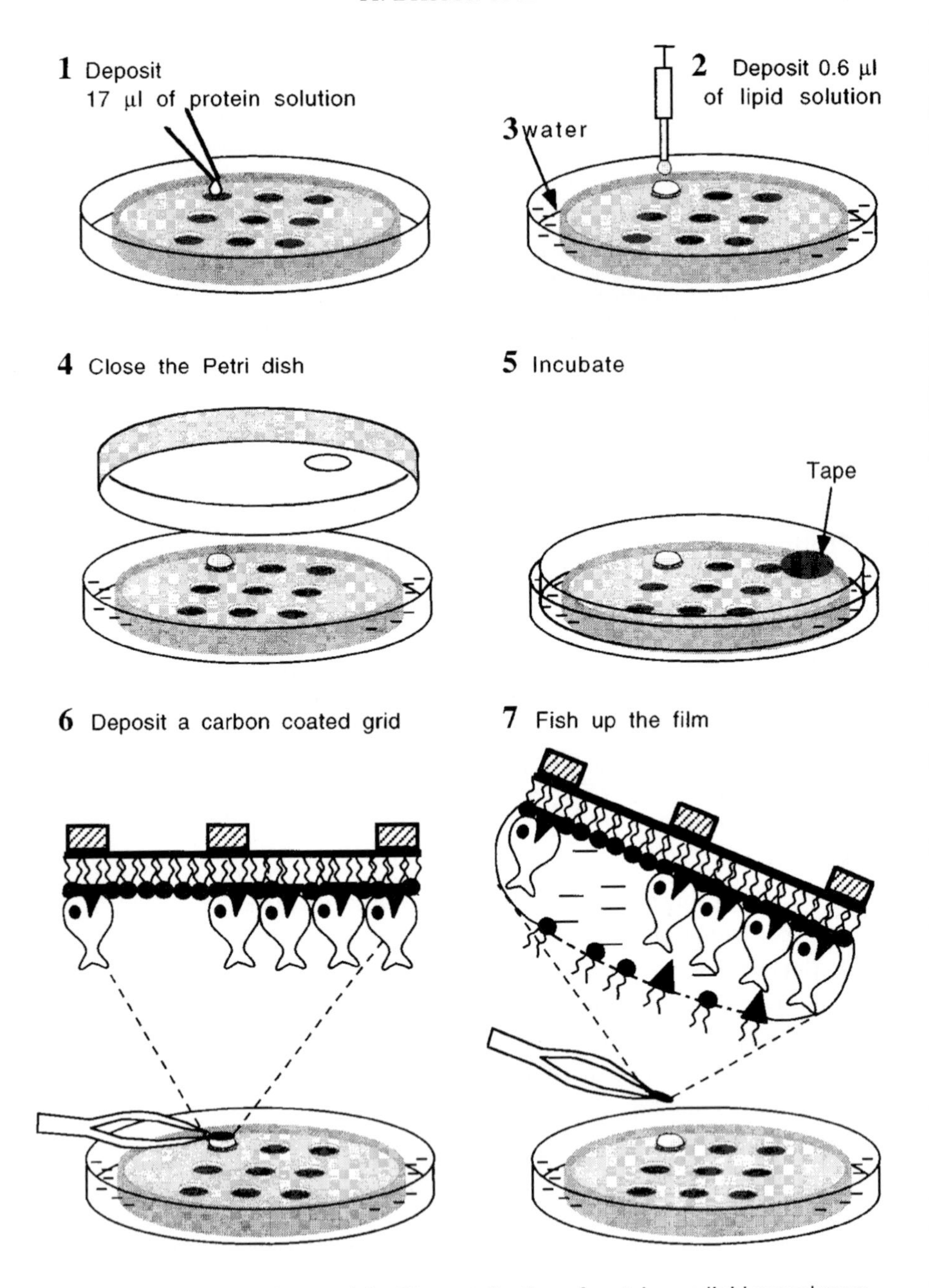

Figure 2. Step-by-step scheme of the 2D crystallization of proteins on lipid monolayers.

4. Transfer of protein–lipid films to an EM grid

4.1 General considerations

The characterization of 2D crystals of proteins, *in situ* at the air–water interface, presents technical difficulties due to the very nature of the specimen. Optical techniques, such as fluorescence microscopy (19), Brewster angle microscopy (20), or ellipsometry (17), as well as mechanical techniques (17), can provide valuable information on the formation of 2D interfacial domains. However, this information is limited to macroscopical features, due to the diffraction limit of light. EM remains the most commonly used method for characterizing and investigating the structure of 2D protein crystals formed by the lipid-layer method. A recent study has demonstrated the potential of atomic force microscopy to follow, *in situ* and in real time, the formation of 2D protein crystals on supported lipid bilayers (21). This study opens new possibilities in various areas and will certainly generate interest for investigating the process of crystal growth in 'live' conditions.

The transfer of protein–lipid films from the air–water interface to an EM grid is certainly the most important step in the whole procedure, as far as the influence of the experimenter is concerned. Most of the practical aspects presented below (*Protocols 3–7*) concern this step of transfer, from which depends the ultimate quality of the structural results.

4.2 Preparation of EM grids

Specimens observed by transmission electron microscopy are in general deposited onto grids coated with a support material. Carbon is the most popular support material because thin carbon films—several nm thick—are easy to prepare, transparent to electrons, mechanically stable, and conducting (22). Two main types of carbon films can be used for transferring protein–lipid films (*Figure 3*):

(a) Continuous carbon films (*Protocol 3*).

(b) Perforated carbon films, presenting holes and commonly named holey carbon films (*Protocols 4* and *5*).

Many methods of preparation of EM grids coated with carbon films have been reported (22). We will only describe here some standard methods which have given satisfactory results for the transfer of protein–lipid interfacial films.

4.2.1 Preparation of EM grids coated with a continuous carbon film

Three successive steps can be distinguished:

(a) Formation of a plastic film onto which grids are deposited (*Protocol 3*, steps 1–8).

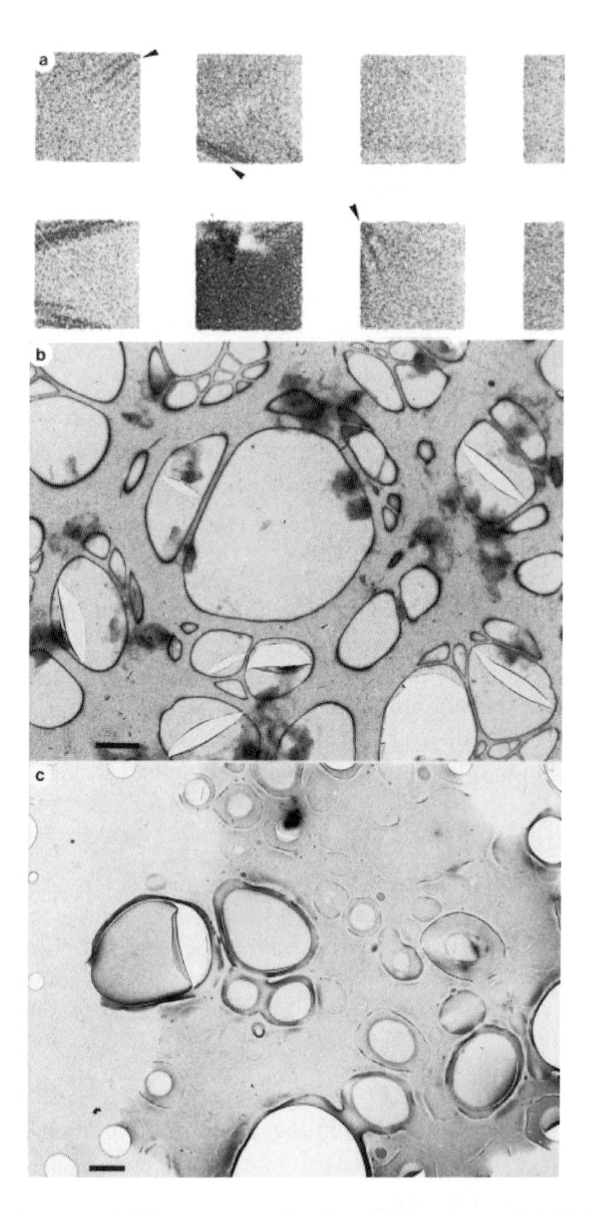

Figure 3. Overall aspect of different types of carbon film supports. (a) Grid coated with a continuous carbon film (adapted from ref. 23). This image was recorded by reflected light microscopy, which constitutes an appropriate method for visualizing the planarity of carbon films. The carbon film is mostly flat, except near some corners, where wrinkles are observed (arrowheads). Each square is 37 μm wide. (b) (c) EM images of holey films prepared by the methods described in *Protocol 4* and *5*, respectively. The films presented here have been used to transfer interfacial films of streptavidin (b) and annexin V (c). Most of the holes are covered with a continuous layer exhibiting a homogeneous greyness, indicating that the transferred material is of uniform thickness. Several holes present breaks. Note the almost complete absence of vesicles and multilayered domains, in comparison with *Figure 4*. Scale bars: (b) 1 μm; (c) 2 μm.

(b) Evaporation of carbon on the plastic film (*Protocol 3*, step 9).

(c) Dissolution of the plastic film (*Protocol 3*, steps 10 and 11).

Protocol 3. Preparation of EM grids coated with a continuous carbon film

Equipment and reagents

- EM grids (300–400 mesh): clean before use by washing them either in acetone, or successively in an 0.1 M H_2SO_4 solution, then water, and then acetone (30 sec each)
- A beaker (10 cm diameter) filled with about 5 cm water
- Glass slides: clean with alcohol and let them dry by placing them vertically on a filter paper
- 0.3% (w/v) nitrocellulose (collodion) solution in amyl acetate

Method

1. Deposit four droplets of the nitrocellulose solution on a clean glass slide, with a Pasteur pipette.
2. Form a continuous and homogeneous liquid film by tilting the slide.
3. Eliminate most of the liquid by holding the slide vertically against a piece of filter paper.
4. Let dry for 10 min.
5. Float off the plastic film on a water surface, by slowly inserting the slide at glancing angle into the water-bath.
6. Deposit EM grids on top of the plastic film.
7. Deposit a piece of absorbent paper or a piece of Parafilm on top of the plastic film covered with the grids, while maintaining the paper/ Parafilm by one edge. Wait until a good contact is formed with the underlying plastic film.
8. Lift up the paper/Parafilm and deposit it on a clean surface. Allow for a complete drying, under a lamp.
9. Evaporate a thin film of carbon on the nitrocellulose side, using standard EM procedures (22).[a]
10. Place the grids, carbon side up, on top of several pieces of filter paper soaked with amyl acetate, in a glass Petri dish. Close the dish and leave for several hours to overnight.
11. Transfer the grids on a dry filter paper.[b]

[a] Grids prepared up to here can be used for negative staining without removal of the plastic film. On the other hand, for cryo-microscopy experiments, it is mandatory to remove the plastic film in order to avoid artefacts.

[b] Carbon films prepared by this method are in general flat (*Figure 3a*) (23). The flatness of carbon films is certainly an important parameter for allowing a good contact with the lipid chains and thus achieving an efficient transfer.

4.2.2 Preparation of EM grids coated with a holey carbon film

The fabrication of holey films has received considerable attention from electron microscopists. Several methods have been developed and more or less modified in many laboratories for particular needs. We describe here two methods (*Protocols 4* and *5*) used routinely in our laboratory, which give two types of films differing by the nature, the size, and the distribution of holes.

Protocol 4. Preparation of EM grids coated with a holey carbon film[a]

Reagents

- 0.3% (w/v) formvar solution in dichloroethane

A. *Formation of a plastic film with 'pseudo'-holes*

1. Bring to the boil a 500 ml beaker filled up with water.
2. Dip a glass slide into a 50 ml beaker filled with the formvar solution for about 10 sec.
3. Pull it gently out of the solution.
4. Keep it vertically against a filter paper to drain most of the liquid.
5. When the slide is still wet, place it horizontally in the stream of water vapour.
6. Take the slide away from the stream when the film at its surface becomes milky. Let it dry.
7. Check the aspect of the network with a light microscope, and mark the good areas[b,c] with a needle.

B. *Formation of holes in the plastic film*

1. Scratch with a needle the borders of the good areas.
2. Float off the plastic film on water, deposit EM grids, and pick up the whole as in *Protocol 3*, steps 5–9.
3. Place a filter paper in a Petri dish and soak it with acetone.
4. Place few (approx. five) grids on a support made of a thin wire mesh, standing 5 mm above the bottom of the Petri dish.
5. Leave the grids in the acetone vapour for 60 sec.
6. Check the aspect of the lacy network with a light microscope, and repeat part B, step 5 by varying the length of the etching time until a convenient aspect is observed.[d]
7. Repeat the procedure for the other grids.
8. Evaporate a thick layer of carbon (as in *Protocol 3*, step 9).[e]

C. *Dissolution of the plastic film*

1. Dissolve the formvar film by depositing the grids on top of several pieces of filter paper soaked with dichloroethane, in a glass Petri dish.
2. Let stand overnight.
3. Transfer the grids on a dry paper.[f]

[a] This method is adapted from Sjöstrand (24).
[b] The selection step is crucial. The network present on the glass slide must be of the 'good' size, as an open network might be too fragile while a dense network might result in too small holes.
[c] As this stage, there is still a thin layer of plastic covering the 'pseudo'-holes.
[d] The time of etching has to be optimized. It is recommended to start with a time of 1 min, which often gives good results, and with a small number (approx. five) of grids.
[e] Due to the lacy structure of the holey film, it is recommended to evaporate a thick layer of carbon, to improve the mechanical stability of the film and to ensure good conductivity.
[f] It is recommended to check grids before use for EM.

Protocol 5. Preparation of EM grids coated with a holey carbon film[a]

Equipment and reagents

- A metal block pre-cooled in a freezer at −20°C
- 0.25% (w/v) cellulose acetate butyrate solution in ethyl acetate (triafol)

A. *Formation of a plastic film with holes*

1. Clean glass slides as in *Protocol 3*.
2. Plunge the slides in a bath of 0.1% (w/v) Tween 20.
3. Let them dry vertically against a filter paper.
4. Take the metal block out of the freezer and deposit a glass slide on it.
5. When moisture appears at the surface of the slide—this takes few seconds—pour a few droplets of the cellulose acetate butyrate solution on the slide.[b]
6. Eliminate the excess of solution by blotting with a filter paper.
7. Let the slide dry vertically.
8. Check the quality of the holey film with an optical microscope.
9. Float off the thin holey plastic film on a water-bath, deposit EM grids onto it, and evaporate a thin carbon layer, following the procedure described in *Protocol 3*, steps 5–10.

B. *Dissolution of the plastic film*

1. Remove the plastic film by placing the grids over a filter paper soaked with chloroform or ethyl acetate, in a closed glass chamber saturated with vapour.

Protocol 5. *Continued*

2. Check each grid with an optical microscope for homogeneity and integrity of the holey film.[c,d]

[a] This method has been adapted from Fukami and Adachi (25) by Chrétien *et al.* (26) (adapted from ref. 26 with permission).
[b] It is possible to adjust the size of the holes at this step. A long exposition on the metal block will produce larger holes. Several trials are necessary to obtain holes with the desired size.
[c] With holey films prepared by this method, about half of the surface is covered with holes and the other half with carbon (*Figure 3c*).
[d] The main advantage of this method, as compared with most other methods of fabrication of holey films, is that floating of the plastic film is easy and reproducible.

4.2.3 Transfer of interfacial films to EM grids

Interfacial films are transferred to EM grids by depositing an EM grid coated with a carbon film on top of a crystallization well. EM grids are deposited with the carbon film facing the lipid tails. This method of transfer is often referred to as the Schaeffer method (15) (*Protocol 6*).

Another method of transfer has recently been proposed, which makes use of a wire loop to pick up the protein–lipid film and to transfer it to an EM grid, of either type mentioned before (27). This method has only been applied to a limited number of specimens and will not be further described here.

Protocol 6. Transfer of a protein–lipid interfacial film to an EM grid: 'fishing' step

Method

1. Hold a carbon coated grid with tweezers, and deposit it horizontally, carbon side facing down, on top of the lipid film.[a]
2. Wait 2–5 min.
3. Lift up the grid with the tweezers. A thin layer of water covering the carbon surface[b] must be visible.
4. The grid is now ready for the specimen preparation step, which must be performed immediately (see *Protocol 7*).

[a] Observe carefully the deposition of the grid on the film. When 2D crystals are present at the interface, the grid 'sticks' to the film upon deposition. When no crystals are present at the interface, as for example with pure fluid lipid layers, the grid moves and rotates freely.
[b] As a large excess of lipid is present at the interface, a new lipid film will spontaneously cover the water surface after the first film has been picked up. The well is in principle ready for a second cycle of crystallization. It must be noted that results obtained with a 'second' or even a 'multiple' fishing are more variable, which may be due to the formation of heterogeneous lipid layers or to some overflowing of lipids on the Teflon surface.

5. Characterization of protein 2D crystals by EM

The observation of biological specimens by EM requires the use of specific methods of preparation, the role of which is to protect these specimens against dehydration and electron radiation. In the case of 2D crystals of proteins, as in general for observations at the molecular level, two preparation methods are best adapted; the classical negative staining method (*Protocol 7*) and the more complex cryo-methods, applied to unstained specimens. For each new study, the initial steps are performed by negative staining, which provides an efficient way for screening various conditions of crystallization. The resolution achievable by negative staining is however limited to ~ 10–20 Å and the structural information is restricted to the molecular envelope. Therefore, further steps of the analysis, abutting ideally to the determination of the molecular structure at high resolution (28), must be performed by cryo-EM on unstained specimens. This latter part is outside the scope of this chapter.

Protocol 7. Negative staining of protein 2D crystals

Equipment and reagents

- 1% (w/v) uranyl acetate solution in water, pH ~ 3.5

Method

1. Immediately after the fishing step (*Protocol 6*, step 4), add a 5 μl droplet of the uranyl acetate solution to the liquid film present on the grid.[a]
2. Wait for 30 sec.
3. Remove the excess liquid by touching a grid border with a filter paper.
4. Allow for complete drying.
5. In the case of holey films, evaporate a thin layer of carbon on the side of the protein–lipid film.[b]
6. The grid is ready for observation in the microscope.

[a] At the beginning of each new study, it is recommended to compare the results obtained with several negative stains. Sodium phosphotungstate (2% (w/v) aqueous solution, pH 7) is another commonly used negative stain.
[b] The deposition of a carbon layer enhances the mechanical stability and improves the conductivity of self-supported interfacial films.

EM of negatively stained specimens provides two types of 'low resolution' structural information:

1. At low magnification (× 2000), the overall aspect of the material transferred on the EM grid is visible. The nature and the aspect of the transferred material depend highly on the type of carbon films used (*Figures 4–6*). As a

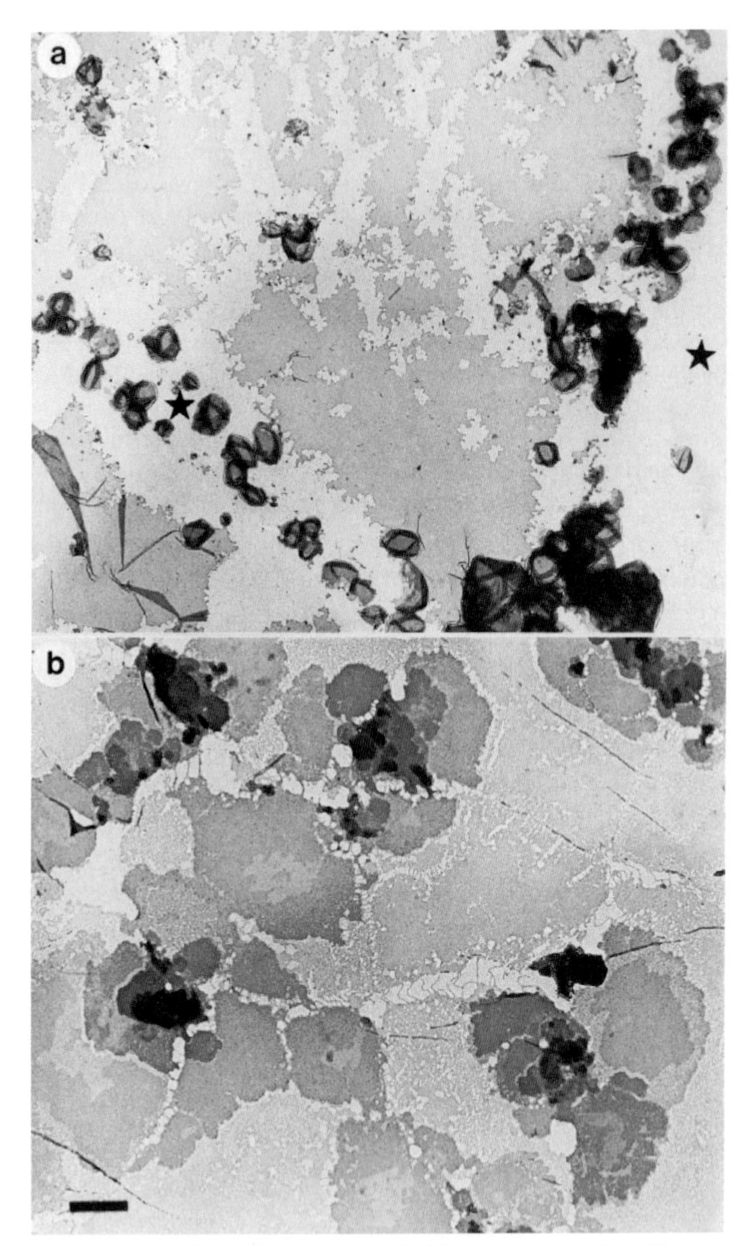

Figure 4. Aspect of interfacial films transferred with continuous carbon films. (a) and (b) correspond to annexin V and streptavidin, respectively. (a) Domains exhibiting a uniform greyness and vesicles cover the carbon film support. The vesicles surround the domain areas and most probably form during the transfer step. Extensive carbon film areas are devoid of domains and vesicles (*). (b) Domains showing multilayered structures are characteristic of films transferred with continuous carbon supports. On these low magnification images (scale bar: 1 μm), the crystalline nature of the domains is not visible.

rule that undoubtedly has exceptions, interfacial films picked up with a continuous carbon film present domains easily distinguishable from the carbon background, together with vesicular material (*Figure 4*). These domains are often folded or overlap in multilayered structures, and present morphologies characteristic of each protein–lipid system. On the other hand, interfacial films transferred with a holey carbon film appear as homogeneous layers of uniform greyness and thus uniform thickness, mostly devoid of domains or large vesicles (*Figures 3b*, *3c*, *5*, and *6*). It is now commonly accepted that the interfacial films are transferred without, or with minor, reorganization when holey films are used (8, 18, 29), while they are submitted to profound reorganization upon transfer/drying with continuous carbon films (13, 18, 30). The annexin V system constitutes an extreme case in this context, as p6 crystals are obtained with holey films, while p3 crystals are observed with continuous carbon films (18). Most strikingly, these p3 crystals do not pre-exist at the air–water interface and their formation is induced by the transfer step. Cholera toxin constitutes another interesting case as highly ordered 2D crystals are obtained after transfer with continuous carbon films while close-packed 2D domains are obtained with holey films, and thus pre-exist at the air–water interface (18). The coherent picture which emerges from these studies is that upon specific binding to ligands incorporated into lipid monolayers at the air–water interface, some proteins form 2D crystals, while many others self-organize in close-packed assemblies. Upon transfer with a continuous carbon film, these close-packed assemblies are 'stressed' and may reorganize into more compact and better ordered 2D crystals.

2. At high magnification (× 50 000), the crystalline nature of the transferred material can be visualized. However, it is in general not possible to get a quantitative evaluation of the crystalline order by a mere 'eye' observation. Even when strongly contrasted stain striations are observed, this information is of low resolution as it represents most often the accumulation of stain between molecules (see for example *Figure 5*). The most objective way to evaluate the crystalline quality of protein–lipid interfacial films is by optical diffraction or Fourier transform calculation (see Section 6).

The characterization of interfacial films by EM is one of the most time-consuming steps in the whole procedure. The main reason is the huge number of areas of potential interest on each grid and the variability of aspect existing between grids and also within different areas of one given grid. It is important to consider that:

(a) In the case of holey films, holes covered by either 2D crystals or close-packed assemblies present the same aspect, and there are of the order of 10^4 to 10^6 holes per grid.

(b) With continuous carbon films, the number of domains of potential interest present on a grid is even larger.

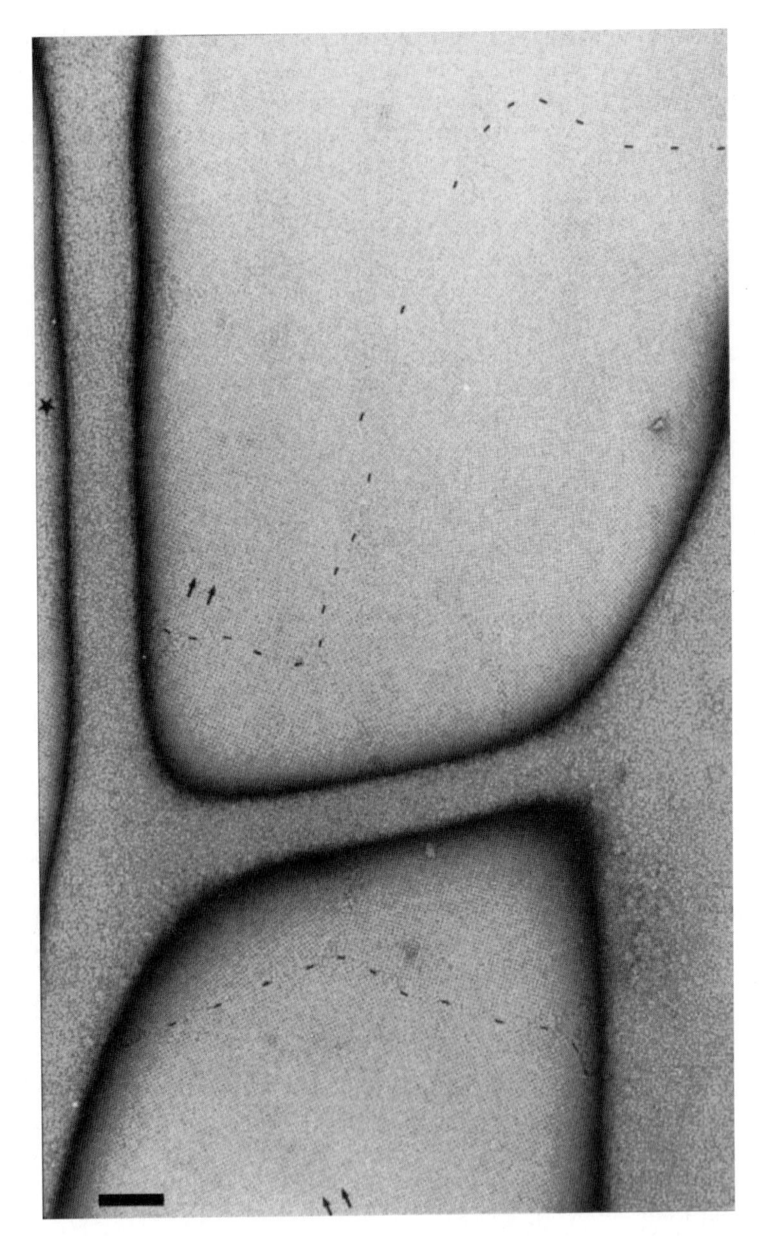

Figure 5. Interfacial film of streptavidin transferred with a holey carbon film. The streptavidin film consists of a mosaic of crystalline domains. Two large crystals, one in each hole, are indicated by two arrows aligned along the main directions of stain striations. Their frontiers with adjacent 2D crystals are delineated with dashed lines. Next to the carbon threads, the streptavidin film is often disordered (*), suggesting that the crystalline structure is disorganized when the carbon film touches the interfacial film or during the 'fishing step'. Scale bar: 1000 Å.

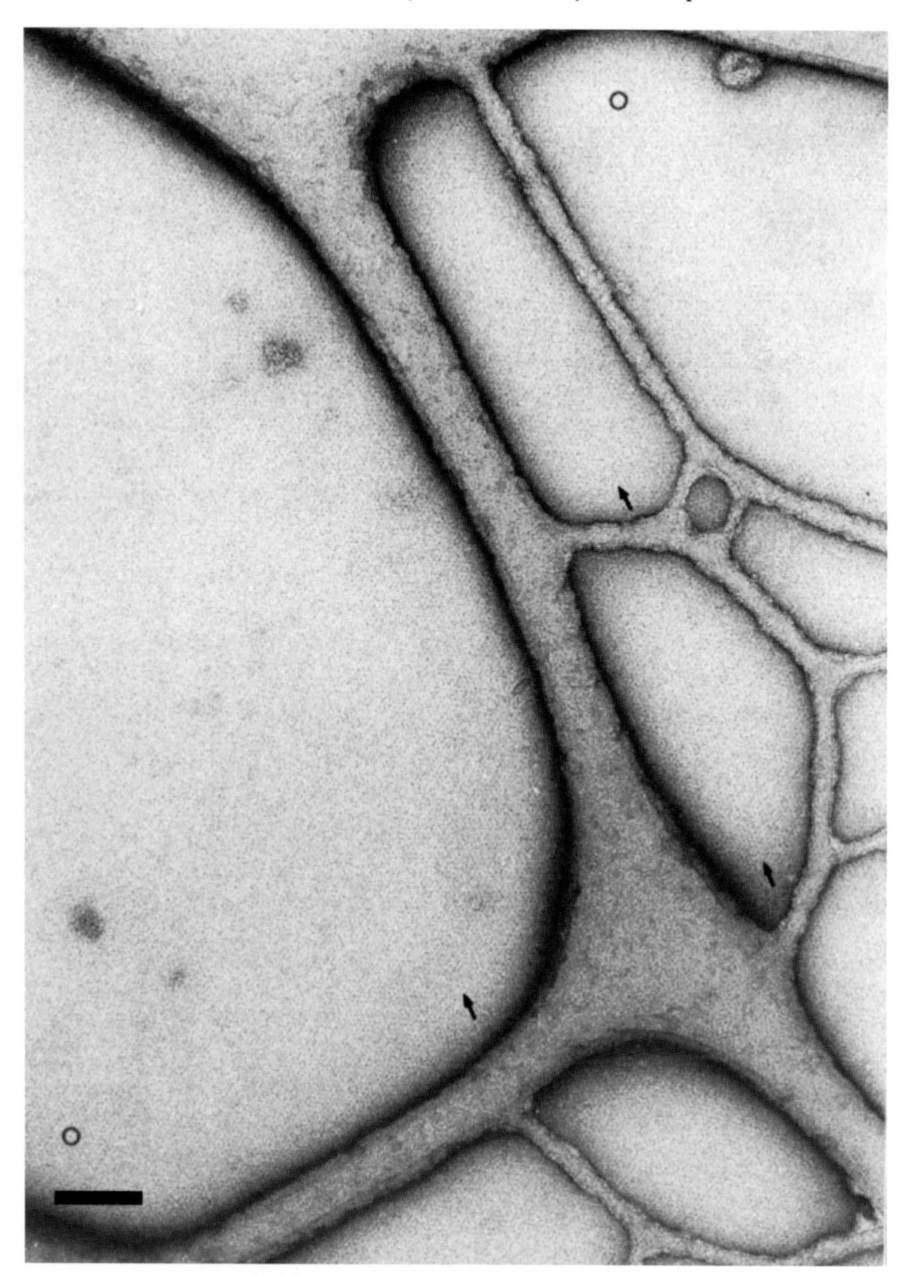

Figure 6. Interfacial film of annexin V transferred with a holey carbon film. Single crystalline domains of annexin V cover the holes. The main orientation of the lattice (arrows) is almost conserved between adjacent holes, suggesting that the interfacial film was a single monocrystal before transfer. The crystal is built up of trimers of annexin V (circles) assembled with p6 symmetry. Scale bar: 1000 Å.

(c) Crystalline domains of 1 μm^2 are large enough to provide high resolution information and the surface of a grid is equivalent to about 10^6 such domains. This explains why screening EM grids for the presence of crystals and optimizing crystallization conditions are extremely time-consuming.

6. Characterization of the protein–lipid crystals by optical diffraction

As mentioned above, the most standard method for characterizing the crystallinity of a specimen is by optical diffraction (*Figure 7*). An optical diffraction set-up consists merely of a laser source, a convergent lens, and an observation screen placed in the diffraction plane of the lens. When an EM negative is placed along the optical system, either before or after the lens, its diffraction image, or diffraction pattern, is displayed on the screen. An optical diffraction bench provides a simple and fast way of calculating on-line Fourier transforms. Optical diffraction gives access to:

(a) The resolution of the crystalline order, from the position of the peaks farthest away from the direct unscattered beam.
(b) The overall crystalline quality, from the sharpness of the diffraction peaks and the completeness of the information.
(c) The optical conditions, like focusing and astigmatism, from the position and shape of the rings of the contrast transfer function.

7. Conclusion

The lipid-layer crystallization method is a rational and general method for growing highly ordered 2D crystals of macromolecules. Until now, it has been

---

Figure 7. Optical diffraction. (a) Scheme of an optical diffraction set-up. The principal components of an optical diffraction set-up are: a laser source, a convergent lens, and a screen placed in the diffraction plane of the lens. In the set-up presented here, the lens, placed at a distance p from the laser source, is illuminated by a non-parallel beam. The rays emerging from the lens converge at the diffraction plane, located at a distance p′ from the lens, such as: $1/p + 1/p' = 1/f$ (f: focal length of the lens). The advantage of using a non-parallel beam illumination is that the size of the diffraction pattern from the EM negative can be easily adjusted by changing the distances between the lens and the source and/or between the EM negative and the lens. Sub-areas of the negatives are evaluated for their crystalline quality and selected for further processing. According to the diffraction theory, a periodic grating of period d, illuminated by a coherent beam of wavelength λ, gives rise to two diffracted beams forming an angle θ with the direct beam, such as: $d \sin \theta = \lambda$. On the screen placed at a distance L from the negative, two diffraction peaks will be observed, located at a distance D from the centre, such as: $D = L\ \mathrm{tg}\ \theta$. As θ is small, $dD = L\lambda = \mathrm{cst}$. (b) Example of a diffraction pattern of a negatively stained 2D crystalline domain of annexin V (adapted from ref. 35). The diffraction peaks are arranged onto a hexagonal lattice. The (0,6) and (6,2) reflections, at 1/13.4 and 1/11.2 $Å^{-1}$, respectively, are circled. Scale: 1 cm = 0.028 $Å^{-1}$.

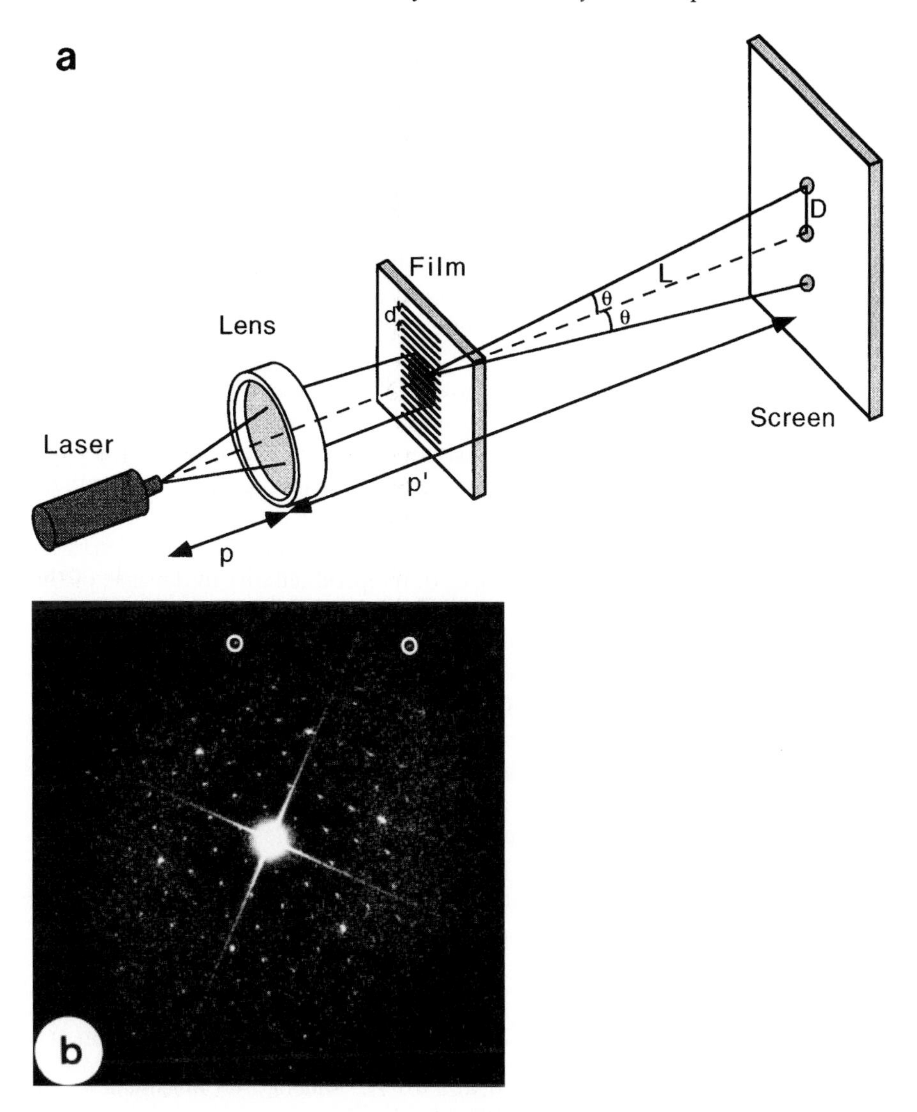

applied almost exclusively to soluble proteins. Its application to the field of membrane proteins is confronted to the presence of detergents required to maintain membrane proteins in a soluble form. These detergents are likely to intercalate and possibly alter the integrity of the planar lipid films. However, one can expect that solutions to this problem will be found, as for example by combining a rapid elimination of the detergent molecules and/or the use of lipids in a solid state during the initial step of binding.

An extension of the lipid-layer crystallization method has recently been developed in our group with the helical crystallization of proteins by specific interaction with lipids forming tubular aggregates (31, 32). In this work, biotinylated lipids were synthesized which self-assembled as tubules in aqueous solutions. Binding of streptavidin molecules to the biotin head groups was followed by their self-organization into helical arrays at the tubular surface. This approach is of particular interest in electron crystallography as the complete 3D structure of macromolecules arranged with helical symmetry can be retrieved by image analysis of individual images of tubular crystals. A variant of this novel approach has been proposed with the incorporation of Ni^{+}-chelating lipids into lipid tubules and the helical crystallization of polyhistidine-containing proteins (33). The use of 2D crystals formed on planar lipid layers as seeds for epitaxial growth of 3D crystals constitutes another potential application of the lipid-layer crystallization method (34).

The main problem associated with the application of the lipid-layer crystallization method is the difficulty in obtaining reproducible results. This problem is also encountered with other crystallization methods, and is due to a lack of understanding and control of some of the parameters involved in crystallization. We believe that the main source of irreproducibility in the case of the lipid-layer crystallization method originates at the transfer step, particularly when continuous carbon films are used for transfer: the presence and number of crystalline domains, their size, or the extent of vesicular material are almost unpredictable. This is certainly related to the lack of planarity of EM grids and/or carbon films; the lipid layer is almost atomically flat, while the surface of the carbon film is far from ideally flat (23). It is therefore highly recommended to use holey films, at least as a control, with which more reproducible results are obtained. On the other hand, the examples of annexin V and cholera toxin demonstrate that the formation of highly ordered crystals can be induced during the transfer step with continuous carbon films (18). Therefore, the variability of the results must not hide the main interest of this method of crystallization: it works!

References

1. Kimura, Y., Vassylyev, D. G., Miyazawa, A., Kidera, A., Matsushima, M., Mitsuoka, K., *et al.* (1997). *Nature*, **389**, 206.
2. Nogales, E., Wolf, S. G., and Downing, K. H. (1998). *Nature*, **391**, 199.
3. Kühlbrandt, W. (1992). *Q. Rev. Biophys.*, **25**, 1.
4. Jap, B. K., Zulauf, M., Scheybani, T., Hefti, A., Baumeister, W., Aebi, U., *et al.* (1992). *Ultramicroscopy*, **46**, 45.
5. Uzgiris, E. E. and Kornberg, R. D. (1983). *Nature*, **301**, 125.
6. Harris, J. R. (1992). *Microsc. Anal.*, 13.
7. Avila-Sakar, A. J., Guan, T. L., Arad, T., Schmid, M. F., Loke, T. W., Yonath, A., *et al.* (1994). *J. Mol. Biol.*, **239**, 689.
8. Kubalek, E. W., Kornberg, R. D., and Darst, S. A. (1991). *Ultramicroscopy*, **35**, 295.

9. Mosser, G., Mallouh, V., and Brisson, A. (1992). *J. Mol. Biol.*, **226**, 23.
10. Celia, H., Hoermann, L., Schultz, P., Lebeau, L., Mallouh, V., Wigley, D. B., *et al.* (1994). *J. Mol. Biol.*, **236**, 618.
11. Avila-Sakar, A. J. and Chiu, W. (1996). *Biophys. J.*, **70**, 57.
12. Kornberg, R. D. and Darst, S. A. (1991). *Curr. Opin. Struct. Biol.*, **1**, 642.
13. Brisson, A., Olofsson, A., Ringler, P., Schmutz, M., and Stoylova, S. (1994). *Biol. Cell*, **80**, 221.
14. Gaines, G. L., Jr. (1966). *Insoluble monolayers at liquid-gas interphases.* Wiley, New York.
15. Roberts, G. (ed.) (1990). *Langmuir-Blodgett films.* Plenum Press, New York.
16. Mosser, G. and Brisson, A. (1991). *J. Struct. Biol.*, **106**, 191.
17. Vénien-Bryan, C., Lenne, P.-F., Zakri, C., Renault, A., Brisson, A., Legrand, J.-F., *et al.* (1998). *Biophys. J.*, **74**, 2649.
18. Brisson, A., Bergsma-Schutter, W., Oling, F., Lambert, O., and Reviakine, I. (1999). *J. Cryst. Growth*, **196**, 456.
19. Blankenburg, R., Meller, P., Ringsdorf, H., and Salesse, C. (1989). *Biochemistry*, **28**, 8214.
20. Frey, W., Schief, W. R., and Vogel, V. (1996). *Langmuir*, **12**, 1312.
21. Reviakine, I., Bergsma-Schutter, W., and Brisson, A. (1998) *J. Struct. Biol.*, **121**, 356.
22. Baumeister, W. and Hahn, M. (1978). In *Principles and techniques of electron microscopy: biological applications* (ed. M. A. Hayat), Vol. 8, p. 1. Van Nostrand Reinhold Co., New York.
23. Schmutz, M., Lang, J., Graff, S., and Brisson, A. (1994). *J. Struct. Biol.*, **112**, 252.
24. Sjöstrand, F. S. (1956). In *Stockholm Conf. Electron Microscopy*, Proc. 20.
25. Fukami, A. and Adachi, K. (1965). *J. Electron Microsc.*, **14**, 112.
26. Chrétien, D., Fuller, S. D., and Karsenti, E. (1995). *J. Cell Biol.*, **129**, 1311.
27. Asturias, F. J. and Kornberg, R. D. (1995). *J. Struct. Biol.*, **114**, 60.
28. Henderson, R., Baldwin, J. M., Ceska, T. A., Zemlin, F., Beckmann, E., and Downing, K. H. (1990). *J. Mol. Biol.*, **213**, 899.
29. Voges, D., Berendes, R., Burger, A., Demange, P., Baumeister, W., and Huber, R. (1994). *J. Mol. Biol.*, **238**, 199.
30. Darst, S. A., Ahlers, M., Meller, P. H., Kubalek, E. W., Blankenburg, R., Ribi, H. O., *et al.* (1991). *Biophys. J.*, **59**, 387.
31. Ringler, P., Müller, W., Ringsdorf, H., and Brisson, A. (1997). *Chem. Eur. J.*, **3**, 620.
32. Huetz, P., van Neuren, S., Ringler, P., Kremer, F., van Breemen, J. F. L., Wagenaar, A., *et al.* (1997). *Chem. Phys. Lipids*, **89**, 15.
33. Wilson-Kubalek, E. M., Brown, R. E., Celia, H., and Millagan, R. A. (1998). *Proc. Natl. Acad. Sci. USA*, **95**, 8040.
34. Edwards, A. M., Darst, S. A., Hemming, S. A., Li, Y., and Kornberg, R. D. (1994). *Nature Struct. Biol.*, **1**, 195.
35. Olofsson, A., Mallouh, V., and Brisson, A. (1994). *J. Struct. Biol.*, **113**, 199.

13

Soaking techniques

E. A. STURA and T. GLEICHMANN

1. Introduction

Once crystals of a macromolecule are obtained there are many circumstances where it is necessary to change the environment in which the macromolecule is bathed. Such changes include the addition of inhibitors, activators, substrates, products, cryo-protectants, and heavy atoms to the bathing solution to achieve their binding to the macromolecule, which may have sufficient freedom to undergo some conformational changes in response to these effectors. In fact, macromolecular crystals have typically a high solvent content which ranges from 27–95% (1, 2). Although, part of this solvent, 'bound solvent' (typically 10%) is tightly associated with the protein matrix consisting of both water molecules and other ions that occupy well defined positions in refined crystal structure it can be replaced in soaking experiments, at a slower rate compared to the 'free solvent'.

In this chapter we will consider the relative merits of various methods for modifying crystals, the restraints that the lattice may impose on the macromolecule, and the relative merits of soaking compared to co-crystallization.

1.1 The crystal lattice

The size and configuration of the channels within the lattice of macromolecular crystals will determine the maximum size of the solute molecules that may diffuse in. The solvent channels are sufficiently large to allow for the diffusion of most small molecules to any part of the surface of the macromolecule accessible in solution except for the regions involved in crystal contacts, although in some cases lattice forces may hinder conformational changes or rearrangements of the macromolecule in crystal. In other cases, the forces that drive the conformational changes can be sufficient to overcome the constraints imposed by the crystalline lattice leading to the disruption of intermolecular and crystal contacts resulting in the cracking and dissolution of the crystals. Some lattices may be more flexible and capable of accommodating conformational changes, and while crystals may crack initially, they may subsequently anneal into a new rearrangement and occasionally improve their crystallinity.

In general small changes are easily accommodated and many macromolecules maintain their activity in the crystalline state. This is exploited in time-resolved crystallography to obtain structural information of transition states of enzymes.

1.2 Reasons for soaking

The major use of soaking is for the introduction of heavy atom substances into crystals for the determination of phases in the techniques of single (SIR) and multiple isomorphous replacement (MIR) which is needed for the determination of macromolecular crystal structures that lack a model suitable for molecular replacement (MR). In SIR and MIR, phase information is obtained by analysing the changes in the intensity of reflections as a result of derivatization with heavy atom containing reactants. The magnitude of the changes depends on the number of electrons in the 'derivative' relative to the 'native' protein. Changes in intensity may also be the result of changes in the unit cell parameters of the crystal. These latter changes, which are referred to as non-isomorphous changes, are undesirable and decrease the resolution to which the intensity differences can be used in phase determination. In soaking we try to minimize such changes while maximizing the incorporation of the heavy-atom into the crystal lattice.

Another method for the determination of phases is multiple anomalous dispersion (MAD). This method obviates the need for heavy-atom derivatization by incorporating anomalous scattering atoms in the crystal and by collecting data sets at three or four different wavelengths to make the best use of the anomalous dispersion. Because data are collected from crystals grown under identical conditions, in some cases the same crystal, this method does not suffer from lack of isomorphism. However, because the anomalous differences are considerably smaller, more accurate data are needed. This is accomplished using longer exposures, and to avoid differences from one crystal and another, data are collected from the same crystal at a synchrotron source. Such experiments are carried out with crystals flash-frozen at close to liquid nitrogen temperatures so that the diffraction is preserved for the longer periods of time needed to collect complete data sets at each of the wavelengths on and on either side of the absorption edge. To avoid loss of crystallinity the solvent used in the crystallization must be exchanged for a cryo-solvent before the crystals are flash-frozen. This is done by soaking or dipping them in an appropriate solution. Because crystals maintain their diffraction for longer periods of time at cryogenic temperatures, this method is extensively utilized to collect high resolution data from crystals of proteins irrespective of the method used for the phasing of their structure, and is becoming common for data collection at synchrotron facilities.

Soaking may also be the method of choice for the determination of ligand binding sites in proteins although co-crystallization is a better alternative when conformational changes occur as a result of ligand binding.

1.3 Soaking of crystals versus co-crystallization

Complexes of macromolecules can be obtained either by co-crystallizing directly from solution or by soaking preformed crystals of the macromolecule in a ligand or reactant solution. Both methods have their own advantages and disadvantages. Soaking of crystals of a macromolecule whose structure has been determined in that crystal form, reduces the complexity of the crystallographic problem to that of determining the positions of the newly introduced atoms by difference Fourier methods. However it may not be possible to ensure that 'true' binding occurs as rearrangements in the macromolecule may be inhibited by the crystal lattice. In co-crystallization, the formation of the complex does not have to contend with lattice forces, but the solubility and the conformation of the complex may be sufficiently different from that of the native molecule that new crystallization conditions may need to be determined. Co-crystallization of complexes can however yield crystals which are totally isomorphous with the uncomplexed protein crystals. This can be encouraged through the use of seeding (Chapter 7). However, the soaking of relatively hydrophobic ligands in aqueous solution may present problems. Such ligands have poor solubility in aqueous solutions. In these cases, a large volume of soaking solution is used so that a one- to five-fold stoichiometric ratio is achieved when the number of molecules in the solution are integrated over the entire volume. Long soak times are commonly used. To increase the water solubility of such ligands, and achieve higher ligand concentrations, organic compounds which are miscible with water are used (*Protocol 1*).

Protocol 1. Preparation of solution for soaking hydrophobic ligands

A crystal suitable for data collection will contain from 5–100 μg of protein, a proportionate amount of ligand must be present in the soaking solution so that a stoichiometric or higher concentration is achieved in an appropriate solution compatible with crystal stability. The exact molarity or stoichiometric ratio required will depend on the affinity of the compound and is likely to vary from case to case.

Equipment and reagents

- Microbalance
- Ligand
- Various solvents
- Crystallization tray
- Spot plate or capillary
- Microscope

Method

1. Measure 1–10 mg of ligand. Make a saturated solution of the ligand in a suitable solvent (see *Table 1*) by adding solvent to the ligand until fully dissolved. Calculate the molarity of the solution obtained.

Protocol 1. *Continued*

2. Test the solubility by adding 1 μl of the saturated solution to 1 ml of the same buffer solution used for the protein. Continue adding until the solution becomes opalescent. The molarity of the resulting solution can be calculated.
3. If the solubility is in the millimolar range, soaking can be done directly in the drops or capillaries (*Protocol 5*). If the solubility is less than 0.2 mM a soaking volume of 300 μl or greater will be needed and spot plates or vials should be used (*Figure 1*).
4. Mix the appropriate volume of ligand saturated solution with the precipitant used in the crystallization experiment so that the desired molarity or so that ligand–protein stoichiometry will be achieved.
5. Test that the precipitant–solvent mixture to be used for soaking is compatible with the crystal. For volatile solvents, just replace the reservoir solution with the precipitant–solvent mixture and allow the crystal to equilibrate with the reservoir by vapour diffusion. Check for cracks and if possible test that crystals equilibrated in such a manner still diffract. Later exchange the mother liquor in the drop with this solution and repeat checks.

An alternative to soaking of hydrophobic compounds is co-crystallization. This presents similar problems. In general it is advisable to mix the ligand with the protein at low protein, low ligand concentrations and subsequently concentrate and purify the complex.

Other situations in which co-crystallization may be needed is to achieve binding of large multi-metal clusters used in the phasing of large macromolecular assemblies ($> 10^6$ Da) as such clusters may be too large to diffuse through the crystal lattice. Since the channels between such assemblies are likely to be proportionately large, soaking experiments should also be attempted. See Thygesen *et al.* (3) for a review of the usage of multi-metal clusters in phasing large assemblies.

1.4 Soaking techniques

While in some cases it is possible to transfer crystals directly from the mother liquor from which they are grown to a fresh soak solution, more gradual changes in the crystal soaking, ending with the desired soak solution, can be effective in slowly annealing the crystal into the new conditions. The number of steps to the final soaking conditions varies from crystal to crystal. Typically, in the first stage the free solvent is replaced without major disruption, and in later stages, pH is changed and ligands are soaked-in or exchanged. The time needed for diffusion of ligands will vary from crystal to crystal. Intuitively, crystals with large channels and high solvent content will equilibrate faster.

Table 1. Organic solvents, additives, and cryo-protectants[a]

Additive	Concentration	Usage
Ethanol	5–20%	Solubilization (steroids) additive for crystallization
Methanol	5–15%	Solubilization (phospholipids combined with MPD)
Hexafluoropropanol	1–5%	Solubilization (very versatile, peptides and mimetics, steroids, etc.)
2-Propanol	5–20%	Solubilization (steroids), (cryo-protectant at > 70% best in combination with others), additive in crystallization
Glycerol	15–45%	Cryo-protectant and additive for crystallization
DMSO	2–20%	Solubilization of ligands and cryo-protectant
Ethylene glycol	15–45%	Cryo-protectant
PEG 200–600	35–50%	Cryo-protectant, precipitant
Sucrose	> 50% (w/v)	Cryo-protectant, best in combination with others
MPD	0.5–55%	Solubilization (phospholipids combined with methanol) Additive for crystallization, cryo-protectant, precipitant
Erythritol	5–35%	Cryo-protectant, best in combination with others
Xylitol	5–35%	Cryo-protectant, best in combination with others
Inositol	5–35%	Cryo-protectant, best in combination with others
Raffinose	5–35%	Cryo-protectant, best in combination with others
Trehalose	5–35%	Cryo-protectant, best in combination with others
Glucose	5–35%	Cryo-protectant, best in combination with others
L-2,3-Butanediol	15–45%	Cryo-protectant (levo isomer, racemic mixture also useful)
Propylene glycol	15–45%	Similar to ethylene glycol

[a] Typical concentrations and usage for organic solvents and additives. As it is suggested throughout the table, combinations of these compounds can be more effective to solubilize ligands and less likely to be incompatible with the crystals. Several organic compounds are suitable both as additives to crystallization set-ups as for use as cryo-solvents. For example the effect of slow equilibration of MPD onto crystals of the multisubstrate adduct complex of glycinamide ribonucleotide transformylase, which under room temperature conditions diffracted to only 2.0 Å was to extend the resolution to 1.96 Å at cryogenic temperature collected with a conventional X-ray source (35). The improvement in resolution may have been due to MPD rather than cryo-cooling.

Experimentally, by soaking the crystals in suitable dyes (for example, the mercury containing dye merbromin and rose Bengal containing iodine, as used by the authors) the time needed for soaking can be determined by observing the crystals to becoming coloured. In practice, 20 minutes to one hour between steps is a good starting point. Longer intervals, and repeated soaks should be used when exchanging one ligand for another. For fragile crystals, it is preferable to add further ligand to the soaking solution than to transfer crystals between solutions. By adding volatile organic solvents or further salt to the reservoir such solvent can be introduced into the crystals or the salt concentration increased to stabilize crystals before soaking. This latter step is necessary if it is known from crystallization trials that the protein–ligand complex has a higher solubility than the protein alone. Flow cells,

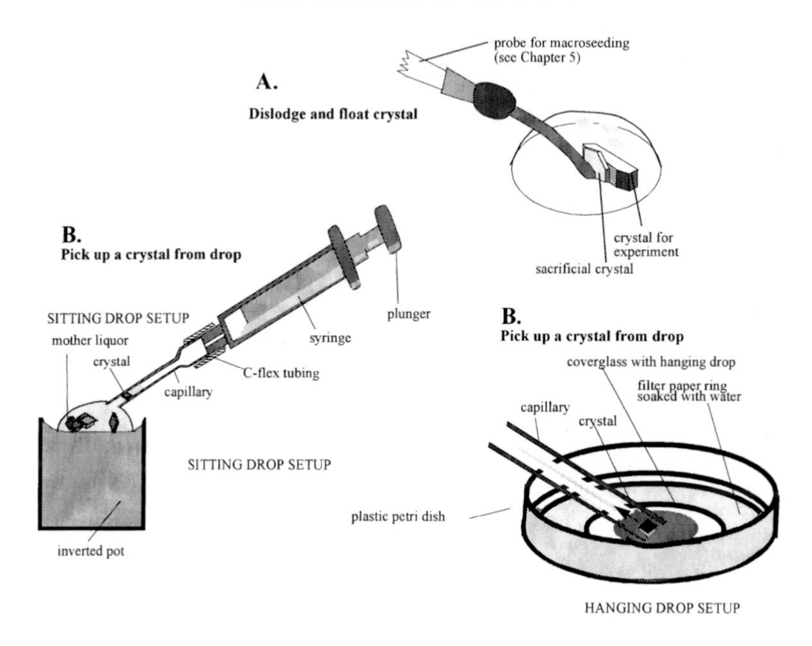

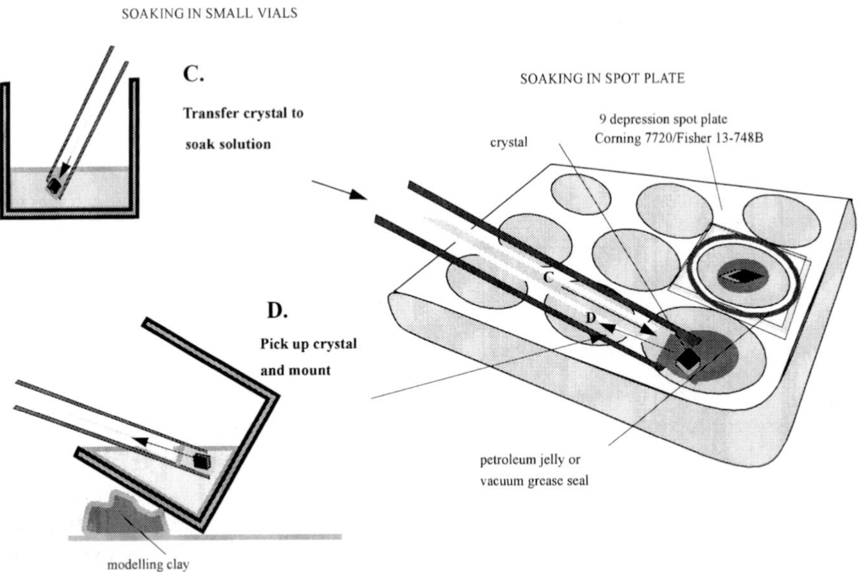

which have been used to change the mother liquor in which the crystals are bathed, such as for the introduction of a substrate, are well described elsewhere (4, 5), pressure cells are used for the incorporation of krypton and xenon into crystals.

Buffers are exchanged in order to change the pH, to analyse pH-induced changes, to favour heavy-atom or drug binding, or to avoid conditions of

Figure 1. Schematic drawing of the transfer of crystals from either a sitting drop set-up, or a hanging drop set-up, to a well for soaking. (A) A crystal is dislodged from the drop using a probe. If the crystal adheres firmly to the glass (or plastic) use a sacrificial crystal placed between the probe and the crystal to push against. The crystal is then floated to the surface to easily be picked up as in *Figure 3* or in (B). (B) Crystals can be transferred directly from a sitting drop to a capillary because the vapour from the reservoir solution (not shown) protects the drop containing the crystal from dehydration. For a hanging drop vapour diffusion experiment the coverslip is placed at the centre of a plastic Petri dish within a ring of filter paper soaked in water. Evaporation of water from the filter paper will ensure that the drop does not dry out while the crystal is picked up into the capillary connected to the syringe. (C) The crystal is transferred into a small vial or a well in a spot plate containing the soak solution. (D) After soaking the crystal is removed from the soak solution for mounting. The walls of the vial used for soaking the crystal should not be high, as this will restrict the angle at which the crystal can be picked up from the soak solution into the capillary, as further restrictions are also imposed by the dissecting microscope, which also limits the working angle. For soak volumes below 1 ml a spot plate may be preferable.

incompatibility between certain heavy-atoms and the crystallization buffer. For example, ammonium sulfate is a poor mother liquor for heavy-atoms binding at pHs above 6 because of the production of NH_3 which acts as a nucleophile. Ammonium sulfate can be replaced by sodium and potassium phosphate, except when uranium and rare earth compounds are used as heavy-atoms as these form insoluble phosphates. Soaking crystals in a cryo-protectant is necessary to favour the formation of vitreous rather than crystalline ice when crystals are flash-frozen. Since the solvent within the crystals does contribute to the diffraction at low resolution it is advisable to collect a new 'native' data set for these crystals after the buffer change to differentiate between those changes induced by the buffer, and those caused either by heavy-atoms, cryo-temperature, or other modifications to the crystals.

2. Soaking of substrates, activators, and inhibitors

Many enzymes remain catalytically active in the crystal (e.g. see refs 6 and 7) and the soaking of substrates may yield a mixture of substrates and products. In fact, with the use of synchrotron radiation, the conversion of heptenitol to heptulose-2-phosphate in the presence of inorganic phosphate, was followed crystallographically (8). Soaking is the method of choice for the determination of the binding sites for analogues, inhibitors, activators, substrates, and products, since it is often difficult to determine the binding of substrates, activators, and inhibitors to crystals when reliable phases for the native crystals have been obtained. Co-crystallization is advisable until then, as a means of confirming the result obtained by soaking, and when soaking fails. When soaking crystals, the development of hair-line cracks can be a good indication of ligand binding. Control experiments in which the native crystals are handled in the same fashion with the same buffer should be performed.

2.1 Soaking techniques for crystals in drops

When sitting drops are used for crystallization (see Chapter 5 and ref. 9), soaking experiments can follow *Protocol 2* with the use of a syringe or *Protocol 4* using a loop. The protocol for sitting drops is easier than that used for hanging drops since the evaporation from the reservoir solution is able to slow down sufficiently the evaporation from the drop containing the crystals during the procedure.

Protocol 2. Soaking of crystals grown in sitting drops (see *Figure 1*)

This work is carried out under a dissecting microscope using a magnification of × 10 to × 100.

Equipment and reagents

- Dissecting microscope
- Glass syringe
- C-flex tubing
- Glass capillary
- Tweezers
- Whisker
- Spot plate or vial
- Modelling clay

Method

1. Connect a glass or quartz capillary tube to a 1 ml glass syringe with a short piece of rubber tubing such as c-flex (Fisher, 14-169-5c) which gives an excellent seal.
2. Snap open the end of the capillary with tweezers or scissors. The glass capillary may be siliconized if the experimental situation can benefit from a diminished adhesion of the solution to the glass wall such as when viscous solutions are handled. After siliconizing it should be extensively washed.
3. To increase the volume available for handling crystals, mother liquor (20–50 μl) can be added to the drop (some crystals require the mother liquor to contain protein for stability). If the crystals adhere to the well, withdraw liquid from the drop and gently eject it onto a chosen crystal. Check that the desired crystal moves in the flow.
4. If flushing with liquid fails to dislodge the crystals the probes for streak seeding (Chapter 5) are used for this purpose. Select a thick whisker with a sharp point or cut a new point if needed. Run the point around the contour of the crystal, this will detach the crystal from precipitated or denatured protein in the depression. Now push gently on the crystal with the wide side of the whisker. Slowly apply pressure and watch for movement. Should the crystal show signs of breaking up or cracking, select a smaller crystal that can be sacrificed and utilize it as shown in

Figure 1. Using a loop to dislodge the crystal is another option (*Protocol 4*). Unfortunately, some crystals have severe adhesion problems and cannot be dislodged without breaking them. For such problematic crystals, glass pots with depressions, or microbridges, coated with a thin film of Corning vacuum silicone grease should be used in the original crystallization set-up.

5. Pick up crystals into the capillary by pulling back the plunger of the syringe.
6. Transfer crystals from the syringe directly into a soak solution from which they are later picked up and mounted for X-ray studies.

Crystals grown by hanging drop (see Chapter 5) can be flushed with mother liquor from the coverslip into a larger container and then picked up as described in *Protocol 2* for sitting drops. Since it can be difficult to find small crystals in a large container, the method described in *Protocol 3* may be preferable.

Protocol 3. Soaking crystals grown by hanging drops

Equipment and reagents

- Petri dish
- Filter paper
- Distilled water
- Microscope
- Syringe with capillary
- Mother liquor

Method

1. Cut a circular piece of filter paper to fit a Petri dish 4–5 cm inside diameter.
2. Cut out a small circle from the centre of the filter paper such that the coverglass from the hanging drop can fit inside this without touching the paper.
3. Soak the filter paper with distilled water.
4. Place the coverglass with the hanging drop in the centre. Mother liquor is added to the drop (20–50 μl) and the crystals for soaking can be picked up and soaked as described in *Protocol 2*, steps 2–6.

This set-up has been used for the stable transportation of crystals to synchrotron facilities, by soaking the filter paper with mother liquor instead of water.

Instead of a capillary and a syringe a loop can be used for the handling of crystals as described in *Protocol 4*. This method is widely used for soaking crystals in cryo-solvents.

Protocol 4. Handling of crystals using loops (see *Figure 3*)

Equipment and reagents

- Cryo-loop
- Microscope
- Tweezers to open
- Crystallization setup

Method

1. Follow *Protocol 2,* steps 1 and 2 if it is necessary to increase the volume of the drop. Select a loop with a diameter about 1.5 times the maximum size of the crystal. Loops can be made with individual fibres from plain dental floss or can be purchased pre-made from Hampton Research.
2. Clean the loop in methanol and wash with water. If the crystal is stuck it can be dislodged as in *Protocol 2,* steps 3 and 4 or by gently pushing with the loop. Tease the crystal to the top of the drop. When close to the surface of the drop, place the loop under the crystal and lift it out of the drop. Keep the loop only slightly above the drop, to avoid it drying out and focus the microscope on the loop to ensure that the crystal is in the loop.
3. Transfer the crystal to the soaking well as rapidly as possible. Drying out of the solvent around the crystal is the main disadvantage of the method.

2.2 Soaking of crystals in capillaries

Once the crystals have been introduced into the thin glass capillary, using either of the two above described procedures, the mother liquor can be removed by allowing the crystal to adhere to the capillary wall and pushing the mother liquor out of the capillary onto a piece of absorbent paper, while the crystal remains *in situ* because of surface tension. Crystals that do not adhere to the capillary wall can be stopped from flowing with the mother liquor by wedging a hair against the crystal while the solution is removed (*Figure 2*).

Alternatively the solution can be removed with a thin glass capillary tube (0.1–0.05 mm outside diameter) or a thin strip of filter paper. Once the mother liquor is removed, with the syringe still connected to the capillary tube, soaking is performed following *Protocol 5*.

This technique is particularly important for soaking compounds which are available only in limited quantities. After data collection, capillaries may be opened and a soak solution added to the crystal with a Hamilton syringe and the capillary sealed for the duration of the soak with paraffin oil. The oil and

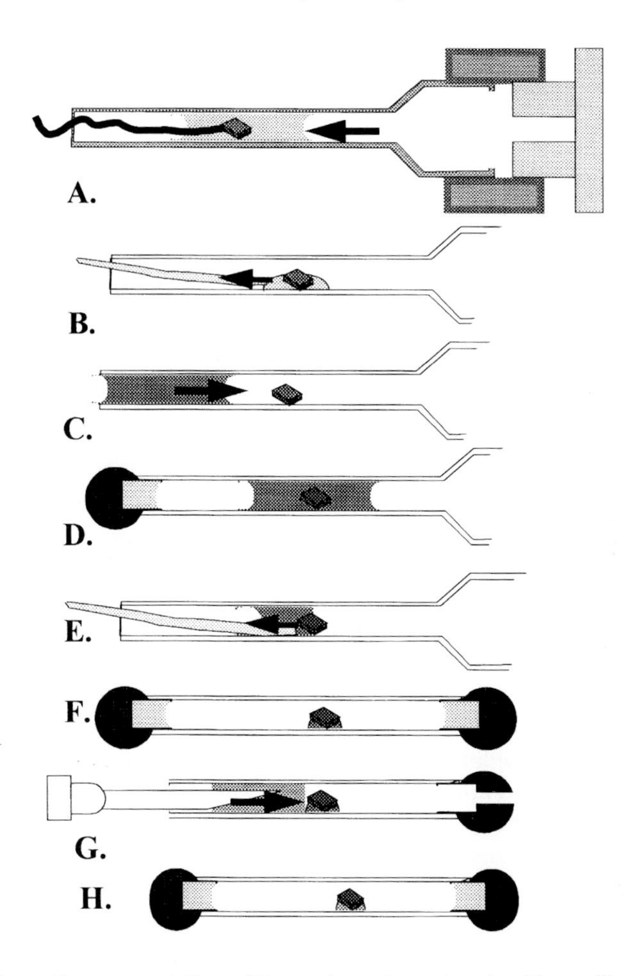

Figure 2. Schematic representation of the various stages involved in capillary soaks. (A) After the crystal is picked up into the capillary (*Figure 1A*) the mother liquor is removed from around the crystal by pushing the liquid out, while holding the crystal in position with a hair (in many cases the surface tension between the crystal and the capillary is sufficient to hold the crystal in place). (B) With a thin strip of filter paper, taking care not to touch the crystal the excess liquid is removed. (C) The soak solution is drawn into the capillary to bathe the crystal. (D) Paraffin oil or buffer is added to the open end and the capillary is sealed with molten wax. (E) After soaking the capillary can be snapped open with thin-nosed forceps and the soak solution is removed using a thin piece of filter paper. (F) The capillary can now be sealed at both ends and the crystal used for X-ray diffraction studies. (G) Crystals that have been used for X-ray work can be soaked by snapping of one end of the sealed capillary with forceps and opening the other end with a hot needle. A piece of wet filter paper is placed over the crystal to prevent the crystal from warming up during this procedure. A solution is then introduced at the broken end of the capillary so that it bathes the crystal. Petroleum jelly is used to seal the experiment as it is easy to remove prior to resealing the thin walled capillary tube with wax. (H) The soaked crystal can be used for X-ray diffraction analysis.

the soak solution are then removed and the crystal used for further X-ray studies such as for collecting an inhibitor complex data set after the native protein data have been measured, if the crystal has survived the damage from the first irradiation.

Protocol 5. Soaking crystals in capillaries (*Figure 2*)

Equipment and reagents

- Syringe with capillary
- Forceps
- Paraffin oil
- Microscope
- Wax
- Filter paper
- Syringe with needle

Method

1. Suck the new solution into the capillary fully immersing the crystal.
2. Add paraffin oil to the open end of the capillary, leaving an air gap between the oil and the soak solution, for the duration of the soak.
3. After the soak period has elapsed the oil and the solution are removed.
4. Remove the excess solution around the crystal with filter paper.
5. Add either soak solution and or oil to the open end of the capillary to maintain a moist environment for the crystal.
6. Seal with wax while still attached to the syringe. A wet strip of filter paper (5 mm wide) can be placed on the outside of the capillary to keep the crystal cool while the ends are sealed. The crystal is now mounted for X-ray diffraction work. Other techniques for mounting crystals can be found in Chapter 14 and elsewhere (10).

2.3 Soaking of crystals in dilute ligand solutions

There are many situations in which it is desirable to soak crystals in diluted solution. High concentration of heavy-atom substances can cause non-isomorphous changes, while prolonged soaking at low concentrations is well tolerated, some ligands are poorly in aqueous solutions or may need to be introduced slowly into the crystals for the crystals to be able to anneal to the changes. *Protocol* 6 gives some suggestions to increase the success with this method.

Protocol 6. Soaking of crystals in dilute ligand solution

Equipment and reagents

- Microscope
- Vacuum grease
- Depression plate or vial
- Syringe with capillary for crystal transfer
- Glass cover to fit depression well or vial

Method

1. Follow *Protocol 2* using a large depression plate (*Figure 1*) or large volume in a soak vial.
2. Soak several crystals for 20 min to several days. Harvest one crystal at a time, mount in a capillary, and test to determine whether the ligand or heavy-atom has bound to the crystal.
3. Remove old soak solution leaving the remaining crystals in the vial or depression plate and add fresh solution. Continue testing the crystals and replacing or adding more ligand or heavy-atom solution.

2.4 Cross-linking of crystals

The use of glutaraldehyde to stabilize crystals is well known from the early work on carboxypeptidase A (11), where it was shown that cross-linked crystals were resilient to changes in mother liquor. The cell dimensions of the cross-linked crystals were shown to remain relatively constant, under a variety of low and high salt conditions as well as extremes of pH, from 5–11. Such cross-linked crystals also retained catalytic activity. Further examples, such as the cross-linking of phosphorylase *a*, with 0.03% glutaraldehyde for 1 h, also indicated that the reagent produces little change in the diffraction pattern of the cross-linked crystals to a resolution of 5.5 Å, while maintaining crystal integrity even after major conformational changes (7).

Most cross-linking reagents link between the ϵ-amino groups of lysine residues (12, 13). Bifunctional diimidates of variable length provide a means of restricting the length of the cross-links from 3.7 Å for dimethyl malonic diimidate to 14.5 Å for dimethyl dodecanoic diimidate as the reacting groups must be within the maximal distance of the reactive groups (14). Dimethyl malonic diimidate was used to study complex crystals of glycogen phosphorylase *b* with the inhibitor glucose-6-phosphate (15). In the above study, a 2 mg/ml solution of dimethyl malonic diimidate in 0.1 M triethanolamine–HCl, 10 mM magnesium acetate pH 7.8, was used to cross-link crystals for 2 h before the reaction was stopped by lowering the pH to 7.1.

3. Soaking application

In this section we will consider some of the more typical applications of soaking: heavy-atom derivatization and the soaking for cryo-crystallography.

3.1 Heavy-atom soaking and isomorphous replacement

The method of isomorphous replacement has been central to X-ray analysis of protein crystals from the initial work on haemoglobin (16). In this procedure a single or a limited number of heavy-atoms per macromolecule are introduced, as an addition or replacement of an endogenous atom, without disrupting or significantly altering the crystal lattice. This addition of electrons in the structure causes significant changes in X-ray recorded intensities which can be used to obtain an estimate of the 'phase' for each reflection, which are then used for the calculation of the electron density and the solution of the structure. Such modifications to crystals of macromolecules are normally carried out by soaking the native crystals in the mother liquor containing the heavy-atom compound. Although obtaining a good heavy-atom derivative is a trial and error process, there are general considerations which give the best chance of success.

It is clear that the preparation of isomorphous derivative crystals will depend on the pH, composition of the mother liquor, and temperature. Many successful pH values for heavy-atom soaking are 6–8. If the pH value is below 6, most reactive groups which could bind the metal ion will be protonated and blocked. Since many heavy-atom compounds are alkaline labile, at high pH they may form insoluble hydroxides. Except at low pH (i.e. below 6), ammonium sulfate is a poor mother liquor for heavy-atom binding due to the production of the good nucleophile NH_3 (17). If possible, the crystals should be transferred to Mg or Na sulfate, or Na and K phosphate. However, an excess of phosphate is undesirable for the binding of uranium and rare earth metals. The temperature can change the rate of reaction and sometimes the degree of binding. In most cases the soaking temperature will be the same as the crystallization temperature.

Other important considerations are the heavy-atom concentration and soaking time. The necessary concentration will depend on the solubility of heavy-metal compound. Typically, 1–2 mM is appropriate as a starting value. The soaking time can vary from 20 minutes to months, but for the initial screening, 4–18 hours is sufficient. The crystal is observed continuously for the first ten minutes of the soak, and hourly for the first four hours, and again at the end of the soak. If the crystal appearance is unchanged from the native crystal, and small changes are found from the data reduction, it is advisable to increase the concentration of the heavy-atom substance and soak time. On the other hand, if the crystal cracks, indicating that the changes have caused non-isomorphism, or if diffraction resolution dramatically decreases, a lower concentration and shorter soaking time should be tried. Sometimes, back-soaking is necessary in order to reduce the cell changes to acceptable values (18). For back-soaking, after the initial soak the crystals are transferred to solutions with a lower concentration of heavy-atoms or to the original mother liquor in order to reduce binding to the lower affinity sites. Another useful criteria for evaluating the soaking conditions is the relative temperature factor of the derivative data compared to the native. When the derivative data have significantly larger temperature factors than the native it is likely that some disorder has been introduced into the derivative crystal. A reasonable temperature factor can often be achieved by lowering the concentration of the heavy-atom compound. As the heavy-atom concentration is lowered the volume of the soak solution should increase in proportion to ensure a good stoichiometric ratio between the heavy-atom compound and the protein.

Since many heavy-atom compounds have a very vigorous photochemistry, soaking should be carried out under low power illumination or in the dark. A drawer is sufficient for this. Finally, freshly prepared soaking solutions should be used whenever possible. It is frequently the case that many heavy-atoms (10 to 50) need to be tried before a good isomorphous derivative is found. There is no substitute for patience and hard work at this stage.

3.2 Selecting a heavy-atom compound

This chapter is not dedicated to the selection of a heavy-atom for structure determination. The frequency with which the various compounds have been used successfully in phasing protein structures are listed in *Tables 2–12* and the table legends provide some chemical guidelines. The frequency can be used in a statistical manner to evaluate the likelihood that these compounds may be useful for the phasing of new proteins. It is suggested that other references (e.g. 10 and 19) should also be consulted. The use of xenon and krypton at high pressure is becoming popular for the phasing of proteins, and equipment for their use is now available at some synchrotron facilities such as LURE and SSRL (20, 21). The number of sites can be varied with pressure, typically from 0.4–2 MPa. Crystals are pressurized for 30 minutes before data collection. Since xenon binding sites are generally different from those for other heavy-atoms this technique can be used as a second resort when the initial soaks in heavy-atoms are not successful.

3.3 Soaking for cryo-crystallography

Cryo-crystallography provides a means of increasing the lifetime of some protein crystals by reducing radiation damage during data collection (22, 23) and allowing a complete data set to be collected from one crystal, often to the same or sometimes higher resolution than crystals analysed at room temperature (24). It has also been suggested that cooling may also increase the internal order of parts of the protein which are mobile at room temperature (25) and also provide ways to observe enzyme substrate complexes and unstable intermediates (26) with the use of Laue X-ray photography. It is now widely used for collecting data at synchrotron facilities (27).

The use of several cryo-protective solvents and combinations of such solvents was pioneered by Petsko (25). The most commonly used cryo-protectants are glycerol (28), ethylene glycol, 2-methyl-2,4-pentane diol (MPD) (also commonly used in crystallization) sometimes in combination with others such as low molecular weight polyethylene glycol (200–600),

Table 2. ^{80}Hg—mercury compounds[a]

Frequency of usage	Name	Abbreviation	Supplier[b]
24	$HgCl_2$		Ac, Al, Af
21	$Hg(CH_3CO_2)_2$	HgAc	Ac, Al, Af, Si
20	$C_2H_5HgPO_4$	EMP	N
17	Ethyl mercury thiosalicylate	EMTS	Ac
15	CH_3HgCl	MMCl	Af, St
13	Mersalyl		Al, Si
12	$CH_3Hg(CH_3CO_2)_2$	MMAc	P
12	K_2HgI_4	PMTI	Af, M
9	*p*-Chloromercuribenzene sulfonate	*p*CMBS	Al, I
8	Tetrakis (acetoxymercury) methane	TAMM	St
8	*p*-Chloromercuribenzoate	*p*CMB	Al, I
4	C_2H_5HgCl	EMCl	Af
4	Baker's dimercurial	Baker's	An
4	*p*-Hydroxymercuribenzoate	*p*HMB	Al, I
4	2-Chloromercuri-4-nitrophenol	CNP	Ac
4	$Hg_2(CO_2)_2$ (oxalate/malonate)	DMMA	P
4	3-Chloromercuri-2 methoxypropyl urea (chlormerodrin)	CMMPU	–
3	Hg-deoxyuridine triphosphate	HgdUTP	Si
3	$K_2Hg(CN)_4$		
3	$(CH_3)_2Hg$	DMHg	Al, Af, St
2	*2*-Chloromercuriphenol	CMP	CS
2	Phenyl mercuriglyoxal	PMG	–
2	$HgBr_2$		A, St
2	HgI_2		St
1	$Hg(NO_3)_2$		St
1	$Hg(CN)_2$		Ac, Al, St
1	HgO		St
1	Dimercuri acetate	DMA	An
	$Hg_2(CH_3CO_2)_2$	DMDA	P
1	CH_3HgOH	MMOH	Af, St
1	CH_3HgBr	MMBr	Af
1	*p*-Hydroxymercuriphenyl sulfonate	*p*HMPS	Si
1	*p*-Chloromercuriphenyl sulfonate	*p*CMPS	Al, Fl

[a] Mercury compounds are targeted to sulfhydryl groups. Short soak times 1–3 h can produce useful derivatives with concentrations as low as 0.01 mM. Mercury has also a strong tendency to bind to zinc sites at a histidine nitrogen. The mercury compounds can be grouped in three classes: the ionic group, the most commonly used are mercury chloride and mercury acetate; the alkyl chain mercury compounds, ethyl mercury phosphate, ethyl mercury chloride, methyl mercury chloride, methyl mercury acetate; and the aromatic group, the most popular being EMTS, mersalyl, *p*CMBS, and *p*CMB. K_2HgI_4 cannot be grouped with the ionic mercurials as it tends to give different results, but it is definitely a compound worth trying. Baker's dimercurial consisting of two mercury atoms, and TAMM, a heavy metal cluster of four mercury atoms which has been used in the phasing of large molecular assemblies (36–39), have good solubility and are definitely worth trying.
[b] See *Table 13*.

Table 3. ^{78}Pt—platinum compounds[a]

Frequency	Name	Supplier[b]
71	K_2PtCl_4	Af, Fl, St
24	*cis*-$Pt(NH_3)_2Cl_2$	Af, St
17	K_2PtCl_6	Ac, Af, St
14	$K_2Pt(NO_2)_4$	Af, Al, St
14	$K_2Pt(CN)_4$	Af, St
12	Di-μ-iodobis (ethylenediamine) diplatinum nitrate (PIP)	St
6	$Pt(NH_2CH_2CH_2NH_2)Cl_2$	Af, Al, St
5	(2,2′:6′,2″)-terpyridinium platinum chloride	Fl
4	$Pt(NH_3)_2(NO_2)_2$	Af, Al, St
4	K_2PtBr_4	St
3	K_2PtI_6	Af
3	$K_2Pt(SCN)_4$	Al
3	$Pt(NH_2CH_2CH_2NH_2)_2Cl_2$	Af, Al, St
2	*trans*-$Pt(NH_3)_2Cl_4$	Af, St
2	K_2PtBr_6	St
2	$Pt(NH_3)_2Cl_2$	Ac, Af, Al
1	$K_2Pt(CN)_6$	Af, Al, St
1	$K_2Pt(SCN)_6$	P
1	$K_2Pt(CN)_2$	Af, St

[a] K_2PtCl_4 is the most widely used compound. Platinum compounds are good ligands for methionine, histidine, and cysteine residues (see refs 5 and 10 for more details).
[b] See *Table 13*.

Table 4. ^{79}Au—gold compounds[a]

Frequency	Name	Supplier[b]
19	$KAu(CN)_2$	Af, Fl, St
12	$KAuCl_4$	Fl, St
5	$NaAuCl_4$	Fl
2	$AuCl_3$	Af
1	$KAuBr_4$	P, St
1	$HAuCl_4$	Af, Fi

[a] Gold compounds have a propensity for binding to sulfhydryl groups, and may provide a good alternative to ionic mercurials with which they often share sites.
[b] See *Table 13*.

Table 5. ^{82}Pb—lead compounds[a]

Frequency	Name	Supplier[b]
14	$(CH_3)_3Pb(CH_3CO_2)_2$	Af, An
9	$Pb(CH_3CO_2)_2$	Ac, Af
5	$Pb(NO_3)_2$	Ac, Af
3	$(CH_3)_3PbCl$	Al, Af
2	$PbCl_2$	Al, Af
1	$(C_2H_5)_3PbCl$	Af
	$(C_2H_5)_3Pb(CH_3CO_2)_2$	An

[a] Outstanding among the lead compounds is trimethyl lead acetate. It is sparsely soluble and long soaks in large volumes may be necessary. It is an excellent reagent for hydrophobic sites, in the α/β TCR structure determination it was found to bind in close proximity to EMTS (40) which binds to the only cysteine not involved in a disulfide bond found in a hydrophobic pocket. Lead can bind at zinc and other divalent metal sites, the acetate and nitrate salts are commonly used.
[b] See *Table 13*.

Table 6. ^{81}Tl—thallium compounds[a]

Frequency	Name	Supplier[b]
3	$TlCl_3$	Af, Al
1	$TlCl$	Af, St
1	$Tl(CH_3CO_2)_3$	Af, St

[a] Thallium is not a heavy-atom of choice due to its **extreme toxicity**. It is wise to experiment with various mercury compounds first before resorting to this option. The relatively good solubility of thallium salts, particularly in phosphate buffers is its main attraction.
[b] See *Table 13*.

Table 7. ^{77}Ir—iridium compounds[a]

Frequency	Name	Supplier[b]
12	K_3IrCl_6	Af
3	$IrCl_3$	Ac, Af
2	Na_3IrCl_6	Af
1	$(NH_4)_3IrCl_6$	St
1	H_2IrCl_6	Af, Fl, I

[a] Iridium shares many properties with platinum, it gives stable anionic (e.g. $IrCl_6^{3-}$) and cationic complexes (e.g. $Ir(NH_3)_6^{3+}$).
[b] See *Table 13*.

Table 8. ^{76}Os—osmium compounds[a]

Frequency	Name	Supplier[b]
10	K_2OsO_4	Af, Al, St
6	K_2OsCl_6	St
2	$(NH_4)_2OsBr_6$	St
1	Na_2OsCl_6	Al, St
1	$OsCl_3$	Ac, Al, St

[a] Osmium, OsO_4 is **extremely toxic**. It is a good reactant for ribose moieties and the 3′ terminus of RNA (41, 42).
[b] See *Table 13*.

Table 9. ^{92}U—uranium compounds[a]

Frequency	Name	Supplier[b]
21	$K_3UO_2F_5$	Sp
16	$UO_2(CH_3CO_2)_2$	Al, Fl
15	$UO_2(NO_3)_2$	Al, Fl, St
6	UO_2SO_4	P
4	UO_2Cl_2	P, St
1	$(NH_4)_2U_2O_7$	P

[a] Uranyl is the third most popular heavy-atom reagent after mercury and platinum compounds. Uranyl nitrate and uranyl acetate give similar results. In the structure determination of the erythropoietin EMP1 complex (43) the best results were obtained with uranyl nitrate in acetate buffer. It should be noted that $K_2UO_2F_5$ is the fifth most popular single compound used for protein structure phasing.
[b] See *Table 13*.

Table 10. ^{74}W—tungsten compounds[a]

Frequency	Name	Supplier[a]
1	Na_2WO_4	Ac, Al
1	$(NH_4)_2WS_4$	Af, Al, St

[a] The low frequency with which tungsten compounds have been used is somewhat surprising. Tungstate is a phosphate mimic and its complex with human protein phosphatase 1 was used for multiple wavelength anomalous dispersion experiments (44). Analysis of the changes induced by tungstate on several enzymes able to bind to phosphate moieties by the method of reverse screening (45), it has been noticed that in general, the solubility of the tungstate complex improved. Indeed, crystals of FIV dUTPase, dissolved when soaked in tungstate solution (Stura and Prasad, unpublished results) and was not be used in its structure determination (46). The tungstate complex with deoxyribose-5-phosphate aldolase was obtained by co-crystallization. This resulted in a crystal form of this enzyme different from that reported in ref. 47. Multi-tungsten clusters have been used for the phasing of fumarase C (48) and riboflavin synthetase (49), see review (3).
[b] See *Table 13*.

Table 11. Lanthanide compounds[a]

Frequency	Name	Supplier[b]
8	$^{62}SmCl_3$	Al, St
3	$^{57}La(NO_3)_3$	Af, St
3	$^{63}EuCl_3$	Af, St
3	$^{64}GdCl_3$	Af
3	$^{62}Sm(CH_3CO_2)_3$	St
2	$^{71}Lu(CH_3CO_2)_3$	Af, St
2	$^{70}YbCl_3$	Af, St
1	$^{66}DyCl_3$	Af, St
1	$^{63}Eu(NO_3)_3$	Af, St
1	$^{62}Sm(NO_3)_3$	St
1	$^{71}LuCl_3$	Af, St
1	$^{59}PrCl_3$	St
1	$^{60}NdCl_3$	Af
1	$^{67}HoCl_3$	Af
1	$^{66}DyI_3$	Af

[a] The combined usage of lanthanide compounds compares with that of the acetate and nitrate salts of uranyl. The overall low frequency of each compound reflects the very similar behaviour of each. $SmCl_3$, $Gd_2(SO_4)_3$, and $Yb_2(SO_4)_2$ all occupy the same site as the magnesium ion at the threefold axis of the FIV dUTPase trimer otherwise occupied by magnesium (46).
[b] See *Table 13*.

Table 12. Other compounds[a]

Frequency of usage	Name	Supplier[b]
1	$^{73}Ta^{35}Br_5$	Af, Al, CS
1	$^{56}BaSO_4$	Al
1	$^{56}BaCl_2$	Al
1	$^{56}Ba^{38}Sr^{41}Nb_4O_{12}$	Af, Al
1	$^{48}CdCl_2$	Ac, Af, Al
3	$I_2 + KI$	Al

[a] Iodination is well described in refs 5 and 50. Certain iodide salts of heavy metals may derivatize proteins due to I^- rather than the metal itself. Tantalum and niobium compounds are of interest for their use in phasing large macromolecular assemblies (3). Brominated and iodinated nucleotides are used in the phasing of protein nucleotide complexes.
[b] See *Table 13*.

Table 13. Suppliers of compounds in *Tables 2–12*[a]

Ac: Acros Chemicals; 711 Forbes Avenue, Pittsburgh, PA 15219, USA. Tel: (800) 227-6701. http://www.fishersci.com/catalogs
Af: Alfa; Johnson Matthey Catalog Company, Inc., PO Box 8247, Ward Hill, MA 01835-0747, USA. Tel: (800) 343-0660.
Al: Aldrich Chemical Co.; 1001 West St Paul Avenue, Milwaukee, WI 53233, USA. Tel: (800) 558-9160.
http://www.sigma.com/SAWS.nsf/Pages/Aldrich?EditDocument
An: Anatrace Inc.; 434 West Dussel Drive, Maumee, OH 43537-1624, USA. Tel: (800) 252-1280. http://www.anaTrace.com
CS: Chem Service; PO Box 3108, West Chester, PA 19381-3108, USA. Tel: (610) 692-3026.
Fi: Fisher Scientific; 711 Forbes Avenue, Pittsburgh, PA 15219, USA. Tel: (800) 766-7000. http://www.fishersci.com/catalogs
Fl: Fluka Chemie AG; Industriestrasse 25, CH-9471 Buchs, Switzerland.
http://www.sigma.aldrich.com/SAWS.nsf/Pages/Fluke?EditDocument
I: ICN Pharmaceutical Inc.; 3300 Hyland Avenue, Costa Mesa, CA 92626, USA. Tel: (714) 545-0100.
M: Mallinckodt; 470 Frontage Road, West Haven, CT 06516, USA. Tel: (203) 933-7064.
N: Noah Technologies; 1 Noah Park, San Antonio, TX 28249, USA. Tel: (210) 691-2000.
P: Pfaltz and Bauer; 172 E Aurora Street, Waterbury, CT, USA. Tel: (203) 574-0075.
Si: Sigma Chemical Company; PO Box 14508, St. Louis, MO 63178, USA.
http://www.sigma.aldrich.com/SAWS.nsf/Pages/Sigma?EditDocument
Sp: SPECS and BioSPECS bv; Koninginnegracht, 94-95, 2514 AK The Hague, The Netherlands. PO Box 85586, 2508 CG The Hague, The Netherlands (mailing address) Tel: 31-70-355-4473. Fax: 31-70-355-8527.
Brandon/SPECS Inc.; (North American sales company), PO Box 1244, Merrimack, New Hampshire 03054, USA. Tel: 603-424-2035. Fax: 603-424-2035.
St: Strem Chemical; 7 Mulliken Way, Newburyport, MA 01950, USA. Tel: (508) 462-3191.

[a] Data for *Tables 2–12* has been compiled from structures reported in: *Macromolecular structures* (1991–4), and *Atomic structures of biological macromolecules* (1990–3) (ed. W. A. Hendrickson and K. Wüthrich). Current Biology Ltd., London.

ethanol, propanol, xylitol, erythriol, inositol, raffinose, trehalose, glucose, L-2,3 butanediol. The methods for soaking of crystals for this application does not vary substantially from that used for other applications. The loop technique is used more often specially in the 'dip and shoot' method, where the crystal is soaked in the cryo-solvent for a few seconds allowing for only the liquid outside the crystal to be replaced. This can be compared to the oil technique, where crystals are transferred to oil prior to freezing them (29). Alternatively, the cryo-solvent is added to the mother liquor in small steps, slowly increasing the summed concentration of all the cryo-solvents to about 35%. All such techniques are aimed at preserving crystal integrity by obtaining a transition from water to vitreous ice, preventing crystallization of the water in the mother liquor.

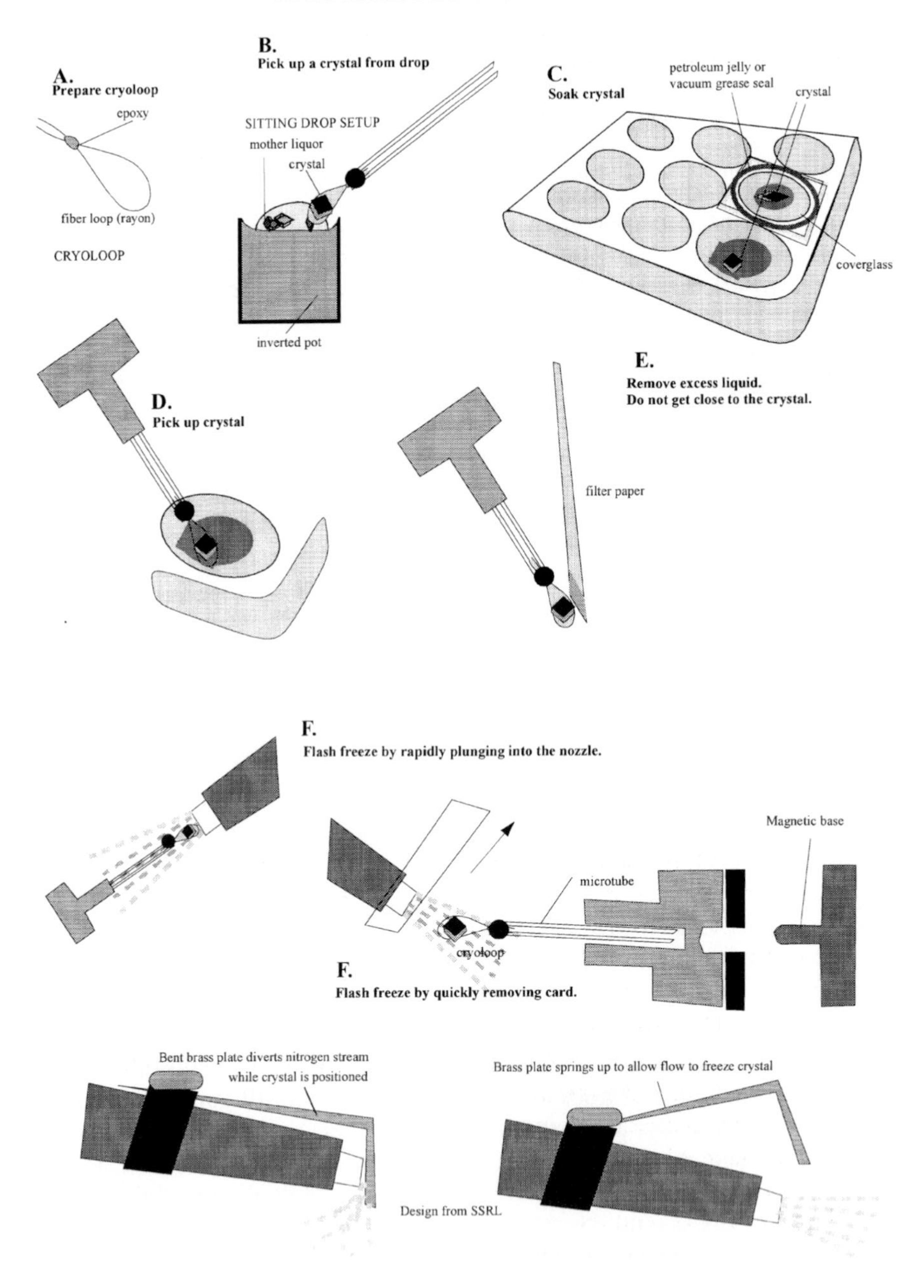
A.
Prepare cryoloop
epoxy
fiber loop (rayon)
CRYOLOOP
B.
Pick up a crystal from drop
SITTING DROP SETUP
mother liquor
crystal
inverted pot
C.
Soak crystal
petroleum jelly or
vacuum grease seal
crystal
coverglass
D.
Pick up crystal
E.
Remove excess liquid.
Do not get close to the crystal.
filter paper
F.
Flash freeze by rapidly plunging into the nozzle.
Magnetic base
microtube
cryoloop
F.
Flash freeze by quickly removing card.
Bent brass plate diverts nitrogen stream
while crystal is positioned
Brass plate springs up to allow flow to freeze crystal
Design from SSRL

Figure 3. (A) Prepare a loop (cryo-loop) from a rayon fibre. This is best done by placing both ends of the fibre into a needle and pulling the two ends until a loop of the correct diameter is obtained. Place a spot of epoxy at the junction of the fibre. (b) Pick up the crystal from the drop using the loop mounted on a wooden or glass rod. (C) Soak crystal in spot plate (easier than in vials). For long soaks, place a ring of petroleum jelly or vacuum grease around the well, and cover with a coverglass. Place the spot plate in a drawer for platinum and other light-sensitive heavy-atom solutions. Warning: loops used for heavy-atom work may retain some heavy-atom soaked in the fibre. (D) After the soak, pick up the crystal from the soak solution. (E) If the drop attached to the loop is very large, the excess liquid can be removed by touching the outside of the loop with a thin strip of filter paper, being careful not to get close to the crystal. (F) The crystal is then plunged into the nitrogen stream or the stream is blocked by a card or ruler and when the crystal is in position the card is rapidly removed. In the design used at SSRL a thin bent brass plate clips onto the nozzle and diverts the beam while the crystal is being positioned. When the crystal is in place the brass plate is made to springs back allowing the stream to flash-freeze the crystal.

Protocol 7. Soaking of crystals for cryo-crystallography

Equipment and reagents

- Cryo-solvents
- Cryo-loops
- Vacuum grease
- Microscope
- Spot plate
- Cover glass

Method

1. Select crystals for freezing of roughly 0.4 mm in each dimension or smaller. Larger crystals are more problematic as they may develop cracks, and form crystalline ice due to slow heat transfer. Since the radiation damage is small or negligible at cryo-temperature, the strength of the X-ray source and exposure times will be able to compensate for the smaller crystal size. Fast data collection on a wiggler line at a synchrotron radiation source is preferable than long exposure times with conventional sources.
2. Select a cryo-solvent. See *Table 1* for suggestions. Garman and Mitchell (28) give the minimum amount of glycerol to be added to 50 typical crystallization conditions. The cryo-buffer typically is a combination of the crystallization reservoir and a cryo-solvent. For crystallizations from small molecular weight PEG (200–600) or MPD no cryo-solvent is required, although the precipitant concentration may need to be increased. If the crystallization is from PEG 4000 or higher molecular weight PEG, replace some of the PEG 4000 by PEG 200 or just add PEG 200 at varying concentrations. The next most popular choices are ethylene glycol or MPD. The alcohol sugars are generally milder, and may be tolerated by those crystals that crack under the other conditions.

Protocol 7. *Continued*

3. A small amount of this buffer is picked up in a loop (*Figure 3*) and shock-frozen in the nitrogen stream. Plunge loop into nozzle to flash-freeze or block stream with a paper card, then remove it quickly. If the buffer stays transparent it has formed vitreous ice, whereas opaqueness indicates formation of crystalline ice which will give a powder diffraction pattern, 'ice-rings'. If there is no loss in resolution or increased mosaicity (30) or anisotropy minimal ice-rings can be tolerated as most integration programs can cope with this problem if not too severe.
4. The next step is to optimize the cryo-solvent. Check whether the cryo-protectant causes lattice damage, this normally results in cracks, very fine hair-line cracks manifest themselves as a brown tinge when the crystal is observed under the microscope. Take a few diffraction images at various concentrations and with different cryo-protectants at room temperature with crystals mounted in capillaries to select the least damaging cryo-solvent and to maximize the resolution limit. Some cryo-solvents may indeed enhance diffraction. Since this process can take a long time, if the crystals are large enough to be analysed on a conventional source, it is best to perform these tests in advance of synchrotron data collection. Often the functional plot of cryo-protectant versus resolution limit has a minimum (30).
5. Pull crystal through cryo-protectant in the loop to transfer the crystal to the cryo-buffer (*Figure 3*). In a stepwise transfer using solutions with increasing amounts of cryo-protectant is important in order to reduce osmotic shock. It is also important to keep the number of operations small to reduce damage and subsequent increase of mosaicity. For a small crystal 20–30 sec between transfers is sufficient and the crystal should be frozen immediately, else the mosaicity might increase. Evaporation of buffer also requires speedy transfer to the cryo-stream.

4. Conclusions

Soaking is most commonly used to obtain heavy-atom derivatives, although crystallization of previously modified proteins either chemically or biologically have also been used (31, 32). Soaking and co-crystallization are two different approaches to achieving complexes of macromolecules. The two procedures are both alternative and complementary to each other. By soaking effectors into preformed crystals it is possible to analyse the structure of complexes, only if crystal lattice constraints permit. The problem of cracking of the crystals which may occur both when binding effectors (15) and heavy-atoms (33), can be often resolved by the use of cross-linking agents. One must however understand that complexes obtained by soaking may differ from complexes obtained by co-crystallization. Flow cells in which a constant supply of substrate is supplied to the enzyme in the crystal and product is washed away may answer the problem in cases where the rate of product formation is significantly slower than the rate of diffusion through the crystal (19, 34).

Acknowledgements

We would like to thank Dr Ping Chen for her contribution to the first edition of this chapter and Dr Ian A. Wilson for reading and support of that work through his grants by the National Institutes of Health Grants AI-23498, GM-38794, and GM-38419. T. G. was supported by BMBF grant 05 641BJA 4 (to Rolf Hilgenfeld) and by the Australian Research Council grant AD 984283 (to B. Kobe). E. S. thanks the French Atomic Energy Commission (CEA) for support during the revision work.

References

1. Matthews, B. W. (1968). *J. Mol. Biol.*, **33**, 419.
2. Cohen, C., Caspar, D. L. D., Parry, D. A. D., and Lucas, R. M. (1971). *Cold Spring Harbor Symp. Quant. Biol.*, **36**, 205.
3. Thygesen, J., Weinstein S., Franceschi, F., and Yonath, A. (1996). *Structure*, **4**, 513.
4. Wyckoff, H. W., Doscher, M. S., Tsernoglou, D., Inagami, T., Johnson, L. N., Hardman, K. D., *et al.* (1967). *J. Mol. Biol.*, **127**, 563.
5. Petsko, G. A. (1985). In *Methods in enzymology* (eds H. W. Wychoff, C. H. W. Hirs, and S. N. Timasheff), Academic Press, London. Vol. 114, p. 141.
6. Remington, S., Wiegand, G., and Huber, R. (1982). *J. Mol. Biol.*, **158**, 111.
7. Kasvinsky, P. J. and Madsen, N. B. (1977). *J. Biol. Chem.*, **251**, 6852.
8. Hadju, J., Acharya, K. R., Stuart, D. I., McLaughlin, P. J., Barford, D., Oikonomakos, N. G., *et al.* (1987). *EMBO J.*, **6**, 539.
9. McPherson, A. (1982). *Preparation and analysis of protein crystals.* Wiley, New York.
10. Blundell, T. L. and Johnson, L. N. (1976). *Protein crystallography.* Academic Press, New York.
11. Quicho, F. A. and Richards, F. M. (1964). *Proc. Natl. Acad. Sci. USA*, **52**, 833.
12. Hunter, M. J. and Ludwig, M. L. (1962). *J. Am. Chem. Soc.*, **84**, 3491.
13. Browne, D. T. and Kent, S. B. H. (1975). *Biochem. Biophys. Res. Commun.*, **67**, 126.
14. Dombradi, U., Hadju, J., Bot, G., and Friedrich, P. (1980). *Biochemistry*, **19**, 2295.
15. Lorek, A., Wilson, K. S., Sansom, M. S. P., Stuart, D. I., Stura, E. A., Jenkins, J. A., *et al.* (1984). *Biochem. J.*, **218**, 45.
16. Green, D. W., Ingram, V. M., and Perutz, M. F. (1954). *Proc. Roy. Soc.*, **A225**, 287.
17. Sigler, P. B. and Blow, D. M. (1965). *J. Mol. Biol.*, **14**, 640.
18. Mondragon, A., Wolberg, C., and Harrison, S. C. (1989). *J. Mol. Biol.*, **205**, 179.
19. Petsko, G. A. (1985). In *Methods in enzymology* ibid ref. 5. Vol. 114, p. 147.
20. Schiltz, M., Prange, T., and Fourme, R. (1994). *J. Appl. Cryst.*, **27**, 950.
21. Stowell, M. H. B., Soltis, S. M., Kisker, C., Peters, J. W., Schindelin, H., Rees, D. C., *et al.* (1996). *J. Appl. Cryst.*, **29**, 608.
22. Rodgers, D. W. (1994). *Structure*, **2**, 1135.

23. Tsernoglou, D., Hill, E., and Banaszack, L. J. (1972). *J. Mol. Biol.*, **69**, 75.
24. Hope, H., Frolow, F., von Bohlen, K., Makowski, I., Kratky, C., Halfon, Y., *et al.* (1989). *Acta Cryst.*, **B45**, 190.
25. Petsko, G. A. (1975). *J. Mol. Biol.*, **96**, 381.
26. Douzou, P., Hui Bon Hoa, G., and Petsko, G. A. (1975). *J. Mol. Biol.*, **96**, 367.
27. Hadju, J., Machin, P. A., Campbell, J. W., Greenhough, T. J., Clifton, I. J., Zurech, S., *et al.* (1987). *Nature*, **329**, 176.
28. Garman, E. F. and Mitchell, E. P. (1996). *J. Appl. Cryst.*, **29**, 584.
29. Hope, H. (1988). *Acta Cryst.*, **B44**, 22.
30. Mitchell, E. P. and Garman, E. F. (1994). *J. Appl. Cryst.*, **27**, 1070.
31. Stura, E. A., Johnson, D. L., Inglese, J., Smith, J. M., Benkovic, S. J., and Wilson, I. A. (1989). *J. Biol. Chem.*, **264**, 9703.
32. Yang, W., Hendrickson, W. A., Crouch, R. J., and Satow, Y. (1990). *Science*, **249**, 1398.
33. Ringe, D., Petsko, G. A., Yakamura, F., Suzuki, K., and Ohmori, D. (1983). *Proc. Natl. Acad. Sci. USA*, **80**, 3879.
34. Farber, G. K., Glasfeld, A., Tiraby, G., Ringe, D., and Petsko, G. A. (1989). *Biochemistry*, **28**, 7289.
35. Klein, C., Chen, P., Arevalo, J. H., Stura, E. A., Marolewski, A., Warren, M. S., *et al.* (1995). *J. Mol. Biol.*, **249**, 53.
36. Deisenhofer, J., Epp, O., Miki, K., Huber, R., and Michel, H. (1984). *J. Mol. Biol.*, **180**, 385.
37. O'Holloran, T. V., Lippard, S. J., Richmond, T. J., and Klug, A. (1987). *J. Mol. Biol.*, **194**, 705.
38. Bentley, G. A., Boulot, G., Riottot, M. M., and Poljak, R. J. (1990). *Nature*, **348**, 254.
39. Reinemer, P., Dirr, H. W., Ladenstein, R., Schaeffer, J., Gallay, O., and Huber, R. (1991). *EMBO J.*, **10**, 1997.
40. Garcia, K. C., Degano, M., Stanfield, R. L., Brunmark, A., Jackson, M. R., Peterson, P. A., *et al.* (1996). *Science*, **274**, 209.
41. Kim, S.-H., Shin, W.-C., and Warrant, R. W. (1985). In *Methods in enzymology* ibid ref. 5. Vol. 114, p. 156.
42. Stout, C. D., Mizuno, H., Rao, S. T., Swaminathan, P., Rubin, J., Brennan, T., *et al.* (1978). *Acta Cryst.*, **B34**, 1529.
43. Livnah, O., Stura, E. A., Johnson, D. L., Mulcahy, L. S., Wrighton, N. C., Dower, W. J., *et al.* (1996). *Science*, **273**, 464.
44. Egloff, M. P., Cohen, P. T., Reinemer, P., and Barford, D. (1995). *J. Mol. Biol.*, **254**, 294.
45. Stura, E. A., Satterthwait, A. C., Calvo, J. C., Kaslow, D. C., and Wilson, I. A. (1994). *Acta Cryst.*, **D50**, 448.
46. Prasad, G. S., Stura, E. A., McRee, D. E., Laco, G. S., Hasselkus-Light, C., Elder, J. H., *et al.* (1996). *Protein Sci.*, **5**, 2429.
47. Stura, E. A., Ghosh, S., Garcia-Junceda, E., Chen, L., Wong, C.-H., and Wilson, I. A. (1995). *Proteins*, **22**, 67.
48. Weaver, T. M., Levitt, D. G., Donnelly, M. I., Stevens, P. P., and Banaszak, L. J. (1995). *Nature Struct. Biol.*, **2**, 654.
49. Ladenstein, R., Bacher, A., and Huber, R. (1987). *J. Mol. Biol.* **195**, 751.
50. Ghosh, D., Enman, M., Sawicki, M., Lala, P., Weeks, D. R., Li, N. *et al.* (1999). Acta Cryst. **D55**, 779.

14

X-ray analysis

L. SAWYER and M. A. TURNER

1. Introduction

This chapter covers the preliminary characterization of the crystals in order to determine if they are suitable for a full structure determination. Probably more frustrating than failure to produce crystals at all, is the growth of beautiful crystals which do not diffract, which have very large unit cell dimensions, or which decay very rapidly in the X-ray beam, though this last problem has been largely overcome by freezing the sample.

It is impossible in one brief chapter to give more than a flavour of what the X-ray crystallographic technique entails and it is assumed that the protein chemist growing the crystals will have contact with a protein crystallographer, who will carry out the actual structure determination and in whose laboratory state-of-the-art facilities exist. However, preliminary characterization can often be carried out with little more than the equipment which is widely available in Chemistry and Physics Departments and so the crystal grower remote from a protein crystallography laboratory can monitor the success of their experiments. The reader should refer to the first edition for protocols useful for photographic characterization but such techniques are seldom used nowadays. It must be remembered, in any case, that X-rays are dangerous and the inexperienced should *not* try to X-ray protein crystals without help.

2. Background X-ray crystallography

It is necessary to provide an overview of X-ray crystallography, to put the preliminary characterization in context. For a general description of the technique the reader should refer to Glusker *et al.* (1) or Stout and Jensen (2). For protein crystallography in particular, the books by McRee (3) and Drenth (4) describe many of the advances since the seminal work of Blundell and Johnson (5). Amongst many excellent introductory articles, those by Bragg (6), published years ago, and Glusker (7) are particularly recommended.

2.1 X-rays

2.1.1 Why use X-rays

The scattering or diffraction of X-rays is an interference phenomenon and the interference between the X-rays scattered from the atoms in the structure produces significant changes in the observed diffraction in different directions. This variation in intensity with direction arises because the path differences taken by the scattered X-ray beams are of the same magnitude as the separation of the atoms in the molecule. Put another way, to 'see' the individual atoms in a structure, it is necessary to use radiation of a similar wavelength to the interatomic distances, typically 0.15 nm or 1.5 Å and radiation of that wavelength lies in the X-ray region of the electromagnetic spectrum. It is also important to realize that it is the *electrons* which scatter the X-rays and so what is in fact observed is the *electron density* of the sample. Because the electrons cluster round the atomic nuclei, regions of high electron density correspond to the atomic positions.

2.1.2 X-ray sources

X-rays are produced in the laboratory by accelerating a beam of electrons into an anode, the metal of which dictates what the wavelength of the resulting X-rays will be. Monochromatization is carried out either by using a thin metal foil which absorbs much of the unwanted radiation or, better, by using the intense low order diffraction from a graphite crystal. To obtain a brighter source, the anode, which is water cooled to prevent it melting, can be made to revolve in what is known as a rotating anode generator. For most work with proteins, the target is copper and the characteristic wavelength of the radiation is 0.1542 nm (1.542 Å).

An alternative source of X-radiation is obtained when a beam of electrons is bent by a magnet. This is the principle behind the synchrotron radiation sources which are capable of producing X-ray beams some thousand times more intense than a rotating anode generator (3). A consequence of this high intensity radiation source is that data collection times have been drastically reduced, making kinetic crystallography feasible (8). A further advantage is that the X-ray spectrum is continuous from around 0.05–0.3 nm, dependent upon the particular machine, and this has distinct advantages for the crystallographer. The use of shorter wavelengths has usually been found to prolong the room temperature lifetime of a crystal in the X-ray beam. The main drawback is that synchrotrons are centralized facilities and consequently access is significantly less convenient, particularly for preliminary work.

2.2 What is a crystal?

A crystal is a regular, repeating array of atoms or molecules in three dimensions. It is convenient to describe such an object with the aid of a lattice, which is a geometric construction defined by three axes and the three angles between them. Along each axis direction; a point will repeat at a distance referred to as the unit translation or unit cell repeat and labelled a, b, and c, respectively. The angles between b and c, a and c, and a and b are α, β, and γ, respectively. The basic building block of a crystal, then, is a parallelepiped described by the dimensions a, b, and c and α, β, and γ and called the *unit cell.*

There are seven crystal systems which arise from the only possible combinations of these unit cell parameters. However, it is sometimes easier to consider a larger unit cell but with a simpler shape, for example with mutually perpendicular axes. This choice can be illustrated in the two-dimensional example shown in *Figure 1*. The choice of the basic building block containing a single 'molecule' can be made in a variety of ways because the lattice is no more than a geometrical construction affording a convenient description of the repeating figure. Crystallographers adopt the convention that the unit cell which is chosen is the one with angles nearest to 90°. Such a cell with only one copy of the molecular structure is called *primitive* but, as noted above, a more convenient cell may have two or even four copies (see *Figure 1*, where the non-primitive, centred cell is at the right). There are 14 so-called Bravais lattices which can be constructed in three dimensions (there are five in 2D). As an example of the limited number of lattices, construct a centred square lattice and it is evident that a smaller, primitive square lattice is also present.

Although the basic building block of a crystal is the unit cell and the lattice produced by its repetition has a characteristic symmetry (see *Table 1*), within the unit cell there may be further symmetry. For example, the molecule itself may have symmetry about an axis which is either a proper rotation of 360°, 180°, 120°, 90°, or 60° only, or an improper one which involves 'inversion' through the point. Both of these can be illustrated with a molecule like methane. A threefold rotation axis (120°) is evident when the molecule is viewed along an H–C bond whereas a fourfold improper rotation axis bisects an H–C–H angle in the plane of the other H–C–H so that a 90° rotation of one

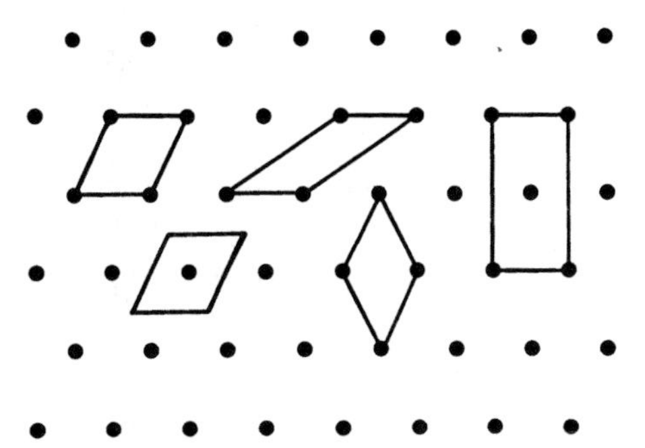

Figure 1. Each of the unit cells shown in this two-dimensional example is a valid choice for the lattice of points. The cell on the right is a centred cell and has twice the contents of the others.

Table 1. The crystal systems and related data for a chiral molecule

System	Necessary cell parameters	Bravais lattice[a]	Class[b]	Numbers[c]	Available space groups[d]	Multiplicity[e]
Triclinic	a, b, c, α, β, γ	P	1	1	P1	1
Monoclinic	a, b, c, β	P	2	3–4	P2, $P2_1$	2
	(α = γ = 90°)	C		5	C2	4
Orthorhombic	a, b, c	P	222	16–19	P222, $P222_1$, $P2_12_12_1$, $P2_12_12$	4
	(α = β = γ = 90°)	C		20–21	C222, $C222_1$	8
		F		22	F222	16
		I		23–24	I222, $I2_12_12_1$	8
Tetragonal	a (= b), c	P	4	75–78	P4, $P4_1$, $P4_2$, $P4_3$	4
	(α = β = γ = 90°)	I		79–80	I4, $I4_1$	8
		P	422	89–96	P422, $P42_12$, $P4_122$, $P4_12_12$, $P4_222$, $P4_22_12$, $P4_322$, $P4_32_12$	8
		I		97–98	I422, $I4_122$	16
Trigonal	a (= b), c, γ = 120°	P	3	143–145	P3, $P3_1$, $P3_2$	3
	(α = β = 90°)	R	3	146	R3	3
		P	312	149, 151, 153	P312, $P3_112$, $P3_212$	3
		P	321	150, 152, 154	P321, $P3_121$, $P3_221$	6
	a (= b = c), α = β = γ ≠ 90°	R	32	155	R32	6
Hexagonal	a (= b), c, γ = 120°	P	6	168–173	P6, $P6_1$, $P6_2$, $P6_3$, $P6_4$, $P6_5$	6
	(α = β = 90°)		622	177–182	P622, $P6_122$, $P6_222$, $P6_322$, $P6_422$, $P6_522$	12
Cubic	a (= b = c)	P	23	195, 198	P23, $P2_13$	12
	(α = β = γ = 90°)	F		196	F23	48
		I		197	I23	24
		P	432	207–8, 212–3	P432, $P4_232$, $P4_332$, $P4_132$	24
		F		209–210	F432, $F4_132$	96
		I		211, 214	I432, $I4_132$	48

[a] The lattice types are: P, primitive; C, C-face centred; F, all faces centred; I, body centred. Alternative lattice types may occasionally be chosen.
[b] The symbols under class refer to the rotational symmetry axes which are a characteristic of it.
[c] Number refers to the number in *International tables*.
[d] The Herman–Mauguin nomenclature for space groups gives the lattice type first, then the symmetry elements in an order which depends upon the crystal system. Refer to *International tables for X-ray crystallography*, Volume A for a fuller explanation of these symbols.
[e] Multiplicity gives the number of copies of the asymmetric unit in the unit cell.

hydrogen about this axis brings it to a point on the opposite side of the C atom to an adjacent H atom. Proteins are made up of L-amino acids and nucleic acids have a chiral ribose unit which preclude centres or mirrors. The combination of these symmetries and the crystal systems leads to the 32 *point groups* or *crystal classes*, of which only 11 can accommodate protein molecules. The rotations referred to above are the only ones allowed in the formation of a crystal but of course other rotations about a point within the molecule are possible as in the case of a spherical virus which has 532 point group symmetry. Only the threefold and twofold axes can be exploited in building up the crystal, leaving the fivefold axis as a *non-crystallographic* symmetry element.

As well as the rotational symmetry possibly present in a unit cell, translational relationships between molecules also exist. The spatial repetition of a crystal is such that a convenient packing involves axes which combine a rotation with a translation. For example, if a rotation of 180° together with a translation of half a unit cell along the axis of rotation is applied twice, it will produce not the initial molecule (as with a pure rotation) but an equivalent one in the next cell. Such an axis is a *screw axis* and several consistent types exist.

It can be shown mathematically that there are only 230 combinations of these symmetry elements possible in three dimensions. Thus, any crystal *must* have a unit cell which conforms to one of these combinations, its *space group*. Further, the presence of symmetry elements within a unit cell means that there are at least two copies of the molecule which are related by an algebraic relationship: if there is an atom at position x, y, z in a cell with a screw axis parallel to the b axis, there *must* be an atom at $-x$, $\frac{1}{2} + y$, $-z$. The effect of this, is to reduce the crystallographer's problem to one of locating the atoms in the *asymmetric unit*, rather than in the whole unit cell. Because all proteins and nucleic acid crystals comprise only one optical isomer, there are only 65 space groups available for such chiral molecules. *Table 1* shows the available crystal systems, classes and space groups for a protein.

2.3 How do X-rays interact with crystals?

The explanation of how X-rays are scattered by crystals is largely the result of a beautiful simplification by Bragg, resulting in the law which bears his name. Consider a crystal lattice, represented in *Figure 2* by the rows of points A, B, C. For X-rays X_2 scattered from row 2 to enhance those scattered from row 1, X_1, there must be an integral number of wavelengths difference. The relationship between the spacing of the rows, d, the wavelength, λ, and the angle at which the emergent ray is observed relative to the direction of the rows, θ, is:

$$n.\lambda = 2d.\sin\theta$$

Thus, as Bragg pointed out, X-ray diffraction can be regarded as the *reflection* of the beam of X-rays from the planes of points in the crystal lattice. Provided

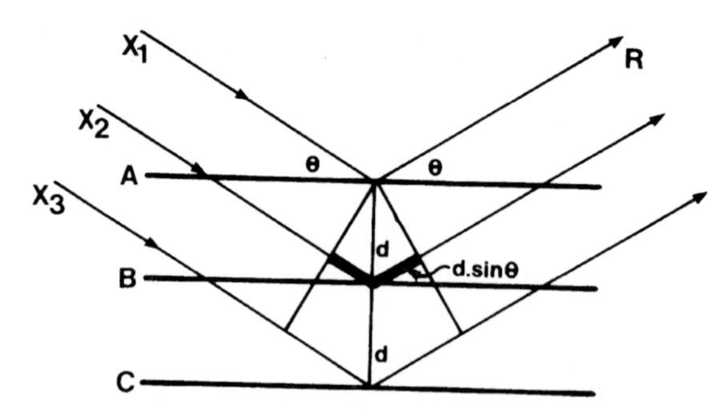

Figure 2. X-rays (X_1, X_2, X_3) reflected from lattice planes A, B, C. To observe a scattered beam of X-rays in direction R, the thickened path must equal a whole number of wavelengths. The ray from plane C travels twice as far as that from B, and so on.

there are a large number of planes contributing to the interference, the position in space at which a given reflection is observed is highly defined. These positions are defined by the crystal lattice and since very few, if any, atoms actually lie on the lattice points, the scattered intensity is modulated by the atomic arrangement within the unit cell. To repeat, the direction of a diffracted ray is defined by the crystal lattice, the intensity of the ray depends upon the atomic arrangement within the unit cell. One further point concerns n, the order of diffraction, which is the number of wavelengths difference between the scattering from adjacent planes; the higher the order, the larger the angle of scattering. Alternatively, the scattering can be considered as arising from planes which are closer together: e.g. using the equation above, it can be seen that a reflection at θ can be considered either as the nth order from planes of spacing d, or the first order from planes of spacing d/n. Crystallographers generally adopt the latter approach.

A diffraction pattern for a protein crystal contains many reflections which must be appropriately indexed and the most convenient system is to use the order of diffraction with respect to each of the unit cell axes. The Miller indices as they are called, which were derived originally to label crystal faces for mineralogical studies, are illustrated in *Figure 3*. Each index along the a, b, and c axes, respectively, is derived by taking the reciprocal of the intercept that the first plane of the set not passing through the origin, makes with each axis in turn. Thus, the 100 planes are the set which have a spacing of $a \times 1/1$ on the x axis, $b \times 1/\infty$ on the y axis, and $c \times 1/\infty$ on the z axis. The 200 planes have a spacing $a \times 1/2$ on the x axis, and so on. Notice that the planes $h00$ are all parallel to one another but the spacing decreases with increasing h. Hence the angle of diffraction increases with increasing h, consistent with Bragg's Law. The letters h, k, and l are used to refer to the indices in general terms.

Each of the many sets of planes defined by the lattice gives rise to one reflection and *Figure 4* shows the relationship in two dimensions of the planes in the crystal (real space) to the points in diffraction space or *reciprocal space*.

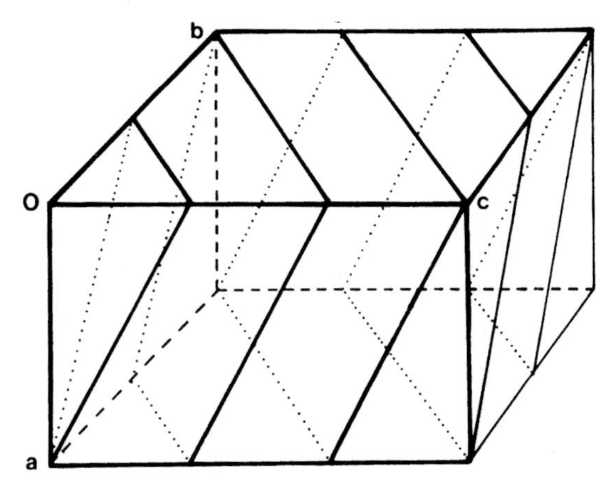

Figure 3. The set of planes 123 are shown as they cut a unit cell. The intercepts on *b* occur every 1/2 and on *c* every 1/3.

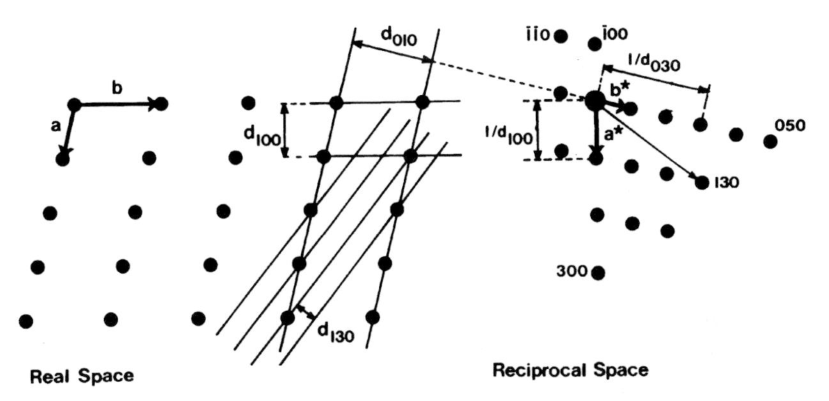

Figure 4. A diagram illustrating the relationship between sets of planes in a crystal in real or direct space and points representing a diffracted X-ray beam in reciprocal (or diffraction) space. Notice that the direction from the origin of reciprocal space (large point) to any point, e.g. 130, is perpendicular to the planes in the crystal and that the length is proportional to the reciprocal of the plane spacing.

The points can be seen to make up another lattice (reciprocal lattice) whose axes and angles are derived from those of the crystal. This idea can be extended to three dimensions. It is important to realize that each reflection contains a contribution from every atom in the crystal and, conversely, each atom in the crystal contributes to every reflection. Thus, as the crystal is moved about in the X-ray beam, reflections flash out and can be recorded when the geometrical arrangement of X-ray beam, crystal orientation, and detector satisfies Bragg's Law.

To help understand diffraction from a crystal, there is a construction intro-

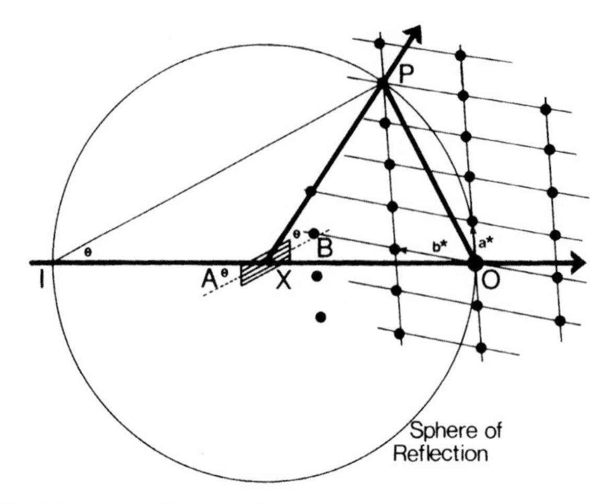

Figure 5. The Ewald construction. For clarity, this is shown as a planar diagram but IXO is the diameter of a sphere of radius $1/\lambda$.

duced by Ewald and shown in *Figure 5*. As we move the crystal, the reciprocal lattice also moves about a fixed origin. With the crystal, X, as centre, a sphere is drawn of radius $1/\lambda$ and the origin, O, of the reciprocal lattice is taken as the point where the X-ray beam leaves the sphere after passing through the crystal. As the crystal is rotated about the z axis (perpendicular to the page) the reciprocal lattice rotates until the point P lies on the surface of the sphere. The point P is the 410 reflection arising from the planes of spacing d_{410}. The angles at IX and XP, i.e. IXA and BXP are equal to θ so that OXP = 2θ and OP is perpendicular to the crystal planes AXB. Now OP = $2 \times$ XO $\times \sin\theta$ = $2 \times (1/\lambda) \times \sin\theta$. However, OP = $1/d_{410}$ and so $1/d_{410} = (2/\lambda) \times \sin\theta$ which is Bragg's Law. Thus, the Ewald sphere gives a readily understandable way of relating the orientation of the crystal to the diffraction pattern observed. In order to collect a set of X-ray data, it is necessary to move the crystal (and in some methods, the detector) in such a way that every reciprocal lattice point passes through the sphere of reflection (*Figure 6*). There are various ways of achieving this, some of which are described in Section 4.

The space group in which a molecule crystallizes may impose certain conditions on the reflections which can be observed so that by looking at the diffraction pattern of the crystal, it is often possible to determine the space group unambiguously. Furthermore, the higher the symmetry of the crystal, the less data is actually required to be collected. A diffraction pattern has a centre of symmetry since reflections in opposite directions from the same planes must have the same intensity ($I(h\,k\,l) = I(\bar{h}\,\bar{k}\,\bar{l})$ is Friedel's Law) (see *Figure 3*). Thus the diffraction symmetry shown in *Table 2* has a centre of symmetry even though the space groups do not. The effects of the lattice type and symmetry elements upon the diffraction pattern are shown in *Table 3*

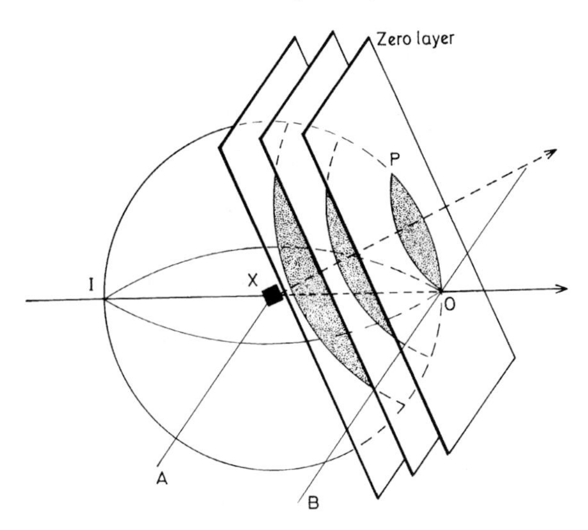

Figure 6. The Ewald sphere intersected by several reciprocal lattice layers. As the crystal is moved, the reciprocal lattice pivots about O but XA and OB remain parallel at all times. As shown, a film placed perpendicular to XO will record a series of concentric circles. As the crystal is rotated through a small angle about AX, the circles will become extended into lunes as the neighbouring spots on each level pass through the sphere of reflection.

and the effect can be explained with reference to *Figure 2*. If the beam X_3 scattered from row C is one wavelength behind X_1 scattered from row A, then X_2 scattered from row B is exactly half a wavelength behind and it will cancel out the reinforcing contributions from rows A and C. Thus, interposing planes midway between the planes separated by the unit cell repeat as is the case for a centred lattice, leads to a systematic absence of reflections. Further, if a twofold screw axis is perpendicular to the planes, there will always be an identical (but rotated) set of scatterers to row A, on row B. Only when the index is even along the axial direction will constructive interference occur and the reflection be observed. Notice that simple rotation axes do not generate any systematic absences.

2.4 How is a protein crystal structure solved?

The formation of a magnified image by a light microscope involves collecting all of the scattered light waves in the objective lens which recombines them in the correct way to produce the magnified image. But what is 'the correct way'? Associated with each wave is not only its amplitude but also its phase relative to the unscattered light. The focusing by the objective lens uses both amplitude and phase to produce the magnified image. In the case of X-rays, the crystal produces a diffraction pattern which needs to be recombined in the correct phase relationship but in this case, no lens exists which is able to perform the task and in recording the pattern as one must, the vital phase

Table 2. Equivalent data for the chiral point groups

System	Class	Laue group	Equivalent reflections[a]
Triclinic	1	−1	$I(hkl)$
Monoclinic	2	2/m	$I(hkl)$, $\underline{I(\bar{h}k\bar{l})}$[b]
Orthorhombic	222	mmm	$I(hkl)$, $\underline{I(\bar{h}\bar{k}l)}$, $\underline{I(\bar{h}k\bar{l})}$, $I(h\bar{k}\bar{l})$
Tetragonal	4	4/m	$I(hkl)$, $\underline{I(\bar{k}hl)}$, $I(\bar{h}\bar{k}l)$, $I(k\bar{h}l)$
	422	4/mmm	$I(hkl)$, $\underline{I(\bar{k}hl)}$, $I(\bar{h}\bar{k}l)$, $I(k\bar{h}l)$
			$\underline{I(h\bar{k}\bar{l})}$, $I(\bar{k}\bar{h}\bar{l})$, $I(\bar{h}k\bar{l})$, $I(kh\bar{l})$
Trigonal[c]	3	−3	$I(hkl)$, $\underline{I(kil)}$, $I(ihl)$[d]
	312	−3m1	$I(hkl)$, $\underline{I(kil)}$, $I(ihl)$, $\underline{I(\bar{k}\bar{h}\bar{l})}$, $I(\bar{i}\bar{k}\bar{l})$, $I(\bar{h}\bar{i}\bar{l})$
	321	−31m	$I(hkl)$, $\underline{I(kil)}$, $I(ihl)$, $\underline{I(kh\bar{l})}$, $I(ik\bar{l})$, $I(hi\bar{l})$
Rhombohedral	3	−3	$I(hkl)$, $\underline{I(klh)}$, $I(lhk)$
	32	−3m	$I(hkl)$, $\underline{I(klh)}$, $I(lhk)$, $\underline{I(\bar{k}\bar{h}\bar{l})}$, $I(\bar{l}\bar{k}\bar{h})$, $I(\bar{h}\bar{l}\bar{k})$
Hexagonal[c]	6	6/m	$I(hkl)$, $\underline{I(kil)}$, $I(ihl)$, $\underline{I(\bar{h}\bar{k}l)}$, $I(\bar{k}\bar{i}l)$, $I(\bar{i}\bar{h}l)$
	622	6/mmm	$I(hkl)$, $\underline{I(kil)}$, $I(ihl)$, $\underline{I(\bar{h}\bar{k}l)}$, $I(\bar{k}\bar{i}l)$, $I(\bar{i}\bar{h}l)$,
			$\underline{I(kh\bar{l})}$, $I(\bar{i}\bar{k}\bar{l})$, $I(hi\bar{l})$, $I(\bar{h}\bar{i}\bar{l})$, $I(\bar{k}\bar{h}\bar{l})$, $I(\bar{i}\bar{k}\bar{l})$
Cubic	23	m−3	$I(hkl)$, $\underline{I(\bar{h}\bar{k}l)}$, $\underline{I(\bar{h}k\bar{l})}$, $I(h\bar{k}\bar{l})$, $\underline{I(klh)}$, $I(lhk)$,
			$I(\bar{k}\bar{l}h)$, $I(k\bar{l}\bar{h})$, $I(\bar{k}l\bar{h})$, $I(\bar{l}\bar{h}k)$, $I(\bar{l}h\bar{k})$, $I(l\bar{h}\bar{k})$
	432	m−3m	$I(hkl)$, $\underline{I(\bar{h}\bar{k}l)}$, $\underline{I(\bar{h}k\bar{l})}$, $I(h\bar{k}\text{-}l)$, $\underline{I(klh)}$, $I(lhk)$,
			$I(\bar{k}\bar{l}h)$, $I(k\bar{l}\bar{h})$, $I(\bar{l}\bar{h}k)$, $I(\bar{l}h\bar{k})$, $I(l\bar{h}\bar{k})$, $I(\bar{k}l\bar{h})$,
			$\underline{I(\bar{k}hl)}$, $I(\bar{h}lk)$, $I(\bar{l}kh)$, $I(h\bar{l}k)$, $I(l\bar{k}h)$, $I(k\bar{h}l)$,
			$I(hlk)$, $I(lk\bar{h})$, $I(kh\bar{l})$, $I(\bar{k}\bar{h}\bar{l})$, $I(\bar{l}\bar{k}\bar{h})$, $I(\bar{h}\bar{l}\bar{k})$

[a] The reflections listed here are identical . If Friedel's Law holds then $I(hkl) = I(\bar{h}\bar{k}\bar{l})$ and this generates an equal number of equivalent reflections. In protein crystallography, anomalous scattering which leads to a breakdown in Friedel's Law, is used to help with phasing the reflections and so the two sets, equivalent to $I(hkl)$ and $I(\bar{h}\bar{k}\bar{l})$ must be kept separate.
[b] The underlined reflections are those which are required to specify the Laue symmetry with the others being generated by repeated application of the symmetry elements.
[c] The axes in the trigonal and hexagonal systems referred to here are $a = b, c, \alpha = \beta = 90°, \gamma = 120°$
[d] When hexagonal axes are being used, $i = \overline{hk}$.

information is lost. It must be calculated and this 'phase problem' is central to crystallography. Ironically, if the positions of the atoms are known, then the phase for each reflection can be calculated. Whilst this phase problem may seem insuperable, if the positions of only a few heavy atoms are known, whether these are added by soaking into crystals in the traditional way, or introduced during protein biosynthesis with selenomethionine (see Chapter 3), their contribution can be calculated and this is generally sufficient to solve the phase problem for a protein. The preparation of heavy metal derivatives of proteins has been dealt with in Chapter 13. It should be pointed out that molecular replacement (9) is applicable where a similar structure already exists and this is increasingly found to be the case.

A phase must be calculated for each reflection to be included in the calculation of the electron density map. The more X-ray reflections that are phased and included, the clearer the map will be and the better will be the resulting model of the protein. Thus the *resolution* of the data is usually reported and this refers to the minimum plane spacing included in the calculation; thus for a

Table 3. Conditions affecting possible reflections

Element	Symbol	Reflection observed for	Notes
Primitive lattice	P		
Lattice centred on the C face	C	hkl with h + k even	The C face is contained by **a** and **b**
Face centred lattice	F	hkl with h, k, and l all odd or all even	
Body centred lattice	I	hkl with h + k + l even	
Rhombohedral lattice	R	–h + k + l = 3n	'= 3n' means divisible by 3
Twofold screw axis ll c	2_1	00l with l even	For an axis along a, the row is h00
Threefold screw axes ll c	3_1, 3_2	00l with l = 3n	The two possible threefold axes have the same pitch but opposite hands
Fourfold screw axes ll c	4_1, 4_3	00l with l = 4n	
	4_2	00l with l even	cf. the twofold screw axis
Sixfold screw axes ll c	6_1, 6_5	00l with l = 6n	
	6_2, 6_4	00l with l = 3n	cf. the threefold screw axes
	6_3	00l with l even	cf. the twofold screw axis

3.5 Å map, all reflections with plane spacings greater than or equal to 3.5 Å will be included. The higher the resolution, the greater the amount of X-ray data which must be measured. Disregarding the symmetry of the reflection data, the total number of reflections is approximately $5V/d^3$ where V is the unit cell volume and d is the resolution.

2.5 Importance of preliminary characterization

There are a number of reasons why the preliminary characterization of a newly crystallized molecule is important. Most obviously, the first point to establish is that the crystal does diffract X-rays. As part of the process of checking that the crystal does diffract, some idea of the crystal lifetime in the X-ray beam will be obtained together with the resolution which can be achieved. Even when a crystal appears perfect, it should not be assumed that it will be suitable for X-ray work. Occasionally, some or all of the crystals in a batch give no discernible diffraction pattern. The reason for this is obscure but possible avenues to explore before abandoning the particular crystallization conditions used, are:

(a) Try crystals from different drops, tubes, or preparations.

(b) Search for crystals with a different morphology and X-ray them; sometimes different forms appear in the same tube.

(c) Cool the crystal before and during the X-ray exposure, possibly down to liquid nitrogen temperatures (10). If the crystal has already been cooled, try it at room temperature.

(d) Use synchrotron radiation with a short wavelength. It has been found to extend the lifetime of sensitive crystals (11).

(e) Attempt to crosslink the molecules in the crystal with a bifunctional reagent such as glutaraldehyde (12).

These ideas may also be worth trying if the crystals produce feeble or rapidly fading diffraction, and (e), in particular, may allow successful handling of crystals which are very fragile.

If the spots obtained on initial images are streaked rather than the well-defined spots illustrated in *Figure 7*, then the crystal is likely to have a degree of disorder which may render successful structure determination impossible. The only recourse then is to re-examine the crystallization procedure.

The aim of the preliminary X-ray investigation should be to determine the unit cell dimensions and the space group. Not only must these be known to solve the crystal structure but also, with the crystal's X-ray lifetime, they dictate the strategy for efficient data collection. The amount of data to be collected is determined by the diffraction symmetry of the crystal and it is often possible to reduce the number of exposures by ensuring that the crystal

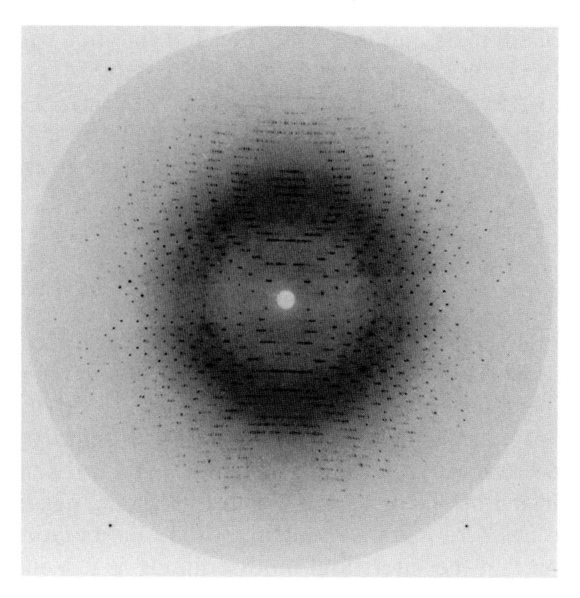

Figure 7. A 3° oscillation photograph of a hexagonal, cytochrome c_4 crystal taken with synchrotron radiation. Lunes from several major zones ([100], [$1\bar{1}0$], and [010]) can be seen clearly. The rotation axis was horizontal.

is mounted in a particular way. For example, it is best to mount a hexagonal crystal with the sixfold axis roughly parallel to the rotation axis of the instrument.

It is normal practice to determine the volume occupied per unit molecular weight (V_m, often called the Matthews' coefficient) since this can be used to determine the number of molecules in the asymmetric unit. V_m has been found to be around 2.4 Å^3/dalton for globular protein crystals, although this value is subject to quite large fluctuations (13). It is obtained by dividing the unit cell volume by the product of the protein molecular weight and the number of equivalent positions (asymmetric units). Unfortunately, it is often found with large unit cells that more than one value of V_m is reasonable and in such a case some biological insight may help resolve any ambiguity. If it is possible to determine the crystal density (see Chapter 2) and the weight loss on drying, the protein molecular weight can be calculated, which when compared with the known value, also gives the number of molecules per asymmetric unit. The approximate solvent content can also be calculated from the formula:

$$V_{sc}(\%) = 100(1 - 1.23/V_m).$$

One final point about the preliminaries is that biological information may emerge about the subunit structure. If it is found that the asymmetric unit contains half of the expected molecular weight, the protein must consist of an even number of subunits and it is probable that the molecular twofold axis coincides with a crystallographic one. This will be consistent with the space group which must possess such a symmetry element. Conversely, if a crystal is found to have three or four molecules in the asymmetric unit of a relatively low symmetry space group, then one should be alerted to the possibility of having missed a higher symmetry space group.

3. Mounting crystals

Mounting a protein crystal is a procedure which requires a reasonable degree of manual dexterity. It is impossible to be dogmatic about the right and wrong way, and each person develops their own technique, modifying it as required from protein to protein depending on the size, strength, temperature behaviour, need to exclude oxygen, or toxicity. Although early workers did dry their crystals (14), drying out of mother liquor in the crystal generally disrupts it such that no useful data can be collected. Mounting methods are therefore designed to maintain the interstitial mother liquor as it is in the drop from which the crystal grew. 'Flash-cooling' is a way of greatly reducing radiation damage (15–17) but it can also help with the problem of fragile crystals by preventing the loss of the interstitial water necessary to maintain crystal integrity. Indeed, nowadays many laboratories routinely freeze their crystals.

3.1 Initial examination with a microscope

3.1.1 Observation

Well-formed protein crystals examined under the light microscope exhibit a symmetric arrangement of edges and faces which are related to the packing of the molecules. Thus, examination of crystal morphology may give a first glimpse of the symmetry of the unit cell. A stereo-zoom dissecting microscope, ideally fitted with a crossed polarizing attachment, with a magnification in the range $\times$ 10 to $\times$ 40 is best for such examination since crystals which cannot be readily seen with such an instrument are probably not going to diffract sufficient X-rays, even with synchrotron radiation. It is important to ensure that the illuminating light source does not heat the microscope stage lest undue evaporation and denaturation occurs. The use of crossed polarizers can indicate the direction of a principal axis. Rotating the crystal on the stage in the dark field (polarizer and analyser at 90°), the crystal appears as a light colour until an optic axis lies along the direction of the polarizer whereupon extinction occurs, depending on the crystal system. During a full rotation, extinction occurs every 90°. This effect will not be observed for cubic crystals, for tetragonal, trigonal, and hexagonal crystals viewed along their unique axis or for non-crystalline material. Note that the crystal should not be contained in a plastic container (like a tissue culture plate) if polarized light is to be used because these containers affect the polarization, usually producing splendid colours. Salt crystals are usually highly coloured, even if they are small and, if a crystal is thought to be salt rather than protein, pressure with a fine probe will produce an audible 'plink' as the tip slips off the hard salt crystal. A protein crystal on the other hand will shatter with very little pressure at all. It is well worth getting to know the crystal habit and its relationship to the axes since this saves considerable time if alignment in the X-ray beam is required.

3.1.2 Selection of a crystal for mounting

Protein crystals with dimensions of 0.2–0.5 mm are most suitable for use in an X-ray diffraction experiment. Use of smaller crystals is possible, however the diffraction pattern tends to be weaker requiring longer exposure times and possibly poorer resolution. On the other hand, crystals much larger than 0.5 mm may not be uniformly bathed in the X-ray beam (depending on the size of beam collimator used) and generate their own problems associated with absorption of X-rays. Larger crystals may also pose problems in a low temperature experiment because of difficulties in freezing them uniformly.

Crystals for X-ray work should be single and should appear transparent (containing no cracks) with well defined edges and faces. Birefringent single crystals, when observed under a polarizer, extinguish light sharply when rotated through 360°. Less obviously twinned or multiple crystals may sometimes be detected if different sections of the crystal extinguish light at different rotations of the microscope stage. Crystals which have grown into one another, or have grown as clumps may be carefully split using a fine probe or a fresh scalpel blade. Gently touch the crystal at the point where the extra piece joins the chosen crystal keeping the blade parallel to the direction in which the crystals are to be separated. A gentle pressure is usually all that is required since crystals will generally cleave readily along the axial directions.

3.2 The basic techniques

The methods suggested take practice and a number of trials, preferably with a batch of old or non-precious crystals, before a decision can be made as to which steps are best suited to maintaining the crystal. The first involves drawing the crystal up into the Lindemann tube by a pipette or syringe attachment, allowing controlled movement of the crystal and buffer in the capillary. For flash-freezing, several methods are possible but the one described in Section 3.2.4 has the advantage of being straightforward and as such is worth practising until perfected.

3.2.1 Mounting for room temperature

This procedure is summarized by *Protocol 1* and illustrated by *Figure 8*.

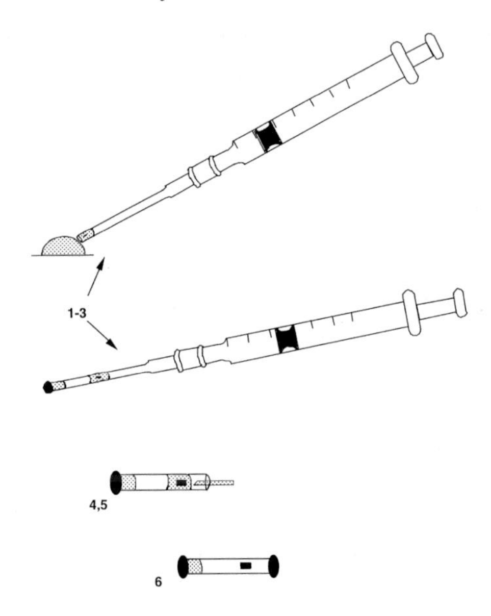

Figure 8. A diagram of the steps involved in mounting a crystal. The numbers refer to the steps described in *Protocol 1* in the text.

Protocol 1. Mounting crystals for room temperature data collection

Equipment and reagents

- Glass slides and coverslips
- Lindemann tubes (thin-walled glass or quartz capillary tubes, 1.0 or 0.7 mm diameter)
- A supply of pipettes sufficiently small to fit inside a 1.0 mm Lindemann tube: use smaller bore Lindémann tubes or Pasteur pipettes drawn to narrow diameter in a flame
- Small glass vials ('pots' made by cutting a vial to give a container perhaps 8 mm in diameter by 6 mm deep are ideal)
- Sealing grease (silicone grease, petroleum jelly)
- Modelling clay (Plasticine)
- Probe with a fine tip (like a sewing needle)
- Scalpel with a new and pointed blade
- Spirit burner, low temperature soldering iron, or Bunsen burner with which to melt wax, wax, forceps, supply of appropriate buffer
- Filter paper strips or cotton thread for soaking up excess mother liquor inside the Lindemann tubes
- Nylon fishing line, fine wire, or glass fibres for manipulating crystals
- Diamond cutting tool for quartz capillaries
- Rubber or plastic tubing to connect Lindemann capillary to syringe
- Disposable tuberculin (1 ml) syringe

Method

1. Attach a 1.5 cm length of rubber tubing to the end of a disposable tuberculin syringe (1 ml). With the diamond tool, score a Lindemann tube near the closed end and break it neatly. Insert the wide end of the

Lindemann tube into the rubber tubing and if necessary, roll back the ends of the tubing to improve the seal. In this way, you have created a narrow-bore pipette.

2. Draw the crystal with a small amount of mother liquor or handling buffer into the Lindemann tube. The tuberculin syringe is small enough that the apparatus can be held in one hand with the thumb available for drawing up on the plunger. Remove the end of the tube from the crystal droplet and continue drawing the crystal further up the Lindemann tube. With the crystal at the desired height in the tube, draw a final small plug of mother liquor into the end of the Lindemann tube.
3. Seal the open end of the Lindemann tube with wax. Soak a small piece of tissue in water and drape the wet paper over the tube at the height of the crystal. This is to protect the crystal from heat conduction up the tube while melted wax is being applied to the end. Seal the end of the tube with wax. Applying a small piece of Plasticine to the wax makes handling easier and allows the capillary to be stuck on a microscope slide or the table top for subsequent manipulation.
4. Score and break with the diamond tool the Lindemann tube a second time — this time 'above' the crystal. If breaking or cutting a glass tube without scoring it, add a drop of wax just to the crystal side of where the break is to be made, but well clear of the crystal, to prevent the tube collapsing when being broken.
5. Dry the remaining buffer from around the crystal with the aid of a shred of filter paper inserted through the open end of the tube. If necessary, larger volumes can be removed with a finely drawn-out Pasteur pipette or small bore Lindemann/syringe assembly as in step 1, before the drying stage. A dry mount is preferred for two reasons. The faces are more easily visible when aligning the crystal and the absence of solvent may reduce the effects of crystal slippage. It should be kept in mind, however, that in this dry atmosphere, the crystal is susceptible to solvent loss, thus the following steps should be performed as quickly as possible.
6. Seal the other end with wax using the wet tissue draped once again over the tube to protect the crystal.

Many variations are possible at the discretion of the mounter. For example, it may be preferable to have two plugs of buffer in the Lindemann tube; one on either side of the crystal. This can be accomplished by adding a small amount of buffer to the top of the tube before the final wax seal is applied. If it is necessary to reposition the crystal, opening up the wax plug is most easily done with a heated needle.

3.2.2 Mounting for low temperatures

It is often necessary to maintain the crystal at temperatures lower than ambient. For example, crystal stability may require working at around 0°C. Working at low (around 0°C) temperatures can be accomplished by housing the X-ray equipment in a cold room but, since such a system is not always available, the alternative is to pass a stream of cooled, dried air or nitrogen over the crystal from a nozzle mounted on the X-ray instrument as close to the crystal as possible (18). Usually the stream is co-axial with the capillary tube and goniometer head (see *Figure 9*) and a plastic collar added to protect the instrument. Below room temperature but above the freezing point of the solution, the crystals should be mounted in the cold room otherwise when the crystal is cooled on the camera or diffractometer, the temperature gradient produced by the cooler will lead to water distilling along the tube and dissolving the crystal. Arrange the cooler to pass cold gas along the tube and ensure that a plug of mother liquor is only at the end of the capillary closest to the cooler so that condensation will occur preferentially at the drop rather than at the crystal.

However, if cooling is required at all, it now makes sense to 'flash-freeze' the crystal and carry out the data collection at around 100 K (near liquid

Figure 9. A typical goniometer head with the crystal sealed in a capillary fixed upon it. The key shown is for adjusting the slides and arcs. It has a fine Allen key at the other end for locking the arcs after adjustment. The threaded ring at the base will screw onto an X-ray camera or diffractometer.

Figure 10. The cryo-loop method of crystal mounting. (a) A goniometer head with the magnetic base and mounted loop. (b) A close-up of the tip of a typical mounted loop of about 0.5 mm diameter. (c) A protein crystal flash-frozen in its cryo-solvent. (d) Typical equipment for handling frozen crystals. Left to right: a goniometer head, a magnetic base, a CrystalCap with mounted cryo-loop, an 18 mm cryo-vial, a cryo-vial in a plastic handling tube, a mounted cryo-loop in a plastic pipette tip for mounting a crystal, a plastic handling tube made from a Pasteur pipette and ideal for filling the cryo-vial with liquid nitrogen.

nitrogen temperature). As noted already, recent developments in the cryo-crystallography of biological molecules have meant that in many laboratories data collection at 100 K is now routine. A general overview of these developments is given by Rodgers (19). If it becomes apparent that very low temperatures will be required (because conventionally mounted crystals have unworkably short lifetimes in the X-ray beam) a different mounting procedure must be applied. Whilst there is some benefit in equilibrating the protein crystals in cryo-protectant, it is not strictly necessary, though it is essential to have a suitable cryo-protectant mother liquor. This can often be obtained by mixing crystal mother liquor with increasing concentrations of glycerol until a capillary containing the solution remains transparent when plunged into liquid nitrogen.

In addition to the equipment mentioned in Section 3.2.1, some special equipment is needed, both for mounting but also for X-ray work. A popular and convenient device for maintaining the crystal at 100 K whilst in the X-ray beam is the Cryostream made by Oxford Cryosystems but most X-ray generator manufacturers provide an equivalent. Much of the equipment for crystal mounting can conveniently be obtained from Hampton Research but it can also be hand-made in the laboratory. *Protocol 2* and *Figure 10* illustrates how the crystal is mounted in a cryo-loop and also shows a convenient and cheap way of handling the mounted loop once frozen in liquid nitrogen.

Protocol 2. Mounting crystals for cryo-crystallography

Equipment and reagents

- Mounted cryo-loops
- A magnetic base
- A CrystalCap: these are convenient both for handling and for long-term storage
- Handlers for the CrystalCap: these can conveniently be made out of disposable Pasteur pipettes or disposable pipette tips, but tweezers can also be used
- A goniometer head with arcs: modified heads (20, or bought from Charles Supper) simplify the placing and removing the crystal on the goniometer head, but are not strictly necessary for horizontal goniometer spindles (e.g. the Mar Image Plate), once the method below has been mastered
- A Dewar of liquid nitrogen

Method

1. Attach a holder (a 1 ml plastic pipette tip is suitable) to the magnetic base end of a mounted cryo-loop. Insert the vial part of the CrystalCap into a small (5 × 5 × 1 cm) expanded polystyrene float with a hole to fit the vial firmly, the open end being up.
2. Select the crystal to be frozen and place on the same slide a drop of cryo-protectant buffer solution.
3. Submerge the float and vial in the liquid nitrogen until it is cold (boiling ceases) and the vial is full of liquid nitrogen.
4. Carefully scoop up the crystal with the cryo-loop in which it will be held by surface tension, and immediately immerse it in the cryo-

protectant. The time required for cryo-protection varies depending on buffer system and cryo-protectant used. This will need to be determined by trial and error and it is sensible to begin with some less good crystals.

5. Carefully scoop up the crystal from the cryo-protectant and plunge it into the liquid nitrogen in the vial. Allow the magnetic base to cool down (boiling ceases) before screwing the cap onto the vial.
6. The crystal is now mounted and frozen—it must now be maintained at approximately this temperature for as long as it is required.
7. Either transfer the vial and cap to a suitable storage Dewar containing liquid nitrogen if the X-ray work is not to proceed immediately or take the crystal in the Dewar to the X-ray laboratory. To transfer the crystal to the X-ray diffractometer, the method will depend upon the exact arrangement of the goniometer spindle. If vertical (e.g. R-Axis Image Plate), a goniometer with an extension to permit the frozen crystal to be positioned so that it points downwards, is essential. If the spindle is horizontal (e.g. Mar Image Plate) proceed as follows.
8. Place the goniometer head with magnetic base attached on the spindle and adjust the z-translation so that the cryo-loop when in position will be in the centre of the cold gas stream. This adjustment is conveniently done with the mounted loop when finding a suitable cryo-protectant. It may be helpful to withdraw the nozzle of the cold stream slightly to allow some extra room for the next stage. Also arrange for the arc with the largest angular displacement to be vertical and at the upper extremity of its travel. The crystal when mounted will then be pointing downwards by some 20–30° depending on the goniometer head.
9. Attach the mounted loop to the base, unscrew the cap and withdraw the vial of nitrogen, holding the vial (re-)filled with liquid nitrogen in a holder so that the metal part of the cap is free to be located on the magnetic base. The crystal should now be in the stream of cold dry nitrogen. Care should be taken not to disturb the X-ray back-stop.
10. Reposition the nozzle, if necessary, to be as close to the crystal as possible without interfering with the X-ray beam. This is conveniently done by a second person as soon as the vial is removed. Ensure that there are no drafts in the laboratory which might deflect the flow from the cryo-cooler and similarly do not breathe at the crystal whilst mounting it.
11. Begin the X-ray measurements.

4. X-ray data

The fundamental data about a crystal which must be known before the structure solution can be attempted are the unit cell dimensions and the space group. Until relatively recently, these data were always determined first, in order that the strategy for data collection could be optimized, a necessary prerequisite for crystals with limited X-ray lifetime. Nowadays however, most data are recorded automatically by the oscillation/rotation method, often before the space group and cell dimensions are known, and the main purpose of examining the first images is to determine that the crystal is single, uncracked, and diffracts X-rays. In addition, some clues about the space group can be obtained from the symmetry of the pattern near the principal zones, but this is not really necessary. The data images are stored on tape or disc and the images further processed (fairly) automatically by computer. The unit cell dimensions are calculated and the space group is determined, once the data have been processed, by plotting out layers as 'mock' precession photos to observe the systematic absences more easily and hence determine the space group. Thus the strategy has now become one of shoot first and ask questions later.

4.1 Oscillation methods for data collection

Most data collection nowadays is done by the oscillation method (21) with some sort of area detector mostly, imaging plate systems. Recently, the development of the charge coupled device (CCD) has provided the sensitivity and dynamic range of the imaging plate but has eliminated the time-consuming scanning step typical of phosphor-imaging plate technology. Whatever device is used for data capture, however, the oscillation method remains the technique of choice. Unlike the precession method, it is not necessary and indeed is undesirable, that the crystal be perfectly aligned before data collection starts. The crystal is rotated through a small angle (0.1–1.5°) about an axis perpendicular to the X-ray beam. As the crystal rotates (about an axis through X perpendicular to the page in *Figure 5*), the successive reciprocal lattice planes (rotating about O in *Figure 5*) cut the Ewald sphere producing extended circles or lunes as shown in *Figures 6* and *7*. Several passes or oscillations through the rotation range minimize the effects of fluctuations of X-ray intensity. Provided the rotation angle is not too large, adjacent levels will not overlap and data from many layers can be collected on each image. The size of the oscillation range is chosen depending on the detector, crystal cell dimensions, Bravais lattice type, properties of the incident X-ray beam, and crystal mosaic spread. An estimate of the maximum permissible rotation angle can be obtained from:

$$\Delta\phi < (\mathrm{dmax}/\mathrm{q}).180/\pi - \Delta$$

where dmax is the maximum resolution for which data are required, q is the spacing of planes perpendicular to the X-ray beam (e.g. *a* when the *a* axis is parallel to the X-ray beam), and Δ, which is typically 0.1–0.3°, is the reflecting range of the crystal, or mosaic spread. The strategy adopted in current practice is to use a relatively large rotation range, or as large a range as possible.

However, the high degree of automation available with area detector and image plate software and the cheapness of disc storage allow oscillation ranges less than or comparable to the actual diffraction spot size to be used which in turn allows integration of the spot as it traverses the reflecting position. Oscillation ranges larger than a typical spot result in the collection of intensity not just of the spot itself but of background 'in front of' and 'behind' the spot as well, thus reducing the effective signal-to-noise ratio for that spot. Larger oscillation ranges, however, are used to minimize the time of overall data collection with image plates because of the relatively large time requirement for scanning each image before the plate can be used for collection of the next frame.

4.2 Optical alignment

Adjust the 'height' (the distance of the crystal from the base of the goniometer head) of the crystal using the adjustment on the instrument and the z-translation on the goniometer head, as necessary. Centring of the crystal is then carried out to ensure that it remains in the X-ray beam during rotation about the spindle. The two bottom sledges on the goniometer head are used to do this (NB: NOT the arcs). First, rotate the crystal through 360°, noting its position in the microscope cross-hairs at 0, 90, 180, and 270°. To centre the crystal, put one sledge perpendicular to the direction of view — this will normally correspond to either the 0/180° or the 90/270° positions. Move the sledge to place the centre of the crystal at the midpoint of the 0/180° (or 90/270°) readings and repeat for the other sledge. The process is repeated until the crystal (not the tube or the cryo-loop) is stationary through a full rotation. Note that the cross-hairs on the telescope may not define the centre of the rotation.

4.3 Crystal characterization with an area detector/image plate

It is assumed that assistance is available to the user getting started on data collection. Two pre-collection files are required for area detector data collection, the flood field and brass plate images, necessary to calibrate respectively intensity and spatial fluctuations on the surface of the detector. Ideally, these calibrations should be done every time the detector is moved to a new distance and they must be done at the distance at which data collection will occur. The calibrations are done using an ^{57}Fe source and cannot be done when the crystal is in place on the goniometer. Thus, initially, the calibrations must be carried out before any knowledge of crystal cell dimensions is available.

Calibration is also required for image plate data collection in order to determine the position of the beam as the centre of the diffraction pattern. This is often carried out by capturing the concentric circle diffraction pattern of wax but may also be recorded on each image if the beamstop has a tiny hole in it allowing a 'centre' spot to be exposed. A typical strategy for data collection is described in *Protocol 3*.

Protocol 3. Data collection

Equipment

- X-ray generator or synchrotron producing monochromatic radiation around 0.1 nm wavelength
- Imaging plate diffractometer
- Appropriate graphics workstation and software

Method

1. With the detector at the desired distance, centre the crystal in the X-ray beam as described above. The shorter the crystal-to-detector distance, the higher is the resolution to which data can be measured but the greater the likelihood of spot overlap. If nothing is known beforehand, it saves time to use the setting already in use.
2. Check the diffraction pattern by exposing a frame for an arbitrary length of time and oscillation range, for example, 120 sec and 0.25° with an area detector, 10 min and 1° with an image plate. If spots are visible to the edges of the image, it may be desirable to swing the detector out to a non-zero 2θ angle, or decrease the crystal-to-detector distance. This will help determine, according to Bragg's equation, the resolution to which the crystal diffracts.
3. Having decided the length of time to be spent exposing each frame and the oscillation range desired to achieve spot separation even at the edges of the detector, begin data collection. Depending on the system and programs used, it is recommended that data processing be started as soon as possible. It may become obvious while attempting to process the data that problems exist with the crystal. If this is the case, the decision can be made to end the measurement and try another crystal without wasting detector time.

Data processing packages are as varied as the hardware used to measure intensities. In general, after the calibrations of the detector face are made, the orientation of the crystal with respect to the laboratory system must be determined. A series of frames is read and a peak search procedure records the positions of strong, well-defined spots to be used in autoindexing. *Figure 11* shows a typical workstation screen with the observed pattern displayed and the predicted pattern superimposed. The unit cell dimensions, the crystal orientation, and the crystal-to-detector distance are modified to obtain the best fit of predicted to observed pattern. The procedure now to be described

(a)

(b)

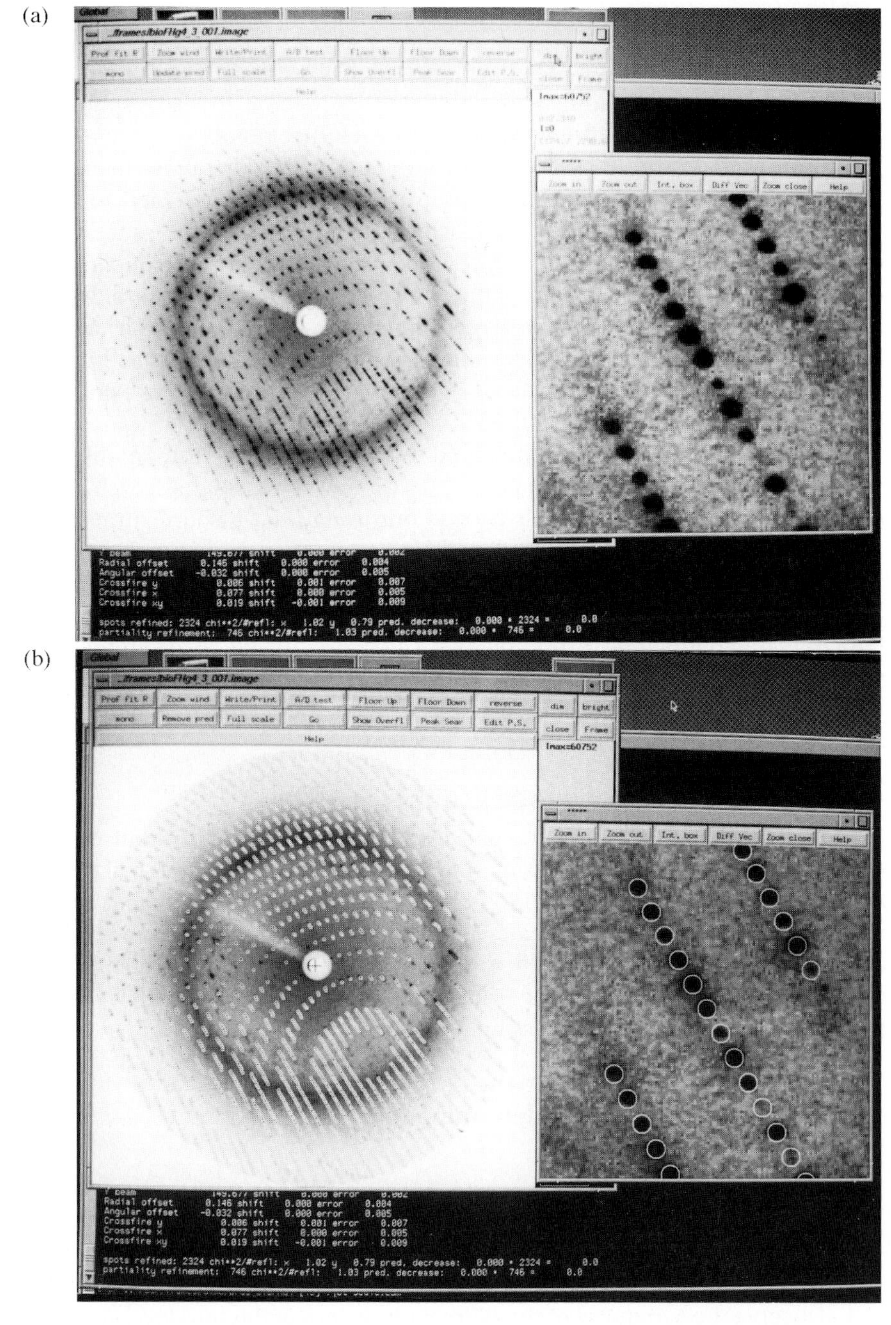

Figure 11. A picture of a typical workstation during the processing of oscillation data from an image plate system. (a) The image recorded is shown on the left with a magnified part of the image shown inset on the right from which it can be seen that the spots are single and not overlapping. (b) The same image as in (a) but with the predicted diffraction pattern superimposed. The inset on the right shows how well the prediction fits the image.

is that used in the program *XDS* developed by Kabsch (22). The autoindexing routine begins by assigning a reciprocal-space vector to each spot. Low resolution differences between these reciprocal lattice points are accumulated in clusters and are sorted by decreasing population. The first two which are at an angular separation $> 45^\circ$ are chosen and indices are assigned to them. These are used as a basis set from which the remaining difference-vector clusters can be indexed. Originally, it was expected that space group and cell dimensions of the crystal were known prior to running the autoindexing routine however contemporary algorithms allow both orientation and unknown cell dimensions to be determined. Alternative choices of cell dimensions are given with associated agreement factor allowing statistical consideration of all the possibilities.

The autoindexing routine employed by *DENZO* (23) uses a different algorithm coined 'real space indexing' whereby a complete search of all possible indices of a reflection is carried out using a Fast Fourier transform. Once the three best linearly independent vectors with minimal unit cell volume are found, the cell is 'reduced' to describe a standard basis for the description of the unit cell. In *DENZO*, a basis set for each of the 14 Bravais lattices is found and a distortion index is calculated for the peaks in the peak search list. The user must then, on the basis of the magnitude of the deviation from ideal Bravais lattice symmetry, decide upon most likely cell dimensions and the space group.

4.4 Determination of space group

As noted above, most software packages which are used to control area detectors provide a means of estimating resolution limits and unit cell dimensions. Thus there is no real need to obtain cell dimensions by an independent method. The software (see *Protocol 4*) also usually gives the Bravais lattice and all that remains to be done is to determine the space group which is best done by displaying the various principal zones (e.g. hk0, h0l, 0kl) on the workstation used for data processing.

Protocol 4. Space group determination

Equipment and reagents

- Appropriate graphics workstation, software and associated printer/plotter

Method

1. Observe the symmetry of the diffraction pattern of the zero level zones ('mock' precession photographs) which must be consistent with the unit cell parameters and lattice type already determined. This gives the diffraction (Laue) symmetry given in *Table 2* for the 11 relevant classes (and helps to ensure that principal zones have indeed been identified).

Diffraction symmetry always has an inversion centre. For example, a triclinic cell, P1, has –1 diffraction symmetry. It is the appearance of extra symmetry which allows labelling of crystal class. At this point the axes can be assigned as *a*, *b*, or *c* such that α, β, and γ are close or equal to 90° (unless a trigonal or hexagonal cell is suspected) and the cell is primitive (see *Table 2*). A zero layer photograph by definition, arises from either the hk0, h0l, or 0kl sets of planes. These can be assigned arbitrarily in the case of certain space groups. The *International tables for X-ray crystallography* (24) will help with the task of assigning axes according to crystallographic convention. In general, the unique axis is *b* for monoclinic cells and *c* for cells of higher symmetry. The upper layer images, which contain no reciprocal axes, must also be assigned as hkn, hnl, or nkl where $n \geq 1$.

2. Index the spots, h, k, l on each image. Be aware that systematically absent reflections also require indexing.
3. Analyse the systematically absent reflections in the diffraction pattern. This pin-points the space group often, but not always, uniquely. Use the axial absences to identify any screw axes.
4. Check that assignments of systematic absences are consistent with upper level images as well. The upper layers also allow, for example, distinction between a sixfold and a threefold axis. (These look the same on a zero level photograph.)
5. Identify, with use of the *International tables*, a list of space groups compatible with the observed diffraction patterns. In some cases there is no ambiguity: e.g. $P2_12_12_1$, whilst in others no distinction is possible until the structure solution is under way, e.g. I222 and $I2_12_12_1$ have identical systematic absences as have the enantiomorphs $P3_121$ and $P3_221$ where only the hand of the screw axis differs.
6. Try to find as high a symmetry space group which is consistent with your observations and work to lower symmetries as need be.
7. Determine the approximate number of molecules in the unit cell from the unit cell dimensions, the molecular weight of the molecule, and V_m. Knowing the crystal system helps in this, e.g. if the crystals are orthorhombic, there must be a multiple of four molecules in the unit cell. (Note that a 'molecule' may also be some identically repeated portion of the protein or polynucleotide.)

4.5 Other techniques for diffraction data collection

Precession photography gives an undistorted image of a reciprocal lattice plane. It does, however, require care and experience to align the crystal and it is fairly slow and as has been said, it is seldom used now. Two other methods which may be used are mentioned here only for completeness. Protein crystallographers do occasionally use the Laue technique but this is exclusively carried out at synchrotron sources since white radiation is required. Speed is the main benefit of this technique which can record a full diffraction pattern in a few seconds. It is not a technique used for the initial characterization of a newly crystallized protein, however. The other method which can be used is diffractometry. Modern four-circle diffractometers are the backbone of small molecule crystal structure laboratories and have sophisticated control programs which allow cell dimensions to be obtained with little if any user intervention. However, most instruments use Mo radiation and the crystal-to-detector distance is often not large enough to allow easy resolution of spots with spacings typical of proteins. The instrument consists of a series of concentric circles, three forming an Eulerian cradle capable of rotating the crystal to (nearly) any angle relative to the X-ray beam, the fourth moving the detector in a horizontal plane. Considering the Ewald construction again (see *Figure 5*), using the three circles of the goniometer, it is possible to orient the normal to any desired set of crystal planes in the horizontal plane in such a way that the normal bisects the angle between the incident X-ray beam and the detector. This satisfies Bragg's Law and the reflection is observed by stepping the crystal from one side of the exact bisecting position to the other, thus moving the crystal through the reflecting position. A plot of detector counts versus angle will then show a peak as the reflection passes through the Ewald sphere. If such an instrument is available, preferably with a Cu tube and an extension on the detector arm fitted with a helium path, it may be worth trying to determine the cell dimensions. Whilst this approach may appear to be the simplest, and with good crystals it is convenient and very accurate, it is very time-consuming since reflections are measured one at a time.

5. Concluding remarks

The object of this brief excursion into X-ray crystallography has been to introduce the ideas and methods required to collect the information necessary for the first publication on a new crystalline material. Such papers should include not only the purification and crystallizing conditions, which should be reproducible, but also the techniques employed to obtain the X-ray diffraction data and the crystal lifetime in the X-ray beam and on the shelf. The unit cell dimensions and space group together with the resolution obtainable from a crystal have been the main concern of this chapter. V_m, the number of molecules in the asymmetric unit, the solvent content, and any comments about the subunit structure are also generally mentioned. Increasingly, molecular replacement techniques will reveal similarities to known structures and so the 'crystallization note' is often superseded by the preliminary structure, obtained rapidly from the complete data set which is collected from the first crystals. Finally, protein crystallographers always enjoy talking about their subject and the number of groups around the world has risen considerably since the first edition of this book. You will have discovered that the technique requires a modicum of dedication and therefore do seek guidance in getting your project under way.

References

1. Glusker, J. P., Lewis, M., and Rossi, M. (1994). *Crystal structure analysis for chemists and biologists*. VCH, New York.
2. Stout, G. H. and Jensen, L. H. (1989). *X-ray structure determination*, 2nd edn. John Wiley, New York.
3. McRee, D. E. (1993). *Practical protein crystallography*. Academic Press Inc., New York.
4. Drenth, J. (1994). *Principles of protein X-ray crystallography*. Springer–Verlag, New York.
5. Blundell, T. L. and Johnson, L. N. (1976). *Protein crystallography*. Academic Press, London.
6. Bragg, W. L. (1968). *Sci. Am.*, **219**, 58.
7. Glusker, J. P. (1994). *Methods Biochem. Anal.*, **37**, 1.
8. Johnson, L. N. and Hajdu, J. (1990). *Eur. J. Biochem.*, **29**, 1669.
9. Rossmann, M. G. (1990). *Acta Cryst.*, **A46**, 73.
10. Hope, H. (1990). *Annu. Rev. Biophys. Biophys.Chem.*, **26**, 107.
11. Acharya, R., Fry, E., Stuart, D., Fox, G., Rowlands, D., and Brown, F. (1989). *Nature*, **337**, 709.
12. Quiocho, F. A. and Richards, F. M. (1964). *Proc. Natl. Acad. Sci. USA*, **52**, 833.
13. Matthews, B. W. (1968). *J. Mol. Biol.*, **33**, 491.
14. Hodgkin, D. C. and Riley, D. P. (1968). In *Structural molecular biology* (ed. A. Rich and N. Davidson), pp. 15–28. Freeman, San Francisco.
15. Henderson, R. (1990). *Proc. Roy. Soc. Lond.*, **B241**, 6.
16. Teng, T. Y. (1990). *J. Appl. Cryst.*, **23**, 387.
17. Garman, E. A. and Mitchell, E. P. (1996). *J. Appl. Cryst.*, **29**, 584.
18. Hajdu, J., McLaughlin, P. J., Helliwell, J. R., Sheldon, J., and Thompson, A. W. (1985). *J. Appl. Cryst.*, **18**, 528.
19. Rodgers, D. W. (1994). *Structure*, **2**, 1135.
20. Engel, C., Wierenga, R., and Tucker, P. A. (1996). *J. Appl. Cryst.*, **29**, 208.
21. Arndt, U. W. and Wonacott, A. (ed.) (1978). *The rotation method in crystallography*. North-Holland Publishers, Amsterdam.
22. Kabsch, W. (1988). *J. Appl. Cryst.*, **21**, 67.
23. Otwinowski, Z. and Minor, W. (1997). *Methods in enzymology* (eds C. W. Carter and R. M. Sweet), Academic Press, London. Vol. 276, pp. 307.
24. Hahn, T. (ed.) (1987). *International tables for X-ray crystallography*. D. Reidel Publishing Co., Dordrecht, Netherlands.

A1

List of suppliers

Aldrich-Chemical Co., Inc., 1001 W. St Paul Avenue, PO Box 355, Milwaukee, WI 53201, USA. (chemicals)

Alpha Laboratories Ltd., Eastleigh, Hampshire, UK. (multiple liquid dispenser)

American Can Company, Greenwich, CT 06830, USA. (Parafilm® 'M', laboratory film)

Amersham

Amersham International plc., Lincoln Place, Green End, Aylesbury, Buckinghamshire HP20 2TP, UK.

Amersham Corporation, 2636 South Clearbrook Drive, Arlington Heights, IL 60005, USA.

Amicon Division, W. R. Grace and Co., 72 Cherry Hill Drive, Beverly, MA 01915, USA. (filters, membranes)

Anderman

Anderman and Co. Ltd., 145 London Road, Kingston-Upon-Thames, Surrey KT17 7NH, UK.

Applied Biosystems, Inc., 850 Lincoln Center Dr., Foster City, CA 94404, USA and Birchwood Science Park North, Warrington, Cheshire WA3 7PB, England. (biochemical instrumentation, chemicals)

Appligene, route du Rhin, BP 72, 67402 Illkirch Cedex, France. (biochemicals)

Bachem, Hauptstrasse 144, CH-4416 Bubendorf, Switzerland. (detergents)

BDH Limited, Broom Road, Poole, BH12 4NN, UK. (electrophoresis products)

Beckman, 4550 Noris Canyon Road, PO Box 5101, San Ramon, CA 94583, USA. (centrifugation, pipetting station)

Beckman Instruments

Beckman Instruments UK Ltd., Progress Road, Sands Industrial Estate, High Wycombe, Buckinghamshire, HP12 4JL, UK.

Beckman Instruments Inc., PO Box 3100, 2500 Harbor Boulevard, Fullerton, CA 92634, USA.

Becton-Dickinson and Co., Clay Adams Div., 299 Webro Road, Parsippany, NJ 07054, USA. (Falcon plasticware)

Becton Dickinson

Becton Dickinson and Co., Between Towns Road, Cowley, Oxford OX4 3LY, UK.

Becton Dickinson and Co., 2 Bridgewater Lane, Lincoln Park, NJ 07035, USA.

Bender and Hobein GmbH, D-8000 Munchen 2, Lindwurmstrasse 71, Germany. (free flow electrophoresis)

Bijhoelt and Heuvelen SV, The Netherlands. (transparent and adhesive plastic foils)

BioBlock Scientific, BP 111, F-67403 Illkirch Cedex, France. (scientific equipments)

BioRad, 1414 Harbour Way South, Richmond CA 94804, USA. (HPLC, IEF equipments, biochemicals)

Boehringer Mannheim, PO Box 310120, D-6800 Mannheim 31, Germany. (biochemicals)

Bio

Bio 101 Inc., c/o Statech Scientific Ltd, 61–63 Dudley Street, Luton, Bedfordshire LU2 0HP, UK.

Bio 101 Inc., PO Box 2284, La Jolla, CA 92038–2284, USA.

Bio-Rad Laboratories

Bio-Rad Laboratories Ltd., Bio-Rad House, Maylands Avenue, Hemel Hempstead HP2 7TD, UK.

Bio-Rad Laboratories, Division Headquarters, 3300 Regatta Boulevard, Richmond, CA 94804, USA.

BioWhittaker, Inc., 8830 Biggs Ford Road, Walkersville, MD 21793, USA.

Boehringer Mannheim

Boehringer Mannheim UK (Diagnostics and Biochemicals) Ltd, Bell Lane, Lewes, East Sussex BN17 1LG, UK.

Boehringer Mannheim Corporation, Biochemical Products, 9115 Hague Road, P.O. Box 504 Indianapolis, IN 46250–0414, USA.

Boehringer Mannheim Biochemica, GmbH, Sandhofer Str. 116, Postfach 310120 D-6800 Ma 31, Germany.

British Drug Houses (BDH) Ltd, Poole, Dorset, UK.

Brookhaven Instrument Corp., 750 Blue Point Road, Holtsville, NY 11743, USA. (light scattering instrumentation)

Bunton Instrument Co., Inc., 615 South Stonestreet Avenue, Rockville, MD 20850, USA. (microgrippers)

Calbiochem Behring Diagnostics, 10933 N. Torrey Pines Road, La Jolla, CA 923037, USA. (biochemicals, detergents)

Cambridge Repetition Engineers Ltd., Green's Road, Cambridge, CB4 3EQ, UK. (dialysis buttons for crystallization)

CEA verken AB, S-152 01 Strängnäs, Sweden. (X-ray films)

Charles Supper Company Inc., 15 Tech Circle, Natick, MA 07160, USA. (crystallographic equipment)

CJB Developments Limited, Airport Service Road, Portsmouth, Hampshire PO35PG, UK. (large-scale preparative electrophoretic apparatus)

Cole and Palmer Instrument Co., 7425 N. Oak Park Avenue, Chicago, IL 60648, USA. (scientific equipments)

Corning, Inc., Science Products, MP-21–5–8, Corning, NY 14831, USA. (glassware, pipettes)

Costar Nucleopore®, One Alewife Center, Cambridge, MA 02140, USA and Costar Europe, Ltd., PO Box 94, 1170 AB Badhoevedrop Sloterweg 305a, 1171 VC Vadhoevedrop, The Netherlands. (titration and crystallization plates, pipettors)

Cruachem Ltd., West of Scotland Science Park, Acre Road, Glasgow G20 0UA.

Difco Laboratories

Difco Laboratories Ltd., P.O. Box 14B, Central Avenue, West Molesey, Surrey KT8 2SE, UK.

Difco Laboratories, P.O. Box 331058, Detroit, MI 48232–7058, USA.

Douglas Instruments Ltd., 255 Thames House, 140 Battersea Park Road, London SW11 4NB, UK. (automatic batch crystallization system)

Dow Corning Corp., Dow Corning Center, Box 0994, Midland, MI 48686–0994, USA. (silicone oil, grease)

Dupont de Nemours and Co., Concord Plaza, Wilmington, DE 19898, USA. (chemicals, instrumentation)

Du Pont

Dupont (UK) Ltd., Industrial Products Division, Wedgwood Way, Stevenage, Herts, SG1 4Q, UK.

Du Pont Ltd., NEN Life Science Products, PO Box 66, Hounslow TW5 9RT, UK.

Du Pont Co. (Biotechnology Systems Division), P.O. Box 80024, Wilmington, DE 19880–002, USA.

Dynatech Laboratories, Inc., 14340 Sullyfield Circle, Chantilly, VA 22021, USA. (titration plates for crystallization robots)

Eastman-Kodak Co., 343 State St., Rochester, NY 14650, USA and Kodak House, Station Road, Hemel Hempstead, Herts HP1 1JU, UK. (chemicals, films)

Enraf Nonius Delft, PO Box 483, 2600 AL Delft, The Netherlands. (diffractometry)

Euromedex, Produits de Recherche, 29 rue Herder, F-67000 Strasbourg, France. (chemicals, protease inhibitors)

European Collection of Animal Cell Culture, Division of Biologics, PHLS Centre for Applied Microbiology and Research, Porton Down, Salisbury, Wilts SP4 0JG, UK.

Everett's Co., Parkgate, Nr, Southampton, UK. (vacuum wax, seals, and lubrifiants)

Falcon (Falcon is a registered trademark of Becton Dickinson and Co.).

Fisher Scientific Company, 711 Forbes Avenue, Pittsburgh, PA 15219–4785, USA. (biochemicals, scientific equipments)

Flow Laboratories, Woodcock Hill, Harefield Road, Rickmansworth, Herts. WD3 1PQ, UK.

Flow Laboratories International SA, via Lambro 23/25, I-20090 Opera (MI), Italy. (biochemical equipments, Linbro plate, CrystalPlate and coverslips)

Fluka

Fluka-Chemie AG, CH-9470, Buchs, Switzerland.

Fluka Chemicals Ltd., The Old Brickyard, New Road, Gillingham, Dorset SP8 4JL, UK.

Fluka Chemie AG, Industriestrasse 25, CH-9470 Buchs, Switzerland. (biochemicals, detergents)

Genset SA, 1, rue Robert et Sonia Delaunay, 75011 Paris, France.

Genzyme Corporation, 75 Kneeland Street, Boston, MA 02111, USA. (protease-free deglycosylation enzymes)

Gibco BRL, Bethesda Research Laboratories, Life Technologies, Inc. PO Box 6009, Gaithersburg, MD 20877, USA. (biochemicals, growth media)

Gibco BRL (Life Technologies Inc.), 3175 Staler Road, Grand Island, NY 14072–0068, USA.

Gibco BRL (Life Technologies Ltd.), Trident House, Renfrew Road, Paisley, Scotland, PA3 4EF, UK.

Gilson Medical Electronics, Inc., 72 rue Gambetta, BP 45, F-95400 Villers-le-Bel, France and 3000 W. Beltine Hwy., PO Box 27, Middleton, WI 53562, USA. (sample changers)

Gow-Mac Inc., PO Box 32, Bound Brook, NJ 08805-0032, USA. (thermal conductivity detectors)

Hamilton Co., PO Box 10030, Reno, NV 89520-0012, USA. (syringes)

Hampton Research, 27632 El Lazo Road, Suite 100, Laguna Beach, CA 92677–3913, USA.

Heraeus Feinchemikalien und Forchungsbedarf GmbH, Alter Weinberg, D-7500 Karlsruhe 41-Ho., Germany. (chemicals, reagents for silanization)

Hewlett-Packard Co., Analytical Group, Mailstop 20B AE, Palo Alto, CA 94403, USA. (robotics)

Hilgenberg Glass Company, D-3509 Malsfeld, Germany. (X-ray glass/quartz capillaries)

Arnold R. Horwell, 73 Maygrove Road, West Hampstead, London NW6 2BP, UK.

Huber Diffraktionstechnik GmbH, D-8219 Rimsting, Germany. (diffractometry)

Hybaid

Hybaid Ltd., 111–113 Waldegrave Road, Teddington, Middlesex TW11 8LL, UK.

Hybaid, National Labnet Corporation, P.O. Box 841, Woodbridge, NJ. 07095, USA.

HyClone Laboratories 1725 South HyClone Road, Logan, UT 84321, USA.

IBF Biotechnics, 35 avenue Jean-Jaurés, 92290 Villeneuve-la-Garenne, France. (chromatographic matrices, biochemicals)

ICI Cambridge Research Chemicals, Gadbrook Park, Northwich, Cheshire CW9 7RA, UK. (chemicals)

ICN Biomedicals, Inc., Micromedica Systems Diagnostic Division, 102 Witmer Road, Horsham, PA 19044–2281, USA. (robotic protein crystallization system II, pipetting stations)

ICN Flow, 330 Hyland Avenue, Costa Mesa, CA 92626, USA. (biochemical equipments, Linbro plate, CrystalPlate® and coverslips)

Imaging Technology Inc., 600 West Cummings Park, Woburn, MA 01801, USA. (digitizers)

Intermec Corp., 4405 Russell Road, PO Box 360602, Lynnwood, WA 98046-9702, USA. (barcode printer)

International Biotechnologies Inc., 25 Science Park, New Haven, Connecticut 06535, USA.

Invitrogen Corporation

Invitrogen Corporation 3985 B Sorrenton Valley Building, San Diego, CA. 92121, USA.

Invitrogen Corporation c/o British Biotechnology Products Ltd., 4–10 The Quadrant, Barton Lane, Abingdon, Oxon OX14 3YS, UK.

Jouan SA, rue Bobby Sands, F-44800 Saint Herblain, France. (laboratory equipments)

Keithley Data Acquisition and Control, 28775 Aurora Road, Cleveland, OH 44139, USA. (instrument interfaces)

Kodak: Eastman Fine Chemicals 343 State Street, Rochester, NY, USA.

Kohyo Trading Company, Kyodo Bldg 4–1, 2 Chome, Iwando-cho, Chiyoda-ky, Tokyo, Japan. (detergents)

Leica SARL, see Wild-Leitz.

Leitz/Leica, 111 Deer Lake Road, Deerfield, IL 60015, USA.

Life Technologies Inc., 8451 Helgerman Court, Gaithersburg, MN 20877, USA.

Marresearch, Grosse Theaterstrasse 42, Postfach 303670, 2000 Hamburg 36, Germany. (image plate)

Memmert GmbH and Co., Aeussere Ritterbacherstrasse 38, D-8540 Schwabach, Germany. (laboratory equipments, thermostated cabinets)

Merck

Merck Industries Inc., 5 Skyline Drive, Nawthorne, NY 10532, USA.

Merck, Frankfurter Strasse, 250, Postfach 4119, D-64293, Germany.

Merck, Frankfurter Strasse 250, D-6100 Darmstadt, Germany. (chemicals and biochemicals)

Microflex Technology, Inc., The Millennium Centre, PO Box 31, Triadelphia, WV 26059, USA. (microgrippers)

Micromedica System, Inc., (see ICN Biomedicals). (pipetting station)

Millipore

Millipore (UK) Ltd., The Boulevard, Blackmoor Lane, Watford, Herts WD1 8YW, UK.

Millipore Corp./Biosearch, P.O. Box 255, 80 Ashby Road, Bedford, MA 01730, USA.

Millipore Waters, PO Box 255, Bedford MA 01730, USA and Zone Industrielle, F-67120 Molsheim, France. (filtration, membranes, HPLC equipments)

NAPS Göttingen GmbH, Nucleic Acids Products Supply, Rudolf-Wissel Str. 28, 37070 Göttingen, Germany.

National Institute of Standards and Technology, (Standard Reference Data) Bldg. 221/A323, Gaithersburg, MD 20899, USA. (Software with crystallization data bank)

National Instruments Corp., 12109 Technology Blvd, Austin, TX 78727–6204, USA. (instrument interfaces, laboratory software)

Neosystem Laboratories, Technopole du Rhin, 21 rue du la Rochelle, F-67100 Strasbourg, France. (peptides)

New England Biolabs (NBL)

New England Biolabs (NBL), 32 Tozer Road, Beverley, MA 01915–5510, USA.

New England Biolabs (NBL), c/o CP Labs Ltd., P.O. Box 22, Bishops Stortford, Herts CM23 3DH, UK.

Nikon Corporation Instrument Div., Fuji Bldg 2–3, 3-Chome, Maranouchi, Chiyoda ku, Tokyo 100, Japan. (stereo microscopes)

Nikon Europe BV, Shipholm weg 321, 1171 AE Badhoevedorp, The Netherlands. (stereo microscopes)

Nunc Inc., 2000, North Aurora Road, Naperville, IL 60566, USA. (plastic tubes and plates)

Ominifit Ltd., 51 Norfolk Street, Cambridge CB1 2LE, UK and 2005 Park Street, Box 56, Atlantic Beach, NY 11509, USA. (valves)

Omnilabo Holland BV, Breda, The Netherlands. (multi-well plates)

Oxyl, Peter Henlein Strasse 11, D-8903 Bobingen, Germany. (detergents)

Panasonic Inc., One Panasonic Way, Secaucus, NJ 07094, USA. (optical disc recorder)

Pentapharm Ltd., Engelgasse 109, CH-4002 Basel, Switzerland. (protease inhibitors)

Peptide Institute, 476 Ina Miush-shi, Osaka 562, Japan. (protease inhibitors)

Perkin-Elmer

Perkin-Elmer Ltd., Maxwell Road, Beaconsfield, Bucks. HP9 1QA, UK.

Perkin-Elmer Ltd., Post Office Lane, Beaconsfield, Bucks, HP9 1QA, UK.

Perkin-Elmer-Cetus (The Perkin-Elmer Corporation), 761 Main Avenue, Norwalk, CT 0689, USA.

Perpetual Systems Corporation, 2283 Lewis Avenue, Rockville, Maryland 20851, USA. (sitting-drop rods for cystallization)

PerSeptive Biosystems, City of Dover, Kent County, DL 19901, USA.

Pfanstiel Laboratory, Inc., 1219 Glen Rock Avenue, Wavkega, IL 60085 0439, USA. (detergents)

Pharmacia Biosystems

Pharmacia Biotech Europe Procordia EuroCentre, Rue de la Fuse-e 62, B-1130 Brussels, Belgium.

Pharmacia Biosystems Ltd. (Biotechnology Division), Davy Avenue, Knowlhill, Milton Keynes MK5 8PH, UK.

Pharmacia LKB Biotechnology AB, Björngatan 30, S-75182 Uppsala, Sweden.

Phenomenex, 6100 Palos Verdes Drive S., Rancho Palos Verdes, CA 90274, USA. (HPLC columns for tRNA)

Pierce, PO Box 1512, 3260 BA Oud-Beijerland, The Netherlands. (laboratory supplies)

Polycrystal book service, PO Box 3439, Dayton, Ohio 45401, USA. (crystallography books)

PolyLabo Paul Block et Cie, BP 36, F-67023 Strasbourg Cedex, France. (scientific equipments)

The Product Integrity Company, Enfield, CT 06082, USA. (programs for factorial analysis)

Prolabo, 12 rue Pelée, F-7511 Paris, France. (chemicals, equipments)

Promega

Promega Ltd., Delta House, Enterprise Road, Chilworth Research Centre, Southampton, UK.

Promega Corporation, 2800 Woods Hollow Road, Madison, WI 53711–5399, USA.

Protein Solutions Incorporated, 2300 Commenwealth Drive, Suite 102, Charlottesville, VA 22901, USA.

Pye Unicam Ltd, York Street, Cambridge CB1 2PX, UK. (Philips X-ray generator)

Qiagen

Qiagen Inc., c/o Hybaid, 111–113 Waldegrave Road, Teddington, Middlesex, TW11 8LL, UK.

Qiagen Inc., 9259 Eton Avenue, Chatsworth, CA 91311, USA.

Radiometer, A/S 49 Krogshojvej, DK 2880 Dagsvaerd, Denmark. (pH-meter, conductimeter)

Rainin Instrument Co. Inc., Mack Road, Woburn, MA 01801, USA. (filters)

Resolution Technology, 26000 Avenida Aeropuerto 22, San Juan Capistrano, CA 92675, USA. (time-lapse VCR)

Rigaku, Monschauer Strasse 7, D-4000 Düsseldorf-Heerdt, Germany & 3 Electronics Avenue, Danvers, MA 01923, USA. (X-ray generators, image plate)

Roucaire, BP 65, F-78143 Velizy-Villacoublay Cedex, France. (scientific equipments)

Schleicher and Schuell

Schleicher and Schuell Inc., Keene, NH 03431A, USA.

Schleicher and Schuell Inc., D-3354 Dassel, Germany.

Schleicher and Schuell Inc., c/o Andermann and Company Ltd.

Seikagaku Kogyo Co. Ltd., 1–5, Nihonbashi-Honcho 2-Chome Chuo-ku, Tokyo, 103, Japan. (biochemicals, glycosylases)

Serva Feinbiochemica GmbH and Co., PO Box 105260, D-6900 Heidelberg, Germany. (biochemicals)

Setaram, 7 rue de l'Oratoire, BP. 34, F-69641 Caluire Cedex, France. (instrumentation, calorimeters)

Shandon Scientific Ltd., Chadwick Road, Astmoor, Runcorn, Cheshire WA7 1PR, UK.

Siemens AG, Mess., Pruf. und Prozesstechnik, Ostl. Rheinbrückenstrasse 50, D-7500 Karlsruje 21, Germany. (diffractometry)

Sigma Chemical Company

Sigma Chemical Company (UK), Fancy Road, Poole, Dorset BH17 7NH, UK.

Sigma Chemical Company, 3050 Spruce Street, P.O. Box 14508, St. Louis, MO 63178–9916, USA.

Societe 3412, 65 avenue de Stalingrad, F-95104 Argenteuil, France. (crystallization boxes)

Sofranel, 59 rue Parmentier, 78500 Sartouville, France. (X-ray glass/quartz capillaries)

Sorvall DuPont Company, Biotechnology Division, P.O. Box 80022, Wilmington, DE 19880–0022, USA.

Speciality Chemicals, PO Box 1466, Gainesville, FL 32602, USA. (Prosil®-28 reagent for silanization)

Spectrum Medical Industries, Inc., 8430 Santa Monica Blvd, Los Angeles, CA 90069, USA. (dialysis membranes-Spectrapore®)

Stratagene

Stratagene Ltd., Unit 140, Cambridge Innovation Centre, Milton Road, Cambridge CB4 4FG, UK.

Strategene Inc., 11011 North Torrey Pines Road, La Jolla, CA 92037, USA.

Tosohaas, 6th and Market Streets, Philadelphia, PA 19105, USA. (HPLC columns)

Transformation Research Inc., PO Box 241, Framington, MA 01701, USA. (protease inhibitors)

United States Biochemical, P.O. Box 22400, Cleveland, OH 44122, USA.

Vegatec S.A.R.L., 7 place des Onze Arpents, F-94800 Villejuif, France. (detergents)

Velmex, Inc., PO Box 38, E. Bloomfield, NY 14443, USA. (stepper motors, motorized slides)

Wellcome Reagents, Langley Court, Beckenham, Kent BR3 3BS, UK.

Whatman Laboratory Sales Ltd, Unit 1, Colred Road, Parkwood, Maidstone, Kent, ME15 9XN, UK. (chromatography supports)

Wild-Leitz (Leica SARL), 86 avenue du 18 juin 1940, F-92563 Rueil-Malmaison Cedex, France and CH-9435 Heerbrugg, Switzerland. (stereo microscopes)

Wolfgang Müller, Reierallee 12, D-1000 Berlin 27, Germany. (X-ray glass/quartz capillaries)

Index